Polyphenols and Flavones from Agricultural Products and Their Health-Promoting Properties

Polyphenols and Flavones from Agricultural Products and Their Health-Promoting Properties

Editors

Jicheng Zhan
Zhigang Liu
Hui-Min David Wang

MDPI

Basel • Beijing • Wuhan • Barcelona • Belgrade • Novi Sad • Cluj • Manchester

Editors
Jicheng Zhan
China Agricultural University
Beijing, China

Zhigang Liu
Northwest A&F University
Yangling, China

Hui-Min David Wang
National Chung Hsing University
Taichung, Taiwan

Editorial Office
MDPI
St. Alban-Anlage 66
4052 Basel, Switzerland

This is a reprint of articles from the Special Issue published online in the open access journal *Antioxidants* (ISSN 2076-3921) (available at: https://www.mdpi.com/journal/antioxidants/special_issues/Polyphenols_Flavones_Health).

For citation purposes, cite each article independently as indicated on the article page online and as indicated below:

Lastname, A.A.; Lastname, B.B. Article Title. *Journal Name* **Year**, *Volume Number*, Page Range.

ISBN 978-3-0365-9965-6 (Hbk)
ISBN 978-3-0365-9966-3 (PDF)
doi.org/10.3390/books978-3-0365-9966-3

Contents

antioxidants

Review

Functional and Nutraceutical Significance of Amla (*Phyllanthus emblica* L.): A Review

Maryam Gul [1,†], Zhi-Wei Liu [2,†], Iahtisham-Ul-Haq [3], Roshina Rabail [1], Fatima Faheem [1], Noman Walayat [4], Asad Nawaz [5], Muhammad Asim Shabbir [1,*], Paulo E. S. Munekata [6,*], José M. Lorenzo [6,7] and Rana Muhammad Aadil [1,*]

1 National Institute of Food Science and Technology, University of Agriculture, Faisalabad 38000, Pakistan; maryamgul1608@gmail.com (M.G.); roshina.rabail@gmail.com (R.R.); fatimafaheem538@gmail.com (F.F.)
2 College of Food Science and Technology, Hunan Agricultural University, Changsha 410128, China; zwliu@hunau.edu.cn
3 Kauser Abdulla Malik School of Life Sciences, Forman Christian College (A Chartered University), Lahore 54600, Pakistan; iahtisham@fccollege.edu.pk
4 College of Food Science and Technology, Zhejiang University of Technology, Hangzhou 310014, China; nomanrai66@zjut.edu.cn
5 Shenzhen Key Laboratory of Marine Microbiome Engineering, Institute for Advanced Study, Shenzhen University, Shenzhen 518060, China; 007298@yzu.edu.cn
6 Centro Tecnológico de la Carne de Galicia, Avd. Galicia No. 4, Parque Tecnolóxico de Galicia, San Cibrao das Viñas, 32900 Ourense, Spain; jmlorenzo@ceteca.net
7 Universidade de Vigo, Área de Tecnoloxía dos Alimentos, Facultade de Ciencias, 32004 Ourense, Spain
* Correspondence: dr.asim@uaf.edu.pk (M.A.S.); paulosichetti@ceteca.net (P.E.S.M.); muhammad.aadil@uaf.edu.pk (R.M.A.)
† These authors contributed equally to this work.

Citation: Gul, M.; Liu, Z.-W.; Iahtisham-Ul-Haq; Rabail, R.; Faheem, F.; Walayat, N.; Nawaz, A.; Shabbir, M.A.; Munekata, P.E.S.; Lorenzo, J.M.; et al. Functional and Nutraceutical Significance of Amla (*Phyllanthus emblica* L.): A Review. *Antioxidants* **2022**, *11*, 816. https://doi.org/10.3390/antiox11050816

Academic Editors: Jicheng Zhan, Zhigang Liu and Hui-Min David Wang

Received: 21 March 2022
Accepted: 20 April 2022
Published: 22 April 2022

Abstract: *Phyllanthus emblica* L. (also popularly known as amla) is a tree native to the India and Southeast Asia regions that produces fruits rich in bioactive compounds that could be explored as part of the increasing interest in naturally occurring compounds with biological activity. Thus, this review aims to highlight the nutritional aspects, rich phytochemistry and health-promoting effects of amla. Scientific evidence indicates that polyphenols are central components in fruits and other sections of the amla tree, as well as vitamin C. The rich composition of polyphenol and vitamin C imparts an important antioxidant activity along with important in vivo effects that include improved antioxidant status and activity of the endogenous antioxidant defense system. Other potential health benefits are the anti-hyperlipidemia and antidiabetic activities as well as the anticancer, anti-inflammatory, digestive tract and neurological protective activities. The promising results provided by the studies about amla bioactive compounds support their potential role in assisting the promotion of health and prevention of diseases.

Keywords: polyphenols; ascorbic acid; antioxidant activity; cardiovascular protection; hyperlipidemia; diabetes; health promotion

1. Introduction

Phyllanthus emblica L. (popular known as amla or Indian gooseberry) is an ephemeral tree belonging to the *Euphorbiaceae* family. Amla fruits are edible and are mainly found in regions of India, Southeast Asia, China, Iran, and Pakistan [1]. Amla has an important role in the traditional medicine of India to reduce anxiety and burning sensation in skin and eyes, improve anemic condition, favor the health of the male reproductive system and reproduction, facilitate digestion, improve liver health, and also exert a tonic effect in the cardiovascular system [2,3].

The fruit of *P. emblica* L. is one of the most popular botanicals, with a wide range of uses in the medicinal, cuisine, and cosmetic industries. This is the first tree to be "produced in the universe", according to ancient Indian mythology [4]. It is a great nutritional supplement with several medicinal benefits [5]. Due to the abundance of phenolic compounds, Emblic fruit could be regarded as a plant source for natural antioxidants and nutraceuticals or medicinal components. Consumers like Emblic fruit because of its unique flavor and pleasant smell. In various animal and human investigations, amla has been proven to have anti-hyperglycemic, hypoglycemic, anti-inflammatory, anti-hyperlipidemic, and antioxidant activities [1]. Amla is rich in antioxidants such as gallic acid, ascorbic acid and phenolic compounds and thus helps the body's immune systems and digestion [6]. Thus, due to the increasing interest and the potential of *P. emblica* L., this review aims to provide an overview of the nutritional composition, phytochemistry and potential health benefits associated with the consumption of phytochemicals naturally found in amla.

2. Nutritional Composition of Amla

Amla fruits are a relevant source of carbohydrates that account for >70 g/100 g dry weight (DW) (Table 1). Fiber is another relevant component (7.2–16.5 g/100 g DW) as well as contents of protein, minerals such as (iron, calcium and phosphorous), and fat (2.0–4.5, 2.1–3.1, and 0.2–0.6 g/100 g DW, respectively) [7–12]. The variability in the composition of amla fruit has been attributed to the cultivar in many studies [10–12].

Table 1. Nutritional constituents of amla fruit from different studies.

Variety	Moisture	Carbohydrate	Fiber	Minerals	Protein	Fat	Vitamin C	Ref.
Local variety (no name)	81 g/100 g	14 g/100 g	3.2 g/100 g	0.3 g/100 g	1 g/100 g	0.5 g/100 g	720 mg/100 g	[8]
Local variety (no name)	82.8 g/100 g	7.6 g/100 g	5.1 g/100 g	2.3 g/100 g	2.0 g/100 g	0.3 g/100 g	573 mg/100 g	[9]
NA-7 NA-9 NA-10 Balwant Chakaiya Hathijhool	84.9–87.5 g/100 g	77.2–81.9 g/100 g DW	11.7–16.0 g/100 g DW	2.1–3.0 g/100 g DW	3.0–4.5 g/100 g DW	0.2–0.5 g/100 g DW	489.9–585.0 mg/100 g	[10]
NA-7, Banarasi, Kanchan, Chakaiya and Desi	81.3–84.6 g/100 g	73.8–87.1 g/100 g DW	7.2–22.4 g/100 g DW	2.2 to 3.1 g/100 g DW	2.0 to 3.2 g/100 g DW	0.4–0.5 g/100 g DW	193–315 mg/100 g	[11]
Krishna, Kanchan, NA-7, Chakaiya	85.6–87.7 g/100 g	70.7–73.8 g/100 g DW	13.9–16.5 g/100 g DW	2.3–2.8 g/100 g DW	2.9–3.6 g/100 g DW	0.5–0.6 g/100 g DW	421–506 mg/100 g	[12]

DW: dry weight.

Another important component found in amla fruit is ascorbic acid (vitamin C). Values between 193 and 720 mg/100 g have been reported in different studies that evaluated a different variety of amla [8–12]. Although the optimum recommended daily intake has not been defined yet due to the emergency of new factors from modern society, many governmental health authorities around the globe established Recommended Dietary Allowance (minimum level to meet the need for a healthy person for a day) that varies between 40 and 110 mg vitamin C/day [13]. Moreover, the Australian and China health authorities have proposed a daily intake of 190–220 mg/day. In this sense, a serving portion of at least 100 g of fresh amla fruits (2–3 pieces) from any of the varieties indicated in Table 1 should suffice the daily need for vitamin C. Comparatively, *P. emblica* L. juice can present more vitamin C content than any other fruits such as apple, lime, pomegranate, and some types of grapes [14,15]. Additionally, other nutritionally relevant compounds found in amla fruits are vitamins A, B1, and E (290 IU, 30 mg/100 g, and 0.17 mg/100 g) as well as calcium and iron (25 and 1 mg/100 g) [8].

3. Phytochemistry of Amla

Amla has been found to possess rich phytochemistry distributed in different sections of the plant (fruits, leaves, and roots). Polyphenols (Figure 1) comprise the main group of secondary metabolites wherein several compounds belonging to phenolic acids, flavonoids, tannins, other phenolics and derivatives compounds have been reported in different studies.

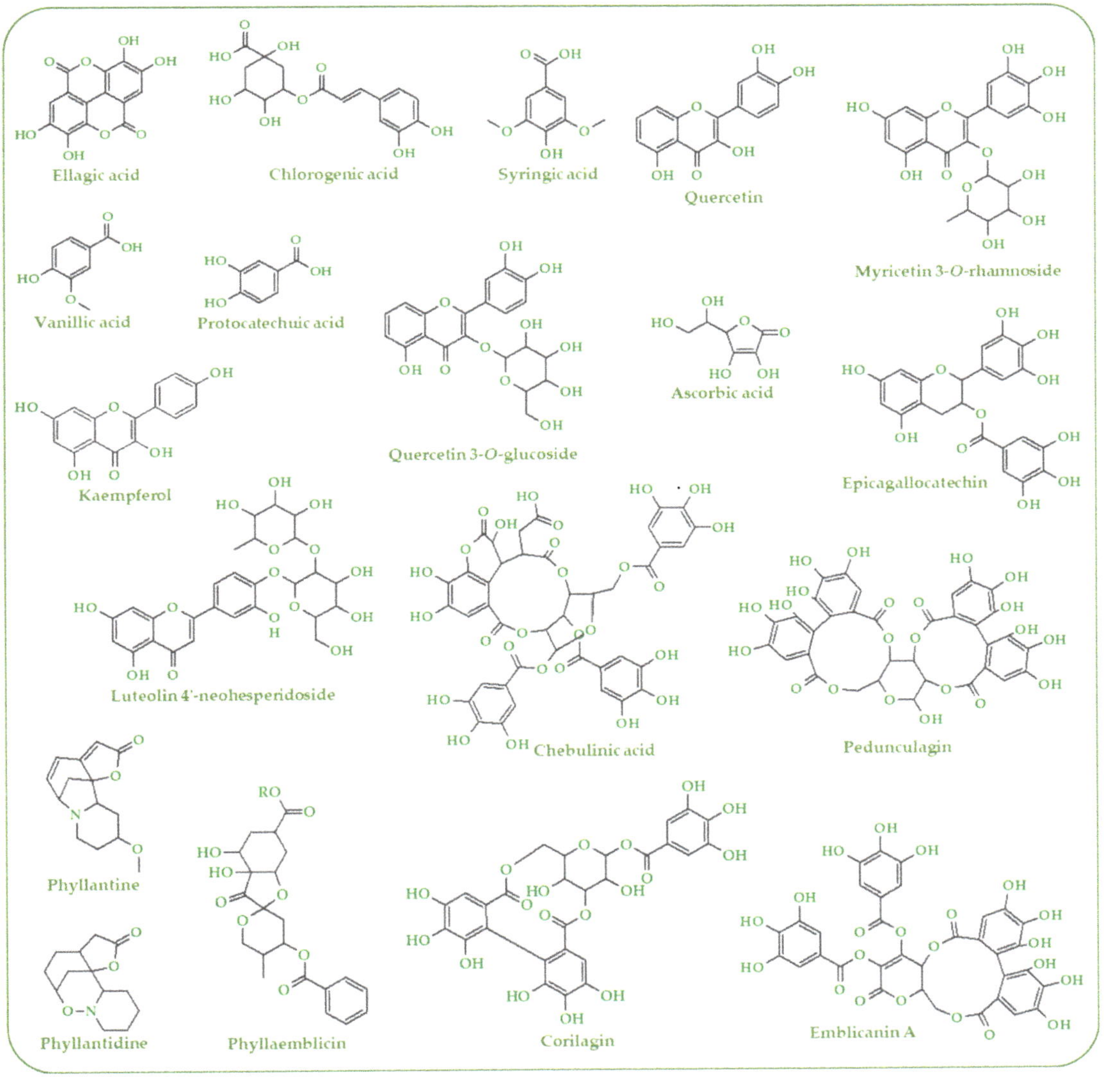

Figure 1. Phytochemicals found in amla.

Regarding the phenolic acids, the presence of hydroxybenzoic acids (4-hydroxybenzoic acid, coumaric acid, gallic acid, protocatechuic acid, syringic acid, and vanillic) acid were identified in the fresh fruit and commercial products prepared from the fruits [16–20]. Gallic acid is the only hydroxybenzoic acid reported in leaves and branches [21]. The presence of hydroxycinnamic acids (caffeic acid and chlorogenic acid) was indicated only in amla fruits [16–18]. Another class of compounds reported in the amla plant is flavonoids (particularly flavonols, flavones, flavanones, and flavan-3-ols). Flavonols are widely distributed in the different sections of the amla plant. Kampferol their derivatives (dihydrokaempferol, kaempferol 3-b-dglucopyranoside, kaempferol 3-o-rhamnoside, kaempferol-3-o-α-l-(6″-ethyl)-rhamnopyranoside, and kaempferol-3-o-α-l-(6″-methyl)-rhamnopyranoside) are found in fruits, leaves branches and shoots [19,21–23]. In a similar way, quercetin and its derivatives (quercetin 3-b-D-glucopyranoside, quercetin 3-O-glucoside, quercetin 3-O-rhamnoside, and rutin) are distributed in fruits, leaves and branches [16–19,21,23].

Regarding flavones, the presence of apigenin, luteolin, and myricetin was indicated in the fresh fruits and fruit commercial products [18,19]. Myricetin 3-O-rhamnoside was reported only in the leaves and branches of the amla tree [21]. Interestingly, flavanones and flavan-3-ols were reported only in leaves and branches [21]. The identified flavanones were eriodictyol, naringenin, and their derivatives ((S)-eriodictyol 7-O-(6″-O-galloyl)-β-D-glucopyranoside, (S)-eriodictyol 7-O-(6″-O-trans-p-coumaroyl)-β-D-glucopyranoside, naringenin 7-O-(6″-O-galloyl)-glucoside, naringenin 7-O-(6″-O-trans-p-coumaroyl)-glucoside, and naringenin 7-O-glucoside). Regarding flavan-3-ols, the detected compounds were epicagallocatechin, epigallocatechin 3-O-gallate, and gallocatechin.

Tannins is another key group of phenolic compounds found in amla fruits, leaves and branches. Many studies indicate the presence of ellagitannins, which includes chebulinic acid, chebulagic acid, corilagin, emblicanin A and B, geraniin, isocorilagin, pedunculagin, phyllanemblinins A–F, and punigluconin [20,23–25]. Ellagic acid and their derivatives (decarboxyellagic acid and 3′-O-methylellagic acid 4-O-α-L-rhamnopyranoside) were also reported in fruits, leaves and brunches [21]. Hydrolysable tannins (1,2,3,4,6-penta-O-galloyl-β-D-glucose, 1,2,3,6-tetra-O-galloyl-β-D-glucose, and 1,2,4,6-tetra-O-galloyl-β-D-glucose) and phlorotannins (2-(2-methylbutyryl)phloroglucinol 1-O-(6″-O-β-D-apiofuranosyl)-β-D-glucopyranoside) are mainly found in leaves and branches of amla [18,21]. The exception is tannic acid, which was reported in amla fruit [21]. Moreover, other phenolics (2,4-di-tert-butylphenol and Phenol, 3,5-bis (1,1-dimethylethyl)) were also reported amla fruit [26]. Additionally, alkaloids (especially phyllantine and phyllantidine) were also reported in amla [7].

4. Potential Health Benefits

4.1. Antioxidant Activity

Diverse in vitro, in vivo, and human studies support the antioxidant activity of *P. emblica* L. components. In the case of in vitro studies, the content of polyphenols in this fruit has also been associated with high antioxidant activity [20,27,28], particularly with the capacity to scavenge free radicals such as the 1,1-diphenyl, 2-picrylhydrazyl (DPPH) radical [9,12,16–18,27,29–32]. Other studies also reported the antioxidant activity of amla phytochemicals by 2,2′-azino-bis(3-ethylbenzothiazoline-6-sulfonic acid) (ABTS) and NO radical scavenging methods, the Ferric Reducing Antioxidant Power (FRAP) [33], and LDL oxidation assay method [19]. Additionally, amla polyphenols can also scavenge superoxide anion and hydroxyl free radicals as well as chelated iron (III) [20].

The observed antioxidant activity observed from extracts and isolated compounds from amla fruit using chemical methods has also been observed in a more complex biological system that includes cells, animals, and clinical trials (Table 2). In this case, the antioxidant defense system, has an important role in the protection against oxidative damage in vivo. This system is composed of non-enzyme compounds (glutathione; GSH) and enzymes (such as catalase (CAT), GSH reductase, glutathione peroxidase (GPx), and superoxide dismutase (SOD)) [34].

Table 2. Antioxidant effect and induction of endogenous antioxidant defense mechanism.

Source	Type of Study	Study Characteristics	Main Outcomes	Ref.
Fruit	In vitro (cell)	PC12 cells; dosage (10–50 μM); and incubation (2 h)	No toxicity; ethyl gallate was the most efficient antioxidant (10–50 μM)	[32]
Fruit	In vitro (cell)	HepG2 cells; dosage (5, 10, 20, 50, and 100 μg/mL); and incubation (4, 8, 12, 16, 20, and 24 h)	No Cytotoxicity (up to 100 μg/mL); reduced lipid hydroperoxides reactive oxygen species levels (50 and 100 μg/mL after 8 h); and increased GSH, total antioxidant capacity, SOD, CAT, GPx, GSH reductase, and GSH S-transferase (50 and 100 μg/mL after 12–24 h)	[35]
Fruit	In vitro (cell)	RAW 264.7 cells; dosage (25, 50, or 100 μg/mL); and incubation (24 h)	No Cytotoxicity (100 μg/mL); increased GSH and SOD activity when challenges with H_2O_2 (50 and 100 μg/mL); and reduced MDA level (100 μg/mL)	[36]
Fruit	In vitro (cell)	C2C12 myoblasts; dosage (100 and 200 μg/mL); and incubation (48 h)	Increased cell survivability (200 μg/mL) and reduced ROS levels with increased oxygen consumption (200 μg/mL)	[37]
Leaves	Animal (mice)	Diabetic wistar mice; 100–400 mg/kg BW; oral administration; and 45 days	Induced GSH, GPx, SOD, and CAT activity (200–400 mg/kg BW) and reduced lipid peroxidation (200–400 mg/kg BW)	[38]
Fruit	Animal (mice thymus)	Balb/c male mice; 500 mg/kg BW; oral administration; and 28 days	Improved cell viability, GSH, CAT, and SOD levels and reduced lipid peroxidation, ROS level	[39]
Fruit	Animal (mice liver)	Wistar mice; 5000 mg/kg BW; oral administration; and 24 days	Reduce lipid peroxidation; preserved CD, CAT, and NPSH; and ameliorated SOD reduction	[40]
Fruits	Animal (mice kidney)	Healthy wistar mice; dosage (50, 100, 150, 200, and 250 μg/mL); single application	Increased SOD and CAT (50–250 μg/mL); reduced lipid peroxidation (50–250 μg/mL); and no effect in GSH	[41]
Commercial supplement	Clinical trial	Male smoker subjects (20–60 y); randomized, double-blind placebo-controlled design; 250 mg (twice a day); and 60 days	Increased antioxidant status (FRAP assay) and reduced lipid peroxidation level	[42]
Commercial supplement	Clinical trial	Female and male subjects with metabolic syndrome (30–68 y); randomized, double-blind, and placebo-controlled; 250 and 500 mg per capsule (twice a day); and 12 weeks	Increased GSH level and reduced lipid peroxidation level	[43]
Commercial supplement	Clinical trial	Female and male healthy subjects (36–67 y); randomized, double-blind, placebo-controlled, and crossover; 125 mg per capsule (4 capsules/day)	A non-significant reduction in lipid peroxidation level	[6]

BW: body weight; CAT: Catalase; CAT Catalase; CD: conjugated dienes; GPx: Glutathione peroxidase; GSH: Glutathione; NPSH: non-protein soluble thiol; ROS: Reactive oxygen species; and SOD: Superoxide dismutase.

One example of the protective effect of amla fruit bioactive compounds against oxidative damage is the study carried out by Shivananjappa and Joshi [35] in HepG2 cells. These authors observed that total antioxidant capacity (ABTS method) was improved after 4 h of exposure to different concentrations of extract (25, 50 and 100 μg/mL). The peroxidation level was significantly reduced after 8 h of exposure to 50 and 100 μg/mL of extract. Moreover, the antioxidant defense system was induced (GSH, SOD, CAT, GPx, GSH reductase, and GSH S-transferase) after 12–24 h of exposure to 50 and 100 μg/mL of extract. Similar results were reported in a study with RAW 264.7 cells with up to 100 μg/mL of the extract [36]. Particularly for isolated compounds, ethyl gallate was indicated as the most efficient antioxidant (10–50 μM) to reduce oxidative damage in PC12 cells [32].

Another relevant outcome obtained from these studies is the non-significant cytotoxicity of extracts in the range of concentrations (up to 100 μg/mL) where the antioxidant activity and induction of the antioxidant defense system were observed [32,35,36]. It is worth mentioning that a recent experiment indicated that a higher concentration (200 μg/mL) of extract would be required to reduce reactive oxygen species levels and improve their survivability in myoblasts [37].

The protective effect of amla fruit compounds against oxidative damage has also been observed at the animal level. In this case, the daily administration of 500 mg/kg body weight (BW) of amla fruit extract during 28 days induced the activity of GSH, CAT, and SOD in the thymus of mice. [39]. Accordingly, these authors also observed that lipid peroxidation and reactive oxygen species (ROS) levels were reduced. A similar experiment with amla fruit extract revealed a significant reduction in the lipid peroxidation levels, simultaneously decreased the levels of conjugated dienes and CAT levels, and ameliorated the reduction in SOD levels in the liver of mice (500 mg/100 g BW) caused by arsenic [40]. In another study, the amla fruit extract (50–250 μg/mL) induced SOD and CAT activities and reduced lipid peroxidation in the kidneys of mice [41].

Additionally, a study using the extract from amla leaves (200–400 mg/kg BW) indicated a similar protective effect in diabetic mice by reducing inducing the activity of GSH, GPx, SOD, and CAT activity and also reducing lipid peroxidation [38]. However, this experiment did not indicate significant effects on the antioxidant status of healthy animals consuming the amla extract. In another experiment carried out by Reddy et al. [44], the protective role of *P. emblica* L. fruit extract was observed in mice subjected to alcohol-induced oxidative stress. The authors indicated that polyphenols (especially tannins and flavonoids) present in this fruit extract significantly reduced oxidative stress by scavenging NOx.

Amla antioxidants have also been associated with improvements in the antioxidant status in humans. A recent clinical trial with smokers (randomized, double-blind placebo-controlled design) also supports the role of amla fruit as a relevant option of natural antioxidants [42]. In this study, a significant reduction in the peroxidation level and increased antioxidant status were observed in subjects that consumed 250 mg (twice a day) for 60 days. Another clinical trial with subjects diagnosed with metabolic syndrome (randomized, double-blind and placebo-controlled design) indicated that consuming either 250 or 500 mg capsules (twice a day) for 12 weeks reduced the lipid peroxidation levels and induced GSH levels [43]. Conversely, the consumption of 125 mg capsules (4 capsules/day) had non-significant effects on the antioxidant status in healthy subjects (randomized, double-blind, placebo-controlled, and crossover design) [6].

These studies indicate that amla phytochemicals can exert antioxidant activity by limiting the formation of oxidation products, increasing antioxidant status, and also inducing the endogenous antioxidant defense system. Particularly for clinical trials, the effect is promising to prevent oxidative induced by lifestyle (smoking) or the management of diseases (metabolic syndrome).

The polyphenols naturally present in amla also exert other biological effects beyond antioxidant activity. Table 3 indicates the phenolic compounds, individually or collectively, associated with biological effects. These studies indicate amla fruit as the most studied source of bioactive compounds (especially polyphenols from different groups indicated in Section 3). In most cases, the biological effect is attributed to more than one polyphenol composing the experimental sample. The biological effects of amla polyphenols are discussed in the following sections.

Table 3. Amla polyphenols and their biological effects beyond antioxidant activity.

Source	Main Active Compounds	Biological Effect	Ref.
Fruit	Gallic acid	Cardioprotective activity	[45]
Fruit	Ellagic acid	Antidiabetic activity	[46]
Fruit	Pyrogallol	Anticancer activity	[47]
Fruit	Emblicanin A and B	Anticancer activity	[48]
Fruit	Emblicanin A and B	Anti-inflammatory activity	[49]
Fruit	Emblicanin A and B	Neuroprotective activity	[50]
Fruit	Myricetin, gallic acid, and kaempferol	Cardioprotective activity	[19]
Fruit	Gallic acid, corilagin, and ellagic acid	Anti-inflammatory activity	[36]
Fruit	Emblicanin A and B, punigluconin, and pedunculagin	Cardioprotective activity	[51–53]
Fruit	Emblicanin A and B, punigluconin, and pedunculagin	Anti-inflammatory activity	[24]
Fruit	Emblicanin A and B, punigluconin and pedunculagin	Neuroprotective activity	[54]
Fruit	Gallic acid, chebulagic acid, geraniin, ellagic acid, and corilagin	Cardioprotective activity	[55]
Fruit	Quercetin, rutin, gallic acid, mucic acid, and beta-glucogallin	Anti-inflammatory activity	[56]
Fruit	Emblicanin A and B, punigluconin, pedunculagin, rutin, and gallic acid	Neuroprotective activity	[57,58]
Fruit	Tannins and gallic acid	Gastrointestinal protective activity	[59]
Fruit	Flavonoids	Antidiabetic activity	[60]
Fruit	Polyphenols	Gastrointestinal protective activity	[61–64]
Fruit	Polyphenols	Neuroprotective activity	[65]
Fruit	Polyphenols	Cardioprotective activity	[66]
Leaves	Gallic acid	Anticancer activity	[67]

4.2. Cardioprotective Activity

Hyperlipidemia is one of the major causes of cardiovascular disorders [68], but amla bioactive compounds may assist in the management of this condition. Different studies have been shown the protective effects of amla and/or its constituents against cardiovascular diseases. One example is the study conducted by Nambiar and Shetty [19] who studied the effect of amla juice (myricetin, gallic acid, and kaempferol as main polyphenols) on low-density lipoprotein (LDL) oxidation. According to the authors, limited the uptake of LDL oxidation in macrophages and LDL cholesterol oxidation was reduced by 90%. Another study indicated that amla polyphenols (emblicanin A and B, punigluconin, and pedunculagin) limited fibrosis formation in cardiovascular tissue of mice subjected to schemia and reperfusion [51].

Madan et al. [69] tested the effect of amla supplementation in beetal kids and observed reductions in the levels of LDL, cholesterol, and blood glucose to the non-supplemented group. Another experiment showed that hydroalcoholic amla extract reduced the arterial mean blood pressure, and serum sodium levels and aided to increase the potassium levels in deoxy-corticosterone acetate salt-induced hypertensive mice [52]. This study also indicated that *P. emblica* L. regulated the endogenous antioxidant system, eNOS, activation of serum nitric oxide (NO), and serum electrolytes level.

The polyphenol-rich extracts of *P. emblica* L. reduced metabolic changes caused by excessive fructose consumption (alteration of triglyceride total cholesterol levels and sterol regulatory element-binding protein 1 (SREBP-1) expression) in an animal model. *P. emblica* L. (containing gallic acid, chebulagic acid, geraniin, ellagic acid, and corilagin) reduced and even inhibited the enhanced mitochondrial COX-2, MDA, and Bax expressions in the liver and regulated Bcl-2 expression, but peroxisome proliferator-activated receptors-α (PPARα) and SREBP-2 expressions were unaffected [55]. In another experiment, polyphenol-rich *P. emblica* L. extract also increased PPARα protein (involved in the regulation of cholesterol and lipid metabolism) expression and decreased cholesterol levels in mice [66]. Similarly, a related study carried out with mice that consumed amla juice (2 mL/kg/day; rich in gallic acid) indicated the activation of PPARα and carnitine palmitoyl transferase (involved in lipid oxidation) [45]. Another interesting outcome of this study was the reduction in the activity of liver enzymes involved in lipogenesis (malic enzyme, fatty acid synthase, and glucose-6-phosphate dehydrogenase). However, another study with mice in a high-fat diet (30%), indicated that *P. emblica* L. ethanolic extract reduced the serum triglycerides but no effects were observed in LDL, very-low-density lipoprotein (VLDL), or high-density lipoprotein (HDL) serum levels [70]. The antihyperlipidemic and cardioprotective potential of amla is represented in Figure 2.

At the human level, a 500 mg dose of *P. emblica* L. extract (twice a day) for three months reduced the high sensitive C-reactive protein (CRP), total cholesterol, and LDL levels in Class I obese subjects [53]. Gopa et al. [71] studied the effect of amla fruit capsules (500 mg/capsule; once a day for 42 days) and reported significant reductions in subjects with hyperlipidemia. According to these authors, significant reductions in total cholesterol, LDL, and VLDL levels along with a significant increase in the serum levels of HDL at the end of the trial period were also observed. Moreover, different studies indicated that *P. emblica* L. fruit and extract reduced the VLDL, cholesterol, and LDL levels in hyperlipidemic patients and healthy persons. When administered for 2–6 months, the extract decreased the level of important inflammatory marker CRP, resulting in an enhanced level of HDL and protection against atherosclerosis [71,72].

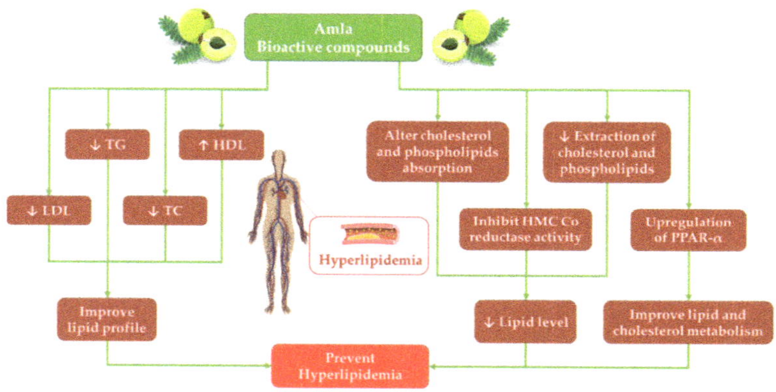

Figure 2. Anti-hyperlipidemic role of amla.

4.3. Antidiabetic Activity

The compounds naturally found in *P. emblica* L. have been associated with protective effects against diabetes. An in vitro study indicated that the activity of the main phytochemicals found in amla (such as ellagic acid and ascorbic acid) reduced the activity of key enzymes involved in glucose digestion (especially amylase and glucosidase) [73].

The protective effects against diabetes have also been reported at the animal level. For instance, a freeze-dried *P. emblica* L. aqueous extract (1.25 g/kg) reduced the levels of serum glucose and triglyceride in diabetic long Evan male mice (induced with streptozotocin) [74]. Likewise, Patel and Goyal [75] observed the antidiabetic potential of *P. emblica* L. juice (1 mL/kg/day) due to the attenuated increase in serum glucose levels of animals with induced diabetes. In another experiment with diabetic mice, the doses of 250 and 500 mg *P. emblica* L. extract/kg (rich in ellagic acid) caused significant reductions in the serum glucose levels and improved the serum insulin levels [46]. Likewise, the experiment carried out by Nain et al. [38] reported a similar outcome from the phytochemical extracted from leaves of *P. emblica* L. According to these authors, the increase in serum insulin and the related reduction in serum glucose were observed for diabetic mice in daily doses between 100 and 400 mg/kg.

Clinical trials also support the health benefits of amla phytochemicals for diabetic patients. For instance, daily doses of up to 3 g of *P. emblica* L. powder extract reduced blood glucose levels in diabetic patients after 21 days of the trial [76]. A similar outcome was reported by Walia et al. [1] who observed significant reductions in blood glucose in diabetic patients after consuming 10 g of amla powder once a day for 90 days. Additionally, flavonoid-rich *P. emblica* L. extract also reduced the risk of neuropathy [60] in diabetic patients. Amla bioactive compounds seem to play a key role in the management of diabetes, particularly in assisting in the restoration of glucose and insulin levels.

4.4. Anticancer Activity

Plant-derived polyphenols have been found to improve the protection against cancer in a variety of nonclinical and clinical investigations [77,78]. Specifically, polyphenols inhibit oxidative stress, produce pro-inflammatory chemicals, prevent DNA damage, and increase apoptosis through various mechanisms [79]. Particularly for amla extracts, DNA fragmentation, increased activity of caspase-3, 7, and 8, and up-regulation of Fas protein were observed in the HeLa cell line, indicating activation of the death receptor pathway for apoptosis, whereas caspase-9 remained unaltered [80]. This study also indicated that *P. emblica* L. decreased the invasiveness of MDA-MB-231 cells (in vitro Matrigel invasion study), and no cytotoxicity was seen in normal lung fibroblasts (MRC5). Likewise, pyrogallol (a polyphenol found in *P. emblica* L.) was investigated in human lung cancer cell lines H441 and H520. Pyrogallol's anti-proliferative effect was achieved via cell arrest in the G2/M phase, which was caused by a drop in cyclin B1, cdc25c, and Bcl-2, as well as an increase in Bax expression [47].

Zhu et al. [81] conducted a study in HeLa cell lines using polyphenolic extract of *P. emblica* L. The extract was capable of inhibiting HeLa cell proliferation by stopping cells in the G2/M phase and promoting apoptosis by inducing apoptotic markers Fas, FasL, and cleaved caspase-8. Huang and Zhong [67] found that gallic acid isolated from *P. emblica* L. leaves induced apoptosis in a hepatocellular cancer cell line (BEL-7404). According to this study, the activity of *P. emblica* L. may be related to the inhibition of the cell cycle in the G2/M phase. Overexpression of Bax and downregulation of Bcl-2 causes a reduction in mitochondrial membrane potential, which activates caspases, resulting in cell death via the apoptotic death-receptor pathway. Additionally, it has also been suggested that *P. emblica* L. play a protective role in chemo- and radiotherapy [48,80,82]. In this sense, the protective effect of amla bioactive compounds seems to have a relevant limiting effect on the progression of cancer progression in different cell lines. However, the evidence supporting the anticancer activity of amla polyphenols is limited and more efforts are still necessary to clarify the mechanisms involved and explored the effects at the animal level and expand the current knowledge.

4.5. Anti-Inflammatory Activity

Amla phytochemistry seems to promote a beneficial effect in the context of inflammation, but current evidence is limited. An example of the anti-inflammatory activity in cell model is the study carried out by Li et al. [36]. According to these authors, RAW 264.7 cells treated with amla extract (rich in gallic acid, corilagin, and ellagic acid) displayed lower levels of inflammatory markers (NO release and production of tumor necrosis factor (TNF-α), interleukin-1β (IL-1β), and interleukin-6 (IL-6)) when an inflammatory response was caused by exposure to lipopolysaccharides.

This modulation of inflammatory markers was also observed at an animal level in a study with arsenic exposure [83]. Animals treated with amla extract (500 mg/kg) showed significantly lower serum levels of TNF-α, IL-1β, and IL-6 than animals exposed only to arsenic. Another relevant outcome obtained from *P. emblica* L. extract was the reduction in edema size in mice's paws. The natural extract induced the production and release of pain and inflammatory mediators. This effect is suggested to be mediated in a similar way to nonsteroidal anti-inflammatory drugs rather than steroidal medicament way [49,56,84]. In another study with mice, Goel et al. [85] demonstrated the analgesic effects and substantial decrease in abdominal writhing of *P. emblica* L. extract with a dose of 600 mg/kg.

In the case of studies involving humans, a randomized crossover clinical trial with type-2 diabetes subjects indicated that *P. emblica* L. fruit extract (500 mg/day; containing punigluconin, emblicanin-A, emblicanin-B, and peduculagin) decreased platelet aggregation in both single and repeated dosage regimens [24].

4.6. Digestive Tract Protection

P. emblica L. polyphenols have also been indicated to protect gastrointestinal organs. One of the potential effects of amla bioactive compounds is the potential inhibition of clarithromycin-resistant *Helicobacter pylori* strains in vitro, since this microorganism is a known cause of gastric ulcers [61].

Relevant outcomes were also reported in studies with animals. Al-Rehaily et al. [62] studied anti-secretory and antiulcer activities of *P. emblica* L. extract in mice with different methods to induce gastrointestinal ulcers: ligating pylorus, administrating indomethacin and necrotizing agents (25% NaCl, 0.2 M NaOH, and 80% ethanol), and inducing hypothermia. Both doses (250 and 500 mg/kg) reduced gastric secretion, ulcer index (pylorus-ligated and necrotizing agent-intoxicated ulcer methods), intraluminal bleeding and gastric lesions (hypothermic restraint-induced ulcer method). Particularly for the indomethacin-induced ulcer method, only the animals in the treatment with 500 mg/kg had a significantly lower ulcer index than animals in the control group (treated only with indomethacin).

The potential of amla phytochemicals to protect the liver was reported in the study carried out by Huang et al. [63] with mice with high fat diet-induced non-alcoholic fatty liver disease. According to the authors, the liver of animals treated with amla extract had significant improvements in adiponectin activity and expression of PPAR-α, which improved steatosis. In another experiment, the use of *P. emblica* L. (200 mg/100 g; rich in tannins and gallic acid) in L-arginine-induced pancreatitis in mice decreased lipase and IL-10 blood concentrations [59]. This study also revealed that animals in the amla group had more suitable nucleic acid content material, pancreatic protein, rate of DNA synthesis, and pancreatic amylase levels and the histological examination indicated an extensively higher share of smooth cells and a lower inflammatory score. Likewise, the methanolic extract of *P. emblica* L. fruit (100 and 200 mg/kg) reduced the histological alterations in the colon of mice from acetic acid-induced colitis [86].

Another interesting potential health benefit from amla consumption was reported in patients with gastroesophageal reflux disease [64]. The daily consumption of amla extract (500 mg/tablet, twice a day) reduced the severity and frequency of regurgitation and heartburn in comparison to placebo group. The aforementioned experiments support the protection of organs involved in digestion with amla phytochemicals (especially polyphenols) consumption. Moreover, seems reasonable to indicate that modern medicine provides partial support to traditional medicine practices with amla.

4.7. Neurological Protection

One of the potential protective effects associated with amla bioactive compound is the attenuation of neurological alterations, particularly the biochemical changes observed in carriers of Alzheimer's disease. For instance, the administration of amla fruit extract (100 mg/kg; rich in emblicanin A and B) for 60 days in mice reduced the neurotoxicity induced by aluminum chloride [50]. Specifically, significant improvement against the triggering of apoptotic mechanisms (involving apoptotic protease activating factor 1, Bax, and cytosolic cyto c proteins) with reduced acetylcholinesterase activity in the cerebellum was observed in the group treated with amla extract. A related experiment indicated a similar outcome in the expression of Bax, caspases-3 and -9, cytochrome c proteins and also indicated the reduction in tau hyperphosphorylation [54]. Moreover, this study also revealed the involvement of GSK-3β/Akt signaling pathway in the reduced phosphorylation of tau protein in animals treated with amla extract (100 mg/kg for 60 days).

Another relevant outcome reported from the administration of amla extract (up to 200 mg/kg; containing emblicanin A and B, punigluconin, pedunculagin, rutin, and gallic acid) in animals with chemically induced neurological impairment is the improvement of memory and learning deficit in different studies [57,58]. The enhancement of neurological functions associated with amla was also observed in healthy animals [87]. In this case, the improvements were observed in animals that consumed extracts from unripe fruits at 100 and 200 mg/kg doses and ripe fruits at 200 mg/kg, which suggests that neuroactive compounds may be found in unripe rather than ripe amla fruits. It is also relevant to mention that the study carried out by Dhingra et al. [65] indicated some antidepressant possible mechanisms of action associated with amla polyphenols. These authors observed partial inhibition of the antidepressant effect of amla extract was observed in animals co-administered with γ-aminobutyric acid, alpha 1-adrenoceptor, and selective D2-receptor antagonists as well as tryptophan hydroxylase inhibitor. The studies reporting the neuroprotective effects of amla phytochemical support the potential role as assisting agent to attenuate biochemical and physiological changes associated with neurological disorders.

5. Conclusions

The rich phytochemistry composition of amla can be seen as a relevant source of compounds with potential health benefits. The antioxidant (from the rich polyphenol composition) is a major property with scientific evidence supporting the direct inhibition of oxidative reactions and the induction of an endogenous antioxidant defense system. Beyond antioxidant activity, seems reasonable to consider with the current level of evidence that amla components (mainly polyphenols) may have a role as supporting source of active compounds to promote health (such as improving antioxidant status in smokers and improving digestive tract protection against stressing agents) and increase the protection against the development of diseases (assisting in the regulation of serum glucose and insulin levels, for instance).

Although a promising scenario can be seen for amla, it is important to promote the progression of studies to strengthen the current evidence with more studies (especially at animal and human levels). Clarifying aspects related to bioaccessibility of bioactive compounds, interaction with gut microbiota, and also exploring technologies and strategies to promote the incorporation into food products (functional foods) are relevant aspects to be explored in future studies.

Author Contributions: Conceptualization, M.G. and R.M.A.; writing—original draft preparation, M.G., Z.-W.L., I.-U.-H., R.R., F.F., N.W., A.N. and M.A.S.; writing—review and editing, M.A.S., P.E.S.M., J.M.L. and R.M.A.; and supervision, M.A.S., P.E.S.M., J.M.L. and R.M.A. All authors have read and agreed to the published version of the manuscript.

Funding: This research received no external funding.

Acknowledgments: Thanks to GAIN (Axencia Galega de Innovación) for supporting this research (grant number IN607A2019/01). P.E.S.M. acknowledges postdoctoral fellowship support from the Ministry of Science and Innovation (MCIN, Spain) "Juan de la Cierva" program (IJC2020-043358-I).

Conflicts of Interest: The authors declare no conflict of interest.

References

1. Walia, K.; Boolchandani, R.; Dhand, S.; Antony, B. Improving glycemic & lipidemic profile with amla powder (*Emblica officinalis*) supplementation in adults with type 2 diabetes mellitus. *Int. J. Basic Appl. Med. Sci.* **2015**, *5*, 251–258.
2. Ramakrishna, N.; Singh, D.R. Ethno-Botanical Studies of Edible Plants Used by Tribal Women of Nirmal District. *Int. J. Sci. Res. Sci. Eng. Technol.* **2020**, *3*, 307–310. [CrossRef]
3. Jaiswal, Y.S.; Williams, L.L. A glimpse of Ayurveda—The forgotten history and principles of Indian traditional medicine. *J. Tradit. Complement. Med.* **2017**, *7*, 50–53. [CrossRef] [PubMed]
4. Pria, F.F.; Islam, M.S. *Phyllanthus emblica* Linn. (Amla)—A Natural Gift to Humans: An Overview. *J. Dis. Med. Plants* **2019**, *5*, 1–9. [CrossRef]
5. Kumar, G.; Madka, V.; Pathuri, G.; Ganta, V.; Rao, C.V. Molecular Mechanisms of Cancer Prevention by Gooseberry (*Phyllanthus emblica*). *Nutr. Cancer*, 2021; *in press*. [CrossRef]
6. Kapoor, M.P.; Suzuki, K.; Derek, T.; Ozeki, M.; Okubo, T. Clinical evaluation of *Emblica officinalis* Gatertn (Amla) in healthy human subjects: Health benefits and safety results from a randomized, double-blind, crossover placebo-controlled study. *Contemp. Clin. Trials Commun.* **2020**, *17*, 100499. [CrossRef]
7. Khan, K.H. Roles of *Emblica officinalis* in medicine—A review. *Bot. Res. Int.* **2009**, *2*, 218–228.

8. Hussain, S.Z.; Naseer, B.; Qadri, T.; Fatima, T.; Bhat, T.A. Anola (*Emblica officinalis*): Morphology, Taxonomy, Composition and Health Benefits. In *Fruits Grown in Highland Regions of the Himalayas*; Hussain, S.Z., Naseer, B., Qadri, T., Fatima, T., Bhat, T.A., Eds.; Springer: Cham, Switzerland, 2021; pp. 193–206. ISBN 30755027_15.
9. KC, Y.; Rayamajhi, S.; Dangal, A.; Shiwakoti, L.D. Phytochemical, Nutritional, Antioxidant Activity and Sensorial Characteristics of Amala (*Phyllanthus emblica* L.) Chutney. *Asian Food Sci. J.* **2020**, *18*, 43–52. [CrossRef]
10. Tewari, R.; Kumar, V.; Sharma, H.K. Physical and chemical characteristics of different cultivars of Indian gooseberry (*Emblica officinalis*). *J. Food Sci. Technol.* **2019**, *56*, 1641–1648. [CrossRef]
11. Parveen, K.; Khatkar, B.S. Physico-chemical properties and nutritional composition of aonla (*Emblica officinalis*) varieties. *Int. Food Res. J.* **2015**, *22*, 2358–2363.
12. Sonkar, N.; Rajoriya, D.; Chetana, R.; Venkatesh Murthy, K. Effect of cultivars, pretreatment and drying on physicochemical properties of Amla (*Emblica officinalis*) gratings. *J. Food Sci. Technol.* **2020**, *57*, 980–992. [CrossRef]
13. Carr, A.C.; Lykkesfeldt, J. Discrepancies in global vitamin C recommendations: A review of RDA criteria and underlying health perspectives. *Crit. Rev. Food Sci. Nutr.* **2021**, *61*, 742–755. [CrossRef] [PubMed]
14. Tarwadi, K.; Agte, V. Antioxidant and micronutrient potential of common fruits available in the Indian subcontinent. *Int. J. Food Sci. Nutr.* **2007**, *58*, 341–349. [CrossRef] [PubMed]
15. Jain, S.K.; Khurdiya, D.S. Vitamin C enrichment of fruit juice based ready-to-serve beverages through blending of Indian gooseberry (*Emblica officinalis* Gaertn.) juice. *Plant Foods Hum. Nutr.* **2004**, *59*, 63–66. [CrossRef] [PubMed]
16. Bansal, V.; Sharma, A.; Ghanshyam, C.; Singla, M.L. Coupling of chromatographic analyses with pretreatment for the determination of bioactive compounds in *Emblica officinalis* juice. *Anal. Methods* **2014**, *6*, 410–418. [CrossRef]
17. Bansal, V.; Sharma, A.; Ghanshyam, C.; Singla, M.L. Rapid HPLC Method for determination of vitamin C, phenolic acids, hydroxycinnamic acid, and flavonoids in seasonal samples of *Emblica officinalis* juice. *J. Liq. Chromatogr. Relat. Technol.* **2015**, *38*, 619–624. [CrossRef]
18. Nambiar, S.S.; Paramesha, M.; Shetty, N.P. Comparative analysis of phytochemical profile, antioxidant activities and foam prevention abilities of whole fruit, pulp and seeds of *Emblica officinalis*. *J. Food Sci. Technol.* **2015**, *52*, 7254–7262. [CrossRef]
19. Nambiar, S.S.; Shetty, N.P. Phytochemical Profiling and Assessment of Low-Density Lipoprotein Oxidation, Foam Cell-Preventing Ability and Antioxidant Activity of Commercial Products of *Emblica officinalis* Fruit. *J. Food Biochem.* **2015**, *39*, 218–229. [CrossRef]
20. Poltanov, E.A.; Shikov, A.N.; Dorman, H.J.D.; Pozharitskaya, O.N.; Makarov, V.G.; Tikhonov, V.P.; Hiltunen, R. Chemical and antioxidant evaluation of Indian gooseberry (*Emblica officinalis* Gaertn., syn. *Phyllanthus emblica* L.) supplements. *Phyther. Res.* **2009**, *23*, 1309–1315. [CrossRef]
21. Zhang, Y.J.; Abe, T.; Tanaka, T.; Yang, C.R.; Kouno, I. Two new acylated flavanone glycosides from the leaves and branches of *Phyllanthus emblica*. *Chem. Pharm. Bull.* **2002**, *50*, 841–843. [CrossRef]
22. Ur-Rehman, H.; Yasin, K.A.; Choudhary, M.A.; Khaliq, N.; Ur-Rahman, A.; Choudhary, M.I.; Malik, S. Studies on the chemical constituents of *Phyllanthus emblica*. *Nat. Prod. Res.* **2007**, *21*, 775–781. [CrossRef]
23. Liu, X.; Cui, C.; Zhao, M.; Wang, J.; Luo, W.; Yang, B.; Jiang, Y. Identification of phenolics in the fruit of emblica (*Phyllanthus emblica* L.) and their antioxidant activities. *Food Chem.* **2008**, *109*, 909–915. [CrossRef] [PubMed]
24. Fatima, N.; Pingali, U.; Muralidhar, N. Study of pharmacodynamic interaction of *Phyllanthus emblica* extract with clopidogrel and ecosprin in patients with type II diabetes mellitus. *Phytomedicine* **2014**, *21*, 579–585. [CrossRef] [PubMed]
25. Zhang, Y.J.; Abe, T.; Tanaka, T.; Yang, C.R.; Kouno, I. Phyllanemblinins A-F, new ellagitannins from *Phyllanthus emblica*. *J. Nat. Prod.* **2001**, *64*, 1527–1532. [CrossRef] [PubMed]
26. Al-Samman, A.M.M.A.; Siddique, N.A. Gas chromatography-mass spectrometry (GC-MS/MS) analysis, ultrasonic assisted extraction, antibacterial and antifungal activity of *Emblica officinalis* fruit extract. *Pharmacogn. J.* **2019**, *11*, 315–323. [CrossRef]
27. Sheoran, S.; Nidhi, P.; Kumar, V.; Singh, G.; Lal, U.R.; Sourirajan, A.; Dev, K. Altitudinal variation in gallic acid content in fruits of *Phyllanthus emblica* L. and its correlation with antioxidant and antimicrobial activity. *Vegetos* **2019**, *32*, 387–396. [CrossRef]
28. Fitriansyah, S.N.; Aulifa, D.L.; Febriani, Y.; Sapitri, E. Correlation of total phenolic, flavonoid and carotenoid content of *Phyllanthus emblica* extract from bandung with DPPH scavenging activities. *Pharmacogn. J.* **2018**, *10*, 447–452. [CrossRef]
29. Bar, F.M.A.; Habib, M.M.A.; Badria, F.A. A new hexagalloyl compound from *Emblica officinalis* Gaertn.: Antioxidant, cytotoxicity, and silver ion reducing activities. *Chem. Pap.* **2021**, *75*, 6509–6518. [CrossRef]
30. Tewari, R.; Kumar, V.; Sharma, H.K. Pretreated Indian Gooseberry (*Emblica officinalis*) Segments: Kinetic, Quality and Microstructural Parameters. *J. Inst. Eng. Ser. A* **2021**, *102*, 523–534. [CrossRef]
31. Pientaweeratch, S.; Panapisal, V.; Tansirikongkol, A. Antioxidant, anti-collagenase and anti-elastase activities of *Phyllanthus emblica, Manilkara zapota* and silymarin: An in vitro comparative study for anti-aging applications. *Pharm. Biol.* **2016**, *54*, 1365–1872. [CrossRef]
32. Zhang, Y.; Zhao, L.; Guo, X.; Li, C.; Li, H.; Lou, H.; Ren, D. Chemical constituents from *Phyllanthus emblica* and the cytoprotective effects on H_2O_2-induced PC12 cell injuries. *Arch. Pharm. Res.* **2016**, *39*, 1202–1211. [CrossRef]
33. Chahal, A.K.; Chandan, G.; Kumar, R.; Chhillar, A.K.; Saini, A.K.; Saini, R.V. Bioactive constituents of *Emblica officinalis* overcome oxidative stress in mammalian cells by inhibiting hyperoxidation of peroxiredoxins. *J. Food Biochem.* **2020**, *44*, e13115. [CrossRef] [PubMed]
34. Birben, E.; Sahiner, U.M.; Sackesen, C.; Erzurum, S.; Kalayci, O. Oxidative stress and antioxidant defense. *World Allergy Organ. J.* **2012**, *5*, 9–19. [CrossRef] [PubMed]

35. Shivananjappa, M.M.; Joshi, M.K. Influence of Emblica officinalis aqueous extract on growth and antioxidant defense system of human hepatoma cell line (HepG2). *Pharm. Biol.* **2012**, *50*, 497–505. [CrossRef] [PubMed]
36. Li, W.; Zhang, X.; Chen, R.; Li, Y.; Miao, J.; Liu, G.; Lan, Y.; Chen, Y.; Cao, Y. HPLC fingerprint analysis of *Phyllanthus emblica* ethanol extract and their antioxidant and anti-inflammatory properties. *J. Ethnopharmacol.* **2020**, *254*, 112740. [CrossRef]
37. Yamamoto, H.; Morino, K.; Mengistu, L.; Ishibashi, T.; Kiriyama, K.; Ikami, T.; Maegawa, H. Amla Enhances Mitochondrial Spare Respiratory Capacity by Increasing Mitochondrial Biogenesis and Antioxidant Systems in a Murine Skeletal Muscle Cell Line. *Oxid. Med. Cell. Longev.* **2016**, *2016*, 1735841. [CrossRef]
38. Nain, P.; Saini, V.; Sharma, S.; Nain, J. Antidiabetic and antioxidant potential of Emblica officinalis Gaertn. leaves extract in streptozotocin-induced type-2 diabetes mellitus (T2DM) rats. *J. Ethnopharmacol.* **2012**, *142*, 65–71. [CrossRef]
39. Singh, M.K.; Yadav, S.S.; Gupta, V.; Khattri, S. Immunomodulatory role of *Emblica officinalis* in arsenic induced oxidative damage and apoptosis in thymocytes of mice. *BMC Complement. Altern. Med.* **2013**, *13*, 193. [CrossRef]
40. Maiti, S.; Chattopadhyay, S.; Acharyya, N.; Deb, B.; Hati, A.K. *Emblica officinalis* (amla) ameliorates arsenic-induced liver damage via DNA protection by antioxidant systems. *Mol. Cell. Toxicol.* **2014**, *10*, 75–82. [CrossRef]
41. Saha, S.; Verma, R.J. Antioxidant activity of polyphenolic extract of *Phyllanthus emblica* against lead acetate induced oxidative stress. *Toxicol. Environ. Health Sci.* **2015**, *7*, 82–90. [CrossRef]
42. Biswas, T.K.; Chakrabarti, S.; Pandit, S.; Jana, U.; Dey, S.K. Pilot study evaluating the use of *Emblica officinalis* standardized fruit extract in cardio-respiratory improvement and antioxidant status of volunteers with smoking history. *J. Herb. Med.* **2014**, *4*, 188–194. [CrossRef]
43. Usharani, P.; Merugu, P.L.; Nutalapati, C. Evaluation of the effects of a standardized aqueous extract of *Phyllanthus emblica* fruits on endothelial dysfunction, oxidative stress, systemic inflammation and lipid profile in subjects with metabolic syndrome: A randomised, double blind, placebo. *BMC Complement. Altern. Med.* **2019**, *19*, 97. [CrossRef] [PubMed]
44. Reddy, V.D.; Padmavathi, P.; Paramahamsa, M.; Varadacharyulua, N.C. Amelioration of alcohol-induced oxidative stress by Emblica officinalis (Amla) in rats. *Indian J. Biochem. Biophys.* **2010**, *47*, 20–25. [PubMed]
45. Variya, B.C.; Bakrania, A.K.; Chen, Y.; Han, J.; Patel, S.S. Suppression of abdominal fat and anti-hyperlipidemic potential of Emblica officinalis: Upregulation of PPARs and identification of active moiety. *Biomed. Pharmacother.* **2018**, *108*, 1274–1281. [CrossRef]
46. Fatima, N.; Hafizur, R.M.; Hameed, A.; Ahmed, S.; Nisar, M.; Kabir, N. Ellagic acid in *Emblica officinalis* exerts anti-diabetic activity through the action on β-cells of pancreas. *Eur. J. Nutr.* **2017**, *56*, 591–601. [CrossRef]
47. Yang, C.J.; Wang, C.S.; Hung, J.Y.; Huang, H.W.; Chia, Y.C.; Wang, P.H.; Weng, C.F.; Huang, M.S. Pyrogallol induces G2-M arrest in human lung cancer cells and inhibits tumor growth in an animal model. *Lung Cancer* **2009**, *66*, 162–168. [CrossRef]
48. Malik, S.; Suchal, K.; Bhatia, J.; Khan, S.I.; Vasisth, S.; Tomar, A.; Goyal, S.; Kumar, R.; Arya, D.S.; Ojha, S.K. Therapeutic potential and molecular mechanisms of *Emblica officinalis* gaertn in countering nephrotoxicity in rats induced by the chemotherapeutic agent cisplatin. *Front. Pharmacol.* **2016**, *7*, 350. [CrossRef]
49. Golechha, M.; Sarangal, V.; Ojha, S.; Bhatia, J.; Arya, D.S. Anti-inflammatory effect of *Emblica officinalis* in rodent models of acute and chronic inflammation: Involvement of possible mechanisms. *Int. J. Inflam.* **2014**, *2014*, 178408. [CrossRef]
50. Bharathi, M.D.; Thenmozhi, A.J. Attenuation of Aluminum-Induced Neurotoxicity by Tannoid Principles of *Emblica officinalis* in Wistar Rats. *Int. J. Nutr. Pharmacol. Neurol. Dis.* **2018**, *8*, 35. [CrossRef]
51. Thirunavukkarasu, M.; Selvaraju, V.; Tapias, L.; Sanchez, J.A.; Palesty, J.A.; Maulik, N. Protective effects of *Phyllanthus emblica* against myocardial ischemia-reperfusion injury: The role of PI3-kinase/glycogen synthase kinase 3β/β-catenin pathway. *J. Physiol. Biochem.* **2015**, *71*, 623–633. [CrossRef]
52. Bhatia, J.; Tabassum, F.; Sharma, A.K.; Bharti, S.; Golechha, M.; Joshi, S.; Akhatar, M.S.; Srivastava, A.K.; Arya, D.S. *Emblica officinalis* exerts antihypertensive effect in a rat model of DOCA-salt-induced hypertension: Role of (p) eNOS, NO and Oxidative Stress. *Cardiovasc. Toxicol.* **2011**, *11*, 272–279. [CrossRef]
53. Khanna, S.; Das, A.; Spieldenner, J.; Rink, C.; Roy, S. Supplementation of a standardized extract from *Phyllanthus emblica* improves cardiovascular risk factors and platelet aggregation in overweight/class-1 obese adults. *J. Med. Food* **2015**, *18*, 415–420. [CrossRef] [PubMed]
54. Thenmozhi, A.J.; Dhivyabharathi, M.; Raja, T.R.W.; Manivasagam, T.; Essa, M.M. Tannoid principles of *Emblica officinalis* renovate cognitive deficits and attenuate amyloid pathologies against aluminum chloride induced rat model of Alzheimer's disease. *Nutr. Neurosci.* **2016**, *19*, 269–278. [CrossRef] [PubMed]
55. Kim, H.Y.; Okubo, T.; Juneja, L.R.; Yokozawa, T. The protective role of amla (*Emblica officinalis* Gaertn.) against fructose-induced metabolic syndrome in a rat model. *Br. J. Nutr.* **2010**, *103*, 502–512. [CrossRef] [PubMed]
56. Middha, S.K.; Goyal, A.K.; Lokesh, P.; Yardi, V.; Mojamdar, L.; Keni, D.S.; Babu, D.; Usha, T. Toxicological evaluation of *Emblica officinalis* fruit extract and its anti-inflammatory and free radical scavenging properties. *Pharmacogn. Mag.* **2015**, *11*, S427–S433. [CrossRef] [PubMed]
57. Thenmozhi, A.J.; Dhivyabharathi, M.; Manivasagam, T.; Essa, M.M. Tannoid principles of *Emblica officinalis* attenuated aluminum chloride induced apoptosis by suppressing oxidative stress and tau pathology via Akt/GSK-3βsignaling pathway. *J. Ethnopharmacol.* **2016**, *194*, 20–29. [CrossRef] [PubMed]

58. Husain, I.; Akhtar, M.; Madaan, T.; Vohora, D.; Abdin, M.Z.; Islamuddin, M.; Najmi, A.K. Tannins enriched fraction of *Emblica officinalis* fruits alleviates high-salt and cholesterol diet-induced cognitive impairment in rats via Nrf2-ARE pathway. *Front. Pharmacol.* **2018**, *9*, 23. [CrossRef] [PubMed]
59. Sidhu, S.; Pandhi, P.; Malhotra, S.; Vaiphei, K.; Khanduja, K.L. Beneficial effects of *Emblica officinalis* in L-arginine-induced acute pancreatitis in rats. *J. Med. Food* **2011**, *14*, 147–155. [CrossRef]
60. Kumar, N.P.; Annamalai, A.R.; Thakur, R.S. Antinociceptive property of *Emblica officinalis* Gaertn (Amla) in high fat diet fed/low dose streptozotocin induced diabetic neuropathy in rats. *Indian J. Exp. Biol.* **2009**, *47*, 737–742.
61. Mehrotra, S.; Jamwal, R.; Shyam, R.; Meena, D.K.; Mishra, K.; Patra, R.; De, R.; Mukhopadhyay, A.; Kumar, A.; Nandi, S.P. Anti-Helicobacter pylori and antioxidant properties of Emblica officinalis pulp extract: A potential source for therapeutic use against gastric ulcer. *J. Med. Plants Res.* **2011**, *5*, 2577–2583.
62. Al-Rehaily, A.J.; Al-Howiriny, T.S.; Al-Sohaibani, M.O.; Rafatullah, S. Gastroprotective effects of "Amla" *Emblica officinalis* on in vivo test models in rats. *Phytomedicine* **2002**, *9*, 515–522. [CrossRef]
63. Huang, C.Z.; Tung, Y.T.; Hsia, S.M.; Wu, C.H.; Yen, G.C. The hepatoprotective effect of *Phyllanthus emblica* L. fruit on high fat diet-induced non-alcoholic fatty liver disease (NAFLD) in SD rats. *Food Funct.* **2017**, *8*, 842–850. [CrossRef] [PubMed]
64. Karkon Varnosfaderani, S.; Hashem-Dabaghian, F.; Amin, G.; Bozorgi, M.; Heydarirad, G.; Nazem, E.; Nasiri Toosi, M.; Mosavat, S.H. Efficacy and safety of amla (*Phyllanthus emblica* L.) in non-erosive reflux disease: A double-blind, randomized, placebo-controlled clinical trial. *J. Integr. Med.* **2018**, *16*, 126–131. [CrossRef] [PubMed]
65. Dhingra, D.; Joshi, P.; Gupta, A.; Chhillar, R. Possible Involvement of Monoaminergic Neurotransmission in Antidepressant-like activity of *Emblica officinalis* Fruits in Mice. *CNS Neurosci. Ther.* **2012**, *18*, 419–425. [CrossRef] [PubMed]
66. Yokozawa, T.; Kim, H.Y.; Kim, H.J.; Okubo, T.; Chu, D.C.; Juneja, L.R. Amla (*Emblica officinalis* Gaertn.) prevents dyslipidaemia and oxidative stress in the ageing process. *Br. J. Nutr.* **2007**, *97*, 1187–1195. [CrossRef]
67. Huang, J.I.; Zhong, Z.G. Study of galic acid extracted from the leaves of *Phyllanthus emblica* on apoptotic mechanism of human hepatocellular carcinoma cells BEL-7404. *J. Chin. Med. Mater.* **2011**, *34*, 246–249.
68. Goyal, M.R.; Suleria, H. Olive Oil Phenols: Chemistry, Synthesis, Metabolism, Fate, And Their Allied Health Claims. In *Human Health Benefits of Plant Bioactive Compounds*; Goyal, M.R., Suleria, H.A.R., Eds.; Apple Academic Press: Palm Bay, FL, USA, 2019; pp. 95–127. ISBN 9780429457913.
69. Madan, J.; Sindhu, S.; Gupta, M.; Poonia, J. Evaluation of *Emblica officinalis* and *Mentha piperata* supplementation on biochemical parameters in growing beetal kids. *J. Cell Tissue Res.* **2015**, *15*, 4811–4814.
70. Kanthe, P.S.; Patil, B.S.; Bagali, S.C.; Reddy, C.R.; Aaithala, M.R.; Das, K.K. Protective effects of ethanolic extract of Emblica officinalis (amla) on cardiovascular pathophysiology of rats, fed with high fat diet. *J. Clin. Diagn. Res.* **2017**, *11*, CC05–CC09. [CrossRef]
71. Gopa, B.; Bhatt, J.; Hemavathi, K.G. A comparative clinical study of hypolipidemic efficacy of Amla (*Emblica officinalis*) with 3-hydroxy-3-methylglutaryl-coenzyme-A reductase inhibitor simvastatin. *Indian J. Pharmacol.* **2012**, *44*, 238–242. [CrossRef]
72. Koshy, S.M.; Bobby, Z.; Hariharan, A.P.; Gopalakrishna, S.M. Amla (*Emblica officinalis*) extract is effective in preventing high fructose diet-induced insulin resistance and atherogenic dyslipidemic profile in ovariectomized female albino rats. *Menopause* **2012**, *19*, 1146–1155. [CrossRef]
73. Nampoothiri, S.V.; Prathapan, A.; Cherian, O.L.; Raghu, K.G.; Venugopalan, V.V.; Sundaresan, A. In vitro antioxidant and inhibitory potential of *Terminalia bellerica* and *Emblica officinalis* fruits against LDL oxidation and key enzymes linked to type 2 diabetes. *Food Chem. Toxicol.* **2011**, *49*, 125–131. [CrossRef]
74. Ansari, A.; Shahriar, M.S.Z.; Hassan, M.M.; Das, S.R.; Rokeya, B.; Haque, M.A.; Haque, M.E.; Biswas, N.; Sarkar, T. *Emblica officinalis* improves glycemic status and oxidative stress in STZ induced type 2 diabetic model rats. *Asian Pac. J. Trop. Med.* **2014**, *7*, 21–25. [CrossRef]
75. Patel, S.S.; Goyal, R.K. Prevention of diabetes-induced myocardial dysfunction in rats using the juice of the *Emblica officinalis* fruit. *Exp. Clin. Cardiol.* **2011**, *16*, 87–91. [PubMed]
76. Akhtar, M.S.; Ramzan, A.; Ali, A.; Ahmad, M. Effect of amla fruit (*Emblica officinalis* Gaertn.) on blood glucose and lipid profile of normal subjects and type 2 diabetic patients. *Int. J. Food Sci. Nutr.* **2011**, *62*, 609–616. [CrossRef] [PubMed]
77. Lorenzo, J.M.; Munekata, P.E.; Putnik, P.; Kovačević, D.B.; Muchenje, V.; Barba, F.J. Sources, Chemistry, and Biological Potential of Ellagitannins and Ellagic Acid Derivatives. *Stud. Nat. Prod. Chem.* **2018**, *60*, 189–221. [CrossRef]
78. Munekata, P.E.S.; Pateiro, M.; Zhang, W.; Dominguez, R.; Xing, L.; Fierro, E.M.; Lorenzo, J.M. Health benefits, extraction and development of functional foods with curcuminoids. *J. Funct. Foods* **2021**, *79*, 104392. [CrossRef]
79. Rodríguez, M.L.; Estrela, J.M.; Ortega, Á.L. Natural Polyphenols and Apoptosis Induction in Cancer Therapy. *J. Carcinog. Mutagen.* **2013**, *6*, 1–10. [CrossRef]
80. Singh, I.; Soyal, D.; Goyal, P. Radioprotective potential of *Emblica officinalis* fruit extract against hematological alterations induced by gamma radiation. In Proceedings of the International Conference on Emerging Frontiers and Challenges in Radiation Biology, Bikaner, India, 24–25 January 2012.
81. Zhu, X.; Wang, J.; Ou, Y.; Han, W.; Li, H. Polyphenol extract of *Phyllanthus emblica* (PEEP) induces inhibition of cell proliferation and triggers apoptosis in cervical cancer cells. *Eur. J. Med. Res.* **2013**, *18*, 46. [CrossRef]
82. Purena, R.; Seth, R.; Bhatt, R. Protective role of *Emblica officinalis* hydro-ethanolic leaf extract in cisplatin induced nephrotoxicity in Rats. *Toxicol. Rep.* **2018**, *5*, 270–277. [CrossRef]

83. Singh, M.K.; Yadav, S.S.; Yadav, R.S.; Chauhan, A.; Katiyar, D.; Khattri, S. Protective effect of *Emblica officinalis* in arsenic induced biochemical alteration and inflammation in mice. *SpringerPlus* **2015**, *4*, 438. [CrossRef]
84. Dang, G.K.; Parekar, R.R.; Kamat, S.K.; Scindia, A.M.; Rege, N.N. Antiinflammatory activity of *Phyllanthus emblica*, *Plumbago zeylanica* and *Cyperus rotundus* in acute models of inflammation. *Phyther. Res.* **2011**, *25*, 904–908. [CrossRef]
85. Goel, B.; Pathak, N.; Nim, D.K.; Singh, S.K.; Dixit, R.K.; Chaurasia, R. Evaluation of analgesic activity of *Emblica officinalis* in albino rats. *Int. J. Basic Clin. Pharmacol.* **2014**, *3*, 365–368. [CrossRef]
86. Deshmukh, C.D.; Bantal, V.; Pawar, A. Protective effect of *Emblica officinalis* fruit extract on acetic acid induced colitis in rats. *J. Herb. Med. Toxicol.* **2010**, *4*, 25–29.
87. Uddin, M.S.; Mamun, A.A.; Hossain, M.S.; Akter, F.; Iqbal, M.A.; Asaduzzaman, M. Exploring the effect of *Phyllanthus emblica* L. on cognitive performance, brain antioxidant markers and acetylcholinesterase activity in rats: Promising natural gift for the mitigation of Alzheimer's disease. *Ann. Neurosci.* **2016**, *23*, 218–229. [CrossRef] [PubMed]

antioxidants

MDPI

Article

The Biphasic Effect of Flavonoids on Oxidative Stress and Cell Proliferation in Breast Cancer Cells

Xiaomin Xi, Jiting Wang, Yue Qin, Yilin You, Weidong Huang and Jicheng Zhan *

College of Food Science and Nutritional Engineering, China Agricultural University, Beijing 100083, China; xxm1113@cau.edu.cn (X.X.); wangjiting@cau.edu.cn (J.W.); jnqinyue@cau.edu.cn (Y.Q.); yilinyou@cau.edu.cn (Y.Y.); weidonghuang@cau.edu.cn (W.H.)

* Correspondence: zhanjicheng@cau.edu.cn

Abstract: Flavonoids have been reported to play an essential role in modulating processes of cellular redox homeostasis such as scavenging ROS. Meanwhile, they also induce oxidative stress that exerts potent antitumor bioactivity. However, the contradiction between these two aspects still remains unclear. In this study, four typical flavonoids were selected and studied. The results showed that low-dose flavonoids slightly promoted the proliferation of breast cancer cells under normal growth via gradually reducing accumulated oxidative products and demonstrated a synergistic effect with reductants NAC or VC. Besides, low-dose flavonoids significantly reduced the content of ROS and MDA induced by LPS or Rosup but restored the activity of SOD. However, high-dose flavonoids markedly triggered the cell death via oxidative stress as evidenced by upregulated ROS, MDA and downregulated SOD activity that could be partly rescued by NAC pretreatment, which was also confirmed by antioxidative gene expression levels. The underlying mechanism of such induced cell death was pinpointed as apoptosis, cell cycle arrest, accumulated mitochondrial superoxide, impaired mitochondrial function and decreased ATP synthesis. Transcriptomic analysis of apigenin and quercetin uncovered that high-dose flavonoids activated TNF-α signaling, as verified through detecting inflammatory gene levels in breast cancer cells and RAW 264.7 macrophages. Moreover, we identified that BRCA1 overexpression effectively attenuated such oxidative stress, inflammation and inhibited ATP synthesis induced by LPS or high dose of flavonoids possibly through repairing DNA damage, revealing an indispensable biological function of BRCA1 in resisting oxidative damage and inflammatory stimulation caused by exogenous factors.

Keywords: breast cancer; flavonoids; oxidative stress; inflammation; transcriptome; BRCA1

Citation: Xi, X.; Wang, J.; Qin, Y.; You, Y.; Huang, W.; Zhan, J. The Biphasic Effect of Flavonoids on Oxidative Stress and Cell Proliferation in Breast Cancer Cells. *Antioxidants* **2022**, *11*, 622. https://doi.org/10.3390/antiox11040622

Academic Editors: José M. Matés and Stanley Omaye

Received: 13 January 2022
Accepted: 23 March 2022
Published: 24 March 2022

1. Introduction

Breast cancer remains the second most common cause of cancer mortality in women worldwide, although many advances in early detection and treatment have been explored. Breast cancer is categorized into three major subtypes based on the presence or absence of molecular biomarkers, namely estrogen or progesterone receptors and human epidermal growth factor 2 (ERBB2, formerly HER2): hormone-receptor-positive/ERBB2-negative (70% of patients), ERBB2-positive (15–20%) and triple-negative (TNBC, tumors lack the above standard molecular markers, 5%) [1]. To date, treatment options for breast cancer mainly include surgical resection, radiotherapy, chemotherapy and endocrine therapy based on hormone receptors. In recent years, immunotherapy represented by PD-1/PD-L1 has also been applied in breast cancer patients [2]. Unluckily, for people presenting metastatic disease, chemotherapy is the only strategy to select, with goals of prolonging life and palliating symptoms. However, most nonmetastatic breast cancer patients gradually obtained drug resistance in the later stage and suffered from great pain during long-term treatment [3]. Therefore, exploring novel therapies that are effective while minimizing side effects is essential in curtailing breast cancer.

Flavonoids, belonging to secondary metabolites broadly distributed in plants, have multiple biological and pharmacological activities in improving body health, emerging as promising options for breast cancer prevention and treatment. These compounds act as free radical scavengers and antioxidants, exhibiting anti-mutagenic, anti-inflammatory and antiviral effects [4]. Moreover, flavonoids have the ability to reduce plasma levels of low-density lipoproteins, inhibit platelet aggregation and reduce cell proliferation. The health benefits of flavonoids have been historically ascribed to their chelating and antioxidant properties led by their innate chemical structure. The presence of multiple hydroxy groups along with a highly conjugated delocalized electron system enhances the free radical scavenging nature of the flavonoids and its interference with the redox activity of the cell [5]. The excellent antioxidant properties of flavonoids are especially evident under the stimulation of external oxidation conditions, such as in protecting skin cells under ultraviolet radiation; alleviating the oxidation products produced by certain chronic diseases such as diabetes and obesity; and maintaining the balanced redox situation in brain neurons, especially in degenerative brain diseases [6–11]. In terms of cancer biology, a previous study found that apigenin and luteolin induced apoptosis and cell cycle arrest in MCF-7 and TNBC cell lines via downregulating PI3K/Akt and upregulating FOXO3a, which correlated with reactive oxygen species (ROS) removal [12]. Besides, quercetin and resveratrol widely existed in wine-inhibited MCF-7 cell line proliferation through neutralizing hydrogen peroxide (H_2O_2) [13]. Butein was shown to induce apoptosis, as evidenced by increased caspase-3 and caspase-9 activities; notably, it reduced the ROS levels with the dose of 10 μg/mL butein for 5 min, which was also confirmed in BT-474 xenografts [14]. Wu et al. proved Ziziphora clinopodioides flavonoids protected against H_2O_2-induced injury in HUVEC cells and this effect was related to the enhancement of antioxidant capacity and suppression of angiogenesis and apoptosis [15].

Nevertheless, flavonoids were noticed to induce ROS accumulation causing programmed cell death (e.g., apoptosis) or directly lead to severe necrosis, indicating extra innovative mechanisms accounting for such cell death via interfering with the redox balance. Apigenin was found to induce apoptosis in mouse macrophage ANA-1 cells through the increased ROS accumulation and high caspase-3 activity followed by activation of the mitogen-activated protein kinase (MAPK) pathway [16]. The flavanone naringenin induced ROS-dependent apoptosis in MDA-MB-468 cells, and myricetin presented inhibition of the growth of MDA-MB-231 and MDA-MB-468 cell lines in a pro-oxidative way [17–19]. Moreover, the flavone 5,7-dihydroxy-8-nitrochrysin dephosphorylated Akt signals and resulted in ROS-dependent cytotoxicity in HER2-positive MDA-MB-453 breast cancer cells [20,21]. The flavonoid silibinin also exerted strong apoptosis through ROS-dependent Notch-1/ERK/Akt pathway in MDA-MB-231 cells. In vivo studies also supported such a conclusion [22]. The myricetin derived from flavonoid oncamex triggered increased superoxide production and cytotoxicity in MDA-MB-231 xenografts [23]. In addition to the function of single-component flavonoids in inducing oxidative stress, extracts derived from natural products that contained mixtures of flavonoids also exerted similar effects, which will not be repeated here [24–27]. Although the above results demonstrated that flavonoids promoted cancer cell death by inducing oxidative stress at the phenotypical level, they did not explain further which factor of flavonoids determined such quite opposite physiological activities on breast cancer cells. Moreover, these experiments were carried out with a single different drug, and the dosage, cancer cell line and culture conditions were quite different, making it difficult to obtain consistent results. Overall, the effect of flavonoids on the alteration of ROS level in breast cancer cells is still unclear. Therefore, in our study, we aimed to determine whether and how the different doses of flavonoids influenced the cellular oxidative stress and cell number with multiple flavonoids in breast cancer cells and tried to determine the underlying mechanism. In our current study, with four different kinds of flavonoids, we expound that a high dose of flavonoids (more than 20 μM) disturbed the cellular redox homeostasis, as indicated by extensive oxidative stress, and destroyed cellular ATP synthesis due to the impaired mitochondrial function, while a low dose of flavonoid (no more than 10 μM) pretreatment could markedly reverse LPS- or Rosup-induced oxidative stress via acting as an ROS scavenger.

2. Materials and Methods

2.1. Reagents and Chemicals

Lipopolysaccharides (LPSs) from *E. coli* 055:B5 were purchased from Solarbio (Cat. No. L8880, Beijing, China) with a purity > 99%. *N*-Acetyl-L-cysteine (NAC) and vitamin C (VC) were purchased from Beyotime (Cat. No. ST1546; Cat. No. ST1434, Shanghai, China) with a purity > 99%. The standard samples of apigenin, quercetin, chrysin and diosmetin were purchased from CHENGDU MUST Biotechnology CO.,LTD (Cat. No. A0113, A0083, A0292, A0927, MUST, Chengdu, China) with HPLC purity > 99%. Rabbit BRCA1 Polyclonal Antibody was purchased from Beyotime (Cat. No. AF6339, Shanghai, China). BeyoClick EdU 488 kit was purchased from Beyotime (Cat. No. C0071S, Shanghai, China).

2.2. Cell Culture

Human breast cancer cell lines MCF-7 and MDA-MB-231 and macrophage RAW 264.7 cells were cultured in DMEM (Gibco, Invitrogen, Carlsbad, CA, USA) with 10% fetal bovine serum (FBS; Gibco, Invitrogen, Carlsbad, CA, USA) and penicillin (100 U/mL)-streptomycin (100 mg/mL). Murine breast cancer 4T1 was cultured in RPMI-1640 (Gibco, Invitrogen, Carlsbad, CA, USA) with 10% fetal bovine serum, 100 U/mL penicillin, and 100 μg/mL streptomycin. Cells were all cultured in an incubator at 37 °C with 5% CO_2 under sterile conditions. All cells were harvested by treatment with 0.25% trypsin–ethylenediaminetetraacetic acid (Trypsin-EDTA; Gibco, Invitrogen, Carlsbad, CA, USA). The information of above cell lines was listed in Supplementary Table S3.

2.3. Cell Number Detection and Colony Formation Assay

The cell counting kit-8 (CCK-8) assay (Cat. No. C0038, Beyotime, Shanghai, China) was used to detect cell numbers as measured by OD 450. Briefly, the cells were seeded in 100 μL of complete medium at a concentration of 3000-4000 cells per well in a 96-well plate overnight and then treated with indicated drugs and conditions. Then, according to the CCK-8 manufacturer's instructions, a microplate reader (Thermo Fisher, Waltham, MA, USA) was used for the absorbance detection at a wavelength of 450 nm as OD value. For colony formation assay, 500-1000 cells were seeded in a 24-well plate overnight, treated with indicated drugs the next day and then cultured for 1–2 weeks. The cells were fixed with 4% paraformaldehyde (Cat. No. P1110, Solarbio, Beijing, China) for 15 min; then, they washed with PBS twice and stained with 1% crystal violet solution (Cat. No. C0121, Beyotime, Shanghai, China) for 10 min. Finally, the number of colonies was counted.

2.4. Detection of ROS, MDA and SOD

The content of reactive oxygen species (ROS) was detected by both fluorescence microplate reader (Sunrise/Infinite F50, Tecan, Mannedorf, Switzerland) and flow cytometry with DCFH-DA Assay (Cat. No. S0033M, Beyotime, Shanghai, China) according to protocol. The activity of superoxide dismutase (SOD) was detected by a commercial kit (Cat. No. S0101M, Beyotime, China). The concentration of malondialdehyde (MDA) was detected by a commercial kit (Cat. No. S0131S, Beyotime, Shanghai, China). All samples were prepared according to the introductions of each kit, and the OD values were assayed by a full wavelength microplate photometer (Thermo Fisher, Waltham, MA, USA).

2.5. Flow Cytometry for Apoptosis and Cell cycle

Apoptotic cells were assessed using FITC Annexin V Apoptosis Detection kit (Cat. No. C1062M, Beyotime, Shanghai, China), and cell cycle was detected by Cell Cycle Analysis Kit with propidium staining (Cat. No. C1052, Beyotime, Shanghai, China) according to protocol.

2.6. Real-Time Quantitative PCR Assay

Total RNA was extracted using TRIzol reagent (Cat. No. 9108, Takara Biotechnology, Dalian, China) from cells according to the manufacturer's instructions, and then the RNA purity and concentration at 260:280 nm was measured. After diluting RNAs to the same concentration with RNase-free water, we used HiScript III 1st Strand cDNA Synthesis Kit (Cat. No. R312, Vazyme, Nanjing, China) to reversely transcribe 1 μg total RNAs and synthesize the first-strand cDNA according to the instructions. Real-time quantitative PCR was performed using ChamQ SYBR qPCR Master Mix (Cat. No. Q341, Vazyme, Nanjing, China). All analyses were performed via CFX Connect Real-Time System (Bio-Rad, Hercules, CA, USA). GAPDH was selected as the housekeeping gene to normalize the data of each target gene. The reaction conditions were as follows: 95 °C for 30 s, followed by 35 cycles of 95 °C for 5 s, 60 °C for 30 s, and melt curve. Results were presented as the fold change relative to the control. Three independent experiments were performed. All primers were listed in Supplementary Table S1.

2.7. Western Blotting

Cells were washed with cold PBS twice, and the total protein of cells was extracted using RIPA Lysis buffer (Cat. No. P0013B, Beyotime, Shanghai, China) premixed with 1 mM proteinase inhibitor PMSF. The protein concentration of cell lysis was detected using a BCA protein assay kit (Cat. No. P0012S, Beyotime, Shanghai, China). Then, samples were diluted to the same concentration and denatured with 5× SDS loading buffer (Cat. No. P0280, Beyotime, Shanghai, China). Protein was subjected to 10% SDS-PAGE by electrophoresis and transferred onto a polyvinylidene fluoride (PVDF) membrane. Membranes were incubated in 5% skimmed milk for 2 h at room temperature, followed by incubation with primary antibodies at 4 °C overnight. After being washed with 1× Tris-buffered saline containing Tween 20 (Cat. No. ST673, Beyotime, Shanghai, China), membranes were incubated with horseradish peroxidase (HRP)-conjugated secondary antibodies. Bands on the membrane were visualized by using BeyoECL Moon (Cat. No. P0018FS, Beyotime, Shanghai, China). For proteins of interest, band intensities were normalized to the housekeeping protein Histone H3. The detailed information of antibodies was listed in Supplementary Table S2.

2.8. Determination of Mitochondrial Membrane Potential

Mitochondrial membrane potential was determined in accordance with the standard procedures by mitochondrial membrane potential assay kit with JC-1 staining (Cat. No. C2006, Beyotime, Shanghai, China).

2.9. RNA-Seq Analysis

MCF-7 cells were treated with DMSO, 50 μM apigenin or quercetin for 48 h. Then, the total RNA was extracted and isolated with Oligo Magnetic Beads and randomly interrupted using divalent cations in NEB Fragmentation Buffer for cDNA synthesis. Libraries were generated using the NEB Next Ultra RNA Library Prep Kit for Illumina (New England BioLabs, Ipswich, MA, USA) following the manufacturer's instructions. Sequencing was conducted using the Illumina NovaSeq 6000 platform.

2.10. Analysis of Enrichment Pathways

For Gene Ontology (GO) analysis, we used the online website tool (https://david.ncifcrf.gov/home.jsp, accessed on 1 November 2019) and then use R studio for visualization. For volcano plot and gene set enrichment analysis (GSEA) analysis, R studio software was used to analyze the data. The selection of differentially expressed genes from the transcriptome was based on the following standard: $\log_2 |\text{fold change}| > 1$, p value < 0.01.

2.11. Quantitative ATP Detection

The ATP detection kit (Cat. No. S0026, Beyotime, Shanghai, China) was used to detect the concentration of ATP according to the manufacturer's instructions. Briefly, cells were seeded in 2 mL of complete culture medium at the same number of 1×10^5 cells per well in a six-well plate overnight, and then treated with indicated drugs according to the indicated design. The culture medium was removed, the cells were washed twice with cold PBS twice and the cells were lysed with 200 μL lysate per well. In order to fully lyse the cells, the cells were pipetted or the plate was shaken repeatedly to make the lysate completely contact the cells. Usually, cells could be lysed immediately when exposed to the lysate. The supernatant was obtained for subsequent determination via centrifuging at 4 °C and 12,000× g for 10 min. Standard curve and ATP detection working buffer were previously prepared according to the protocol. First, 100 μL ATP detection working buffer was added into the test holes of a 96-well plate, which was placed at room temperature for 5 min so that the background ATP was consumed. Then, 20 μL sample or standard was added into the test hole, which was quickly mixed with a micropipette, and the RLU value was measured with a luminometer after an interval of at least two seconds.

2.12. The Construction of BRCA1-Overexpressing Cells

The BRCA1 gene was subcloned from the cDNA of MCF-7 cells and inserted into the pLenti-CMV-Puro plasmid (Addgene, Watertown, MA, USA) and then verified by Sanger sequencing. HEK 293T cells were used to produce lentivirus with Lipo8000 Transfection Reagent (Cat. No. C0533, Beyotime, Shanghai, China). After cancer cells were seeded and cultured overnight, the fresh medium containing 5 μg/mL polybrene (Cat. No. ST1380, Beyotime, Shanghai, China) together with filtered lentivirus was used to culture cells for 24 h. Then, the cells were selected by puromycin (Cat. No. ST551, Beyotime, Shanghai, China) at an indicated concentration for a week to retain the successfully infected cells.

2.13. Flow Cytometry for CellROX Deep Red Probes

CellROX Deep Red probes (Cat. No. C10491, Thermo Fisher, Waltham, MA, USA) were used to detect the cellular oxidative stress by flow cytometry. Briefly, cells were seeded overnight and treated with indicated drugs the next day; then, they were collected and stained with probes according to protocol.

2.14. Mitochondrial Superoxide Staining

MitoSOX Red Indicator (Cat. No. 40778ES50, YEASEN, Shanghai, China) was applied to detect the cellular mitochondrial superoxide. Briefly, cells were seeded overnight and treated with indicated drugs the next day; then, they were stained with MitoSOX Red Indicator according to protocol. Then, cells were fixed with 4% paraformaldehyde and stained with DAPI (Cat. No. C1005, Beyotime, China) and then observed by confocal laser microscopy (Nikon, Tokyo, Japan).

2.15. DNA Damage Biomarker γ-H2AX Foci Detection

Briefly, cells treated with indicated flavonoid were washed with PBS and fixed for 10 min at room temperature with 4% paraformaldehyde (Cat. No. P0099, Beyotime, China) then permeabilized for 20 min on ice with 0.25% triton-X 100. Then, cells were washed twice with PBS and blocked with blocking buffer for 1 h at room temperature. Next, cells were incubated with the rabbit monoclonal antibody against phospho-histone H2AX (serine 139) (γ-H2AX, Cat. No. AF5836, Beyotime, China) at a 1:500 dilution for 1 h at room temperature. Cells were then washed twice with PBS and incubated for 1 h at room temperature with the secondary antibody Alexa Fluor 488 tagged anti-rabbit IgG (Cat. No. A0423, Beyotime, China). Cells were then washed twice and stained for 15 min with DAPI (Cat. No. C1005, Beyotime, China). Cells were then twice with PBS, and the slide was sealed. Fluorescence was observed by confocal laser microscopy (Nikon, Tokyo, Japan).

2.16. Statistical Analysis

All experiments were repeated three times independently with triplicates in each treatment. All data in this study were analyzed by unpaired Student's t-test or one-way

ANOVA (GraphPad Prism 8.0, La Jolla, CA, USA). Results are presented as the mean and standard error of the mean (mean ± SD). * $p < 0.05$, ** $p < 0.01$, *** $p < 0.001$, **** $p < 0.0001$. ns indicates no significant difference.

3. Results

3.1. Low-Dose Flavonoids Slightly Promoted the Growth of Breast Cancer Cells by Scavenging Accumulated Oxidative Products

Previously, we found that low-dose apigenin (5 μM) could promote breast cancer cell growth. Therefore, four flavonoids with similar chemical structures were detected. Among the cell lines, MCF-7 and MDA-MB-231 belong to human beings and 4T1 comes from mice. Compared with the DMSO group, 5 μM flavonoids for 60 h, but not 12 or 24 h treatment, stimulated cell proliferation in all cell lines, suggesting that the growth promotion induced by low-dose flavonoids requires a relative long treating time (Figure 1A–C). We hypothesized that such a phenomenon was correlated with the accumulated oxidative products such as ROS during the long-period growth. Hence, we obtained the old medium by culturing MCF-7 cells for two days. Then, we added the old medium into the new seeded MCF-7 cells (old medium). The above old medium supplemented with VC (500 μM) was added into the new MCF-7 seeded cells altogether (A_VC). We also prepared another medium by simultaneously adding VC (500 μM) to the medium for 48 h cotreatment in MCF-7 cells, which was subsequently collected and added into the new seeded MCF-7 cells (S_VC). As shown in Figure 1D,E, the growth rate of A_VC was significantly higher than that of S_VC, indicating that the reductants were metabolized during normal cell growth and the post-addition of reductant was better. These results were in line with NAC (5 μM) groups. Moreover, low-dose flavonoids (apigenin, 5 μM) also displayed a similar function of acting as a reductant by alleviating the harmful oxidants (Figure 1F). To determine the cellular toxicity of four flavonoids on cancer cells, the IC_{50} was detected. Generally, chrysin showed the lowest IC_{50} with the strongest inhibitory effect in MCF-7 cells, followed by apigenin, diosmetin and quercetin, which was basically higher than 50 μM (Figure 1G).

3.2. Different Pharmacological Doses of Flavonoids Determined the Biphasic Biological Function in MCF-7 Breast Cancer Cells

To explore the correlation between flavonoid doses and cell proliferation, four flavonoids with a series of low or high doses were used to treat MCF-7 breast cancer cells. We found that a low dose of flavonoids ranging from 1 to 8 μM for 60 h slightly enhanced the cell growth, especially at 8 μM, and some of them started to work at 1 μM. Besides, the pretreatment with VC or NAC for 6 h synergistically promoted the cell growth, in line with previous results (Figure 2A–D). Conversely, a high dose of flavonoids ranging from 20 to 120 μM for 48 h markedly inhibited cell proliferation and induced cell death due to the excessive oxidative stress since it could be partly rescued by VC or NAC pretreatment (Figure 2E–H). Colony formation assay was also applied to evaluate the colonization ability of cancer cells. The results showed that pretreatment with 5 μM chrysin markedly restored the LPS or Rosup inhibited colony numbers (Figures 2I and S3C). However, among high-dose groups, 50 μM chrysin significantly downregulated the colony numbers, and the NAC or VC supplement attenuated such suppression. Taken together, these results showed that flavonoids had dual effects on cancer cells depending on the drug doses and treating time, which was relevant to oxidative stress.

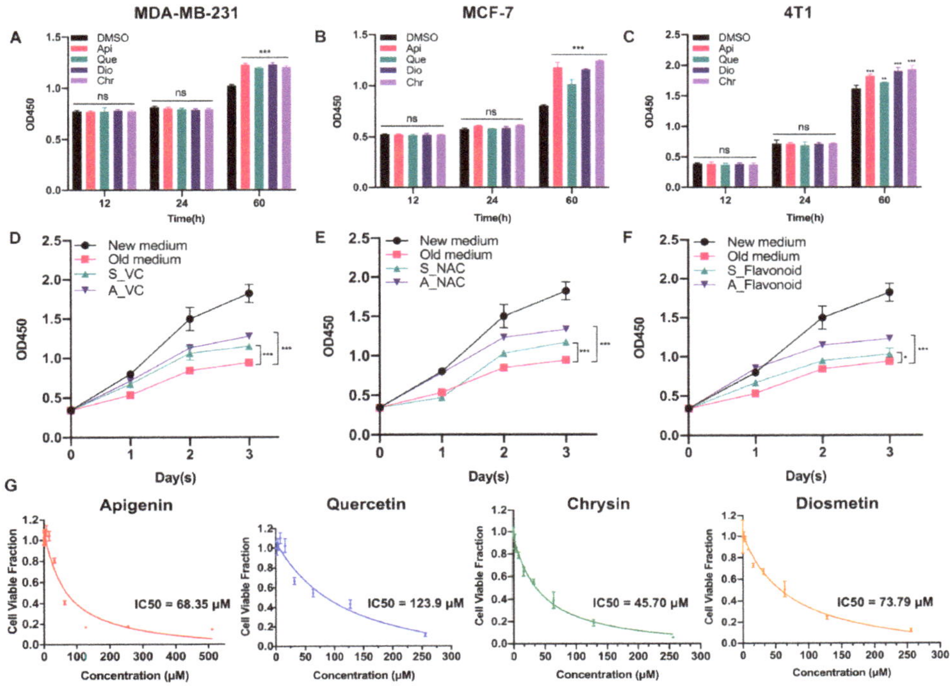

Figure 1. Low-dose flavonoids promoted cell growth by reducing oxidative reactions under long-term treatment. (**A**) The cell number of MDA-MB-231 treated with DMSO or flavonoids under 12, 24 or 60 h was detected. (**B**) The cell number of MCF-7 treated with DMSO or flavonoids under 12, 24 or 60 h was detected. (**C**) The cell number of 4T1 treated with DMSO or flavonoids under 12, 24 or 60 h was detected. (**D**) The growth rate of four groups. New medium: culturing the new seeded cells with fresh medium; Old medium: culturing the new seeded cells with old medium that had already cultured MCF-7 cells for 48 h. A_VC: culturing the new seeded cells with the above old medium but with VC supplement. S_VC: culturing the new seeded cells with collected medium in which VC had been simultaneously added for 48 h cotreatment in MCF-7 cells. VC: 500 μM. (**E**) The growth rate of four groups. The treatment was similar to (**D**). NAC: 5 μM. (**F**) The growth rate of four groups. The treatment was similar to (**D**). Flavonoid: apigenin, 5 μM. (**G**) The IC_{50} of four flavonoids was detected in MCF-7 cells. Data was shown as the mean ± SD. * $p < 0.05$, ** $p < 0.01$, *** $p < 0.001$. ns indicates no significant difference.

3.3. Dual Effects of Flavonoids on Physiological Redox Biomarkers and Relevant Gene Expression Levels

To elucidate the role of flavonoids involved in the oxidative response, we measured the content of ROS, MDA and SOD activity in MCF-7 breast cancer cells. In general, a low dose of chrysin (5 μM) significantly alleviated the oxidative stress induced by LPS or Rosup, characterized by reduced ROS and MDA content but increased SOD activity, performing like reductants NAC/VC (Figure 3A–C). However, a high dose of chrysin (50 μM) showed the effect of promoting oxidative stress, which can be partially reversed by reductants (Figure 3D–F). We also detected the fluorescence intensity of ROS by flow cytometry (Figures S1B and S2A) under different doses of chrysin. Next, the expression levels of antioxidant and inflammatory genes were detected (Figure 3G–I). LPS for 24 h significantly increased the expression of antioxidant genes such as SOD and NAD(P)H quinone dehydrogenase 1 (NQO1), indicating the perturbed redox status. NAC pretreatment alleviated such oxidative stress and restored the imbalance to some extent. Similarly, low-dose chrysin also attenuated the expression of antioxidant genes stimulated by LPS, suggesting its intracellular antioxidant function; however, high-dose chrysin directly activated the expression of these genes, indicating the disorder of cellular redox state. Oxidative stress is usually accompanied by a rapid inflammatory response. We also observed a similar effect on the expression level of inflammatory genes, especially cytokines such as IL-6, IFN-γ and TNF-α, which was examined in macrophage RAW 264.7 cells for further verification (Figure S3A). To further explore which type of ROS was increased following treatment with high-dose flavonoids, MitoSOX Red and CellROX Deep Red probes were used for detection. MitoSOX Red indicated that mitochondria superoxide around the nucleus was significantly enhanced after 50 μM chrysin treatment for 24 h, illustrating that a high dose of flavonoids widely induced oxidative stress both in mitochondria and cytoplasm (Figure S1A). Moreover, the fluorescence intensity of CellROX Deep Red demonstrated that high-dose apigenin triggered strong oxidative stress in the cytoplasm, in line with the above results (Figure S1C). Considering the accumulation of mitochondrial superoxide, we next detected mitochondrial membrane potential (MMP) and ATP production. LPS and 50 μM chrysin induced a higher percentage of JC-1 monomers, indicating an impaired mitochondria function (Figure 3J). Not surprisingly, ATP synthesis was inhibited after high-dose chrysin treatment (Figure 3K). NAC or low-dose chrysin alleviated such energy inhibition, consistent with JC-1 monomers. Taken together, these results presented that flavonoids have a dual effect on the redox of cancer cells, depending on drug doses.

3.4. High Dose of Apigenin or Quercetin Induced Cell Apoptosis and Cell Cycle Arrest to Restrain the Proliferation of MCF-7 Breast Cancer Cells

To investigate the inhibitory effect of high-dose flavonoids on the growth of breast cancer cells, 50 μM apigenin and quercetin were used to treat MCF-7 cells for 48 h. As shown in Figure 4A, the proportions of early apoptosis and late apoptosis in the DMSO treatment group were 0.87% and 3.37%; the proportions of early apoptosis and late apoptosis in the apigenin treatment group were 2.40% and 4.34%, respectively, increasing by 1.53% and 0.97%, showing that 50 μM apigenin induced mild apoptosis in the early stage. Annexin V, being the biomarker of apoptosis, was applied for staining to observe the apoptosis. Here, TNF-α was used as a positive control to induce apoptosis. Obviously, 50 μM apigenin was able to enhance the fluorescence intensity of annexin V, indicating the surge of apoptosis (Figure 4B). In terms of the cell cycle, 50 μM apigenin enhanced the cell distribution of both S and G2/M phases and markedly decreased the proportion of the G1 phase (Figure 4C). To intuitively observe the changes, immunofluorescence was confirmed via EdU and PCNA (proliferating cell nuclear antigen) staining. EdU (5-ethynyl-2′-deoxyuridine) is based on the incorporation of a thymidine analog in the process of DNA synthesis. PCNA is an auxiliary protein of DNA polymerase delta and is involved in the control of eukaryotic DNA replication by increasing the polymerase's processibility during elongation of the leading strand. Both biomarkers reflect the speed of cell replication and proliferation. Compared with control, the amounts of EdU- and PCNA -labeled cells treated with 50 μM apigenin were markedly decreased, presenting the inhibition of cell replication (Figure 4D). Overall, the proliferation inhibition induced by a high concentration of apigenin mainly resulted from cell cycle arrest, since the percentage of induced apoptosis was relatively low. Similarly, 50 μM quercetin triggered apoptosis and cell cycle arrest as well, leading to the same proliferative inhibition as apigenin (Figure 5).

3.5. The Transcriptome Analysis of High Dose of Apigenin or Quercetin in MCF-7 Breast Cancer Cells

To further explore the mechanism of growth inhibition in MCF-7 breast cancer cells, we systematically compared the transcriptomes between the DMSO- and flavonoid-treated groups (apigenin and quercetin, 50 μM for 48 h). The heat map revealed five distinct clusters differing between DMSO- and flavonoid-treated cells, suggesting that apigenin and quercetin resulted in significant alterations of gene expression patterns (Figure 6A). Importantly, the results of GO analysis presented that a high dose of apigenin contributed to the enhanced expression of genes regulated by TNF-α signaling, further supporting the previous results of inflammation response (Figure 6B). We also noticed that apigenin and quercetin inhibited the expression of genes correlated with cholesterol homeostasis, suggesting that apigenin may be able to interrupt the biosynthesis or metabolism of cholesterol in cancer cells (Figures 6C and 7B). Moreover, the GSEA plots presented that TNF-α and unfolded protein response signaling pathways, both reflecting the activation of oxidative stress, were markedly upregulated after apigenin treatment (Figure 6D,E). Besides, multiple genes specifically regulated by apigenin at the transcriptional level were demonstrated in the volcano plot (Figure 6F). Interestingly, quercetin showed similar results to those of apigenin in GO pathways, in which TNF-α signaling and unfolded protein response were largely stimulated in quercetin-treated cells (Figure 7A–D). Finally, the volcano plot of most altered genes was summarized (Figure 7E). Taken together, the transcriptome analysis of apigenin and quercetin displayed highly similar patterns, and both patterns revealed that high doses of flavonoids triggered the activation of oxidative stress and inflammation reaction.

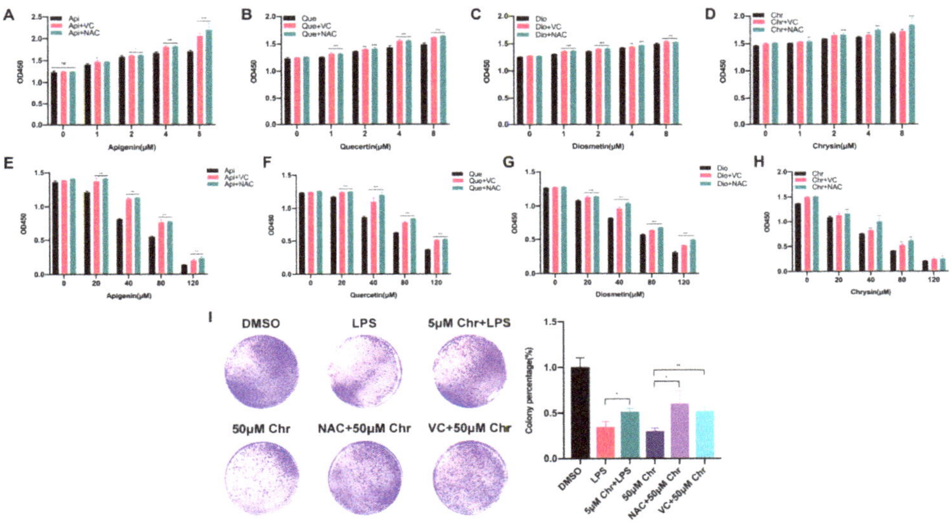

Figure 2. The dual effects of flavonoids on the proliferation of MCF-7 breast cancer cells depending on different drug doses. (**A**) The cell number of MCF-7 cells pretreated with NAC or VC for 6 h and thereafter receiving the indicated dose of apigenin for 60 h. (**B**) The cell number of MCF-7 cells pretreated with NAC or VC for 6 h and thereafter receiving the indicated dose of quercetin for 60 h. (**C**) The cell number of MCF-7 cells pretreated with NAC or VC for 6 h and thereafter receiving the indicated dose of diosmetin for 60 h. (**D**) The cell number of MCF-7 cells pretreated with NAC or VC for 6 h and thereafter receiving the indicated dose of chrysin for 60 h. (**E–H**) The cell number of MCF-7 cells pretreated with NAC or VC for 6 h and thereafter receiving the indicated high dose of four flavonoids for 48 h, respectively. (**I**) (**Left**) Colony formation assay of MCF-7 cells treated with 5 μg/mL LPS and/or indicated dose of chrysin. LPS: cells were treated with 5 μg/mL LPS for 48 h, then provided with fresh medium for one week; 5 μM Chr + LPS: cells were pretreated with 5 μM chrysin for 6 h and then 5 μg/mL LPS for 48 h, then provided with fresh medium for one week; 50 μM Chr: cells were treated with 50 μM chrysin for 48 h, then provided with fresh medium for one week; NAC + 50 μM Chr: cells were pretreated with 5 μM NAC for 6 h and then 50 μM chrysin for 48 h, then provided with fresh medium for one week; VC + 50 μM Chr: cells were pretreated with 500 μM VC for 6 h and then 50 μM chrysin for 48 h, then provided with fresh medium for one week. (**Right**) Quantitative analysis. Data was shown as the mean ± SD. * $p < 0.05$, ** $p < 0.01$, *** $p < 0.001$. ns indicates no significant difference.

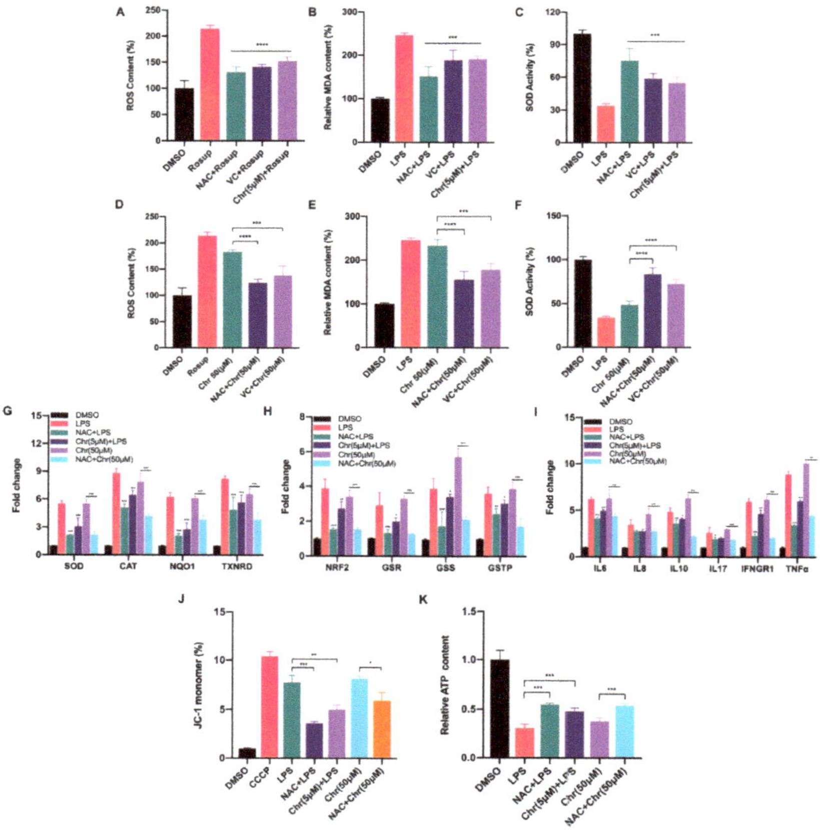

Figure 3. Dual effects of flavonoids on cellular redox biomarkers and relevant gene expression levels. (**A**) The ROS content of different groups. Rosup: 50 μg/mL for 24 h. NAC + Rosup: pretreated with 5 μM NAC for 6 h and then 50 μg/mL Rosup for 18 h. VC + Rosup: pretreated with 500 μM VC for 6 h and then 50 μg/mL Rosup for 18 h. Chr (5 μM) + Rosup: pretreated with 5 μM chrysin for 6 h and then 50 μg/mL Rosup for 18 h. (**B**) The MDA content of different groups. LPS: 5 μg/mL for 24 h. NAC + LPS: pretreated with 5 μM NAC for 6 h and then 5 μg/mL LPS for 18 h. VC + LPS: pretreated with 500 μM VC for 6 h and then 5 μg/mL LPS for 18 h. Chr (5 μM) + LPS: pretreated with 5 μM chrysin for 6 h and then 5 μg/mL LPS for 18 h. (**C**) The SOD activity of different groups. The treatment was similar to (**B**). (**D**) The ROS content of different groups. Rosup: 50 μg/mL for 24 h. Chr (50 μM): 50 μM chrysin for 24 h. NAC + Chr (50 μM): pretreated with 5 μM NAC for 6 h and then 50 μM chrysin for 18 h. VC + Chr (50 μM): pretreated with 500 μM VC for 6 h and then 50 μM chrysin for 18 h. (**E**) The MDA content of different groups. LPS: 5 μg/mL for 24 h. NAC + Chr (50 μM): pretreated with 5 μM NAC for 6 h and then 50 μM chrysin for 18 h. VC + Chr (50 μM): pretreated with 500 μM VC for 6 h and then 50 μM chrysin for 18 h. (**F**) The SOD activity of different groups. The treatment was similar to (**E**). (**G**,**H**) The qPCR results of antioxidative genes. Chr (5 μM) + LPS: pretreated with 5 μM chrysin for 6 h and then 5 μg/mL LPS for 18 h. Chr (50 μM): 50 μM chrysin for 24 h. NAC + Chr (50 μM): pretreated with 5 μM NAC for 6 h and then 50 μM chrysin for 18 h. (**I**) The qPCR results of inflammatory genes. (**J**) The percentage of JC-1 monomers in different groups. Higher proportion of JC-1 indicates more grievous mitochondrial membrane potential collapse. Cells were treated the same as described above. (**K**) The relative ATP content of different groups. Cells were treated the same as described above. Data was shown as the mean ± SD. * $p < 0.05$, ** $p < 0.01$, *** $p < 0.001$, **** $p < 0.0001$. ns indicates no significant difference.

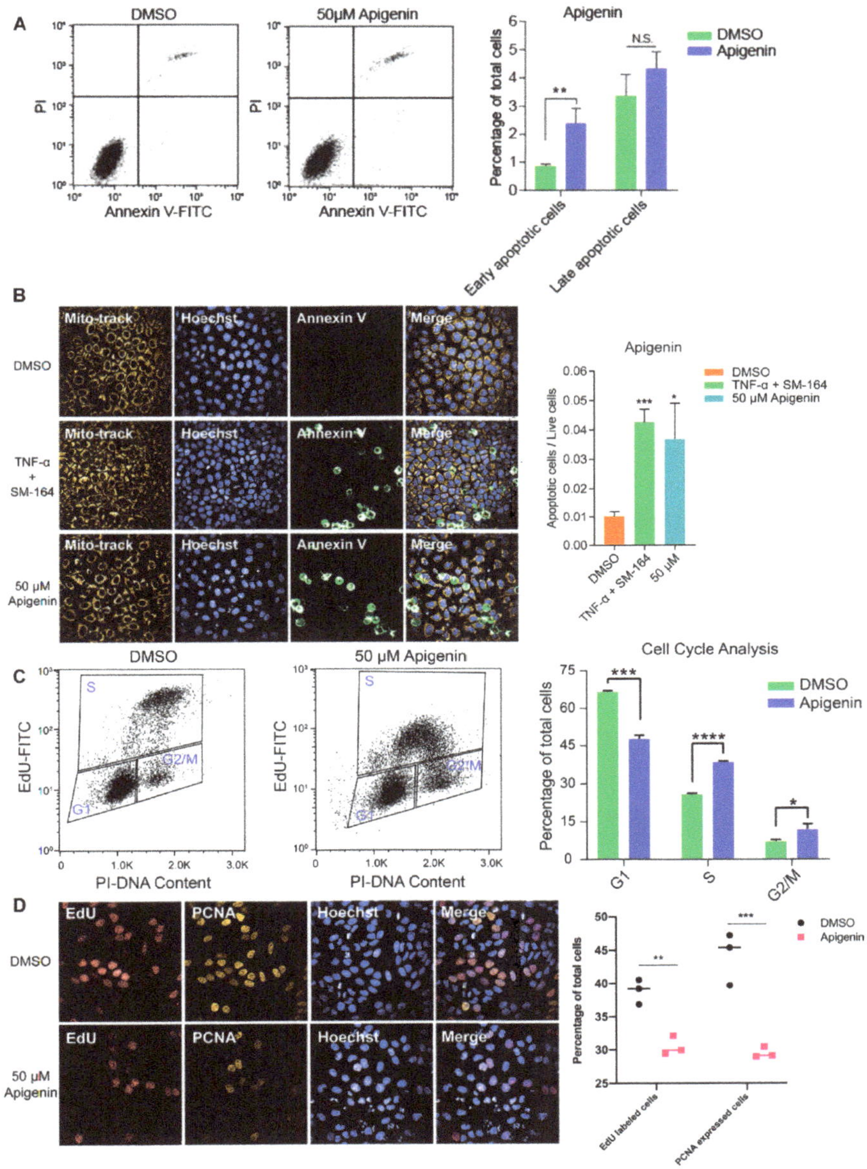

Figure 4. High dose of apigenin induced apoptosis and cell cycle arrest in MCF-7 cells. (**A**) Flow cytometry and statistical results of apoptosis in MCF-7 cells treated with DMSO or 50 μM apigenin for 48 h. TNF-α+SM-164 was positive control. (**B**) Immunofluorescence staining of annexin V and statistical results of apoptosis in MCF-7 cells treated with DMSO or 50 μM apigenin for 48 h. (**C**) Flow cytometry and statistical results of cell cycle arrest in MCF-7 cells treated with DMSO or 50 μM apigenin for 48 h. (**D**) Immunofluorescence results of indicators EdU and PCNA in MCF-7 cells treated with DMSO or 50 μM apigenin for 48 h. Cells were observed under 40× oil laser confocal microscope. Data was shown as the mean ± SD. * $p < 0.05$, ** $p < 0.01$, *** $p < 0.001$, **** $p < 0.0001$. ns indicates no significant difference.

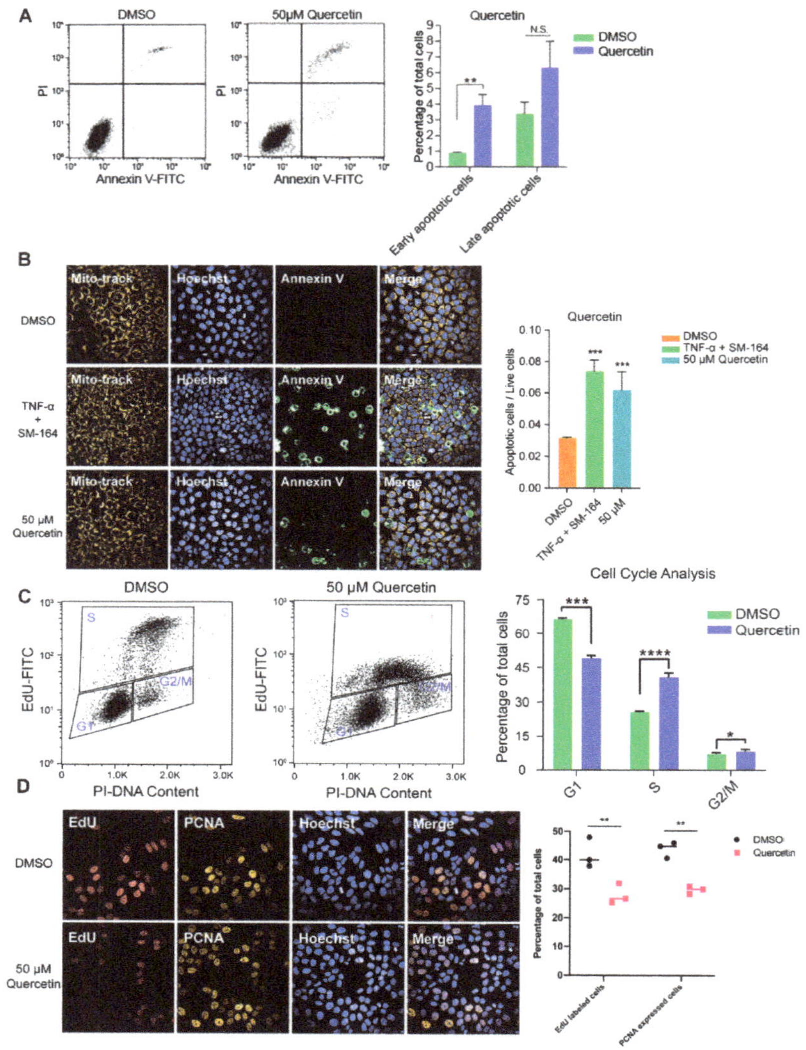

Figure 5. High dose of quercetin induced apoptosis and cell cycle arrest in MCF-7 cells. (**A**) Flow cytometry and statistical results of apoptosis in MCF-7 cells treated with DMSO or 50 μM quercetin for 48 h. TNF-α+SM-164 was positive control. (**B**) Immunofluorescence staining and statistical results of apoptosis in MCF-7 cells treated with DMSO or 50 μM quercetin for 48 h. (**C**) Flow cytometry and statistical results of cell cycle arrest in MCF-7 cells treated with DMSO or 50 μM quercetin for 48 h. (**D**) Immunofluorescence results of indicators EdU and PCNA in MCF-7 cells treated with DMSO or 50 μM quercetin for 48 h. Cells were observed under 40× oil laser confocal microscope. Data was shown as the mean ± SD. * $p < 0.05$, ** $p < 0.01$, *** $p < 0.001$, **** $p < 0.0001$. ns indicates no significant difference.

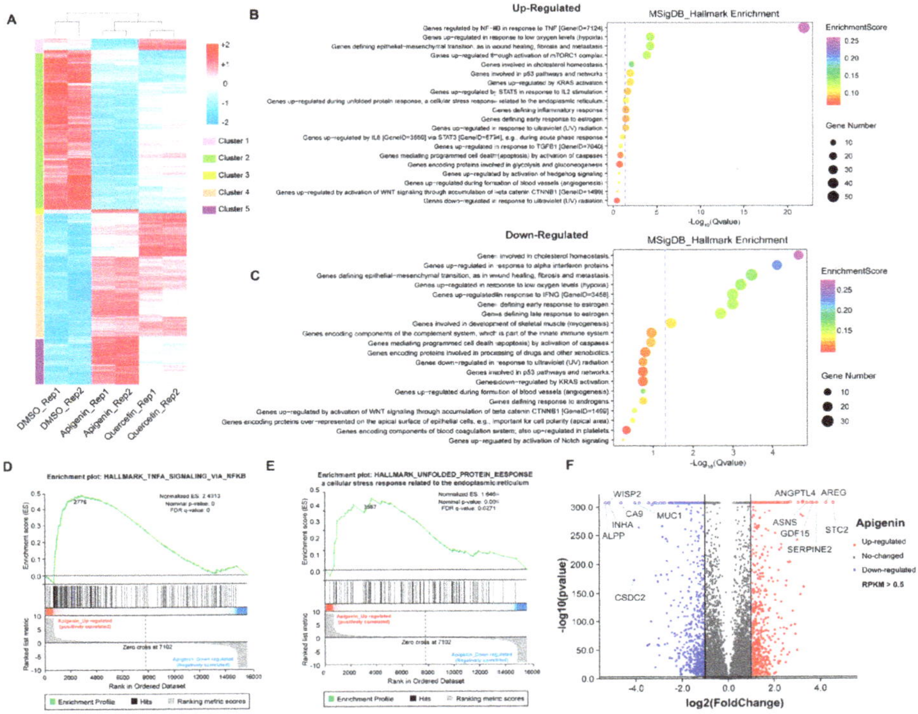

Figure 6. Transcriptome analysis of MCF-7 cells treated with high dose of flavonoids. (**A**) The heat map of cells treated with DMSO or 50 µM apigenin/quercetin for 48 h. (**B**) The GO analysis of cells treated with 50 µM apigenin for 48 h (upregulated). (**C**) The GO analysis of cells treated with 50 µM apigenin for 48 h (downregulated). (**D**) The GSEA analysis of cells treated with 50 µM apigenin for 48 h. TNF-α signaling was enriched and upregulated. (**E**) The GSEA analysis of cells treated with 50 µM apigenin for 48 h. Unfolded protein response signaling was enriched and upregulated. (**F**) The volcano plot revealing the most differentially expressed genes of cells treated with 50 µM apigenin for 48 h. The selection of differentially expressed genes from the transcriptome was based on $\log_2|\text{fold change}| > 1$, p value < 0.01.

3.6. BRCA1 Alleviated the Oxidative Stress and Inflammation Induced by High Dose of Flavonoids

Breast cancer susceptibility gene 1 (BRCA1) is one of the most frequently mutated tumor suppressor genes in human breast cancers. In recent years, BRCA1 has been identified as an essential player in regulating intracellular oxidative stress [28]. Bae et al. found BRCA1 upregulated the expression of multiple genes involved in the cytoprotective antioxidant response, including glutathione S-transferases, oxidoreductases and other antioxidant genes. Meanwhile, BRCA1 overexpression conferred resistance while BRCA1 deficiency conferred sensitivity to oxidizing agents in vitro and in vivo [29–32]. To investigate whether BRCA1 modulates oxidative stress induced by high-dose flavonoids, BRCA1-overexpressing cells were constructed and verified in MCF-7 cells (Figure 8A). Without drug treatment, no difference in the growth rate of BRCA1-overexpressing cells (OE) and wild-type cells (WT) was observed. The survival rate of OE cells was much higher than that of WT cells under stimulation, indicating that BRCA1 overexpression effectively alleviated such oxidative stress and thereby attenuated cell death (Figure 8B). With the NAC pretreatment, no difference in cell number was observed (Figure 8C). Interestingly, the cell number of OE cells treated with NAC and chrysin was still higher than that of WT, suggesting that NAC did not completely offset the chrysin-induced oxidative stress, which indicated that chrysin may trigger cell death via other non-oxidative-stress pathways. Next, the levels of SOD, MDA and ROS were detected. BRCA1 overexpression significantly rescued the oxidative stress stimulated by LPS or high-dose chrysin, evidenced as decreased ROS and MDA but elevated SOD activity in comparison to WT cells (Figure 8D–F). Nevertheless, such difference disappeared with NAC intervention. Moreover, BRCA1 improved the ATP production inhibited by LPS or 50 μM chrysin that was eliminated by the addition of NAC or 5 μM chrysin (Figure 8G). Finally, the expression level of antioxidant and inflammatory genes further supported the conclusion (Figure 8H,I). For many years, BRCA1 has been well established as a tumor suppressor, and it functions primarily by maintaining genome integrity [33]. We then hypothesized that BRCA1 attenuates the oxidative stress induced by high-dose flavonoids by repairing DNA damage. H2AX phosphorylation at Ser139 (γ-H2AX) is a typical indicator of DNA double-strand breaks after different genotoxic stresses, including ionizing radiation, environmental agents and chemotherapy drugs. The fluorescence intensity of γ-H2AX foci was significantly reduced in OE cells when exposed to a high dose of chrysin compared with WT cells (Figure 9). Taken together, these results presented that BRCA1 effectively resisted the oxidative stress induced by a high dose of flavonoids, possibly by repairing DNA damage.

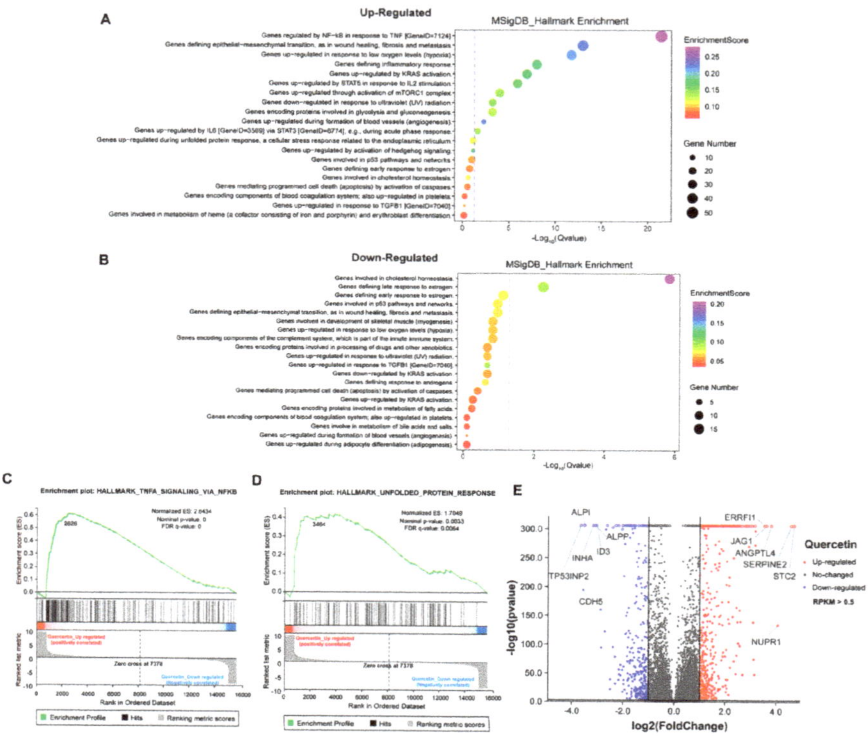

Figure 7. Transcriptome analysis of MCF–7 cells treated with high dose of quercetin. (**A**) The GO analysis of cells treated with 50 μM quercetin for 48 h (upregulated). (**B**) The GO analysis of cells treated with 50 μM quercetin for 48 h (downregulated). (**C**) The GSEA analysis of cells treated with 50 μM quercetin for 48 h. TNF-α signaling was enriched and upregulated. (**D**) The GSEA analysis of cells treated with 50 μM quercetin for 48 h. Unfolded protein response signaling was enriched and upregulated. (**E**) The volcano plot revealing the most differentially expressed genes of cells treated with 50 μM quercetin for 48 h. The selection of differentially expressed genes from the transcriptome was based on $\log_2 |\text{fold change}| > 1$, p value < 0.01.

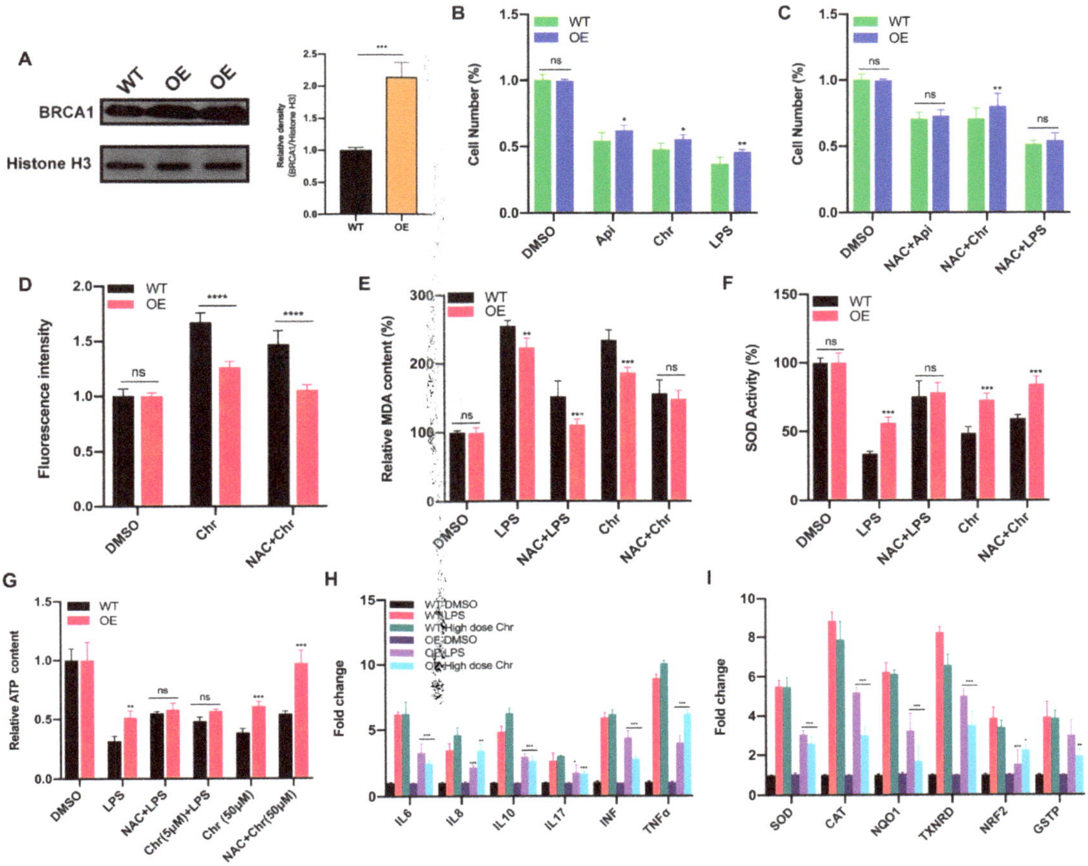

Figure 8. BRCA1 overexpression attenuated the oxidative stress induced by LPS or high dose of flavonoids. (**A**) The Western blot of wild type (WT) and BRCA1-overexpressing cells (OE). (**B**) The cell number of WT/OE cells treated with LPS or 50 μM apigenin/chrysin for 48 h. (**C**) The cell number of WT/OE cells pretreated with 5 μM NAC for 6 h and then 50 μM apigenin/chrysin or 5 μg/mL LPS for 48 h. (**D**) The fluorescence intensity of ROS detected by DCFH-DA assay. Chr: 50 μM chrysin for 24 h. NAC + Chr: pretreated with 5 μM NAC for 6 h and then 50 μM chrysin for 18 h. (**E**) The MDA content of WT/OE cells. LPS: 5 μg/mL LPS for 24 h. NAC + LPS: pretreated with 5 μM NAC for 6 h and then LPS for 18 h. (**F**) The SOD activity of WT/OE cells. (**G**) The relative ATP content of different groups. (**H**) The qPCR results of inflammatory genes in WT/OE cells. LPS: 5 μg/mL LPS for 24 h. High dose Chr: 50 μM chrysin for 24 h. The marked significance of OE groups was obtained by comparing the WT groups with the same treatment, respectively. (**I**) The qPCR results of antioxidative genes in WT/OE cells. The marked significance of OE groups was obtained by comparing the WT groups with the same treatment, respectively. Data was shown as the mean ± SD. * $p < 0.05$, ** $p < 0.01$, *** $p < 0.001$, **** $p < 0.0001$. ns indicates no significant difference.

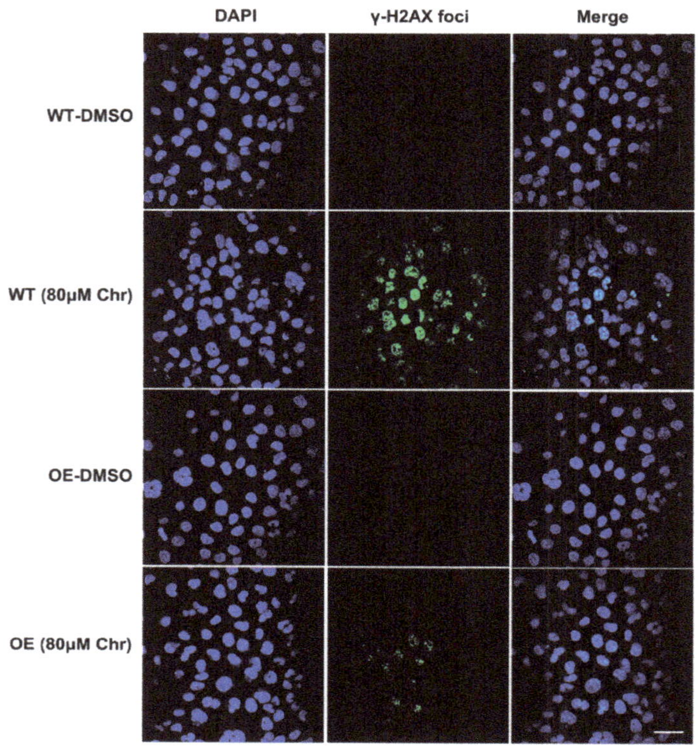

Figure 9. Immunofluorescence results of γH2AX foci. Wild type (WT) or BRCA1-overexpressing cells (OE) were treated with DMSO or high dose of chrysin (80 μM) for 24 h, respectively. DNA damage biomarker γ-H2AX (green) and nuclear DNA stained with DAPI (blue) was presented. Cells were observed under 60× oil laser confocal microscope.

4. Discussion

Free radicals such as ROS in the human body are gradually accumulated spontaneously under normal conditions and trigger cell death or carcinogenesis depending on a certain level. Here, a low dose of flavonoids slightly stimulated the proliferation of cancer cells under 60 h treatment, but not in short periods such as 24 h. Therefore, we proposed a hypothesis that a low dose of flavonoids possibly eliminated the accumulated free radicals and other oxidants during normal growth, and the subsequent medium exchanging experiments verified this conjecture. Although flavonoids have been identified to display remarkable antioxidant effects in vitro and in vivo, the option of appropriate concentration was still suspicious since a single dose for a certain purpose was mostly used. Besides, extracts of natural products were usually tested rather than single components. The chemicals of extracts are complex and interact with each other, making it difficult to determine the specific biofunction. More importantly, most studies focused on the antioxidant ability of substances in a single dose instead of exploring various concentrations to confirm the possible functional activity at the beginning. Recently, the anti-inflammation and antioxidant activities of natural flavonoids and extracts have been extensively investigated in keratinocytes, melanocytes and fibroblasts when exposed to ultraviolet (UV) [34,35]. Meanwhile, some studies focused on the antioxidant function of flavonoids in relieving chronic oxidative stress in mice with specific diseases of diabetes, drug-induced cancer or polycystic ovary syndrome. Different from the external oxidative stress caused by UV or animal models, our results showed that the oxidation products formed during the normal growth could seriously impede the proliferation of cells as well, aside from nutrition exhaustion. One of the main results we obtained is that a high dose of flavonoids (usually more than IC_{50}) can induce severe oxidative stress in cancer cells and increase ROS levels. ROS are important in regulating normal cellular processes and deregulated ROS lead to the development of diseased states in humans, including cancers. Several studies have been found to be marked with increased ROS production which activates protumorigenic signaling, enhances cell survival and proliferation and drives DNA damage and genetic instability [36]. In general, the level of ROS in cancer cells is higher than that in normal cells; this is because tumor cells develop a mechanism of adapting to the high ROS condition by expressing enhanced levels of antioxidant genes while developing neoplastic signaling. However, ROS exceeding a certain threshold can lead to cell disaster and benefits antitumorigenic signaling through initiating oxidative stress, which could be used as a potential target for cancer therapies [37]. Interestingly, the elevated level of ROS induced by high-dose flavonoids was quite sufficient to trigger oxidative stress, thereby promoting cancer cell death. Transcriptome analysis presented that the unfolded protein reaction (UPR) was positively enriched by high doses of apigenin and quercetin. Whether such UPR activation was dependent on enhanced ROS has not yet been determined. Studies have shown that UPR is necessary to maintain the integrity of the endoplasmic reticulum and prevent oxidative stress [38]. Moreover, oxidative stress can induce mitochondrial dysfunction and unfolded protein response in retinal pigment epithelial cells, consistent with our results [39]. However, excessive protein synthesis and unfolded protein accumulation in ER also lead to oxidative stress [40]. In yeasts, UPR and oxidative stress regulate each other, constructing a complicated regulatory network [41,42]. Therefore, the enriched UPR could be the result or cause of oxidative stress.

Previous studies have verified that quercetin and apigenin promoted cell death at high concentrations, and the death mechanisms mainly involve apoptosis, cell cycle arrest and autophagy [43–48]. Our results further reveal that a high dose of flavonoids triggered mitochondrial damage and ATP reduction, thus illustrating the cell death from another perspective. Mitochondria are one of the most prominent sources of ROS that contribute to oxidative stress. The electron transport chain located on the inner mitochondrial membrane generates the majority of mitochondrial ROS during the process of oxidative phosphorylation (OXPHOS). Several lines of evidence suggested that oxidative stress as a vital cause or consequence of mitochondrial dysfunction is one of the leading drivers of this cell death program [49]. In addition to producing a large amount of ROS, damaged mitochondria can also be exacerbated by oxidative stress, as seen in Parkinson's and Alzheimer's diseases. Aside from the decisive role in modulating endogenous apoptosis, mitochondria are also the main manufactory producing ATP [50]. Mitochondrial damage caused by imbalanced membrane potential directly leads to inhibited energy production and then cell death. NAC or VC pretreatment significantly recovered such effect, further proving that oxidative stress participated in ATP regulation. Indeed, flavonoids (baicalein, luteolin, naringenin and quercetin) were found to suppress the Fenton reaction of the iron–ATP complex to inhibit ATP synthesis [51]. Besides, a study showed that the pro-oxidant activity of flavonoids can contribute to their health-promoting activity by inducing important detoxifying enzymes via a supposed toxic chemical reaction, which was another explanation for the toxicity of high-dose flavonoids [52]. Although the antioxidative properties of flavonoids have been investigated for many years, there are queries about what reactions they will perform in protecting or killing cancer cells. Our findings provide new perspectives for better use of flavonoids as antitumor agents.

Intriguingly, inflammation was markedly activated by high-dose flavonoids, as measured by upregulated TNF-α signaling from transcriptome analysis. TNF-α is a cytokine that binds to TNFRSF1A/TNFR1 and TNFRSF1B/TNFBR, is secreted by macrophages and induces tumor cell death and other inflammatory mediators and proteases that orchestrate inflammatory responses. TNF-α could also be produced by tumor cells and act as an endogenous tumor promoter or suppressor. The role of TNF-α has been linked to all steps involved in tumorigenesis, including transformation, proliferation, invasion, angiogenesis and metastasis [53]. In the past, most flavonoids or natural products were found to exert an inhibitory effect on the inflammatory response via reducing the secretion of cytokines or inhibiting transcription factors such as NF-kB, PPAR, AP-1 and Nrf2 and downstream gene expression [54–57]. However, our results showed that similar to the oxidative stress induction, a high dose of flavonoids activated the inflammatory response, as supported by transcriptome and qPCR results. To be more accurate, macrophages as a cell model were used for investigation, and upregulated expression levels of the above inflammatory genes were observed. Compared with macrophages, the fold change of the above genes was found lower in MCF-7 cancer cells, indicating that macrophages were more sensitive to the inflammatory response. We speculated that the regulatory network of inflammation in cancer cells could be more refined due to the genetic mutations and evolutionary escape from immune systems. Huang et al. suggested that quercetin and catechin were of benefit for diabetic vascular complications via promoting inflammation in human monocytes and antioxidant abilities against AGE-mediated oxidative stress [58]. Similarly, preincubation with 2 μM β-carotene tended to inhibit the inflammatory reaction, whereas 20 μM β-carotene significantly increased the secretion of proinflammatory mediators [59]. The secretion of inflammatory cytokines or related protein expression was not measured in this research and should be further detected if possible.

Cholesterol homeostasis plays an essential role in proper cellular and systemic functions. The cellular cholesterol level reflects the dynamic balance between biosynthesis, uptake, export and esterification, a process in which cholesterol is converted to neutral cholesteryl esters either for storage in lipid droplets or for secretion as constituents of lipoproteins. Disturbed cholesterol balance underlies not only cardiovascular disease but also an increasing number of other diseases such as cancers [60]. Commonly, the roles played by cholesterol in cancer development and the potential of therapeutically targeting cholesterol homeostasis is a controversial area in the cancer community. Several epidemiologic studies reported an association between cancer and serum cholesterol levels or statin use, while others suggest that there is no such association. In melanoma, an analysis demonstrated that enhanced expression of cholesterol synthesis genes was associated with decreased patient survival [61]. Loss of CYP27A1 has been determined to dysregulate cholesterol homeostasis in prostate cancer; the same was found for the gene ABCA1 [62,63]. Apolipoprotein E gene was shown to regulate aggressive behaviors in prostate cancer cells by deregulating cholesterol homeostasis [64]. Moreover, cholesterol homeostasis directly affected the drug sensitivity to platinum in ovarian cancer [65]. In our results, both apigenin- and quercetin-treated groups have been observed to significantly inhibit the cholesterol homeostasis pathway, indicating that these flavonoids could possibly modulate intracellular cholesterol biosynthesis or metabolism, thus breaking the energy balance to induce cell death. Previously, red grape juice with multiple flavonoids was found to be able to alter cholesterol homeostasis and increase LDL-receptor activity in human cells in vitro [66]. Likewise, the secretion of hepatocyte ApoB was inhibited by two flavonoids, naringenin and hesperetin, via reduced activity and expression of ACAT2 and MTP [67]. Alcohol-free red wine concentrates were more effective than lovastatin for decreasing total cholesterol in vitro and inhibited cholesterol biosynthesis at a transcriptional level [68]. Interestingly, different flavonoids presented selective modulation for liver X receptor (LXR) activity, which has been found to be involved in adjusting cholesterol homeostasis; among these flavonoids, both apigenin and quercetin activated LXR-β [69]. Nevertheless, Deyhim et al. found that cranberry juice increased antioxidant status without affecting cholesterol homeostasis in orchidectomized rats, possibly because a higher dose was required in vivo for regulating cholesterol condition [70]. Although flavonoids showed diverse mechanisms in the cholesterol maintenance in different cell lines, combined with our results, we speculated that the phenotype of inhibiting cholesterol homeostasis promoted the disorder of energy metabolism and cell death of cancer cells, which was conducive to the aggravation of oxidative stress injury. Taken together, these results revealed that high doses of flavonoids are of great potential in regulating intracellular cholesterol homeostasis, which further enriches our conclusion.

Mutations of BRCA1 are found in a high percentage of hereditary breast and ovarian cancers. Women carrying germline mutations in BRCA1 have a 50–80% lifetime risk of developing breast cancer and a 20–40% lifetime risk of developing ovarian cancer [71]. Generally, BRCA1 has been well established as a tumor suppressor and functions primarily by maintaining genome integrity through DNA repair, especially when cells are exposed to oxidative stress. More and more evidence has suggested that BRCA1 regulates oxidative stress and antioxidant signaling transduction [28]. BRCA1 upregulated the expression of multiple genes involved in the cytoprotective antioxidant response, including glutathione S-transferases, oxidoreductases and other antioxidant genes. Besides, BRCA1 deficiency conferred sensitivity to oxidizing agents (hydrogen peroxide and paraquat) [32]. BRCA1 activated p21 expression and then enhanced the transcriptional level of antioxidant genes [72]. It is physically associated with Nrf2, a master transcription factor modulating antioxidant enzymes, to prevent ubiquitin-dependent degradation by KEAP1 and exert an antioxidant effect [73]. BRCA1 was also found to bind and stabilize hypoxia-inducible factor-1α (HIF-1α) and promote tumor survival [74]. Here, our results showed that BRCA1 suppressed the oxidative stress induced by high-dose flavonoids, possibly by repairing DNA damage, as manifested by fewer γ-H2AX foci. Genome stability is key for normal physiological activities, and the main function of BRCA1 is to keep the genome stable. Considering the fact that ROS directly cause DNA damage, it is plausible that BRCA1 functions like this to regulate oxidative stress. Our results supported and further elucidated the possible mechanism from different levels that BRCA1 restrains oxidative stress via repairing DNA double-strand breaks, highlighting a new perspective that a daily diet containing relatively high doses of flavonoids may be more conducive to tumor inhibition for breast cancer patients, especially for those with BRCA1 mutations or silence, thus leading to synthetic lethality, which could be more beneficial in prolonging prognosis.

Supplementary Materials: The following supporting information can be downloaded at: https://www.mdpi.com/article/10.3390/antiox11040622/s1, Figure S1: (A) The staining of mitochondria superoxide with MitoSOX Red Indicator. Cells were treated with DMSO or 50 μM chrysin for 24 h, stained and observed under a laser confocal microscope. Mitochondria superoxide (red) and nuclear DNA stained with DAPI (blue) was showed. (B) Fluorescence intensity of ROS detected by flow cytometry with DCFH-DA assay. Rosup: 50 μg/mL for 24 h. NAC + Rosup: pretreated with 5 μM NAC for 6 h thereafter 50 μg/mL Rosup for 18 h. Chr (5 μM) + Rosup: pretreated with 5 μM chrysin for 6 h thereafter 50 μg/mL Rosup for 18 h. Chr (50 μM): treated with 50 μM chrysin for 24 h. NAC + Chr (50 μM): pretreated with 5 μM NAC for 6 h thereafter 50 μM chrysin for 18 h. (C) Fluorescence intensity of oxidative stress detected by flow cytometry with CellROX Deep Red probe. TBHP: treated with 250 μM TBHP for 2 h (positive control). NAC + TBHP: pretreated with 1 mM NAC for 1 h thereafter 250 μM TBHP for 2 h. Api: treated with 100 μM apigenin for 2 h. NAC + Api: pretreated with 1 mM NAC for 1 h thereafter 100 μM apigenin for 2 h; Figure S2: The original flow cytometry plots of Figure S1B; Figure S3: (A) The qPCR results of relevant inflammatory genes in RAW 264.7 macrophages treated with four flavonoids, respectively. LPS: 5 μg/mL LPS for 24 h. Low dose means 5 μM and high dose means 50 μM. NAC pretreatment was 6 h thereafter 18 h of flavonoids treatment. (B) The original flow cytometry plots of Figure 8D. (C) The colony formation of Rosup and Chr (5 μM) + Rosup in MCF-7 cells. The treatment was described before. (D) The original WB gels for BRCA1 overexpression. The gels were cropped according to the targeted protein molecular, respectively. Table S1: The qPCR primers of detected genes; Table S2: The information of flavonoids and antibodies; Table S3: The information of used cell lines.

Author Contributions: J.Z. and X.X. designed the study; X.X., J.W. and Y.Q. performed the experiments; X.X. wrote the manuscript; W.H. and J.Z., funding acquisition; J.Z. and Y.Y., supervision. All authors have read and agreed to the published version of the manuscript.

Funding: This study was supported by the National Natural Science Foundation of China Youth Fund under Grant [81900777].

Institutional Review Board Statement: Not applicable.

Informed Consent Statement: Not applicable.

Data Availability Statement: The data presented in this study are available the article and supplementary materials.

Acknowledgments: The authors thank Hongbo Hu from the College of Food Science and Nutritional Engineering, China Agricultural University, for generously gifting MCF-7 and MDA-MB-231 cells and Junying Jia, Shu Meng from the Institute of Biophysics, Chinese Academy of Sciences, for providing technical help with flow cytometry and the Laboratory of Laser Confocal Microscope of the College of Animal Science and Technology, China Agricultural University, for providing technical help on microscope.

Conflicts of Interest: The authors declare no conflict of interest.

References

1. Waks, A.G.; Winer, E.P. Breast Cancer Treatment: A Review. *JAMA-J. Am. Med. Assoc.* **2019**, *321*, 288–300.
2. Maughan, K.L.; Lutterbie, M.A.; Ham, P.S. Treatment of breast cancer. *Am. Fam. Physician* **2010**, *81*, 1339–1346. [PubMed]
3. Sledge, G.W.; Mamounas, E.P.; Hortobagyi, G.N.; Burstein, H.J.; Goodwin, P.J.; Wolff, A.C. Past, present, and future challenges in breast cancer treatment. *J. Clin. Oncol.* **2014**, *32*, 1979–1986. [PubMed]
4. Tapas, A.R.; Sakarkar, D.M.; Kakde, R.B. Flavonoids as nutraceuticals: A review. *Trop. J. Pharm. Res.* **2008**, *7*, 1089–1099.
5. Romagnolo, D.F.; Selmin, O.I. Flavonoids and cancer prevention: A review of the evidence. *J. Nutr. Gerontol. Geriatr.* **2012**, *31*, 206–238. [PubMed]
6. Agrawal, A.D. Pharmacological activities of flavonoids: A review. *Int. J. Pharm. Sci. Nanotechnol.* **2011**, *4*, 1394–1398.
7. Stahl, W.; Sies, H. Carotenoids and flavonoids contribute to nutritional protection against skin damage from sunlight. *Mol. Biotechnol.* **2007**, *37*, 26–30.
8. Bonina, F.; Lanza, M.; Montenegro, L.; Puglisi, C.; Tomaino, A.; Trombetta, D.; Castelli, F.; Saija, A. Flavonoids as potential protective agents against photo-oxidative skin damage. *Int. J. Pharm.* **1996**, *145*, 87–94.
9. Gravandi, M.M.; Fakhri, S.; Zarneshan, S.N.; Yarmohammadi, A.; Khan, H. Flavonoids modulate AMPK/PGC-1α and interconnected pathways toward potential neuroprotective activities. *Metab. Brain Dis.* **2021**, *36*, 1501–1521.
10. Rufino, A.T.; Costa, V.M.; Carvalho, F.; Fernandes, E. Flavonoids as antiobesity agents: A review. *Med. Res. Rev.* **2021**, *41*, 556–585.
11. Lukačínová, A.; Mojžiš, J.; Beňačka, R.; Keller, J.; Maguth, T.; Kurila, P.; Vaško, L.; Racz, O.; Ništiar, F. Preventive effects of flavonoids on alloxan-induced diabetes mellitus in rats. *Acta Vet. Brno* **2008**, *77*, 175–182.
12. Lin, C.H.; Chang, C.Y.; Lee, K.R.; Lin, H.J.; Chen, T.H.; Wan, L. Flavones inhibit breast cancer proliferation through the Akt/FOXO3a signaling pathway. *BMC Cancer* **2015**, *15*, 958.
13. Damianaki, A.; Bakogeorgou, E.; Kampa, M.; Notas, G.; Hatzoglou, A.; Panagiotou, S.; Gemetzi, C.; Kouroumalis, E.; Martin, P.; Castanas, E. Potent inhibitory action of red wine polyphenols on human breast cancer cells. *J. Cell. Biochem.* **2000**, *78*, 429–441. [PubMed]
14. Cho, S.G.; Woo, S.M.; Ko, S.G. Butein suppresses breast cancer growth by reducing a production of intracellular reactive oxygen species. *J. Exp. Clin. Cancer Res.* **2014**, *33*, 51.
15. Wu, Y.; Wang, Y.; Nabi, X. Protective effect of Ziziphora clinopodioides flavonoids against H2O2-induced oxidative stress in HUVEC cells. *Biomed. Pharmacother.* **2019**, *117*, 109156.
16. Liao, Y.; Shen, W.; Kong, G.; Lv, H.; Tao, W.; Bo, P.; Rajesh, M. Apigenin Induces the Apoptosis and Regulates MAPK Signaling Pathways in Mouse Macrophage ANA-1 Cells. *PLoS ONE* **2014**, *9*, e92007.
17. Taabodi, M.; May, E.B.; Mack, K.M.; Squibb, K.S.; Ishaque, A.B. Oxidative stress pathways of flavonoid toxicity in human breast tumor cells. *Int. J. Clin. Exp. Pathol.* **2017**, *10*, 2554–2567.
18. Chobot, V.; Hadacek, F. Exploration of pro-oxidant and antioxidant activities of the flavonoid myricetin. *Redox Rep.* **2011**, *16*, 242–247.
19. Knickle, A.; Fernando, W.; Greenshields, A.L.; Rupasinghe, H.P.V.; Hoskin, D.W. Myricetin-induced apoptosis of triple-negative breast cancer cells is mediated by the iron-dependent generation of reactive oxygen species from hydrogen peroxide. *Food Chem. Toxicol.* **2018**, *118*, 154–167.
20. Samarghandian, S.; Azimi-Nezhad, M.; Borji, A.; Hasanzadeh, M.; Samini, M. Inhibitory and Cytotoxic Activities of Chrysin on Human Breast Adenocarcinoma Cells by Induction of Apoptosis. *Pharmacogn. Mag.* **2016**, *12*, S436–S440.
21. Zhao, X.C.; Tian, L.; Cao, J.G.; Fei, L. Induction of apoptosis by 5,7-dihydroxy-8-nitrochrysin in breast cancer cells: The role of reactive oxygen species and Akt. *Int. J. Oncol.* **2010**, *37*, 1345–1352. [PubMed]
22. Kim, T.H.; Woo, J.S.; Kim, Y.K.; Kim, K.H. Silibinin Induces Cell Death through Reactive Oxygen Species-Dependent Downregulation of Notch-1/ERK/Akt Signaling in Human Breast Cancer Cells. *J. Pharmacol. Exp. Ther.* **2014**, *349*, 268–278. [PubMed]
23. Martínez-Pérez, C.; Ward, C.; Turnbull, A.K.; Mullen, P.; Cook, G.; Meehan, J.; Jarman, E.J.; Thomson, P.I.; Campbell, C.J.; Mcphail, D. Antitumour activity of the novel flavonoid Oncamex in preclinical breast cancer models. *Br. J. Cancer* **2016**, *114*, 905–916. [PubMed]

24. Cao, P.; Xia, Y.; He, W.; Zhang, T.; Hong, L.; Zheng, P.; Shen, X.; Liang, G.; Cui, R.; Zou, P. Enhancement of oxaliplatin-induced colon cancer cell apoptosis by alantolactone, a natural product inducer of ROS. *Int. J. Biol. Sci.* **2019**, *15*, 1676–1684.
25. Yu, J.; Gao, H.; Wu, C.; Xu, Q.M.; Lu, J.J.; Chen, X. Diethyl blechnic, a novel natural product isolated from salvia miltiorrhiza bunge, inhibits doxorubicin-induced apoptosis by inhibiting ROS and activating JNK1/2. *Int. J. Mol. Sci.* **2018**, *19*, 1809.
26. He, G.; He, G.; Zhou, R.; Pi, Z.; Zhu, T.; Jiang, L.; Xie, Y. Enhancement of cisplatin-induced colon cancer cells apoptosis by shikonin, a natural inducer of ROS in vitro and in vivo. *Biochem. Biophys. Res. Commun.* **2016**, *469*, 1075–1082.
27. Sun, X.; Wang, W.; Chen, J.; Cai, X.; Yang, J.; Yang, Y.; Yan, H.; Cheng, X.; Ye, J.; Lu, W. The natural diterpenoid isoforretin A inhibits thioredoxin-1 and triggers potent ROS-mediated antitumor effects. *Cancer Res.* **2017**, *77*, 926–936.
28. Yi, Y.W.; Kang, H.J.; Bae, I. BRCA1 and Oxidative Stress. *Cancers* **2014**, *6*, 771–795.
29. Nikitina, D.; Chen, Z.; Vallis, K.; Poll, A.; Ainsworth, P.; Narod, S.A.; Kotsopoulos, J. Relationship between Caffeine and Levels of DNA Repair and Oxidative Stress in Women with and without a BRCA1 Mutation. *J. Nutr. Nutr.* **2015**, *8*, 174–184.
30. Cao, L.; Xu, X.; Cao, L.L.; Wang, R.H.; Coumoul, X.; Kim, S.S.; Deng, C.X. Absence of full-length Brca1 sensitizes mice to oxidative stress and carcinogen-induced tumorigenesis in the esophagus and forestomach. *Carcinogenesis* **2007**, *28*, 1401–1407.
31. Martinez-Outschoorn, U.E.; Balliet, R.; Lin, Z.; Whitaker-Menezes, D.; Birbe, R.C.; Bombonati, A.; Pavlides, S.; Lamb, R.; Sneddon, S.; Howell, A. BRCA1 mutations drive oxidative stress and glycolysis in the tumor microenvironment: Implications for breast cancer prevention with antioxidant therapies. *Cell Cycle* **2012**, *11*, 4402–4413. [PubMed]
32. Bae, I.; Fan, S.; Meng, Q.; Rih, J.K.; Kim, H.J.; Kang, H.J.; Xu, J.; Goldberg, I.D.; Jaiswal, A.K.; Rosen, E.M. BRCA1 induces antioxidant gene expression and resistance to oxidative stress. *Cancer Res.* **2004**, *64*, 7893–7909. [PubMed]
33. Moynahan, M.E.; Chiu, J.W.; Koller, B.H.; Jasin, M. Brca1 controls homology-directed DNA repair. *Mol. Cell* **1999**, *4*, 511–518.
34. Oh, S.; Kim, Y.J.; Lee, E.K.; Park, S.W.; Yu, H.G. Antioxidative effects of ascorbic acid and astaxanthin on arpe-19 cells in an oxidative stress model. *Antioxidants* **2020**, *9*, 833.
35. Di Francesco, S.; Savio, M.; Bloise, N.; Borroni, G.; Stivala, L.A.: Borroni, R.G. Red grape (Vitis vinifera L.) flavonoids down-regulate collagen type III expression after UV-A in primary human dermal blood endothelial cells. *Exp. Dermatol.* **2018**, *27*, 973–980. [PubMed]
36. Wang, Y.; Qi, H.; Liu, Y.; Duan, C.; Liu, X.; Xia, T.; Chen, D.; Piao, H.; Liu, H.-X. The double-edged roles of ROS in cancer prevention and therapy. *Theranostics* **2021**, *11*, 4839–4857. [PubMed]
37. Arfin, S.; Jha, N.K.; Jha, S.K.; Kesari, K.K.; Ruokolainen, J.; Roychoudhury, S.; Rathi, B.; Kumar, D. Oxidative stress in cancer cell metabolism. *Antioxidants* **2021**, *10*, 642.
38. Kaufman, R.J.; Back, S.H.; Song, B.; Han, J.; Hassler, J. The unfolded protein response is required to maintain the integrity of the endoplasmic reticulum, prevent oxidative stress and preserve differentiation in β-cells. *Obes. Metab.* **2010**, *12*, 99–107.
39. Cano, M.; Wang, L.; Wan, J.; Barnett, B.P.; Ebrahimi, K.; Qian, J.; Handa, J.T. Oxidative stress induces mitochondrial dysfunction and a protective unfolded protein response in RPE cells. *Free Radic. Biol. Med.* **2014**, *69*, 1–14.
40. Malhotra, J.D.; Miao, H.; Zhang, K.; Wolfson, A.; Pennathur, S.; Pipe, S.W.; Kaufman, R.J. Antioxidants reduce endoplasmic reticulum stress and improve protein secretion. *Proc. Natl. Acad. Sci. USA* **2008**, *105*, 18525–18530.
41. Kimata, Y.; Ishiwata-Kimata, Y.; Yamada, S.; Kohno, K. Yeast unfolded protein response pathway regulates expression of genes for anti-oxidative stress and for cell surface proteins. *Genes Cells* **2006**, *11*, 59–69. [PubMed]
42. Guerra-Moreno, A.; Ang, J.; Welsch, H.; Jochem, M.; Hanna, J. Regulation of the unfolded protein response in yeast by oxidative stress. *FEBS Lett.* **2019**, *593*, 1080–1088. [PubMed]
43. Liu, M.M.; Ma, R.H.; Ni, Z.J.; Thakur, K.; Wei, Z.J. Apigenin 7-O-glucoside promotes cell apoptosis through the PTEN/PI3K/AKT pathway and inhibits cell migration in cervical cancer HeLa cells. *Food Chem. Toxicol.* **2020**, *146*, 111843. [PubMed]
44. Li, Y.-W.; Xu, J.; Zhu, G.-Y.; Huang, Z.-J.; Lu, Y.; Li, X.-Q.; Wang, N.; Zhang, F.-X. Apigenin suppresses the stem cell-like properties of triple-negative breast cancer cells by inhibiting YAP/TAZ activity. *Cell Death Discov.* **2018**, *4*, 105.
45. Ashrafizadeh, M.; Bakhoda, M.R.; Bahmanpour, Z.; Ilkhani, K.; Zarrabi, A.; Makvandi, P.; Khan, H.; Mazaheri, S.; Darvish, M.; Mirzaei, H. Apigenin as tumor suppressor in cancers: Biotherapeutic activity, nanodelivery, and mechanisms with emphasis on pancreatic cancer. *Front. Chem.* **2020**, *8*, 829.
46. Chien, M.-H.; Lin, Y.-W.; Wen, Y.-C.; Yang, Y.-C.; Hsiao, M.; Chang, J.-L.; Huang, H.-C.; Lee, W.-J. Targeting the SPOCK1-snail/slug axis-mediated epithelial-to-mesenchymal transition by apigenin contributes to repression of prostate cancer metastasis. *J. Exp. Clin. Cancer Res.* **2019**, *38*, 246.
47. Chen, X.; Xu, H.; Yu, X.; Wang, X.; Zhu, X.; Xu, X. Apigenin inhibits in vitro and in vivo tumorigenesis in cisplatin-resistant colon cancer cells by inducing autophagy, programmed cell death and targeting m-TOR/PI3K/Akt signalling pathway. *J. BUON* **2019**, *24*, 488–493.
48. Tang, Y.-S.; Zhao, Y.-H.; Zhong, Y.; Li, X.-Z.; Pu, J.-X.; Luo, Y.-C.; Zhou, Q.-L. Neferine inhibits LPS-ATP-induced endothelial cell pyroptosis via regulation of ROS/NLRP3/Caspase-1 signaling pathway. *Inflamm. Res.* **2019**, *68*, 727–738.
49. Starkov, A.A. The role of mitochondria in reactive oxygen species metabolism and signaling. *Ann. N. Y. Acad. Sci.* **2008**, *1147*, 37–52.
50. Cano Sanchez, M.; Lancel, S.; Boulanger, E.; Neviere, R. Targeting oxidative stress and mitochondrial dysfunction in the treatment of impaired wound healing: A systematic review. *Antioxidants* **2018**, *7*, 98.
51. Cheng, I.F.; Breen, K. On the ability of four flavonoids, baicilein, luteolin, naringenin, and quercetin, to suppress the Fenton reaction of the iron-ATP complex. *Biometals* **2000**, *13*, 77–83. [PubMed]

52. Lu, K.; Cheng, Y.; Li, Y.; Li, W.; Zeng, R.; Song, Y. Phytochemical flavone confers broad-spectrum tolerance to insecticides in Spodoptera litura by activating ROS/CncC-mediated xenobiotic detoxification pathways. *J. Agric. Food Chem.* **2021**, *69*, 7429–7445. [PubMed]
53. Cruceriu, D.; Baldasici, O.; Balacescu, O.; Berindan-Neagoe, I. The dual role of tumor necrosis factor-alpha (TNF-α) in breast cancer: Molecular insights and therapeutic approaches. *Cell. Oncol.* **2020**, *43*, 1–18.
54. Ahmed, S.; Ahmed, N.; Rungatscher, A.; Linardi, D.; Kulsoom, B.; Innamorati, G.; Meo, S.A.; Gebrie, M.A.; Mani, R.; Merigo, F. Cocoa flavonoids reduce inflammation and oxidative stress in a myocardial ischemia-reperfusion experimental model. *Antioxidants* **2020**, *9*, 167.
55. Chen, L.; Teng, H.; Jia, Z.; Battino, M.; Miron, A.; Yu, Z.; Cao, H.; Xiao, J. Intracellular signaling pathways of inflammation modulated by dietary flavonoids: The most recent evidence. *Crit. Rev. Food Sci. Nutr.* **2018**, *58*, 2908–2924.
56. Owor, R.O.; Bedane, K.G.; Openda, Y.I.; Zühlke, S.; Derese, S.; Ong'amo, G.; Ndakala, A.; Spiteller, M. Synergistic anti-inflammatory activities of a new flavone and other flavonoids from Tephrosia hildebrandtii vatke. *Nat. Prod. Res.* **2021**, *35*, 4486–4493.
57. Ożarowski, M.; Karpiński, T.M. Extracts and Flavonoids of Passiflora Species as Promising Anti-inflammatory and Antioxidant Substances. *Curr. Pharm. Des.* **2021**, *27*, 2582–2604.
58. Huang, S.; Wu, C.; Yen, G. Effects of flavonoids on the expression of the pro-inflammatory response in human monocytes induced by ligation of the receptor for AGEs. *Mol. Nutr. Food Res.* **2006**, *50*, 1129–1139.
59. Yeh, S.-L.; Wang, H.-M.; Chen, P.-Y.; Wu, T.-C. Interactions of β-carotene and flavonoids on the secretion of pro-inflammatory mediators in an in vitro system. *Chem. Biol. Interact* **2009**, *179*, 386–393.
60. Luo, J.; Yang, H.; Song, B.L. Mechanisms and regulation ofcholesterol homeostasis. *Nat. Rev. Mol. Cell Biol.* **2020**, *21*, 225–245.
61. Kuzu, O.F.; Noory, M.A.; Robertson, G.P. The Role of Cholesterol in Cancer. *Cancer Res.* **2016**, *76*, 2063–2070. [PubMed]
62. Alfaqih, M.A.; Nelson, E.R.; Liu, W.; Safi, R.; Jasper, J.S.; Macias, E.; Geradts, J.; Thompson, J.W.; Dubois, L.G.; Freeman, M.R. CYP27A1 loss dysregulates cholesterol homeostasis in prostate cancer. *Cancer Res.* **2017**, *77*, 1662–1673. [PubMed]
63. Lee, B.H.; Taylor, M.G.; Robinet, P.; Smith, J.D.; Schweitzer, J.; Sehayek, E.; Falzarano, S.M.; Magi-Galluzzi, C.; Klein, E.A.; Ting, A.H. Dysregulation of Cholesterol Homeostasis in Human Prostate Cancer through Loss of ABCA1. *Cancer Res.* **2013**, *73*, 1211–1218. [PubMed]
64. Ifere, G.O.; Desmond, R.; Demark-Wahnefried, W.; Nagy, T.R. Apolipoprotein E gene polymorphism influences aggressive behavior in prostate cancer cells by deregulating cholesterol homeostasis. *Int. J. Oncol.* **2013**, *43*, 1002–1010.
65. Criscuolo, D.; Avolio, R.; Calice, G.; Laezza, C.; Esposito, F. Cholesterol Homeostasis Modulates Platinum Sensitivity in Human Ovarian Cancer. *Cells* **2020**, *9*, 828.
66. Davalos, A.; Fernandez-Hernando, C.; Cerrato, F.; Martínez-Botas, J.; Gomez-Coronado, D.; Gomez-Cordoves, C.; Lasuncion, M.A. Red grape juice polyphenols alter cholesterol homeostasis and increase LDL-receptor activity in human cells in vitro. *J. Nutr.* **2006**, *136*, 1766–1773.
67. Wilcox, L.J.; Borradaile, N.M.; Dreu, L.; Huff, M.W. Secretion of hepatocyte apoB is inhibited by the flavonoids, naringenin and hesperetin, via reduced activity and expression of ACAT2 and MTP. *J. Lipid Res.* **2001**, *42*, 725–734.
68. Lee, D.-H.; Choi, S.-S.; Kim, B.-B.; Kim, S.-Y.; Kang, B.-S.; Lee, S.-J.; Park, H.-J. Effect of alcohol-free red wine concentrates on cholesterol homeostasis: An in vitro and in vivo study. *Process Biochem.* **2013**, *48*, 1964–1971.
69. Fouache, A.; Zabaiou, N.; De Joussineau, C.; Morel, L.; Silvente-Poirot, S.; Namsi, A.; Lizard, G.; Poirot, M.; Makishima, M.; Baron, S. Flavonoids differentially modulate liver X receptors activity—Structure-function relationship analysis. *J. Steroid Biochem. Mol. Biol.* **2019**, *190*, 173–182.
70. Deyhim, F.; Patil, B.S.; Villarreal, A.; Lopez, E.; Garcia, K.; Rios, R.; Garcia, C.; Gonzales, C.; Mandadi, K. Cranberry juice increases antioxidant status without affecting cholesterol homeostasis in orchidectomized rats. *J. Med. Food* **2007**, *10*, 49–53.
71. King, M.-C.; Marks, J.H.; Mandell, J.B. Breast and ovarian cancer risks due to inherited mutations in BRCA1 and BRCA2. *Science* **2003**, *302*, 643–646. [PubMed]
72. Ouchi, T.; Monteiro, A.N.A.; August, A.; Aaronson, S.A.; Hanafusa, H. BRCA1 regulates p53-dependent gene expression. *Proc. Natl. Acad. Sci. USA* **1998**, *95*, 2302–2306. [PubMed]
73. Chen, W.; Sun, Z.; Wang, X.-J.; Jiang, T.; Huang, Z.; Fang, D.; Zhang, D.D. Direct interaction between Nrf2 and p21Cip1/WAF1 upregulates the Nrf2-mediated antioxidant response. *Mol. Cell* **2009**, *34*, 663–673. [PubMed]
74. Kang, H.J.; Kim, H.J.; Rih, J.-K.; Mattson, T.L.; Kim, K.W.; Cho, C.-H.; Isaacs, J.S.; Bae, I. BRCA1 plays a role in the hypoxic response by regulating HIF-1α stability and by modulating vascular endothelial growth factor expression. *J. Biol. Chem.* **2006**, *281*, 13047–13056. [PubMed]

antioxidants

MDPI

Article

Lipophilic Grape Seed Proanthocyanidin Exerts Anti-Cervical Cancer Effects in HeLa Cells and a HeLa-Derived Xenograft Zebrafish Model

Changhong Li [1], Linli Zhang [1], Chengmei Liu [1], Xuemei He [2,3], Mingshun Chen [1,*] and Jun Chen [1]

1 State Key Laboratory of Food Science and Technology, Nanchang University, Nanchang 330047, China; 357900210034@email.ncu.edu.cn (C.L.); zll231797@126.com (L.Z.); liuchengmei@ncu.edu.cn (C.L.); chenjun@ncu.edu.cn (J.C.)

2 Agro-Food Science and Technology Research Institute, Guangxi Academy of Agricultural Sciences, Nanning 530007, China; xuemeihegx@gxaas.net

3 Guangxi Key Laboratory of Fruits and Vegetables Storage-Processing Technology, Nanning 530007, China

* Correspondence: chenshun1221@ncu.edu.cn; Tel.: +86-0791-88305871

Abstract: Lipophilic grape seed proanthocyanidin (LGSP) synthesized from GSP and lauric acid exhibits an excellent antioxidant and anti-inflammatory effect. However, its anti-cervical cancer activity is still unknown. In this study, the in vitro anti-cervical cancer activity of LGSP on HeLa cell lines was investigated by MTT assay, flow cytometry and Western blot analysis, and its effect was explored by a HeLa-derived xenograft zebrafish model. LGSP exhibited an excellent anti-proliferative effect on HeLa cells by increasing the level of reactive oxygen species, which further induced cell apoptosis and blocked cell cycle progression in the G2/M phase. LGSP-treated HeLa cells showed a reduction in mitochondrial membrane potential, upregulation of the Bax/Bcl-2 ratio, release of cytochrome *c* into the cytoplasm, and activation of cleaved caspase-9/3 and cleavage of PARP, thus indicating that LGSP induced apoptosis through the intrinsic mitochondrial/caspase-mediated pathway. In the zebrafish model, LGSP effectively suppressed the growth of a HeLa xenograft tumor. These data suggest that LGSP may be a good candidate for the prevention or treatment of cervical cancer.

Keywords: lipophilic grape seed proanthocyanidin; cervical cancer; HeLa cell; apoptosis; zebrafish

Citation: Li, C.; Zhang, L.; Liu, C.; He, X.; Chen, M.; Chen, J. Lipophilic Grape Seed Proanthocyanidin Exerts Anti-Cervical Cancer Effects in HeLa Cells and a HeLa-Derived Xenograft Zebrafish Model. *Antioxidants* **2022**, *11*, 422. https://doi.org/10.3390/antiox11020422

Academic Editors: Jicheng Zhan, Zhigang Liu and Hui-Min David Wang

Received: 22 January 2022
Accepted: 17 February 2022
Published: 19 February 2022

1. Introduction

Cervical cancer is the fourth most common cancer in women with approximately 570,000 new cases and 311,000 deaths reported worldwide in 2018 [1]. Although significant progress has been made in the treatment of cervical cancer, they are costly and unbearable. The prevention of cancer through vaccination has become very important [2], and increasing attention has been paid to its health effects. A large number of epidemiologic studies have investigated the relation between dietary polyphenols and cancer [3], and these have demonstrated that phenolic compounds can reduce the risk of chronic diseases, including cancer, diabetes, cardiovascular and neurodegenerative diseases [4,5].

Phenolic compounds, which are widely distributed in food, are regarded as an important source of antioxidants in the daily diet. Grape seed proanthocyanidin (GSP), the major polyphenol component in grape seed, has been widely used as a dietary supplement and food additive due to its various biological activities [6]: antioxidant [7], anti-tumor [8], anti-inflammatory [9] and anti-cardiovascular disease [10]. However, the hydrophilicity of GSP with its low solubility in lipid systems limits its application in lipidic food matrices and hinders its bioactivity [11].

In recent years, lipophenols have received much attention due to their high antioxidant effects on lipidic food matrices [12,13]. Tea polyphenol palmitate has been allowed as a food additive to fat and oils by the China Food and Drug Administration [14]. Lipophilic grape seed proanthocyanidin (LGSP), synthesized from GSP and lauric acid, has shown antioxidant and anti-inflammatory activity higher than that of GSP [15,16]. Chronic inflammation and cancer in organisms are regulated and stimulated by oxidative stress; thus, the good antioxidant and anti-inflammatory activity of LGSP may also have certain anti-cancer activity [17]. However, the anti-cervical cancer effect of LGSP is still unclear.

In this study, we examined the hypothesis that LGSP has an anti-cervical cancer effect on HeLa cells, and we explored the underlying mechanisms by MTT assay, flow cytometry and Western blot analysis. We then probed the anti-cervical cancer effect of LGSP in vivo, which provided scientific evidence for its future application in cervical cancer prevention or therapy.

2. Materials and Methods

2.1. Materials

GSP (proanthocyanidin content, 95%) was purchased from Xi'an Realin Biotechnology Co. (Xi'an, Shanxi, China). Dulbecco's modified Eagle medium (DMEM), trypsin/EDTA and trypsin were obtained from Gibco (Grand Island, NY, USA). Fetal bovine serum (FBS) was purchased from Biological Industries (Shanghai, China). Thiazolyl blue tetrazolium bromide (MTT) was purchased from Aladdin Biotechnology (St. Louis, MO, USA). The Cell cycle staining kit and Annexin V-FITC/PI apoptosis kit were obtained from MultiSciences (Shanghai, China). Mitochondrial membrane potential (MMP) assay kit, reactive oxygen species (ROS) assay kit and bicinchoninic acid (BCA) protein assay kit were purchased from Beyotime Biotechnology (Shanghai, China). The antibodies against cleaved PARP, PARP, caspase-3, cleaved caspase-3, caspase-9, cleaved caspase-9, cytochrome *c*, Bax, Bcl-2 and GADPH were purchased from Cell Signaling Technology (Beverly, MA, USA).

2.2. Preparation of LGSP

LGSP was prepared as previously described [15]: GSP, lauric acid, immobilized lipase TLIM (Novozymes, Shanghai, China) and ethanol were mixed into a screw-capped glass bottle. After incubating for 12 h with magnetic stirring at 50 °C, the mixture was filtrated to remove the enzyme and stop the reaction. The product was condensed using a rotatory evaporator (RE-52AA, Yarong Co. Ltd., Shanghai, China) to obtain LGSP, the major compositions of which are shown in Figure S1: 3′-*O*-lauroyl catechin, 3′-*O*-lauroyl epicatechin, 3′,5′-2-*O*-lauroyl epigallocatechin, 3′,3″,5″-3-*O*-lauroyl epicatechin gallate, 3′,3″-2-*O*-lauroyl procyanidin A2,3′,3″-2-*O*-lauroyl procyanidin B1 and 3′, 3″,3‴-3-*O*-lauroyl procyanidin C1.

2.3. Cell Culture

HeLa cells, obtained from the Chinese Academy of Sciences (Shanghai, China), were cultured in DMEM supplemented with 10% fetal bovine serum and 1% penicillin–streptomycin at 37 °C under a 5% CO_2 humidified atmosphere.

2.4. Cell Proliferation Analysis

The effect of LGSP on the proliferation of HeLa cells was measured by MTT assay according to the literature with a slight modification [18]. Briefly, the cells (5×10^3) were seeded into 96-well plates and treated with various concentrations (25, 50, 100, 150 and 200 μg/mL) of LGSP for 24 and 48 h. Then, 0.5 mg/mL MTT solution was added and incubated for 4 h. After that, 200 μL of DMSO was added to dissolve the formazan crystals, and the absorbance was measured at 490 nm by a microplate reader (Thermo Scientific, Waltham, MA, USA).

2.5. Cell Apoptosis Analysis

Cell apoptosis was analyzed by flow cytometry using the Annexin V-FITC cell apoptosis kit [18]. HeLa cells (4×10^5) were seeded into 6-well plates and incubated with various concentrations (50, 100 and 200 μg/mL) of LGSP for 48 h. Then, floating and adherent cells were harvested by trypsinization, washed twice with pre-cooling PBS, and resuspended in a 500 μL binding buffer containing 5 μL Annexin V-FITC and 10 μL PI. After incubating for 30 min in the dark at room temperature, the cells were determined by a flow cytometer (BD, Franklin Lakes, NJ, USA). The apoptotic HeLa cells were analyzed by FlowJo v10 software (BD, Franklin Lakes, NJ, USA).

2.6. Cell Cycle Distribution Analysis

The cell cycle distribution was analyzed by flow cytometry as described in the literature [19]. After treatment with LGSP at concentrations of 50, 100 and 200 μg/mL for 48 h, HeLa cells were harvested, washed twice with pre-cooling PBS, and resuspended in 1 mL DNA staining solution with PI and 10 μL permeabilization in the dark at room temperature for 30 min. The DNA content of the stained cells was measured by a flow cytometer (BD, Franklin Lakes, NJ, USA). All results were analyzed using FlowJo v10 software (BD, Franklin Lakes, NJ, USA).

2.7. Determination of ROS Production

ROS production was measured using the ROS assay kit with DCFH-DA [20]. The HeLa cells were seeded into 6-well plates at the density of 4×10^5 cells/well. After attachment, the cells were treated with different concentrations (50, 100 and 200 μg/mL) of LGSP for 48 h. Then, they were harvested, incubated with 10 μM DCFH-DH in the dark at 37 °C for 30 min and washed three times with a serum-free medium. ROS production was determined by a flow cytometer (BD, Franklin Lakes, NJ, USA).

2.8. Measurement of MMP

MMP was measured according to the instruction of the MMP assay kit with JC-1. HeLa cells (4×10^5) were seeded into 6-well plates. After treatment with different concentrations (50, 100 and 200 μg/mL) of LGSP for 48 h, the cells were obtained by trypsinization, incubated with JC-1 working solution for 30 min in the dark at room temperature, and washed twice with JC-1 buffer. The MMP of HeLa cells was observed under an inverted fluorescence microscope (Nikon Eclipse Ti-U, Melville, NY, USA) and analyzed by a flow cytometer (BD, Franklin Lakes, NJ, USA).

2.9. Western Blot Analysis

HeLa cells (4×10^5) were seeded into 6-well plates. After treatment with various concentrations of LGSP (50, 100 and 200 μg/mL) for 48 h, cells were harvested and lysed in a RIPA buffer. The total protein concentration was detected using an BCA protein assay kit [21]. The cell lysates were denatured in a SDS sample buffer and subjected to SDS-PAGE gel. The separated proteins were subsequently transferred onto the polyvinylidene difluoride membrane (Beyotime Biotechnology, Shanghai, China), blocked for 1 h at room temperature, incubated with specific primary antibody (1:1000, v/v) overnight at 4 °C, and followed by secondary antibody (1:3000, v/v) for 2 h at room temperature. The target proteins were detected using an ECL kit (Beyotime, Shanghai, China) by a chemiluminescent gel imaging system (Bio-Rad, Hercules, CA, USA).

2.10. Zebrafish Maintenance

The zebrafish wild-type AB strain was purchased from the China Zebrafish Resource Center (Wuhan, China). Adult zebrafish were maintained in a recirculating zebrafish housing system with a 14 h light/10 h dark photoperiod at 28 °C. After natural pair-mating and reproduction, the embryos were collected and incubated in fish water at 28 °C (0.2% instant ocean salt, pH 6.9–7.2, conductivity 480–510 μS/cm and hardness 53.7–71.6 mg/L $CaCO_3$). The use and manipulations of zebrafish were approved by the ethical review committee of Nanchang University (Nanchang, China) on 25 March 2021.

2.11. Embryo Acute Toxicity Test

The acute toxicity test was performed on zebrafish embryos as described in the literature [22]. Briefly, embryos at 10 h post-fertilization (hpf) were randomly distributed to 6-well culture plates (30 embryos/well) and exposed to LGSP with indicated concentrations (4, 8, 16 and 32 μg/mL). The hatched larvae were counted and the death rate was calculated at different time points (24, 48, 72, 96 and 120 hpf).

2.12. HeLa-Derived Xenograft Zebrafish Model

The zebrafish xenograft model was established by transplantation of stable fluorescent cervical cells expressing GFP protein (HeLa-GFP) into the yolk sac of zebrafish embryos (48 hpf). An average of 30 zebrafish embryos were used for each treatment. After successful transplantation, the zebrafish embryos were maintained at 35 °C for 1 day to promote tumor-cell proliferation and embryo recovery. After exposure to LGSP (4 and 8 μg/mL) for 48 h, the larvae were placed on agarose, and the fluorescent tumor cells were observed under a Leica KL300 LED inverted fluorescence microscope (Leica, Wetzlar, Germany). The fluorescence intensity of tumor cells was quantified by Image J (Rawak Software, Inc., Stuttgart, Germany) [23].

2.13. Statistical Analysis

Data were represented as mean ± standard deviation (SD) from at least three independent experiments. The differences between the experimental and control groups were compared using one-way ANOVA followed by Dunnett's multiple comparisons test using SPSS software (IBM version 16.0, Chicago, IL, USA). A p value < 0.05 was considered statistically significant.

3. Results

3.1. Effect of LGSP on the Proliferation of HeLa Cells

The anti-proliferative effect of LGSP on HeLa cells was determined using MTT assay. As shown in Figure 1A,B, the inhibition rate of LGSP on HeLa cells increased in a dose-dependent manner at 24–48 h. The inhibition rate increased rapidly from 11.67 to 56.69% after treatment with LGSP at concentrations of 25–50 μg/mL for 48 h, followed by enhancing gradually at concentrations of 50–200 μg/mL. The IC_{50} values of LGSP and GSP for anti-proliferative effect on HeLa cells were 57.97 ± 1.23 and 111.16 ± 1.02 μg/mL, respectively, after 48 h of treatment, indicating that LGSP had stronger anti-cervical cancer activity compared to GSP. The morphology of HeLa cells after treatment with LGSP and GSP at concentrations of 50, 100 and 200 μg/mL for 48 h was shown in Figure 1C. The number of HeLa cells clearly decreased after treatment with LGSP. Meanwhile, HeLa cells became round and lost their characteristic stretched appearance. The morphological changes in HeLa cells treated with LGSP were more obvious than those treated with GSP, which was consistent with the results of the anti-proliferative effect.

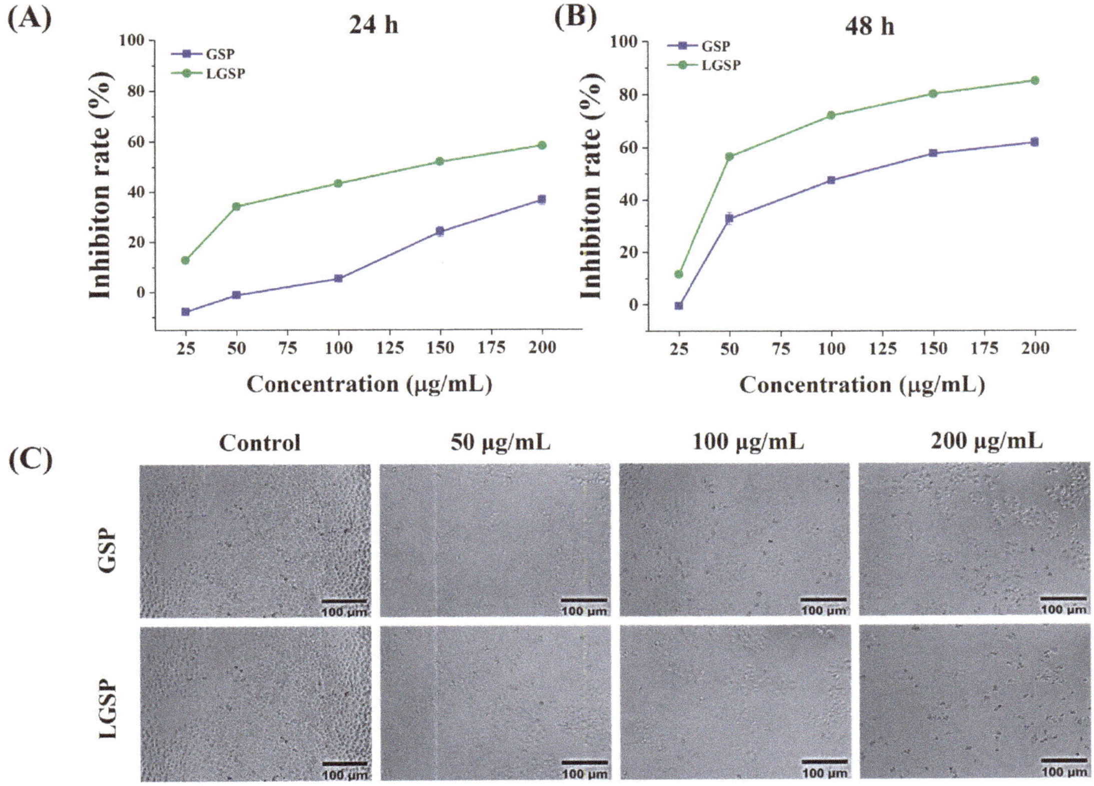

Figure 1. Anti-cervical cancer effect of GSP and LGSP on HeLa cell lines. HeLa cells were treated with various doses of GSP and LGSP for (**A**) 24 h and (**B**) 48 h. (**C**) Morphological image of HeLa cells after treatment with GSP and LGSP for 48 h. Data were mean ± SD (n = 3).

3.2. LGSP-Induced HeLa Cells Apoptosis

To explore the effect of LGSP on HeLa cell apoptosis, the cells (Annexin V+/PI− fraction subpopulations and Annexin V+/PI+ fraction subpopulations) were detected by flow cytometry. As shown in Figure 2A, both LGSP and GSP displayed a dose-dependent increase in apoptosis induction in HeLa cells. LGSP resulted in 18.77, 31.25 and 43.36% apoptotic rates at concentrations of 50, 100 and 200 μg/mL, respectively, which were higher than that of GSP at the same doses.

3.3. Effect of LGSP on the Cell Cycle of HeLa Cells

The effect of LGSP on the cell cycle distribution was analyzed by flow cytometry. As shown in Figure 2B, compared with 18.8% G2/M phase cells of control cells, an increase in the percentage of cells in the G2/M cells (20.1, 24.3 and 36.5%) was observed in cells after treatment with LGSP at concentrations of 50, 100 and 200 μg/mL, respectively. Similar to LGSP, GSP also induced the occurrence of G2/M cell-cycle arrest, but its effect (25.6%) was weaker than that of LGSP at the dose of 200 μg/mL.

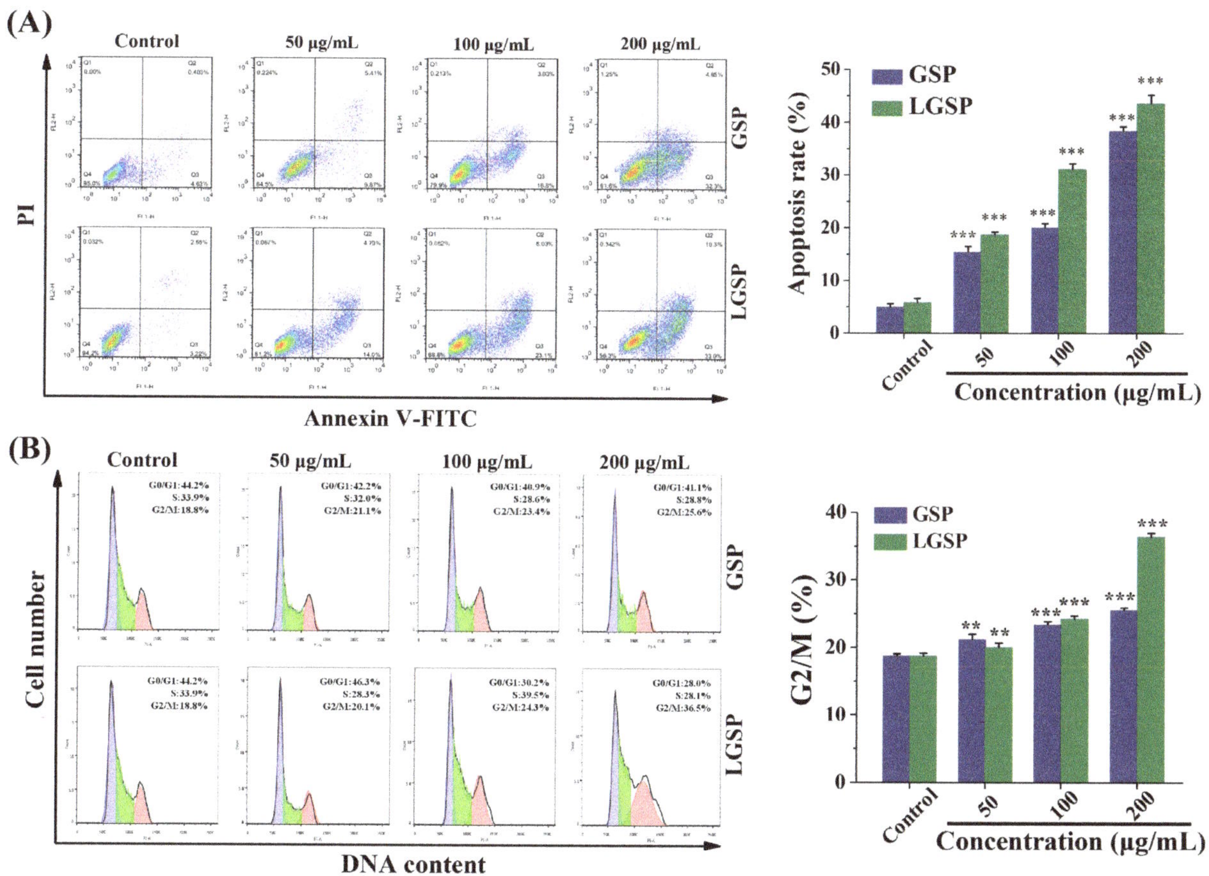

Figure 2. (**A**) Apoptotic effect of GSP and LGSP on HeLa cells double-stained with Annexin V-FITC/PI and analyzed by flow cytometry. (**B**) The cell cycle distribution of HeLa cells treated with GSP and LGSP. HeLa cells were treated with different concentrations of GSP and LGSP for 48 h. Data were mean ± SD ($n = 3$). ** $0.001 < p < 0.01$; *** $p < 0.001$.

3.4. LGSP Enhanced the ROS Levels in HeLa Cells

ROS generation plays a vital role in apoptosis [24]. To investigate the effect from LGSP, on the intracellular ROS level, the production of intracellular ROS was examined by flow cytometry. As shown in Figure 3, treatment of HeLa cells with LGSP at concentrations of 50, 100, and 200 μg/mL for 48 h resulted in a significant dose-dependent increase in the fluorescence intensity. Compared with the control group, the fluorescence intensity of HeLa cells treated with LGSP and GSP at the concentration of 200 μg/mL increased 3.01- and 2.12-fold, respectively, indicating the ROS production of HeLa cells after treatment with LGSP was higher than that with GSP.

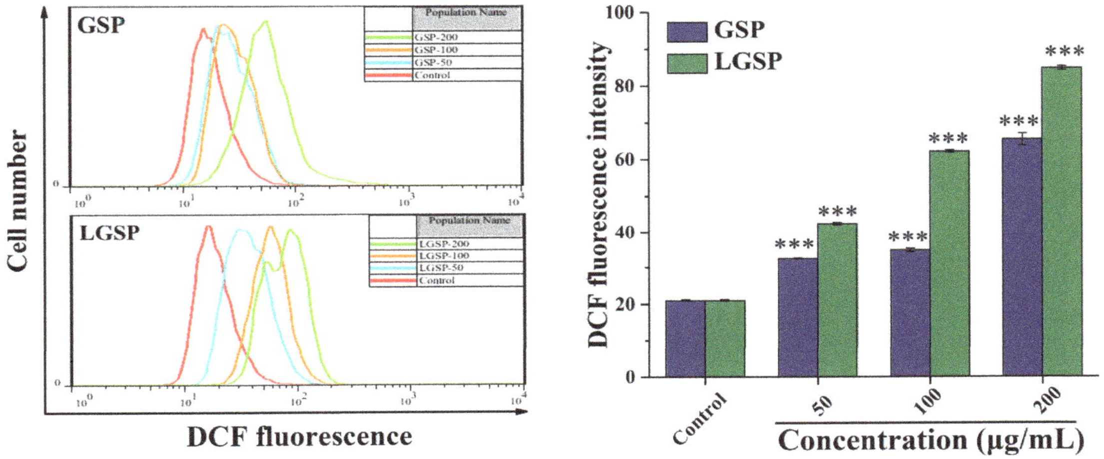

Figure 3. Effect of GSP and LGSP on the ROS generation of HeLa cells treated with different concentrations of GSP and LGSP for 48 h. ROS production was analyzed by flow cytometry using a DCFH-DA probe, and data were mean ± SD (n = 3): *** $p < 0.001$.

3.5. LGSP Induced Mitochondrial Dysfunction in HeLa Cells

To examine whether LGSP-induced apoptosis in HeLa cells was involved in the mitochondria-mediated pathway, the MMP of HeLa cells treated with LGSP was observed under an inverted fluorescence microscope with fluorescence probe JC-1. The change of JC-1 from red to green fluorescence indicated a decrease in membrane potential. As shown in Figure 4A, red fluorescence intensity decreased while green fluorescence intensity increased, suggesting that LGSP treatment triggered mitochondrial damage. For further quantitative analysis, the fluorescence intensity of HeLa cells was determined by flow cytometry. As shown in Figure 4B, the green fluorescence intensity of HeLa cells treated with 200 μg/mL LGSP increased to 49.22%, which was higher than that for GSP (8.54%).

3.6. Effect of LGSP on Bax, Bcl-2 and Cytochrome c Expression in HeLa Cells

To explore the effect of LGSP on apoptosis-regulated proteins, the levels of Bcl-2, Bax and cytochrome *c* on HeLa cells were examined. As shown in Figure 5A, LGSP up-regulated Bax expression and down-regulated Bcl-2 expression, leading to a significant increase in the Bax/Bcl-2 ratio. Moreover, the Bax/Bcl-2 ratio of cells treated with 200 μg/mL LGSP was 29.24, which was higher than that for GSP (8.60) at the same concentration. The Bcl-2 protein family, pro-apoptotic Bax and anti-apoptotic Bcl-2, could strictly control the release of cytochrome *c* from mitochondria into the cytoplasm [25]. The results showed that the level of cytochrome *c* increased in a dose-dependent manner after treatment with LGSP, indicating that LGSP increased the release of cytochrome *c* into the cytoplasm.

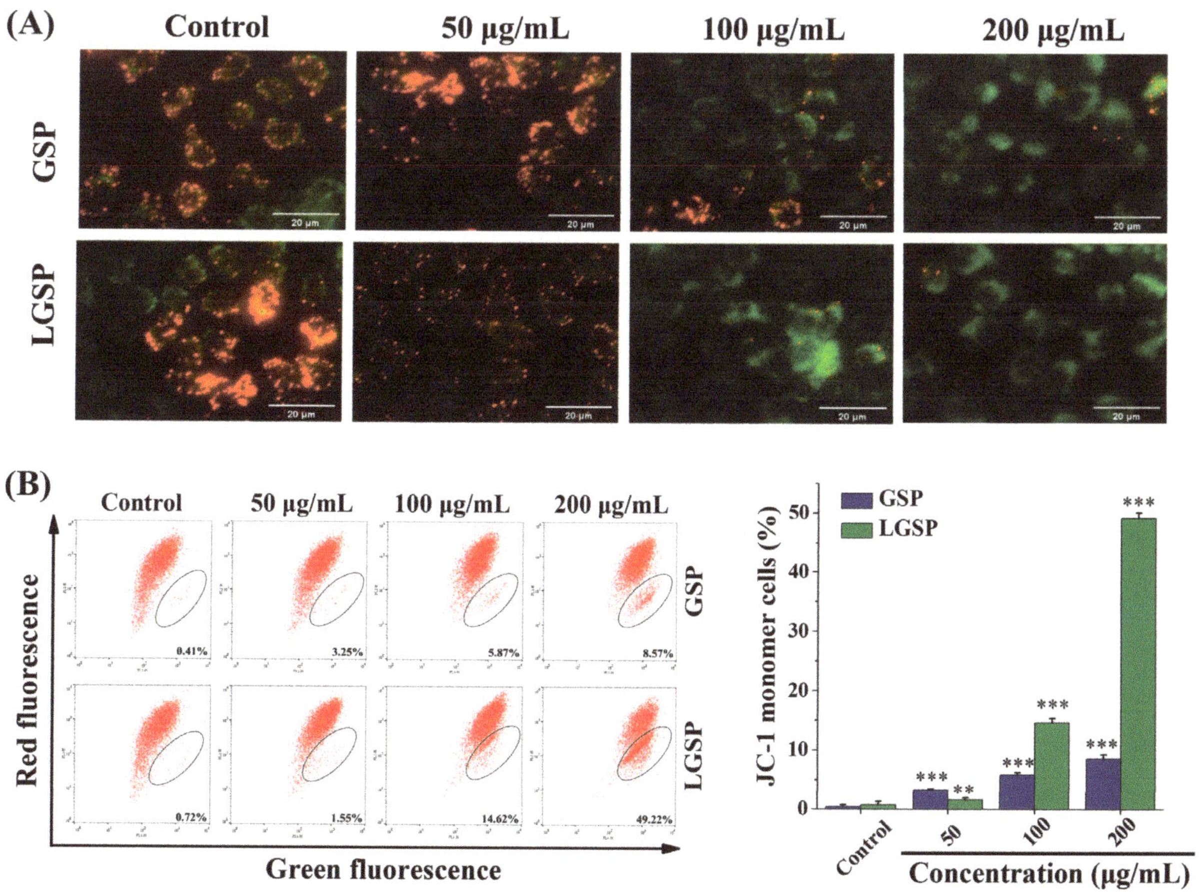

Figure 4. Effect of GSP and LGSP on the mitochondrial membrane potential (MMP) of HeLa cells. HeLa cells were treated with GSP and LGSP for 48 h, and the MMP was (**A**) observed under an inverted fluorescence microscope and (**B**) determined by flow cytometer with JC-1 dye. Data were mean ± SD ($n = 3$): ** $0.001 < p < 0.01$; *** $p < 0.001$.

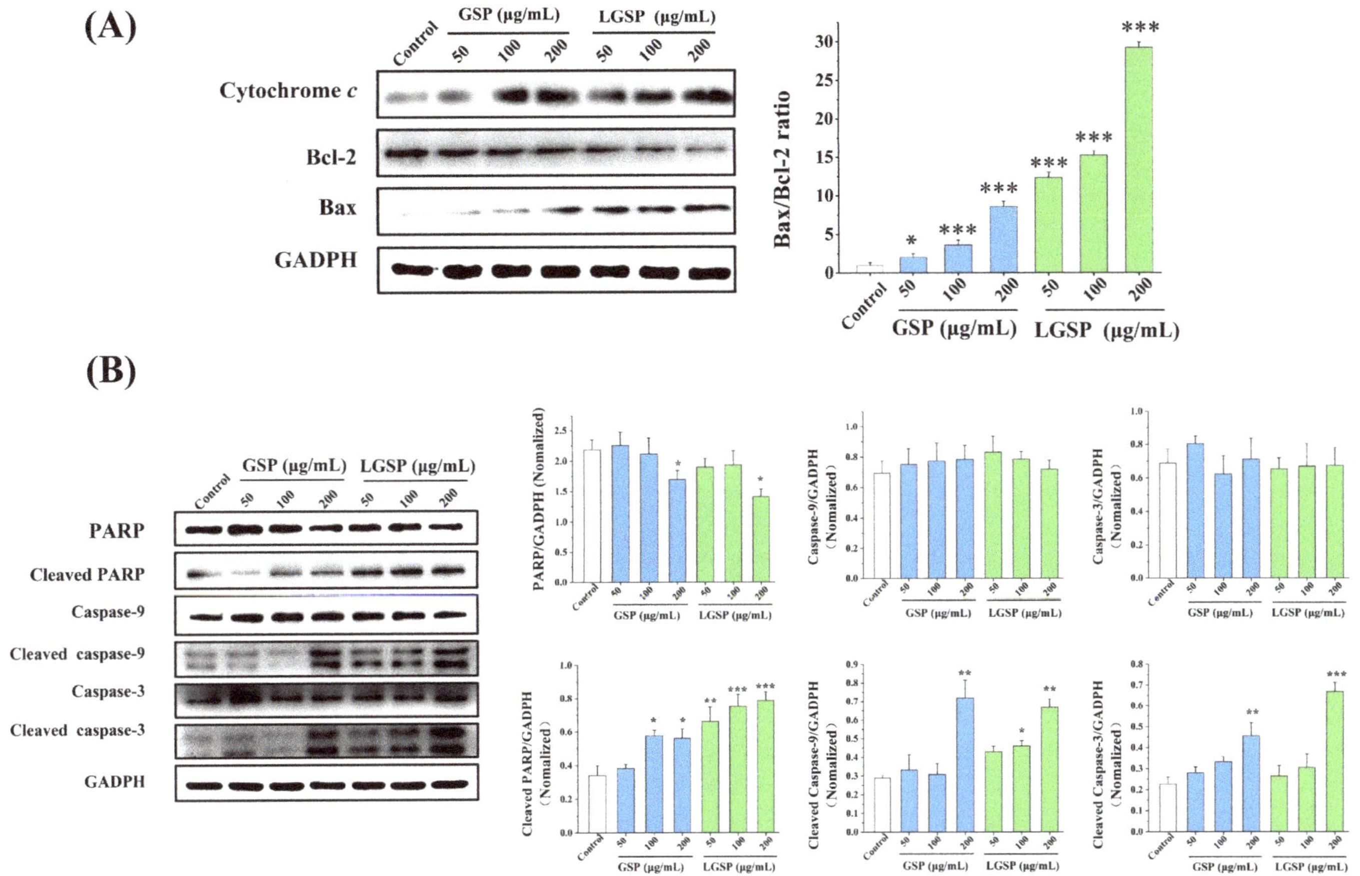

Figure 5. Western blot analysis of the expressions of (**A**) Bax, Bcl-2, cytochrome *c*, (**B**) PARP, cleaved PARP, caspase-9, cleaved caspase-9, caspase-3 and cleaved caspase-3 in HeLa cells treated with different concentrations of GSP or LGSP for 48 h. Data were mean ± SD ($n = 3$): * $0.01 < p < 0.05$; ** $0.001 < p < 0.01$; *** $p < 0.001$.

3.7. LGSP-Induced Activation of Caspases and the Cleavage of PARP Protein in HeLa Cells

The cytochrome *c* released from the mitochondria activated the caspase apoptosis pathway, and we then explored whether LGSP activated it. As shown in Figure 5B, after treatment with LGSP, the levels of cleaved caspase-3, cleaved caspase-9 and cleaved PARP in HeLa cells increased in a dose-dependent manner, indicating that LGSP led to a clear cleavage of caspase-9, caspase-3 and PARP.

3.8. LGSP Blocked Cervical Cancer Growth in a HeLa-Derived Xenograft Zebrafish Model

To test whether LGSP can suppress cervical cancer in vivo, we treated zebrafish carrying HeLa xenograft tumors with LGSP. First, we evaluated the toxicity of GSP and LGSP to zebrafish larvae. The results showed that GSP at doses of 4–8 μg/mL and LGSP at doses of 4–16 μg/mL exhibited non-toxic effects (Figure 6A). Hence, the doses of 4 and 8 μg/mL were selected for further study in the HeLa-derived xenograft zebrafish model. As shown in Figure 6B,C, the circled green fluorescence represented the cervical tumor mass, indicating that the HeLa xenograft model was successfully established in zebrafish. LGSP and GSP at doses of 4 and 8 μg/mL markedly decreased the green fluorescence intensity in a dose-dependent manner, and the inhibition rate of LGSP (84.32%) was higher than that for GSP (68.89%) at the concentration of 8 μg/mL.

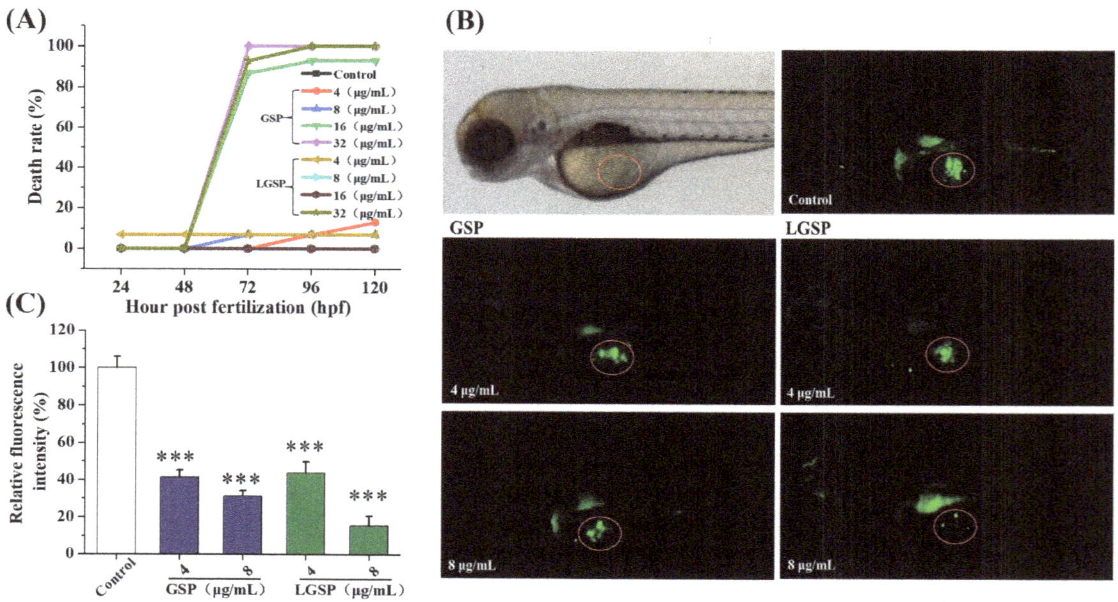

Figure 6. Effect of GSP and LGSP on the HeLa-derived xenograft tumor growth. (**A**) Mortality curve of zebrafish embryos treated with different doses of LGSP and GSP. (**B**) Tumor growth in HeLa-derived xenograft zebrafish embryos treated with GSP and LGSP for 48 h was observed under a microscope. (**C**) The fluorescence intensity of the tumor in Figure 6B. Data were mean ± SD (n = 3): *** $p < 0.001$.

4. Discussion

GSP, the main polyphenol in grape seed, has been proven to inhibit cervical cells by inducing apoptosis and blocking cell cycles [26]. LGSP, synthesized by enzymatic modification of GSP, showed higher antioxidant and anti-inflammatory activity compared to GSP [15,16]. Therefore, we hypothesized that LGSP has a potential chemopreventive effect on cervical cancer.

Our results showed that LGSP exhibited an anti-proliferative effect on HeLa cells, and its effect was higher than that for GSP, which confirmed our hypothesis. Similarly, the oleate derivative of quercetin-3-O-glucoside showed a greater anti-proliferative effect on HepG2 cells compared to quercetin-3-O-glucoside [27]. As we previously reported, the lipophilicity of LGSP was higher than that for GSP [15], which may lead to higher cell membrane permeability and better bioactivity in cell models [12].

Generally speaking, the induction of apoptosis or cell cycle arrest or a combination of these two modes is the common mechanism for inhibiting cancer-cell proliferation [28]. In this study, LGSP increased the apoptosis rate and the percentage of G2/M phase on HeLa cells, indicating LGSP exerted an anti-proliferative effect on HeLa human cervical cancer cells by inducing apoptosis and blocking G2/M cell cycle. It is reported that intracellular ROS was associated with cell apoptosis [29], and could also cause a DNA damage response that could impact the cell cycle progression [30]. In this study, the level of ROS increased dramatically in the HeLa cells treated with LGSP, suggesting that LGSP may induce the production of ROS, leading to cell apoptosis and cycle arrest. In addition, we found that the anti-proliferative effect of GSP on HeLa cells was also achieved by inducing apoptosis and blocking the G2/M cell cycle, which was consistent with the report of Chen et. al [26].

There are two principal apoptosis pathways: the extrinsic pathway and the intrinsic mitochondrial pathway [31], which is the key signaling pathway in the induction of apoptosis [32]. It has been reported that the loss of MMP is a typical phenomenon in mitochondrial-dependent apoptosis [33]. In this work, LGSP decreased the MMP of HeLa cells, indicating that LGSP-induced apoptosis may be through the mitochondrial pathway. The mitochondrial apoptosis pathway is strictly regulated by the two important members of the Bcl-2 protein family: Bax and Bcl-2 [34]. The oligomerization of pro-apoptotic Bax and the reduction of anti-apoptotic Bcl-2 could lead to the release of cytochrome *c* from the mitochondria into the cytoplasm [25]. Our results showed that LGSP clearly increased the Bax/Bcl-2 ratio and promoted the release of cytochrome *c* into the cytoplasm, which confirmed that LGSP induced apoptosis through the mitochondrial pathway.

The cytochrome *c* released from the mitochondria was reported to activate caspase-9 and finally induce apoptotic occurrence [35]. To further study the downstream apoptosis mechanism of LGSP on HeLa cells, we evaluated some relevant proteins of the caspase pathway. Our findings showed that LGSP up-regulated the expression of cleaved caspase-9 caspase-3 and PARP to activate the caspase pathways. The activated cleaved caspases and PARP were considered to be relevant biomarkers for inducing apoptosis [36]. Therefore, our findings suggested that LGSP-induced apoptosis was mediated through the mitochondrial/caspase pathway, which involved the up-regulation of the Bax/Bcl-2 ratio, release of cytochrome *c*, and activation of cleavage of caspase 9, caspase-3 and PARP.

The above results indicated that LGSP has an anti-cervical cancer effect in vitro. It is necessary to evaluate its anti-cervical cancer effect in vivo further. In recent years, zebrafish xenograft models have been generated more frequently to study malignancies due to their special characteristics, such as transparent embryos, large-scale generation, rapid organogenesis and no immune rejection [37,38]. Therefore, we then assessed the effect of LGSP on the growth of HeLa cells in a zebrafish xenograft model. In this study, we found that LGSP remarkably decreased the growth of a HeLa xenograft tumor, and the inhibition rate of LGSP was higher than that for GSP, indicating that LGSP had better anti-cervical cancer activity in vivo, which was consistent with our in vitro studies.

5. Conclusions

In summary, this study first indicated that LGSP possessed an anti-cervical cancer effect on HeLa cells in vitro and in vivo. LGSP exhibited an excellent anti-proliferative effect on HeLa cells by inducing cell apoptosis and blocking cell-cycle progression in the G2/M phase. The apoptosis of HeLa cells triggered by LGSP was through the mitochondrial/caspase-mediated pathway, characterized by Bax/Bcl-2 ratio up-regulation, the release of cytochrome *c*, the loss of mitochondrial membrane potential, and the cleaved caspase-9/3 and cleavage of PARP activation. In the zebrafish model, LGSP effectively suppressed the growth of a HeLa xenograft tumor. Furthermore, LGSP presented better anti-cervical cancer effects than GSP in HeLa cells and zebrafish models. LGSP could be considered a potential supplement for preventing and treating human cervical cancer.

Supplementary Materials: The following supporting information can be downloaded at: https://www.mdpi.com/article/10.3390/antiox11020422/s1, Figure S1: Synthesis of LGSP.

Author Contributions: Conceptualization, C.L. (Chengmei Liu); methodology, M.C. and C.L. (Changhong Li); software, X.H.; validation, C.L. (Chengmei Liu), M.C. and J.C.; formal analysis, L.Z.; investigation, C.L. (Changhong Li); resources, M.C.; data curation, X.H.; writing—original draft preparation, C.L. (Changhong Li); writing—review and editing, C.L. (Changhong Li), L.Z. and M.C.; visualization, M.C.; supervision, M.C.; project administration, J.C.; funding acquisition, M.C. and J.C. All authors have read and agreed to the published version of the manuscript.

Funding: This research was funded by the first batch of high-end talents (youth) projects of science and technology innovation in Jiangxi Province (2019201076) and Natural Science Foundation of Jiangxi Province (20202BAB215014).

Institutional Review Board Statement: The use and manipulations of zebrafish were approved by the ethical review committee of Nanchang University (Nanchang, China) on 25 March 2021.

Informed Consent Statement: Not applicable.

Data Availability Statement: The data presented in this study are available in this manuscript.

Acknowledgments: The authors would like to thank the State Key Laboratory of Food Science and Technology for their expert technical assistance.

Conflicts of Interest: The authors declare no conflict of interest.

References

1. Arbyn, M.; Weiderpass, E.; Bruni, L.; de Sanjose, S.; Saraiya, M.; Ferlay, J.; Bray, F. Estimates of incidence and mortality of cervical cancer in 2018: A worldwide analysis. *Lancet Glob. Health* **2020**, *8*, E191–E203. [CrossRef]
2. Small, W., Jr.; Bacon, M.A.; Bajaj, A.; Chuang, L.T.; Fisher, B.J.; Harkenrider, M.M.; Jhingran, A.; Kitchener, H.C.; Mileshkin, L.R.; Viswanathan, A.N.; et al. Cervical Cancer: A Global Health Crisis. *Cancer* **2017**, *123*, 2404–2412. [CrossRef] [PubMed]
3. Rothwell, J.A.; Knaze, V.; Zamora-Ros, R. Polyphenols: Dietary assessment and role in the prevention of cancers. *Curr. Opin. Clin. Nutr. Metab. Care* **2017**, *20*, 512–521. [CrossRef] [PubMed]
4. Shahidi, F.; Yeo, J. Bioactivities of Phenolics by Focusing on Suppression of Chronic Diseases: A Review. *Int. J. Mol. Sci.* **2018**, *19*, 1573. [CrossRef] [PubMed]
5. Roleira, F.M.F.; Tavares-da-Silva, E.J.; Varela, C.L.; Costa, S.C.; Silva, T.; Garrido, J.; Borges, F. Plant derived and dietary phenolic antioxidants: Anticancer properties. *Food Chem.* **2015**, *183*, 235–258. [CrossRef]
6. Unusan, N. Proanthocyanidins in grape seeds: An updated review of their health benefits and potential uses in the food industry. *J. Funct. Foods* **2020**, *67*, 103861. [CrossRef]
7. Long, M.; Yang, S.H.; Han, J.X.; Li, P.; Zhang, Y.; Dong, S.; Chen, X.L.; Guo, J.Y.; Wang, J.; He, J.B. The Protective Effect of Grape-Seed Proanthocyanidin Extract on Oxidative Damage Induced by Zearalenone in Kunming Mice Liver. *Int. J. Mol. Sci.* **2016**, *17*, 808. [CrossRef]
8. Cai, X.X.; Bao, L.; Ren, J.W.; Li, Y.; Zhang, Z.F. Grape seed procyanidin B2 protects podocytes from high glucose-induced mitochondrial dysfunction and apoptosis via the AMPK-SIRT1-PGC-1 alpha axis in vitro. *Food Funct.* **2016**, *7*, 805–815. [CrossRef]
9. Liu, W.; Zhao, S.Q.; Wang, J.Q.; Shi, J.; Sun, Y.K.; Wang, W.Q.; Ning, G.; Hong, J.; Liu, R.X. Grape seed proanthocyanidin extract ameliorates inflammation and adiposity by modulating gut microbiota in high-fat diet mice. *Mol. Nutr. Food Res.* **2017**, *61*, 1601082. [CrossRef]
10. Huang, L.L.; Pan, C.; Wang, L.; Ding, L.; Guo, K.; Wang, H.Z.; Xu, A.m.; Gao, S. Protective effects of grape seed proanthocyanidins on cardiovascular remodeling in DOCA-salt hypertension rats. *J. Nutr. Biochem.* **2015**, *26*, 841–849. [CrossRef]
11. Zhu, S.; Li, Y.; Ma, C.Y.; Lou, Z.X.; Chen, S.W.; Dai, J.; Wang, H.X. Optimization of lipase-catalyzed synthesis of acetylated EGCG by response surface methodology. *J. Mol. Catal. B Enzym.* **2013**, *97*, 87–94. [CrossRef]
12. Mellou, F.; Loutrari, H.; Stamatis, H.; Roussos, C.; Kolisis, F.N. Enzymatic esterification of flavonoids with unsaturated fatty acids: Effect of the novel esters on vascular endothelial growth factor release from K562 cells. *Process Biochem.* **2006**, *41*, 2029–2034. [CrossRef]
13. Chen, M.; Yu, S. Lipophilic grape seed proanthocyanidin exerts anti-proliferative and pro-apoptotic effects on PC3 Human Prostate cancer cells and suppresses PC3 xenograft tumor growth in vivo. *J. Agric. Food Chem.* **2019**, *67*, 229–235. [CrossRef] [PubMed]

14. Luo, S.Z.; Hu, X.F.; Pan, L.H.; Zheng, Z.; Zhao, Y.Y.; Cao, L.L.; Pang, M.; Hou, Z.G.; Jiang, S.T. Preparation of camellia oil-based W/O emulsions stabilized by tea polyphenol palmitate: Structuring camellia oil as a potential solid fat replacer. *Food Chem.* **2019**, *276*, 209–217. [CrossRef]
15. Chen, M.S.; Yu, S.J. Lipophilized grape seed proanthocyanidin derivatives as novel antioxidants. *J. Agric. Food Chem.* **2017**, *65*, 1598–1605. [CrossRef]
16. Zhang, L.L.; Chen, J.; Liao, H.J.; Li, C.H.; Chen, M.S. Anti-inflammatory effect of lipophilic grape seed proanthocyanidin in RAW 264.7 cells and a zebrafish model. *J. Funct. Foods* **2020**, *75*, 104217. [CrossRef]
17. Reuter, S.; Gupta, S.C.; Chaturvedi, M.M.; Aggarwal, B.B. Oxidative stress, inflammation, and cancer How are they linked? *Free Radic. Biol. Med.* **2010**, *49*, 1603–1616. [CrossRef]
18. Zhang, T.T.; Yang, L.; Jiang, J.G. Effects of thonningianin A in natural foods on apoptosis and cell cycle arrest of HepG-2 human hepatocellular carcinoma cells. *Food Funct.* **2015**, *6*, 2588–2597. [CrossRef]
19. Wang, J.; Liao, A.M.; Thakur, K.; Zhang, J.G.; Huang, J.H.; Wei, Z.J. Licochalcone B extracted from glycyrrhiza uralensis fisch induces apoptotic effects in human hepatoma cell HepG2. *J. Agric. Food Chem.* **2019**, *67*, 3341–3353. [CrossRef]
20. Wu, Q.; Deng, J.J.; Fan, D.D.; Duan, Z.G.; Zhu, C.H.; Fu, R.Z.; Wang, S.S. Ginsenoside Rh4 induces apoptosis and autophagic cell death through activation of the ROS/JNK/p53 pathway in colorectal cancer cells. *Biochem. Pharmacol.* **2018**, *148*, 64–74. [CrossRef]
21. Chen, J.; Zhang, L.L.; Li, C.H.; Chen, R.C.; Liu, C.M.; Chen, M.S. Lipophilized epigallocatechin gallate derivative exerts anti-proliferation efficacy through induction of cell cycle arrest and apoptosis on DU145 human prostate cancer cells. *Nutrients* **2020**, *12*, 92. [CrossRef] [PubMed]
22. Xiong, G.H.; Zou, L.F.; Deng, Y.Y.; Meng, Y.L.; Liao, X.J.; Lu, H.Q. Clethodim exposure induces developmental immunotoxicity and neurobehavioral dysfunction in zebrafish embryos. *Fish Shellfish Immunol.* **2019**, *86*, 549–558. [CrossRef] [PubMed]
23. Bae, H.; Song, G.; Lee, J.-Y.; Hong, T.; Chang, M.-J.; Lim, W. Laminarin-Derived from brown algae suppresses the growth of Ovarian cancer cells via mitochondrial dysfunction and ER stress. *Mar. Drugs* **2020**, *18*, 152. [CrossRef] [PubMed]
24. Aggarwal, V.; Tuli, H.S.; Varol, A.; Thakral, F.; Yerer, M.B.; Sak, K.; Varol, M.; Jain, A.; Khan, A.; Sethi, G. Role of reactive oxygen species in cancer progression: Molecular mechanisms and recent advancements. *Biomolecules* **2019**, *9*, 753. [CrossRef]
25. Fan, Y.Y.; Zhang, Y.; Liu, Y.T.; Xu, W.P.; Yang, Y.; Hao, Y.W.; Tao, L.M. A natural product enhances apoptosis via mitochondria/caspase-mediated pathway in HeLa cells. *J. Cell. Biochem.* **2019**, *120*, 16811–16823. [CrossRef]
26. Chen, Q.; Liu, X.F.; Zheng, P.S. Grape seed proanthocyanidins (GSPs) inhibit the growth of cervical cancer by inducing apoptosis mediated by the mitochondrial pathway. *PLoS ONE.* **2014**, *9*, e107045.
27. Sudan, S.; Rupasinghe, H.P.V. Antiproliferative activity of long chain acylated esters of quercetin-3-O-glucoside in hepatocellular carcinoma HepG2 cells. *Exp. Biol. Med.* **2015**, *240*, 1452–1464. [CrossRef]
28. Sun, X.; Liao, W.; Wang, J.; Wang, P.; Gao, H.; Wang, M.; Xu, C.; Zhong, Y.; Ding, Y. CSTMP induces apoptosis and mitochondrial dysfunction in human myeloma RPMI8226 cells via CHOP-dependent endoplasmic reticulum stress. *Biomed. Pharmacother.* **2016**, *83*, 776–784. [CrossRef]
29. Ye, X.C.; Zuo, D.D.; Yu, L.; Zhang, L.; Tang, J.; Cui, C.C.; Bao, L.; Zan, K.; Zhang, Z.H.; Yang, X.X.; et al. ROS/TXNIP pathway contributes to thrombin induced NLRP3 inflammasome activation and cell apoptosis in microglia. *Biochem. Biophys. Res. Commun.* **2017**, *485*, 499–505. [CrossRef]
30. Wu, G.S.; Lu, J.J.; Guo, J.J.; Li, Y.B.; Tan, W.; Dang, Y.Y.; Zhong, Z.F.; Xu, Z.T.; Chen, X.P.; Wang, Y.T. Ganoderic acid DM, a natural triterpenoid, induces DNA damage, G1 cell cycle arrest and apoptosis in human breast cancer cells. *Fitoterapia* **2012**, *83*, 408–414. [CrossRef]
31. Pfeffer, C.M.; Singh, A.T.K. Apoptosis: A Target for Anticancer Therapy. *Int. J. Mol. Sci.* **2018**, *19*, 448. [CrossRef] [PubMed]
32. Yu, P.; Shi, L.F.; Song, M.Y.; Meng, Y. Antitumor activity of paederosidic acid in human non-small cell lung cancer cells via inducing mitochondria-mediated apoptosis. *Chem. Biol. Interact.* **2017**, *269*, 33–40. [CrossRef] [PubMed]
33. Ben Othmene, Y.; Monceaux, K.; Karoui, A.; Ben Salem, I.; Belhadef, A.; Abid-Essefi, S.; Lemaire, C. Tebuconazole induces ROS-dependent cardiac cell toxicity by activating DNA damage and mitochondrial apoptotic pathway. *Ecotoxicol. Environ. Saf.* **2020**, *204*, 111040. [CrossRef]
34. McArthur, K.; Whitehead, L.W.; Heddleston, J.M.; Li, L.; Padman, B.S.; Oorschot, V.; Geoghegan, N.D.; Chappaz, S.; Davidson, S.; Chin, H.S.; et al. BAK/BAX macropores facilitate mitochondrial herniation and mtDNA efflux during apoptosis. *Science.* **2018**, *359*, 883. [CrossRef] [PubMed]
35. Youle, R.J.; Strasser, A. The BCL-2 protein family: Opposing activities that mediate cell death. *Nat. Rev. Mol. Cell Biol.* **2008**, *9*, 47–59. [CrossRef]
36. Bi, Y.L.; Min, M.; Shen, W.; Liu, Y. Genistein induced anticancer effects on pancreatic cancer cell lines involves mitochondrial apoptosis, G(0)/G(1) cell cycle arrest and regulation of STAT3 signaling pathway. *Phytomedicine* **2018**, *39*, 10–16. [CrossRef]
37. Zhang, B.B.; Xuan, C.; Ji, Y.X.; Zhang, W.M.; Wang, D.G. Zebrafish xenotransplantation as a tool for in vivo cancer study. *Fam. Cancer* **2015**, *14*, 487–493. [CrossRef]
38. Jin, W.D.; Zhou, L.; Yan, B.; Yan, L.; Liu, F.C.; Tong, P.J.; Yu, W.H.; Dong, X.Q.; Xie, L.; Zhang, J.; et al. Theabrownin triggers DNA damage to suppress human osteosarcoma U2OS cells by activating p53 signaling pathway. *J. Cell. Mol. Med.* **2018**, *22*, 4423–4436. [CrossRef]

antioxidants

MDPI

Article

Identification of Crucial Polymethoxyflavones Tangeretin and 3,5,6,7,8,3′,4′-Heptamethoxyflavone and Evaluation of Their Contribution to Anticancer Effects of *Pericarpium Citri Reticulatae 'Chachi'* during Storage

Yexing Tao [1,2,3], Qian Yu [1,2,3], Yuting Huang [1,2,3], Ruiting Liu [1,2,3], Xiwen Zhang [1,2,3], Ting Wu [1,2,3], Siyi Pan [1,2,3] and Xiaoyun Xu [1,2,3,*]

1 College of Food Science and Technology, Huazhong Agricultural University, Wuhan 430070, China
2 Key Laboratory of Environment Correlative Dietology (Ministry of Education), Huazhong Agricultural University, Wuhan 430070, China
3 Hubei Key Laboratory of Fruit & Vegetable Processing & Quality Control, Huazhong Agricultural University, Wuhan 430070, China
* Correspondence: xuxiaoyun@mail.hzau.edu.cn; Tel.: +86-27-87671056; Fax: +86-27-87288373

Abstract: *Pericarpium Citri Reticulatae 'Chachi'* (PCR-C), rich in polymethoxyflavones (PMFs), has potential anticancer bioactivity and its quality will be improved during storage. However, the main factors influencing the PCR-C quality during its storage remain unclear. In this study, multivariate analysis was performed to investigate free and bound PMFs of PCR-C during storage. The anticancer effects of purified PCR-C flavonoid extracts (PCR-CF) and the important PMFs were evaluated using A549 cells. The results showed that PCR-C samples exhibited remarkable differences in free PMFs during storage, which fell into three clusters: Cluster 1 included fresh (fresh peel) and PCR-C01 (year 1); Cluster 2 consisted of PCR-C03 (year 3) and PCR-C05 (year 5); and PCR-C10 (year 10) was Cluster 3. 3,5,6,7,8,3′,4′-heptamethoxyflavone, tangeretin, and isosinensetin were identified as the most important PMFs distinguishing the various types of PCR-C according to its storage periods. Moreover, PCR-CF inhibited A549 cell proliferation and induced cell cycle arrest at G2/M phase, cell apoptosis, and ROS accumulation, and all anticancer indices had an upward tendency during storage. Additionally, tangeretin and 3,5,6,7,8,3′,4′-heptamethoxyflavone exhibited anticancer effects on A549 cells, whereas isosinensetin displayed no anticancer effect, indicating that tangeretin and 3,5,6,7,8,3′,4′-heptamethoxyflavone jointly contributed to anticancer activity of PCR-C during storage. PCR-CF and the most important PMFs killed cancer cells (A549 cells) but had no cytotoxicity to normal lung fibroblast cells (MRC-5 cells). Overall, the high quality of long-term stored PCR-C might be due to the anticancer effects of tangeretin and 3,5,6,7,8,3′,4′-heptamethoxyflavone.

Keywords: *Pericarpium Citri Reticulatae 'Chachi'*; polymethoxyflavones; storage periods; non-small cell lung cancer; apoptosis

Citation: Tao, Y.; Yu, Q.; Huang, Y.; Liu, R.; Zhang, X.; Wu, T.; Pan, S.; Xu, X. Identification of Crucial Polymethoxyflavones Tangeretin and 3,5,6,7,8,3′,4′-Heptamethoxyflavone and Evaluation of Their Contribution to Anticancer Effects of *Pericarpium Citri Reticulatae 'Chachi'* during Storage. *Antioxidants* **2022**, *11*, 1922. https://doi.org/10.3390/antiox11101922

Academic Editors: Jicheng Zhan, Zhigang Liu and Hui-Min David Wang

Received: 25 July 2022
Accepted: 23 September 2022
Published: 28 September 2022

1. Introduction

Pericarpium Citri Reticulatae 'Chachi' (PCR-C, Guangchenpi in China), one of the first food materials promulgated by Chinese Ministry of Health owing to its outstanding efficacy, is a dried and matured pericarp of *Citrus reticulata Blanco* or its cultivars cultivated in Xinhui, Guangdong, China that can not only be used as a dietary supplement but also a traditional Chinese medicine [1]. PCR-C has pharmacological effects as the adjuvant therapy for indigestion, physical weakness, cough accompanied with expectoration of phlegm, and other digestive and respiratory diseases [2,3]. Long-term stored medicine refers to that materials that are stored and maintained through using corresponding methods and then could be more applicable to the traditional medicinal utilization [4]. There is a folk proverb about the aging of PCR-C which is "the longer storage periods, the better", which reflects the effect of storage period on the market price and application value of PCR-C [5]. PCR-C contains abundant flavonoids, which are divided into two categories: flavanone glycosides such as hesperidin and polymethoxylated flavones (PMFs) such as tangeretin and nobiletin [6]. Although the contents of PMFs are quite low in PCR-C, they are characteristic bioactive components with extremely potent activities [7,8]. PCR-C quality is mainly attributed to the accumulation of specific chemical bioactive components during storage. The phenolic acids in PCR-C have been reported to exhibit an increasing trend, and the antioxidant activity of PCR-C is enhanced during storage, indicating that phenolic acids might be the main contributors to the improvement of the quality of long-term stored PCR-C [9]. One previous study has revealed that the contents of all the polymethoxyflavones (PMFs such as sinensetin, 4′,5,7,8-tetramethoxyflavone, nobiletin, tangeretin, and 5-O-desmethyl nobiletin) and their antioxidant activities increase during PCR-C storing, suggesting the positive correlation between the quality of PCR-C during storage and the PMFs contents [5]. On the contrary, one metabolomic analysis that was based on UPLC-QTOF-MS indicated that the contents of PMFs were not positively correlated with the storage periods of PCR-C [10]. The dynamic change of specific chemical bioactive components in PCR-C during storage is still contradictory due to the limited free fraction of components. Thus, the comprehensive evaluation on the dynamic changes of free and bound PMFs in PCR-C during storage is necessary. Among the chemical compounds, the main contributors to the quality of PCR-C during storage also remain largely unknown.

Lung cancers including non-small cell lung cancer (NSCLC, almost accounting for 85%) and small cell lung cancer are a prevailing oncological disease worldwide with high incidence rate and mortality rate [11]. The common treatments that include surgery, radiation, chemotherapy, and targeted therapy are based on the development stage of the cancer [12]. Although a variety of NSCLC treatment strategies have been developed in recent decades, the prognosis of patients is far from satisfactory, and the resistance, side effects, and recurrence are fairly common [13]. Therefore, developing potentially effective agents and exploring novel NSCLC therapeutic strategies are urgent. Natural products with the advantages of low cost, safety, and easy accessibility are considered to be the potentially effective anticancer drugs. The increasing evidence has demonstrated that PCR-C has antioxidant, anti-inflammatory, anti-asthmatic, neuroprotective, and antitumor activities [14–16]. However, during storage, the influential contributors to the anticancer effect of PCR-C remain unclear.

Therefore, it is worth investigating the dynamic change of free and bound PMFs in PCR-C during storage and explore the influential PMFs that mainly contribute to the improvement of anticancer effect of PCR-C during storage. In the present study, the multivariate statistical analysis was performed to reveal dynamic changes of free and bound PMFs and to identify the important PMFs that could distinguish PCR-C during storage. In addition, the anticancer effect of PCR-CF and the important PMFs were evaluated to screen the crucial bioactive components of PCR-CF during storage. The results of dynamic changes of free and bound PMFs and anticancer activity assay revealed that free PMFs mainly contributed to the difference and the formation of unique quality of PCR-C during storage, and the free fraction of 3,5,6,7,8,3′,4′-heptamethoxyflavone, tangeretin, and isosinensetin were the determinant PMFs for discriminating the types of PCR-C at different storage periods. The accumulation of free tangeretin and 3,5,6,7,8,3′,4′-heptamethoxyflavone during storage contributed to the improvement of the quality of PCR-C. Our findings will provide a theoretical basis for the scientific utilization, rational storage, and quality control of PCR-C.

2. Materials and Methods

2.1. Samples and Reagents

The fresh peel of *Citrus reticulata cv. 'Chachiensis'* (Fresh) and PCR-C aged 1, 3, 5, 10 years (PCR-C01, PCR-C03, PCR-C05, PCR-C10) were obtained from Ganze Garden, Shuangshui Town, Xinhui District, Jiangmen City, Guangdong Province, China and authenticated by Hubei university of Chinese Medicine. PCR-CF at the fresh peel and aged 1, 3, 5, 10 years (Fresh-F, PCR-C01F, PCR-C03F, PCR-C05F, PCR-C10F) were obtained from our laboratory according to our previous study [17]. The flavonoid standards tangeretin, isosinensetin, and 3,5,6,7,8,3′,4′-heptamethoxyflavone were purchased from Yuanye Bio-Technology Co., Ltd. (Shanghai, China). Fetal bovine serum (FBS) was purchased from Germini (Woodland, CA, USA). Dulbecco's Modified Eagle Medium/Nutrient Mixture F-12 (DMEM/F12) was obtained from HYcezmbio Biotechnology Co., Ltd. (Wuhan, China). The 3-(4,5-dimethylthiazol-2-yl)-2,5-diphenyltetrazolium bromide (MTT) was provided by Gen-View Scientific Inc. (Calimesa, CA, USA). Trypsin was obtained from Lanheng Shidai Biotechnology Co., Ltd. (Wuhan, China). Penicillin-Streptomycin Solution was purchased from Cienry Biotechnology Co., Ltd. (Huzhou, China). Annexin V FITC/PI apoptosis and cell cycle staining kit were obtained from Multi-sciences (Lianke) Biotechnology Co., Ltd. (Hangzhou, China). 2′,7′-dichlorofluorescein diacetate (DCFH-DA) and Dimethyl sulfoxide (DMSO) in cell culture grade were obtained from Sigma Chemical Co. (St. Louis, MO, USA).

2.2. Cell Culture

Human non-small lung cancer A549 cells and human normal fetal lung fibroblast cell line (MRC-5) were purchased from the Cell bank of Chinese Academy of Sciences in Shanghai, China and cultured in DMEM/F-12 medium that was supplemented with 10% FBS, penicillin (100 U/mL), and streptomycin (100 μg/mL) under a stable atmosphere of 5% CO_2, at 37 °C constant temperature and a relatively high humidity of 95%. When they grew to approximately 80–90% confluence, the cells were passaged until the logarithmic growth phase and used for subsequent experiments.

2.3. MTT Assay

During the logarithmic growth phase, A549 cells and MRC-5 cells were grown in 96-well plates at a density of 3000 cells/well overnight. Then, the cells were treated with the indicated PMFs and PCR-CF diluted in DMEM/F12 culture medium. After 24-h treatment, 0.5 mg/mL MTT solution was added and incubated for 2–4 h. The medium was removed, and 200 uL DMSO was added to each well, followed by 10-min incubation. After shaking for 5 min at medium speed, the absorbance at 490 nm was measured using a microplate reader (MultiScan Go, Thermo Scientific Co., Ltd., Waltham, MA, USA). The cell viability percentage (%) was determined by OD treatment group/OD control group × 100%. IC_{50} was calculated by the Logit method using GraphPad Prism 7.0.

2.4. Cell Cycle Analysis

A549 cells were grown in 12-well plates at a density of 1×10^5 cells/well overnight and treated with PMFs and PCR-CF for 24 h. The cells were harvested and fixed with ethanol (pre-cooled overnight at −20 °C), followed by centrifugation (1000 rpm, 3 min). Subsequently, the cells were stained with 1 mL DNA staining solution (PI containing RNase A, Lianke Biotech, CCS012; Hangzhou, China) for 30 min. The red fluorescence was detected by CytoFLEX S flow cytometer (Beckman Coulter) at the excitation wavelength of 488 nm within one hour and analyzed by FlowJo software 7.6.

2.5. Cell Apoptosis Analysis

The A549 cells were grown in 12-well plates at a density of 1×10^5 cells/well overnight and treated with PMFs and PCR-CF for 24 h. The cells were washed with precooled PBS twice and harvested. After being resuspended in 500 μL binding buffer (1×) and diluted with PBS, the cells were double stained with 5 μL annexin V-FITC and 10 μL PI in the cell suspension for 5 min using the Annexin V-FITC/PI apoptosis kit (Lianke Biotech, AP101). At least 10,000 living cells were analyzed on a CytoFLEX S flow cytometer (Beckman Coulter) within one hour. The green fluorescence of annexin V-FITC was detected by the FITC channel (Ex = 488 nm, Em = 530 nm) and the red fluorescence of PI was detected by the PI channel (Ex = 535 nm, Em = 615 nm). The apoptosis rate was analyzed by FlowJo software 7.6.

2.6. Detection of Intracellular ROS Accumulation

The A549 cells were seeded in 12-well plates at a density of 1×10^5 cells/well overnight and treated with PMFs and purified PCR-CF for 24 h. Afterwards, the cells were incubated for 45 min at 37 °C and washed with PBS twice. After DCFH-DA (10 μM) staining, intracellular ROS accumulation was detected by flow cytometry. The fluorescence intensity was measured by a flow cytometer at an excitation wavelength of 488 nm and an emission wavelength of 525 nm within one hour. The ROS level was analyzed by FlowJo software 7.6.

2.7. Data Processing and Statistical Analysis

All the data were expressed as the means ± SD (standard deviation) of the triplicates (n = 3) from independent experiments. The data were processed using GraphPad Prism 7.0 (GraphPad Software Inc., La Jolla, CA, USA). The IC_{50} values were obtained from nonlinear regression. One way ANOVA analysis and subsequent Duncan's test were performed for comparison among the samples, and $p < 0.05$ was considered as statistically significant. The hierarchical clustering analysis (HCA) was performed by TBtools software. Principal component analysis (PCA) was conducted using Origin software (version 2021; Origin Lab, Northampton, MA, USA). PLS-DA analysis was performed by SIMCA software (version 14.1).

3. Results and Discussion

3.1. Multivariate Statistical Analysis of Bound and Free PMFs during PCR-C Storage

The bound and free PMFs in positive and negative ion modes during PCR-C storage were examined qualitatively and quantitatively by HPLC-Q-TOF-MS/MS based on our previous research, and the contents of eight PMFs with high response intensity including isosinensetin (C1), sinensetin (C2), 5,7,8,4′-tetramethoxyflavone (C3), nobiletin (C4), 5,6,7,4′-tetramethoxyflavone (C5), 3,5,6,7,8,3′,4′-heptamethoxyflavone (C6), tangeretin (C7), and 5-demethylnobiletin (C8) were determined by HPLC [17].

The multivariate statistical analysis was performed to investigate the dynamic changes and the contribution of these eight main PMFs to the diversity of the PCR-C at different storage periods. A cluster heatmap of the bound and free PMFs were plotted, respectively. As shown in Figure 1a, with the extended storage periods, the bound PMFs in PCR-C at different storage periods were not clearly clustered, which indicated that the bound PMFs were similar without significant differences during PCR-C aging. Then, a PCA score plot showed a similar result (Figure 1b). In contrast, Figure 1c showed that the free PMFs in PCR-C10 were clearly separated from those in other PCR-C (Fresh, PCR-C01, PCR-C03, PCR-C05), indicating a remarkable difference in the free PMFs in the late storage period (year 10) of PCR-C. In addition, Figure 1d showed that in terms of the free PMFs, PCR-C fell into three clusters: Cluster 1 was fresh and PCR-C01; Cluster 2 included PCR-C03 and PCR-C05; and Cluster 3 was PCR-C10. And as shown in Figure 1c, with the extended PCR-C storage time, the contents of the free PMFs from Cluster 1 to Cluster 3 exhibited an increasing trend. Thus, it could be concluded that the free PMFs in PCR-C varied with the storage periods, and they mainly contributed to the difference and the formation of unique quality of PCR-C during storage. The contents of the free PMFs had an increasing trend with the extended storage periods of PCR-C. The above results demonstrated that the free PMFs were markedly affected by the storage periods of PCR-C.

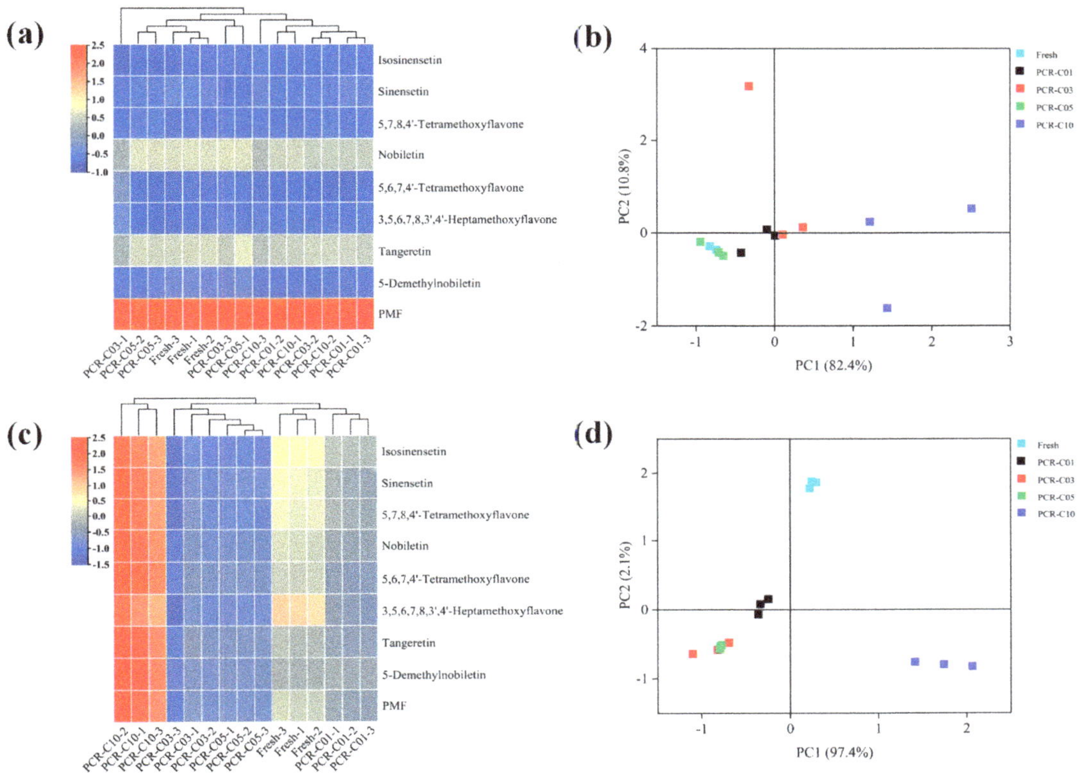

Figure 1. Multivariate statistical analysis of the PMFs in the PCR-C during storage. (**a**) HCA and (**b**) PCA of the bound PMFs in PCR-C (Fresh, PCR-C01, PCR-C03, PCR-C05, PCR-C10); (**c**) HCA and (**d**) PCA of the free PMFs in PCR-C (Fresh, PCR-C01, PCR-C03, PCR-C05, PCR-C10).

Our results were consistent with one previous report that the contents of the 15 types of flavonoid metabolites in PCR-C (year 0, 1, 2, 3, 4, and 29) were increased with the extended storage periods [18]. In addition, the contents of the free phenolic acids in 3-year PCR-C clearly differed from those at 1-year PCR-C [19]. When the whole phenolics profile including phenolic acids and flavonoids are taken into consideration, the free fraction showed a similarity and a slight variation between PCR06 and PCR13. On the contrary, the bound fractions in PCR-C at the later stored period (year 13) were distinct from those in the early time of PCR06, PCR03, and fresh [20]. These previous findings were not in line with our study results, and the possible reason for such an inconsistency might be that we mainly investigated the dynamic changes of PMFs rather than the entire phenolics profiles.

3.2. *Screening of Important PMFs in PCR-C during Storage*

The clustering analysis and PCA successfully divided the PCR-C in different storage periods into three clusters in terms of the free PMFs. In order to screen the potential chemical biomarkers that could distinguish the three clusters of PCR-C, the partial least squares linear discriminant analysis (PLS-DA) was performed to establish the discriminatory models [21]. One previous study has shown that *Citrus reticulata "Chachi"* and *Citrus reticulata Blanco* samples could be separated from each other based on the GC-MS data of volatile compounds by employing PCA, HCA, and orthogonal partial least-squares-discrimination analysis (OPLS-DA), and seven potential chemical markers were identified to be responsible for the special quality control of *Citri reticulatae pericarpium* (CRP) [22].

Considering this, a PLS-DA regression model was established in the present study. R2X, R2Y, and Q2 of the model were 1, 0.727, and 0.659, respectively, which showed that our established model was good. As shown in Figure 2a, in terms of free PMFs, fresh was close to PCR-C01; PCR-C03 was close to PCR-C05; and PCR-C10 was an independent cluster, which was consistent with our PCA results. In the PLS-DA model, the variable importance in the projection (VIP) value was a parameter to evaluate the contributions of the variables to screening biomarkers. Generally, the variables were considered as differential biomarkers when the VIP scores were higher than 1 (VIP > 1) [23]. In our study, the VIP scores of three chemical components including 3,5,6,7,8,3′,4′-Heptamethoxyflavone (C6), Tangeretin (C7), and Isosinensetin (C1) were greater than 1 (VIP > 1) (Figure 2b), which indicated that these three compounds were important variables to discriminate the three clusters of PCR-C at different storage periods, namely, they were the determinant PMFs for discriminating the types of PCR-C at different storage periods. Our results were in line with the previous findings that phenolic acids, flavonol glycosides, fatty acids, and alkyl glycosides were identified as marker compounds by untargeted metabolomics analysis [24].

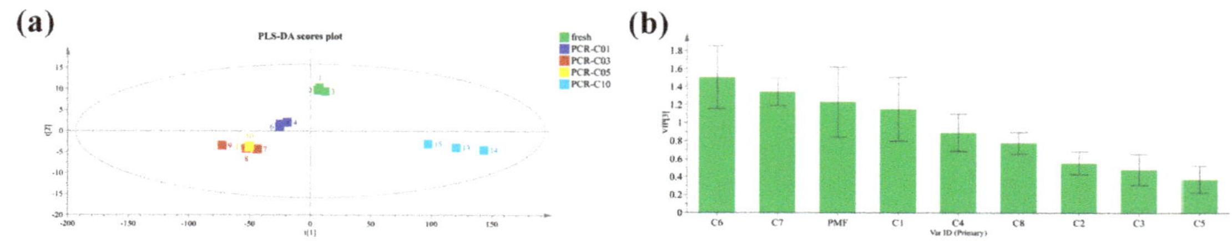

Figure 2. (**a**) PLS-DA of PCR-C samples and (**b**) variable importance in projection (VIP) score based on the free PMFs in PCR-C (Fresh, PCR-C01, PCR-C03, PCR-C05, PCR-C10). C1, isosinensetin; C2, sinensetin; C3, 5,7,8,4′-tetramethoxyflavone; C4, nobiletin; C5, 5,6,7,4′-tetramethoxyflavone; C6, 3,5,6,7,8,3′,4′-heptamethoxyflavone; C7, tangeretin; and C8, 5-demethylnobiletin.

3.3. *PCR-CF Inhibits Cell Growth of A549*

Our previous study has confirmed that the content of PMFs in PCR-C that was stored for 10 years was higher than that in the PCR-C at early stored periods (fresh peel, 1, 3, and 5 years), and that PCR-C10F had excellent antioxidant activity which was positively correlated with nobiletin and 5,6,7,4′-tetramethoxyflavone [17]. Therefore, PMFs were the main active components in PCR-C, and their accumulation might contribute greatly to the efficacy of PCR-C during storage.

Since the free PMFs were the most influential factors discriminating the types of PCR-C during storage, and they showed an increasing trend, we further purified the free PMFs in PCR-C (PCR-CF) using HPD300 resin [17] to investigate their anticancer effect. Recently, A549 cells, human alveolar epithelial cells, have been widely and universally applied to investigate the anticancer effect of natural products in NSCLC [25–27]. Firstly, the morphology changes of A549 cells that were treated with PCR-CF at different storage periods were observed under the inverted phase-contrast microscopy. As shown in Figure 3a, morphology change in A549 cells was observed after treatment with PCR-CF at different storage periods at the dose of 200 μg/mL. With the extended storage periods, the A549 cell number was decreasing and the cell displayed a shrunken shape and detached. In addition, plasma membrane blebbing was observed when the cells were treated with PCR-C10F (indicated by arrows). To determine whether PCR-CF at different storage periods exhibited an anticancer effect against NSCLC and whether the anticancer effect was enhanced with aging years, the cell viability of NSCLC cells (A549) was detected by an MTT assay. As shown in Figure 3b, the results of the MTT assay indicated that 24-h treatment with PCR-CF at different storage periods significantly inhibited the growth of A549 cells in a dose-dependent manner and the cytotoxicity of cells that were treated with PCR-C10F was higher than those that were treated with PCR-CF at early storage periods. Notably, the inhibition effect of PCR-C10F on cell proliferation was better than that of fresh-F, PCR-C01F, PCR-C03F, and PCR-C05F at the IC_{50} values (half maximal inhibitory concentration) of 190.2 μM, 379.4 μM, 213.7 μM, 253.6 μM, and 208.2 μM, respectively (Table 1). Our results were consistent with previous reports that citrus fruit-specific flavonoids nobiletin and 5-demethylnobiletin significantly inhibited the proliferation of human non-small cell lung cancer cells [28]. The anti-proliferative effect of PCR-CF was similar to flavonoids that were isolated from Korean *Citrus aurantium* L. with 230 μg/mL IC_{50} value [29,30]. In addition, PCR-CF at different storage periods had no significant effect on the growth of MRC-5 cells (Figure 3c), indicating that PCR-CF had an inhibitory effect on the growth of cancer cells but a little effect on normal lung cells with no toxicity. These results demonstrated that PCR-CF inhibited the growth of A549 cells in vitro and exhibited an enhanced inhibitory effect when storage years was extended, which might explain why the long-term stored PCR-C had high quality.

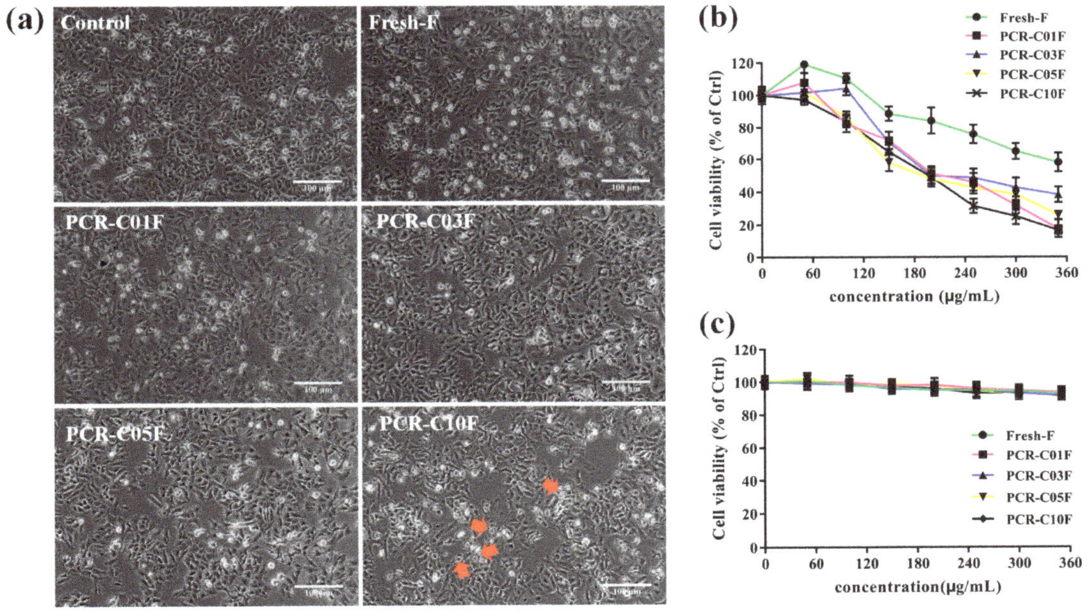

Figure 3. Growth inhibition effect of PCR-CF during storage on A549 cells. (**a**) Morphological changes of A549 cells that were treated with PCR-CF for 24 h under inverted phase-contrast microscopy (scale bar = 100 μm). (**b**) Viability of A549 cells and (**c**) MRC-5 cells that were treated with PCR-CF at the gradient concentrations (50, 100, 150, 200, 250, 300, and 350 μg/mL) for 24 h by MTT assay.

Table 1. The IC_{50} values of A549 cells that were treated with PCR-CF during storage.

		IC_{50} Values (μg/mL)			
Samples	Fresh-F	PCR-C01F	PCR-C03F	PCR-C05F	PCR-C10F
A549	379.4	213.7	253.6	208.2	190.2

3.4. *PCR-CF Induces Cell Cycle Arrest in A549 Cells*

In order to determine whether PCR-CF inhibited cell growth via cell cycle arrest, the effect of PCR-CF at different storage periods on cell cycle arrest of A549 cells was investigated by flow cytometry with PI staining. As shown in Figure 4a, compared to the control group, the PCR-CF treatment group exhibited a significant induction on cell cycle arrest at the G2/M phase, which agreed with the previous report that tangeretin induced G2/M cell cycle arrest and apoptosis to suppress cell growth and to induce cell death in glioma cells [31]. The cell number at the G2/M cell cycle arrest phase ranked in descending order as follows: PCR-C10F > PCR-C05F > PCR-C03F > PCR-C01F > Fresh-F (Figure 4b). Such a change trend demonstrated that the cell cycle arrest that was induced by PCR-CF was enhanced during storage. The cell number at G2/M cell cycle arrest phase of PCR-C10F and PCR-C05F treatment group was significantly higher than that of the other PCR-CF treatment groups. In summary, PCR-CF inhibited cell growth by inducing G2/M phase cell cycle arrest of A549 cells and exhibited an enhanced effect with storage periods extended.

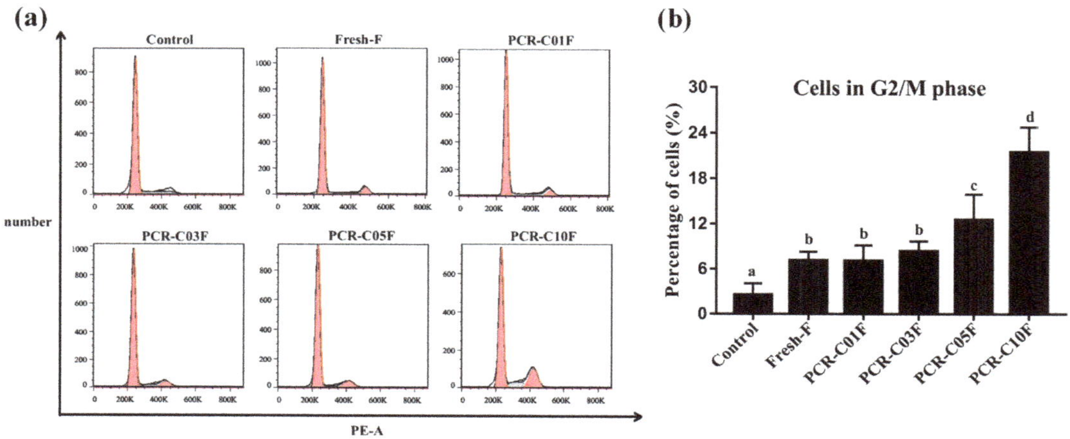

Figure 4. PCR-CF-induced G2/M phase cell cycle arrest in A549 cells. (**a**) The effect of PCR-CF on cell cycle distribution of A549 cells. Fluorescence intensity of A549 cells with or without 24-h PCR-CF (200 μg/mL) treatment by PI staining was measured by flow cytometry. (**b**) Percentage of cells at G2/M phase of cell cycle. The data were expressed as the means ± SD of the triplicates of each independent experiment. The different lowercase letters above the bars indicated a significant difference ($p < 0.05$) between the different groups through one-way ANOVA and Duncan's multiple comparisons.

3.5. PCR-CF Induces Apoptosis in A549 Cells

One previous study has revealed that the ethyl acetate extracts from sweet orange peel exert an antiproliferative effect on human hepatoma cells through cell cycle arrest and apoptosis induction [32]. In order to explore whether PCR-CF induced A549 cell apoptosis, flow cytometry assay with Annexin V-FITC/PI double staining was conducted. As shown in Figure 5a, compared to the control group, the PCR-CF treatment group exhibited a significant induction of apoptosis. The percentage of apoptosis that was induced by PCR-CF treatment ranked in descending order as follows: PCR-C10F > PCR-C05F > PCR-C03F > Fresh-F > PCR-C01F (Figure 5b). Compared with the PCR-CF at early storage periods, PCR-C10F significantly induced apoptosis of A549 cells, indicating higher apoptosis induction of PCR-C10F.

Taken together, our results demonstrated that PCR-CF inhibited the cell growth by inducing cell arrest at the G2/M phase and apoptosis of A549 cells in vitro, and there was an enhanced inhibition trend with extended storage time, which could explain the high quality of the long-term stored PCR-C.

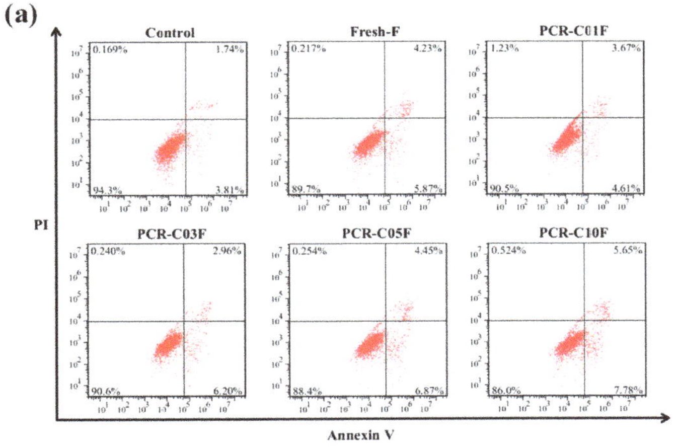

Figure 5. PCR-CF-induced apoptosis of A549 cells. (**a**) The effect of PCR-CF on apoptosis of A549 cells. The apoptosis rate of A549 cells with or without 24-h PCR-CF (200 μg/mL) treatment followed by Annexin V-FITC/PI double staining was measured by flow cytometry. (**b**) The apoptosis rate of A549 cells was calculated (early and late apoptosis). The data were expressed as the means ± SD of the triplicates of each independent experiment. The different lowercase letters above the bars indicated a significant difference ($p < 0.05$) between the different groups through one-way ANOVA and Duncan's multiple comparisons.

3.6. PCR-CF Induces ROS Accumulation in A549 Cells

The excessive accumulation of ROS leads to the suppression of cancer angiogenesis, metastasis, and cancer cell survival [33]. Multiple chemotherapeutic agents have been reported to induce cell apoptosis and cell death through ROS accumulation [34,35]. One previous study has found that Kaempferol can inhibit the Nrf2 signaling pathway, thus inducing ROS accumulation, eventually resulting in apoptosis [36]. In order to determine whether ROS was involved in PCR-CF-induced apoptosis of A549 cells, ROS generation in A549 cells was detected by flow cytometry with DCFH-DA staining. As shown in Figure 6a,b, the level of ROS was increased with the extended storage periods, and the level of ROS in the A549 cells that were treated with the PCR-C10F was significantly higher than those that were treated with PCR-CF at early storage periods when A549 cells were treated with 200 μg/mL PCR-CF for 24 h. The ROS accumulation in A549 cells might be associated with PCR-CF-induced cytotoxicity. In conclusion, these results suggested that ROS accumulation that was induced by the PCR-CF showed an increasing trend with extended storage periods and might mediate the cell viability inhibition, cell cycle arrest, and apoptotic induction in A549 cells.

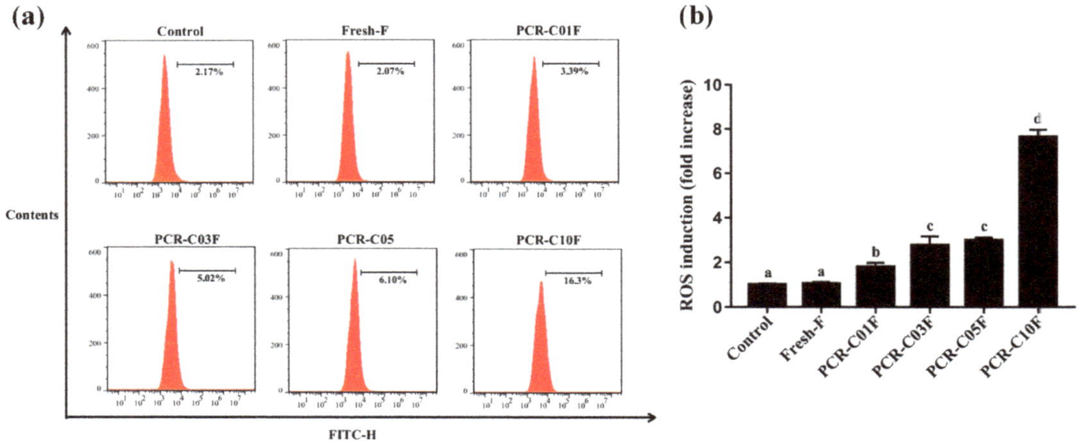

Figure 6. ROS accumulation that was induced by PCR-CF in A549 cells. (**a**) ROS generation in A549 cells was measured by flow cytometry after A549 cells were exposed to 200 μg/mL PCR-CF for 24 h followed by DCFH-DA staining. (**b**) Fluorescence intensity ratio of the treatment group to control group. The data were expressed as the means ± SD of the triplicates of each independent experiment. The different lowercase letters above the bars indicate a significant difference ($p < 0.05$) between the different groups through one-way ANOVA and Duncan's multiple comparisons.

3.7. Important PMFs Inhibit the Cell Growth and Induce Apoptosis of A549 Cells

Since isosinensetin, 3,5,6,7,8,3′,4′-heptamethoxyflavone, and tangeretin were the important factors influencing the quality of PCR-C during storage, and they exhibited antioxidative [37,38], anti-inflammatory [39], neuroprotective [40,41], and anti-tumor [42–44] bioactivities according to the previous studies, this study further evaluated their anticancer effect against NSCLC (A549 cells) in vitro.

To validate whether the important PMFs had an anticancer effect against NSCLC, the effect of PMFs on cell viability of NSCLC cells (A549) was detected by an MTT assay. The MTT assay results indicated that 24-h treatment with the important PMFs significantly inhibited the growth of A549 cells in a dose-dependent manner (Figure 7a). Notably, the inhibitory effect on cell growth of tangeretin was better than isosinensetin and 3,5,6,7,8,3′,4′-heptamethoxyflavone at the IC_{50} of 118.5 μM, 197.6 μM, and 208.6 μM, respectively (Table 2), indicating that they could inhibit A549 cell growth even at the low concentration, and thus they had the good anticancer effect on A549 cells. In addition, isosinensetin, 3,5,6,7,8,3′,4′-heptamethoxyflavone, and tangeretin had no significant effect on the growth of MRC-5 cells (Figure 7b), indicating that the important PMFs had an inhibitory effect on the growth of cancer cells but a little effect on normal lung cells with no toxicity. Collectively, these results demonstrated that all the important PMFs could inhibit the proliferation of A549 cells in vitro and had no cytotoxicity to normal lung MRC-5 cells.

Table 2. The IC_{50} values of A549 cells that were treated with the important PMFs.

	IC_{50} Values (μM)		
Samples	Tangeretin	Isosinensetin	3,5,6,7,8,3′,4′-Heptamethoxyflavone
A549	118.5	197.6	208.6

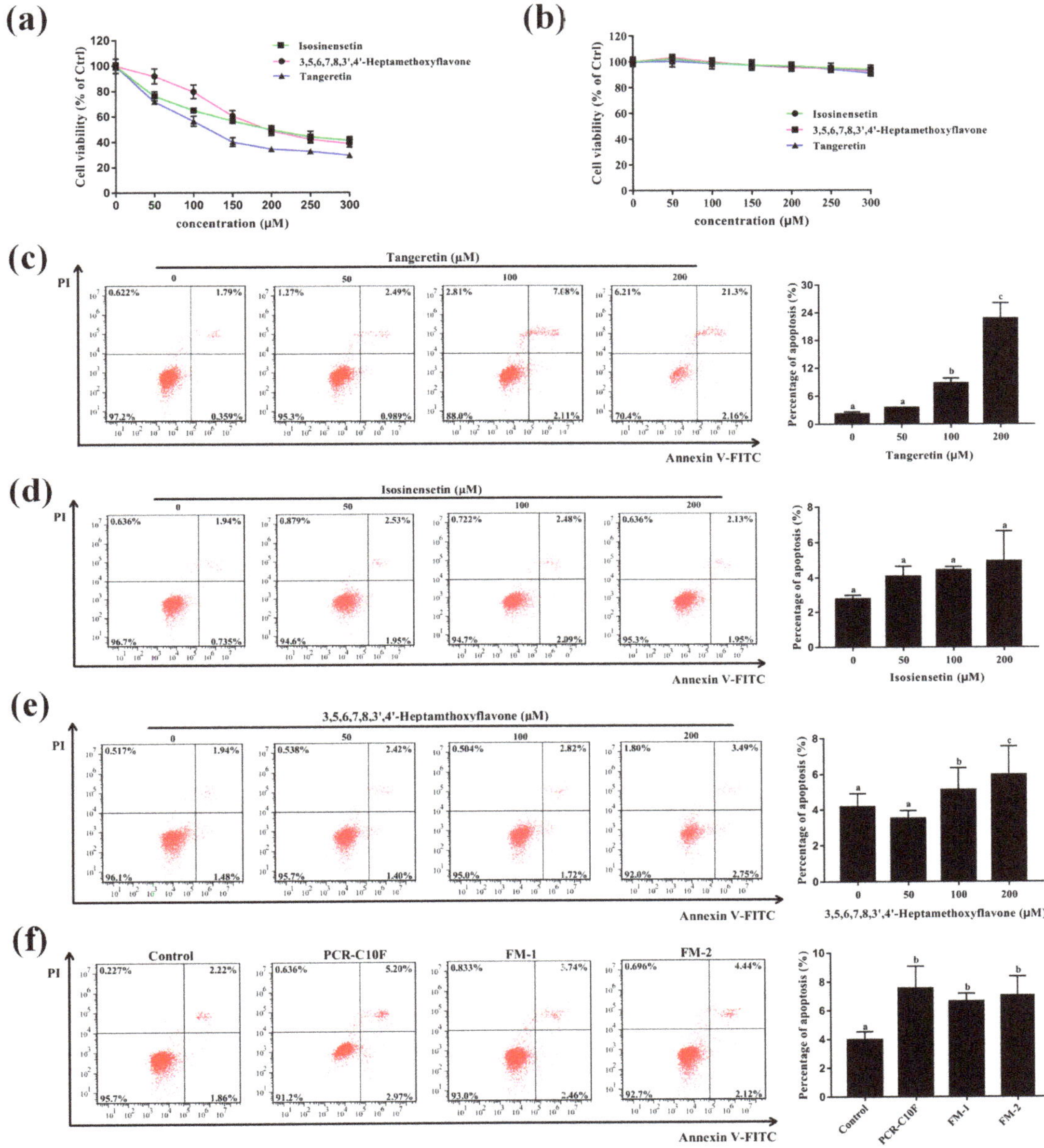

Figure 7. Growth of inhibition and apoptosis induction effect of the important PMFs in A549 cells. (**a**) Viability of A549 cells and (**b**) MRC-5 cells treated with the important PMFs at gradient concentrations (50, 100, 150, 200, 25C, and 300 μM) for 24 h by MTT assay. The apoptosis induction effect of tangeretin (**c**), 3,5,6,7,8,3′,4′-Heptamethoxyflavone (**d**), and Isosinensetin (**e**) in A549 cells and the corresponding apoptosis rate of A549 cells (the early and late apoptosis). The apoptosis rate of A549 cells with or without 24-h PMFs (50, 100, and 200 μM) treatment followed by Annexin V-FITC/PI staining was measured by flow cytometry. (**f**) The apoptosis induction effect of PCR-C10F and the major active flavonoid mixture (FM) with the equivalent amount of PMFs to PCR-C10F. The data were expressed as the means ± SD of the triplicates of each independent experiment. The different lowercase letters above the bars indicate a significant difference ($p < 0.05$) between the different groups through one-way ANOVA and Duncan's multiple comparisons.

Further, flow cytometry assay with Annexin V-FITC/PI double staining was performed to determine whether the important PMFs could induce apoptosis in A549 cells. Compared to the control group, tangeretin and 3,5,6,7,8,3′,4′-heptamethoxyflavone significantly induced apoptosis in a dose-dependent manner at the gradient concentrations of 50 μM, 100 μM, and 200 μM, and the apoptosis induction effect of tangeretin was better than 3,5,6,7,8,3′,4′-heptamethoxyflavone. However, isosinensetin did not significantly induce cell apoptosis (Figure 7c–e). In order to explore the contribution of isosinensetin to the anticancer effect of PCR-CF during aging, an apoptosis assay was performed using the equivalent amount of flavonoid mixture (FM1, 16.5 μM tangeretin, 2.2 μM isosinensetin, and 1.7 μM 3,5,6,7,8,3′,4′-heptamethoxyflavone; FM2, 16.5 μM tangeretin and 1.7 μM 3,5,6,7,8,3′,4′-heptamethoxyflavone) corresponding to the important PMFs in 200 μg/mL PCR-C10F. As shown in Figure 7f, the apoptosis effect of FM2 showed no difference from the FM1 and PCR-C10F, implying that tangeretin and 3,5,6,7,8,3′,4′-heptamethoxyflavone were the crucial active components to predict the anticancer effect of PCR-C during storage.

Taken together, our results demonstrated that tangeretin and 3,5,6,7,8,3′,4′-heptamethoxyflavone could inhibit the cell growth and induce apoptosis in a dose-dependent manner in A549 cells. In spite of the low contents of these two PMFs in PCR-C, they showed strong anticancer effect against NSCLC. Our results were consistent with the previous report that tangeretin was the most crucial active ingredient in long-term stored citrus peel, and it could inhibit cell proliferation, induce the cell cycle arrest and cell apoptosis so as to fight against oral squamous cell carcinoma [45]. Based on these findings, we concluded that the increasing content of the PMFs during storage might improve the quality of PCR-C, and that the tangeretin and 3,5,6,7,8,3′,4′-heptamethoxyflavone might be the contributors to the improvement of the efficacy and quality of long-term stored PCR-C, and thus they might be crucial active indicators for predicting the anticancer effect of PCR-C during storage against NSCLC.

3.8. Crucial Active PMFs Induces ROS Accumulation in A549

The increasing evidence indicated that ROS plays a pivotal role in mediating the viability of cancer cells [40,41]. Our data showed that ROS accumulation induced by the PCR-CF could inhibit the cell growth, induce cell cycle arrest and apoptosis in A549 cells. Therefore, ROS generation that was induced by the crucial PMFs in A549 cells was detected by flow cytometry with DCFH-DA staining. As shown in Figure 8a,b, the ROS level was dramatically increased in a dose-dependent manner when A549 cells were treated with tangeretin and 3,5,6,7,8,3′,4′-heptamethoxyflavone at the gradient concentrations (50, 100, 200 μM) for 24 h. In conclusion, our results further confirmed that the crucial PMFs induced ROS accumulation, thus possibly inhibiting the cell viability and inducing apoptosis of A549 cells.

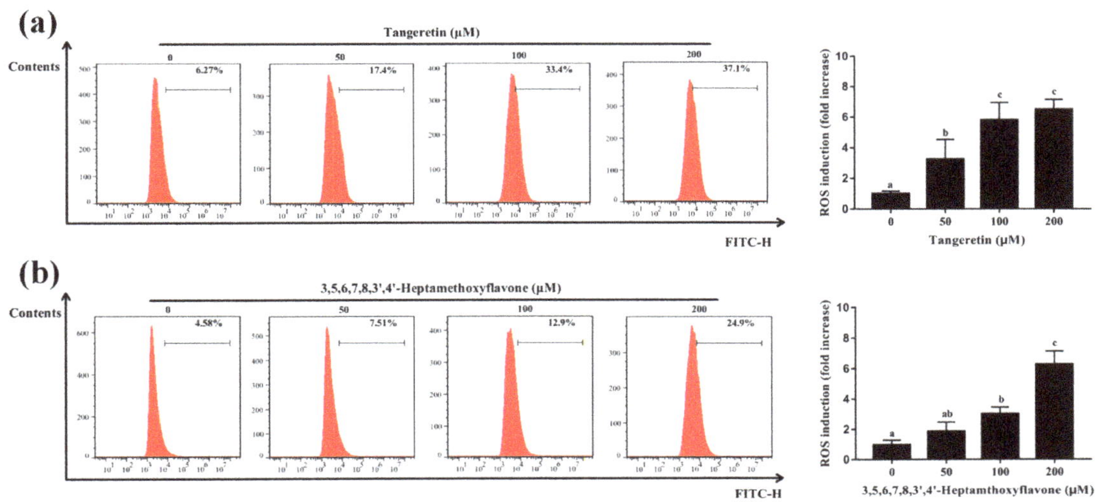

Figure 8. Crucial active PMFs-induced ROS accumulation of A549 cells. ROS generation was detected by flow cytometry after 24-h exposure to tangeretin (**a**) and 3,5,6,7,8,3′,4′-heptamethoxyflavone (**b**) followed by DCFH-DA staining. The fluorescence intensity ratio of the treatment group to the control group was calculated. The data were expressed as the means ± SD of the triplicates of each independent experiment. The different lowercase letters above the bars indicate a significant difference ($p < 0.05$) between the different groups through one-way ANOVA and Duncan's multiple comparisons.

4. Conclusions

In this study, cluster analysis and PCA showed that the free PMFs in PCR-C during storage fell into three clusters, namely, fresh and PCR-C01 (Cluster 1), PCR-C03, and PCR-C05 (Cluster 2), and PCR-C10 (Cluster 3). The PLS-DA results showed that three PMFs including 3,5,6,7,8,3′,4′-heptamethoxyflavone, tangeretin, and isosinensetin were identified as the important PMFs that could distinguish types of PCR-C during storage. Furthermore, our data showed that the anticancer effect against NSCLC of PCR-CF was enhanced with the extension of storage periods, and that tangeretin and 3,5,6,7,8,3′,4′-heptamethoxyflavone exhibited an anticancer effect on A549 cells, but isosinensetin failed to induce the apoptosis of A549 cells. The apoptosis induction effect of FM2 showed no difference from that of FM1 and PCR-C10F, indicating that tangeretin and 3,5,6,7,8,3′,4′-heptamethoxyflavone collectively contributed to the anticancer effect of PCR-C. Our results confirmed that tangeretin and 3,5,6,7,8,3′,4′-heptamethoxyflavone contributed to the high quality of the long-term stored PCR-C, which provides a theoretical basis for scientific utilization, rational storage, and quality control of PCR-C. However, the anticancer effect of PCR-C needs further research on other NSCLC cell lines such as H1299 and H1975 and in vivo study in the future.

Author Contributions: Conceptualization, Y.T. and Q.Y.; methodology, Y.T. and Y.H.; software, R.L.; validation, X.X.; formal analysis, Y.T.; investigation, Y.T. and X.Z.; resources, Q.Y.; data curation, Y.T.; writing—original draft preparation, Y.T.; writing—review and editing, Y.T.; visualization, T.W.; supervision, S.P.; project administration, S.P. and X.X.; funding acquisition, X.X. All authors have read and agreed to the published version of the manuscript.

Funding: This research was funded by the National Natural Science Foundation of China (32172140).

Institutional Review Board Statement: Not applicable.

Data Availability Statement: Data is within the article.

Acknowledgments: Minfeng Zhou is thanked for his assistance with experiments. Yuxin Cheng, Mei Qiu, and Wenjuan Zhu are thanked for their assistance with revision.

Conflicts of Interest: The authors declare no conflict of interest.

References

1. Liu, H.; Qiu, N.; Ding, H.; Yao, R. Polyphenols contents and antioxidant capacity of 68 Chinese herbals suitable for medical or food uses. *Food Res. Int.* **2008**, *41*, 363–370. [CrossRef]
2. Wei, D.; Ci, X.; Chu, X.; Wei, M.; Hua, S.; Deng, X. Hesperidin Suppresses Ovalbumin-Induced Airway Inflammation in a Mouse Allergic Asthma Model. *Inflammation* **2012**, *35*, 114–121. [CrossRef]
3. Tang, X.; Zhao, H.; Jiang, W.; Zhang, S.; Guo, S.; Gao, X.; Yang, P.; Shi, L.; Liu, L. Pharmacokinetics and pharmacodynamics of citrus peel extract in lipopolysaccharide-induced acute lung injury combined with Pinelliae Rhizoma Praeparatum. *Food Funct.* **2018**, *9*, 5880–5890. [CrossRef] [PubMed]
4. Wang, J.; Huang, Q.; Zhang, Y.; Su, J.; Liu, J. Current Situation of the Long-stored TCM Research and Exploration of the Ideas. *China Pharm.* **2012**, *23*, 1433–1435.
5. Fu, M.; Xu, Y.; Chen, Y.; Wu, J.; Yu, Y.; Zou, B.; An, K.; Xiao, G. Evaluation of bioactive flavonoids and antioxidant activity in Pericarpium Citri Reticulatae (Citrus reticulata 'Chachi') during storage. *Food Chem.* **2017**, *230*, 649–656. [CrossRef] [PubMed]
6. Luo, M.; Luo, H.; Hu, P.; Yang, Y.; Wu, B.; Zheng, G. Evaluation of chemical components in Citri Reticulatae Pericarpium of different cultivars collected from different regions by GC-MS and HPLC. *Food Sci. Nutr.* **2018**, *6*, 400–416. [CrossRef] [PubMed]
7. Youn, K.; Lee, S.; Jun, M. Discovery of Nobiletin from Citrus Peel as a Potent Inhibitor of beta-Amyloid Peptide Toxicity. *Nutrients* **2019**, *11*, 2684. [CrossRef]
8. Yao, X.; Zhu, X.; Pan, S.; Fang, Y.; Jiang, F.; Phillips, G.O.; Xu, X. Antimicrobial activity of nobiletin and tangeretin against Pseudomonas. *Food Chem.* **2012**, *132*, 1883–1890. [CrossRef]
9. Bian, X.; Xie, X.; Cai, J.; Zhao, Y.; Miao, W.; Chen, X.; Xiao, Y.; Li, N.; Wu, J.-L. Dynamic changes of phenolic acids and antioxidant activity of Citri Reticulatae Pericarpium during aging processes. *Food Chem.* **2022**, *373*, 131399. [CrossRef] [PubMed]
10. Luo, Y.; Zeng, W.; Huang, K.-E.; Li, D.-X.; Chen, W.; Yu, X.-Q.; Ke, X.-H. Discrimination of Citrus reticulata Blanco and Citrus reticulata 'Chachi' as well as the Citrus reticulata 'Chachi' within different storage years using ultra high performance liquid chromatography quadrupole/time-of-flight mass spectrometry based metabolomics approach. *J. Pharm. Biomed. Anal.* **2019**, *171*, 218–231. [CrossRef]
11. Duma, N.; Santana-Davila, R.; Molina, J.R. Non-Small Cell Lung Cancer: Epidemiology, Screening, Diagnosis, and Treatment. *Mayo Clin. Proc.* **2019**, *94*, 1623–1640. [CrossRef] [PubMed]
12. Zarogoulidis, K.; Zarogoulidis, P.; Darwiche, K.; Boutsikou, E.; Machairiotis, N.; Tsakiridis, K.; Katsikogiannis, N.; Kougioumtzi, I.; Karapantzos, I.; Huang, H.; et al. Treatment of non-small cell lung cancer (NSCLC). *J. Thorac. Dis.* **2013**, *5*, S389–S396. [CrossRef] [PubMed]
13. Lemjabbar-Alaoui, H.; Hassan, O.U.; Yang, Y.-W.; Buchanan, P. Lung cancer: Biology and treatment options. *Biochim. Et Biophys. Acta-Rev. Cancer* **2015**, *1856*, 189–210. [CrossRef]
14. Duan, L.; Dou, L.-L.; Yu, K.-Y.; Guo, L.; Chen, B.-Z.; Li, P.; Liu, E.H. Polymethoxyflavones in peel of Citrus reticulata 'Chachi' and their biological activities. *Food Chem.* **2017**, *234*, 254–261. [CrossRef]
15. Fu, M.; Zou, B.; An, K.; Yu, Y.; Tang, D.; Wu, J.; Xu, Y.; Ti, H. Anti- asthmatic activity of alkaloid compounds from Pericarpium Citri Reticulatae (Citrus reticulata 'Chachi'). *Food Funct.* **2019**, *10*, 903–911. [CrossRef]
16. Chen, X.-M.; Tait, A.R.; Kitts, D.D. Flavonoid composition of orange peel and its association with antioxidant and anti-inflammatory activities. *Food Chem.* **2017**, *218*, 15–21. [CrossRef]
17. Yu, Q.; Tao, Y.; Huang, Y.; Zogona, D.; Wu, T.; Liu, R.; Pan, S.; Xu, X. Aged Pericarpium Citri Reticulatae 'Chachi' Attenuates Oxidative Damage Induced by tert-Butyl Hydroperoxide (t-BHP) in HepG2 Cells. *Foods* **2022**, *11*, 273. [CrossRef]
18. Liang, S.; Wen, Z.; Tang, T.; Liu, Y.; Dang, F.; Xie, T.; Wu, H. Study on flavonoid and bioactivity features of the pericarp of Citri Reticulatae 'chachi' during storage. *Arab. J. Chem.* **2022**, *15*, 103653. [CrossRef]
19. Choi, M.-Y.; Chai, C.; Park, J.H.; Lim, J.; Lee, J.; Kwon, S.W. Effects of storage period and heat treatment on phenolic compound composition in dried Citrus peels (Chenpi) and discrimination of Chenpi with different storage periods through targeted metabolomic study using HPLC-DAD analysis. *J. Pharm. Biomed. Anal.* **2011**, *54*, 638–645. [CrossRef]
20. Wang, H.; Chen, G.; Fu, X.; Liu, R.-H. Effects of aging on the phytochemical profile and antioxidative activity of Pericarpium Citri Reticulatae 'Chachiensis'. *Rsc Adv.* **2016**, *6*, 105272–105281. [CrossRef]
21. Barker, M.; Rayens, W. Partial least squares for discrimination. *J. Chemom.* **2003**, *17*, 166–173. [CrossRef]
22. Zheng, Y.; Zeng, X.; Peng, W.; Wu, Z.; Su, W. Study on the Discrimination between Citri Reticulatae Pericarpium Varieties Based on HS-SPME-GC-MS Combined with Multivariate Statistical Analyses. *Molecules* **2018**, *23*, 1235. [CrossRef]
23. Lloyd, G.R.; Stone, N. Method for Identification of Spectral Targets in Discrete Frequency Infrared Spectroscopy for Clinical Diagnostics. *Appl. Spectrosc.* **2015**, *69*, 1066–1073. [CrossRef]
24. Yang, M.; Jiang, Z.; Wen, M.; Wu, Z.; Zha, M.; Xu, W.; Zhang, L. Chemical Variation of Chenpi (Citrus Peels) and Corresponding Correlated Bioactive Compounds by LC-MS Metabolomics and Multibioassay Analysis. *Front. Nutr.* **2022**, *9*, 825381. [CrossRef]

25. Liu, X.; Jiang, Q.; Liu, H.; Luo, S. Vitexin induces apoptosis through mitochondrial pathway and PI3K/Akt/mTOR signaling in human non-small cell lung cancer A549 cells. *Biol. Res.* **2019**, *52*, 7. [CrossRef]
26. Ren, H.; Zhang, Y.-Y.; Li, Y.-L.; Bai, M.; Yan, Q.-L.; Huang, X.-X.; Cui, W.; Zhao, H.; Gu, L.; Liu, Q.; et al. Semisynthesis and Non-Small-Cell Lung Cancer Cytotoxicity Evaluation of Germacrane-Type Sesquiterpene Lactones from Elephantopus scaber. *J. Nat. Prod.* **2022**, *85*, 352–364. [CrossRef]
27. Wang, J.; Li, X. Chamaejasmine induces apoptosis in human non-small-cell lung cancer A549 cells through increasing the Bax/Bcl-2 ratio, caspase-3 and activating the Fas/FasL. *Minerva Med.* **2021**, *112*, 413–415. [CrossRef]
28. Song, M.; Charoensinphon, N.; Wu, X.; Zheng, J.; Gao, Z.; Xu, F.; Wang, M.; Xiao, H. Inhibitory Effects of Metabolites of 5-Demethylnobiletin on Human Nonsmall Cell Lung Cancer Cells. *J. Agric. Food Chem.* **2016**, *64*, 4943–4949. [CrossRef]
29. Park, K.I.; Park, H.S.; Nagappan, A.; Hong, G.E.; Lee, D.H.; Kang. S.R.; Kim, J.A.; Zhang, J.; Kim, E.H.; Lee, W.S.; et al. Induction of the cell cycle arrest and apoptosis by flavonoids isolated from Korean Citrus aurantium L. in non-small-cell lung cancer cells. *Food Chem.* **2012**, *135*, 2728–2735. [CrossRef]
30. Park, K.-I.; Park, H.-S.; Kim, M.-K.; Hong, G.-E.; Nagappan, A.; Lee, H.-J.; Yumnam, S.; Lee, W.-S.; Won, C.-K.; Shin, S.-C.; et al. Flavonoids identified from Korean Citrus aurantium L. inhibit Non-Small Cell Lung Cancer growth in vivo and in vitro. *J. Funct. Foods* **2014**, *7*, 287–297. [CrossRef]
31. Ma, L.-L.; Wang, D.-w.; Yu, X.-D.; Zhou, Y.-L. Tangeretin induces cell cycle arrest and apoptosis through upregulation of PTEN expression in glioma cells. *Biomed. Pharmacother.* **2016**, *81*, 491–496. [CrossRef] [PubMed]
32. Chin-Chen, C.; Shih-Ying, C.; Charng-Cherng, C.; Pin-Der, D. Antiproliferative effect of sweet orange peel and its bioactive compounds against human hepatoma cells, in vitro and in vivo. *J. Funct. Foods* **2017**, *33*, 363–375. [CrossRef]
33. Woelfle, U.; Seelinger, G.; Bauer, G.; Meinke, M.C.; Lademann, J.; Schempp, C.M. Reactive Molecule Species and Antioxidative Mechanisms in Normal Skin and Skin Aging. *Ski. Pharmacol. Physiol.* **2014**, *27*, 316–332. [CrossRef] [PubMed]
34. Menon, S.; Devi, S.K.S.; Santhiya, R.; Rajeshkumar, S.; Kumar, V.S. Selenium nanoparticles: A potent chemotherapeutic agent and an elucidation of its mechanism. *Colloids Surf. B-Biointerfaces* **2018**, *170*, 280–292. [CrossRef]
35. Bossis, G.; Sarry, J.-E.; Kifagi, C.; Ristic, M.; Saland, E.; Vergez, F.; Salem, T.; Boutzen, H.; Baik, H.; Brockly, F.; et al. The ROS/SUMO Axis Contributes to the Response of Acute Myeloid Leukemia Cells to Chemotherapeutic Drugs. *Cell Rep.* **2014**, *7*, 1815–1823. [CrossRef]
36. Fouzder, C.; Mukhuty, A.; Kundu, R. Kaempferol inhibits Nrf2 signalling pathway via downregulation of Nrf2 mRNA and induces apoptosis in NSCLC cells. *Arch. Biochem. Biophys.* **2021**, *697*, 108700. [CrossRef]
37. Liang, F.; Fang, Y.; Cao, W.; Zhang, Z.; Pan, S.; Xu, X. Attenuation of tert-Butyl Hydroperoxide (t-BHP)-Induced Oxidative Damage in HepG2 Cells by Tangeretin: Relevance of the Nrf2-ARE and MAPK Signaling Pathways. *J. Agric. Food Chem.* **2018**, *66*, 6317–6325. [CrossRef]
38. Kim, H.-I.; Jeong, Y.-U.; Kim, J.-H.; Park, Y.-J. 3,5,6,7,8,3,4-Heptamethoxyflavone, a Citrus Flavonoid, Inhibits Collagenase Activity and Induces Type I Procollagen Synthesis in HDFn Cells. *Int. J. Mol. Sci.* **2018**, *19*, 620. [CrossRef]
39. Lee, Y.Y.; Lee, E.-J.; Park, J.-S.; Jang, S.-E.; Kim, D.-H.; Kim, H.-S. Anti-Inflammatory and Antioxidant Mechanism of Tangeretin in Activated Microglia. *J. Neuroimmune Pharmacol.* **2016**, *11*, 294–305. [CrossRef]
40. Guo, X.-Q.; Cao, Y.-L.; Hao, F.; Yan, Z.-R.; Wang, M.-L.; Liu, X.-W. Tangeretin alters neuronal apoptosis and ameliorates the severity of seizures in experimental epilepsy-induced rats by modulating apoptotic protein expressions, regulating matrix metalloproteinases, and activating the PI3K/Akt cell survival pathway. *Adv. Med. Sci.* **2017**, *62*, 246–253. [CrossRef]
41. Sawamoto, A.; Okuyama, S.; Amakura, Y.; Yoshimura, M.; Yamada, T.; Yokogoshi, H.; Nakajima, M.; Furukawa, Y. 3,5,6,7,8,3′,4′-Heptamethoxyflavone Ameliorates Depressive-Like Behavior and Hippocampal Neurochemical Changes in Chronic Unpredictable Mild Stressed Mice by Regulating the Brain-Derived Neurotrophic Factor: Requirement for ERK Activation. *Int. J. Mol. Sci.* **2017**, *18*, 2133. [CrossRef] [PubMed]
42. Lin, J.-J.; Huang, C.-C.; Su, Y.-L.; Luo, H.-L.; Lee, N.-L.; Sung, M.-T.; Wu, Y.-J. Proteomics Analysis of Tangeretin-Induced Apoptosis through Mitochondrial Dysfunction in Bladder Cancer Cells. *Int. J. Mol. Sci.* **2019**, *20*, 1017. [CrossRef] [PubMed]
43. Iwase, Y.; Takemura, Y.; Ju-ichi, M.; Yano, M.; Ito, C.; Furukawa, H.; Mukainaka, T.; Kuchide, M.; Tokuda, H.; Nishino, H. Cancer chemopreventive activity of 3,5,6,7,8,3′,4′-heptamethoxyflavone from the peel of citrus plants. *Cancer Lett.* **2001**, *163*, 7–9. [CrossRef]
44. Chen, Y.-K.; Wang, H.-C.; Ho, C.-T.; Chen, H.-Y.; Li, S.; Chan, H.-L.; Chung, T.-W.; Tan, K.-T.; Li, Y.-R.; Lin, C.-C. 5-Demethylnobiletin promotes the formation of polymerized tubulin, leads to G2/M phase arrest and induces autophagy via JNK activation in human lung cancer cells. *J. Nutr. Biochem.* **2015**, *26*, 484–504. [CrossRef] [PubMed]
45. Yu, Z.; Wu, Y.; Ma, Y.; Cheng, Y.; Song, G.; Zhang, F. Systematic analysis of the mechanism of aged citrus peel (Chenpi) in oral squamous cell carcinoma treatment via network pharmacology, molecular docking and experimental validation. *J. Funct. Foods* **2022**, *91*, 105012. [CrossRef]

antioxidants

MDPI

Article

The Peanut Skin Procyanidins Attenuate DSS-Induced Ulcerative Colitis in C57BL/6 Mice

Na Wang [1,2,3,4,†], Weixuan Chen [4,5,†], Chenxu Cui [2,4], Yuru Zheng [2,4], Qiuying Yu [2,3,4], Hongtao Ren [2,4], Zhigang Liu [4,5], Chao Xu [2,4,*] and Gaiping Zhang [1,3,*]

1 College of Veterinary Medicine, Henan Agricultural University, Zhengzhou 450002, China
2 College of Food Science and Technology, Henan Agricultural University, Zhengzhou 450002, China
3 International Joint Research Center for Animal Immunology, Zhengzhou 450002, China
4 Key Laboratory of Nutrition and Healthy Food of Zhengzhou, Zhengzhou 450002, China
5 College of Food Science and Engineering, Northwest A&F University, Yangling 712100, China
* Correspondence: chaoxu@henau.edu.cn (C.X.); zhanggaiping@pku.edu.cn (G.Z.)
† These authors contributed equally to this work.

Abstract: Polyphenols from peanut skin have been reported to possess many beneficial functions for human health, including anti-oxidative, antibacterial, anticancer, and other activities. To date, however, its anti-inflammatory effect and the underlying mechanism remain unclear. In this study, the anti-inflammatory effect of peanut skin procyanidins extract (PSPE) and peanut skin procyanidins (PSPc) were investigated by a dextran sodium sulfate (DSS)-induced colitis mouse model. The results showed that both PSPE and PSPc supplementation reversed the DSS-induced body weight loss and reduced disease activity index (DAI) values, accompanied by enhanced goblet cell numbers and tight junction protein claudin-1 expression in the colon. PSPE and PSPc treatment also suppressed the inflammatory responses and oxidative stress in the colon by down-regulating IL-1β, TNF-α, and MDA expressions. Meanwhile, PSPE and PSPc significantly altered the gut microbiota composition by increasing the relative abundance of *Clostridium XlVb* and *Anaerotruncus*, and inhibiting the relative abundance of *Alistipes* at the genus level. PSPE and PSPc also significantly elevated the production of short-chain fatty acids (SCFAs) in mice with colitis. The correlation analysis suggested that the protective effects of PSPE and PSPc on colitis might be related to the alteration of gut microbiota composition and the formation of SCFAs. In conclusion, the current research indicates that supplementation of PSPE and PSPc could be a promising nutritional strategy for colitis prevention and treatment.

Keywords: peanut skin; procyanidins; DSS-induced; ulcerative colitis

Citation: Wang, N.; Chen, W.; Cui, C.; Zheng, Y.; Yu, Q.; Ren, H.; Liu, Z.; Xu, C.; Zhang, G. The Peanut Skin Procyanidins Attenuate DSS-Induced Ulcerative Colitis in C57BL/6 Mice. *Antioxidants* **2022**, *11*, 2098. https://doi.org/10.3390/antiox11112098

Academic Editor: Alessandra Napolitano

Received: 2 September 2022
Accepted: 21 October 2022
Published: 25 October 2022

1. Introduction

Inflammatory bowel disease (IBD), including Crohn's disease and ulcerative colitis (UC), begins in the rectum and extends to the proximal colon [1]. The main symptoms are intestinal mucosal injury, relapsing, abdominal pain, diarrhea, mucinous bloody stools, and weight loss [2]. UC is more prevalent than Crohn's disease globally. Based on current knowledge, it is possible that a combination of genetic and environmental variables contributes to the development of UC, and these modifications lead to abnormalities in the gut microbiome and dysregulation of the mucosal immune system [3]. Pharmacological therapy is used to treat the majority of UC patients. Oral and rectal 5-aminosalycilates are frequently used for mild to moderate UC. Thiopurines, biological agents that target tumor necrosis factor and integrins, as well as small-molecule Janus kinase inhibitors, are among the drug classes used to treat mild to severe colitis. However, up to 15% of individuals will require surgery if medical therapy is ineffective or if dysplasia develops as a result of their protracted colitis [4]. Therefore, it is of great significance to seek new and effective therapies for UC.

Nutritional intervention is an alternative treatment for UC [5]. Numerous studies suggest that polyphenols and their metabolites with potent anti-inflammatory, antioxidant, anticancer, and immunomodulatory properties may be promising contenders for combating UC [6]. In rodent models of colitis, polyphenol-rich diets and substances have been utilized as a therapy, including green tea polyphenols [7], daidzein-rich isoflavones extract [8], polyphenol-enriched cocoa extract [9], curcuma longa extract [10], pomegranate extract and urolithin A [11]. Peanut skin has been shown to be a promising source of phenolic compounds, including phenolic acids, flavonoids, procyanidins, tannic acid, resveratrol, etc. [12]. Peanut skins (testae or seed coat) are the seed coat of the legume family, as a by-product of peanut processing, the output of peanut skins may reach 1 million tons worldwide in recent years [13], most of which are used as animal feed for animal husbandry or wasted [14]. Peanut skin procyanidins (PSPc), which amount to 128 g/kg [15], are one of the most pivotal and functional polyphenols in peanut skin. A study reported that the procyanidins obtained from peanut skins were mainly A-type dimers, different from the B-type dimers that largely exist in grape seeds [16]. PSPc might possess higher bioavailability than grape seed procyanidins [17]. Moreover, according to many in vitro and in vivo studies, PSPc are shown to have many potentially beneficial functions for human health, including anti-oxidative activities, antibacterial, anticancer, hyperlipidemia prevention, anti-allergy, hypoglycemic capability, and inhibiting acrylamide production [18–24]. However, whether PSPc has a therapeutic effect on UC and its mechanism are still unclear.

Genetic variety, gut barrier failure, proinflammatory cytokine over-release, and gut dysbiosis are all identified as risk factors for UC [25]. Studies have shown that polyphenol-enriched cocoa extract reduces colon damage, with significant reductions in both the extent and the severity of the inflammation as well as in crypt damage and leukocyte infiltration in the mucosa [9]. TNF and IL-6, two inflammatory biomarkers, rose in colitic animals and significantly decreased after treatment with green tea polyphenols [7]. Resveratrol could increase the number of lactic acid bacteria and bifidobacteria, and reduce the number of enterobacteria upon DSS treatment [26]. Interestingly, previous studies also have shown that Peanut skin procyanidins can improve gut barrier integrity, restrain the inflammation reaction and regulate gut microbiota. A study found that peanut skin procyanidins improved the gut barrier integrality by restoring gut morphology and enhancing tight junction protein expression including claudin-1 and occludin in the colon in type 2 diabetic mice [27]. Another study found that both acetone-extracted peanut skin extracts and ethanol-extracted peanut skin extract significantly inhibit COX-2 protein expression, exhibiting similar anti-inflammatory effects [28]. In a study of peanut skin extract to ameliorate atherosclerosis, the results showed that peanut skin extract could significantly change the composition of intestinal microbiota, and its anti-atherosclerosis effect might be related to the changes in the composition and function of intestinal microbiota [29]. Especially, various studies have demonstrated that gut microbiota is crucial in the development of UC [30]. Fecal microbiota transplantation and supplementation with short-chain fatty acids (SCFAs) such as butyrate have been successfully identified as therapies for patients with UC. Although a significant amount of polyphenols are not absorbed along the digestive tract, they may build up in the large intestine, where the intestinal microbiota extensively metabolizes the majority of them. A balanced gut microbiome may play an important role in maintaining human health by producing beneficial microbial metabolites and SCFAs that improve host nutrient supply [31] or by preventing pathogen colonization and maintaining normal mucosal immunity [32]. According to the above, peanut skin procyanidins may exert their therapeutic effects on UC by affecting proinflammatory cytokines and improving gut microbiota diversity and intestinal barrier function.

In the present study, the major components from peanut skin were extracted and analyzed. A dextran sodium sulfate (DSS)-induced colitis mouse model was then used to evaluate the effects of PSPE (peanut skin procyanidin extract) and PSPc (PSPE purified product) regimens on colitis development. By detecting the mucosal damage, junction proteins, inflammatory responses, SCFAs, and gut microbiota diversity, this study explored whether or not the PSPE and PSPc could alleviate UC development by reconstructing the microbiota compositions.

2. Materials and Methods

2.1. Extraction and Determination of PSPc

The skins of peanuts were provided by Zhengyang Xindi Peanut Group Co., Ltd. (Zhengyang, China), and were frozen at −20 °C until use. PSPE and PSPc were provided by the Key Laboratory of Nutrition and Healthy Food of Zhengzhou, China.

The extraction of PSPE and PSPc was conducted according to the methods optimized in our laboratory. Peanut skin was ground and the oil was removed by petroleum ether (National Pharmaceutical Group Co, Beijing, China), followed by drying, the addition of 70% ethanol (National Pharmaceutical Group Co, Beijing, China), and then extracted jointly by microwave and ultrasonic. The crude extract was separated by Centrifugation (8000 rpm, 5 min) prior to adsorption by activated carbon to attain the filtrate. The concentrated liquor was frozen and dried after extraction, and the resulting PSPE purity was at about 67%.

On the other hand, PSPc were obtained through the following steps. PSPE was chromatographed on macroreticular resins (AB-8, Solarbio Co., Ltd., Beijing, China). The effective ingredients were washed with 40% alcohol, and the elutriant was filtered through a 0.22 μm filter membrane. The PSPc were obtained after the concentrated liquor was frozen and dried. PSPc were determined by HPLC (Thermo Scientific UltiMate 3000 U, Phenomenex Kinetex C18) with a purity of ca. 95%. The composition of PSPc was detected by ultrahigh-performance liquid chromatography–tandem mass spectrometry (UPLC-QTOF-MS/MS) (Agilent Technologies Technology Co., Ltd., Beijing, China). The operating conditions of UPLC-QTOF-MS/MS were as follows: waters BEH C18 column (2.1 × 100 mm, 1.7 μm), mobile phase A: 0.1% formic acid in water; mobile phase B: acetonitrile solution, gradient elution: 0 min, 95% A, 5% B; 30 min, 10% A, 90% B; 45 min, 0% A, 100% B; 50 min, 0% A, 100% B; 51 min, 95% A, 5% B; 60 min, 95% A, 5% B. The real-time flow rate was 0.3 mL/min, the injection volume was 5 μL, the real-time mass spectrometry scanning range was 50~1200 m/z, the sheath gas temp 350 °C, the sheath gas flow 12 L/min, the ESI-mode and the voltage were 3200 V.

2.2. Animal Experiment

The male mice (C57BL/6J, 8 weeks old, n = 40) were purchased from Xi'an Jiaotong University (Xi'an, Shaanxi, China). Mice in groups of ten were raised in a rectangular cage under a controlled environment (25 ± 2 °C temperature, 50% ± 5% humidity) and were fed with a standard diet (AIN-93) with a 12 h light/dark cycle. The laboratory animal production license number was 81803231.

The mice were divided into four groups randomly: the control group (n = 10), the DSS group (n = 10), the DSS + PSPE group (n = 10), and the DSS + PSPc group (n = 10). The control group and the DSS group were treated with physiological saline for 17 days by gavage before euthanasia. The DSS + PSPE group (200 mg/kg of PSPE dissolved in saline) and the DSS + PSPc group (200 mg/kg of PSPc dissolved in saline) were administered with corresponding doses by gavage once daily for 17 days [33,34]. The UC mouse model was established by replacing the drinking water in the DSS group, DSS + PSPE group, and DSS + PSPc group with a 3% DSS solution on day 11, and the mice were treated for 7 days. Mice were sacrificed while being anesthetized with 3.5% chloral hydrate (10 mL/kg). Following sacrifice, the colon, blood, cecal contents, and feces were collected and instantly frozen with liquid nitrogen. The remaining colon tissue was kept at −80 °C for additional biochemical and immunoblot analysis, while a portion of the colon tissues was preserved in 4% paraformaldehyde/PBS (v/v) for histopathological study at room temperature. Experimental protocols complied with Guidelines for the Care and Use of Experimental Animals: 8th edition (ISBN-10: 0-309-15396-4). The Animal Ethics Committee of Northwest A&F University and the BGI Institutional Review Board on Bioethics and Biosafety (BGI-IRB) authorized the animal experiment procedure.

2.3. Disease Activity Index (DAI) Evaluation

The DAI score was used to evaluate the development of UC [35]. The score was recorded starting from the first day of DSS treatment. DAI was evaluated by loss of body weight, blood in stools, and formation of stools. The maximum DAI score was 10, and it was made up of body weight loss (0–4), formation of stools (0–2), and blood in stools (0–4). The mice's body weight loss was scored as follows: body weight loss of less than 1% received a score of 0, body weight loss of 1–5%, 5–10%, 10–15%, and greater than 15% received scores 1, 2, 3, and 4, respectively. The formation of stools was scored as follows: solid and granular stool received score 0, soft and granular stool received score 1, and loose and showing signs of liquid received score 2. An occult blood test kit (BaSO, Guangdong, China) and the manufacturer's instructions were used to evaluate the presence of blood in the stool. The results were scored as follows: 0 for no evidence of occult blood; 1 for occult blood that was only weakly expressed; 2 for occult blood that was strongly expressed; 3 for occult blood that was strongly expressed and 4 for occult blood that was strongly expressed and visible.

2.4. Colon Histopathology Analysis

The tissues were fixed in 4% paraformaldehyde/PBS (*v*/*v*), embedded in paraffin, and stained with hematoxylin and eosin (H&E). Light microscopy (Olympus, Tokyo, Japan) (×160) was used to examine the histopathologic characteristics of a section of colon tissues with a thickness of 5 μm. The histological scores were assessed by experienced staff in the laboratory according to the extent of infiltration of inflammatory cells and mucosa damage, and these staffers had a lot of hands-on experience, but they were blinded to the samples' treatments. The scores for the infiltration of inflammatory cells were classified using 5 grades: score 0, none; score 1, infiltrate around crypt bases; score 2, infiltrate in muscularis mucosa; score 3, extensive infiltrate in muscularis mucosa with edema; score 4, infiltration of the submucosa. The scores for the extent of mucosa damage were classified into 4 grades: score 0, intact mucosa; score 1, ≤1/3 disruption of mucosa; score 2, 1/3–2/3 disruption of mucosa; score 3, >2/3 disruption of the mucosa. The sum of the two scores was considered the histological score.

2.5. Alcian Blue Staining

Goblet cells and mucus layers have frequently been visualized using alcian blue staining. [36]. The tissue was embedded vertically in paraffin and cut into 5 μm sections. Sections were stained with Alcian blue/Nuclear Fast Red. The colon tissue sections were observed at ×160 magnification.

2.6. Immunohistochemistry Analysis

The immunohistochemistry (IHC) analysis followed the guidelines previously described [37]. The colon tissue sections were dewaxed, washed three times with PBS, and treated in 3% hydrogen peroxide for 10 min. Then sections were blocked for 20 min by a normal goat serum blocking solution. After blocking, sections were incubated with primary antibody (Anti-Claudin 1 antibody, Abcam, Cambridge, UK, 5μg/mL) overnight at 4 °C. After rinsing with PBS 3 times, sections were incubated with 50 μL secondary antibody (Biotin conjugated Goat Anti-Mouse IgG, ZSGB-Bio, Beijing, China) for 30 min and then was coupled with horseradish peroxidase for 30 min at room temperature. Tissues were rinsed with PBS 3 times before visualizing by chromogen DAB (DAB kit, Zhongshan Golden Bridge Biotechnology Co., Ltd., Beijing, China). The final steps were clearing colon sections in xylene, dehydrating them in ethanol, and mounting them with Permount TM Mounting Medium.

2.7. ELISA Assay

Colonic tissues were homogenized and centrifuged, and the supernatant was taken for cytokine analysis. Proinflammatory cytokine concentrations (MDA, TNF-, and IL-1) were quantified by using ELISA kits (Xinle Biotechnology, Shanghai, China), and measured at an optical density of 450 nm (Bio-Rad, Hercules, USA).

2.8. *qRT-PCR*

The protocol of the qRT-PCR was as previously described with the following modifications [38]. The total RNA of the colon tissue was extracted with TRIzol (Jingcai Biotechnology, Xi'an, China) reagent and quantified using NanoDropOne (Thermo Fisher Scientific, MA, USA). The cDNA samples (50 ng/μL) were synthesized using the HiFiScript cDNA Synthesis Kit by RNA reverse transcription. Subsequently, a Real-Time PCR reaction system was prepared by UltraSYBR Mixture (Cowin Biosciences, Beijing, China) for quantitative analysis. The reaction mixture was made by mixing 1 μL upstream primer (6 μM), 1 μL downstream primer (6 μM), 2 μL diluted five-fold sample (10 ng/μL), 6 μL ddH_2O, and 10 μL mixture (10×). The primers used in qRT-PCR were listed in Table 1, which were laboratory designed and cited in previous research [38]. The level of the RNA expressions was quantified using real-time detection, performed by CFX96™ real-time system (Bio-Rad, Hercules, CA, USA). Thermal cycling conditions were as recommended by the manufacturer for 40 cycles. Cycle threshold (Ct) values were recorded. Data were normalized using GAPDH and transformed using the $2^{-\Delta\Delta CT}$ method.

Table 1. Primer sequences used for qRT-PCR analysis.

	Forward Primers (5′-3′)	Reverse Primer (5′-3′)
Occludin	ACGGACCCTGACCACTATGA	TCAGCAGCAGCCATGTACTC
IL-6	CTCTGGCGGAGCTATTGAGA	AAGTCTCCTGCGTGGAGAAA
iNOS	GGGCTGACCTGTTTCCTACT	GGAGGTTGAGACCCAATGGA
COX-2	CCCATTAGCAGCCAGTTGTC	CAGGATGCAGTGCTGAGTTC
Claudin-1	AGCTGCCTGTTCCATGTACT	CTCCCATTTGTCTGCTGCTC
MUC2	AGGGCTCGGAACTCCAGAAA	CCAGGGAATCGGTAGACATCG
Gapdh	TGGAGAAACCTGCCAAGTATGA	TGGAAGAATGGGAGTTGCTGT

2.9. *Analysis of 16S rRNA Sequencing*

The cecal contents samples were collected from mice in a sterile environment after being sacrificed. The 16S rRNA sequence analyses of the collected samples were conducted by the method reported previously [31]. The ribosomal RNA gene's 16S rDNA V3-V4 region was amplified by PCR using the primers 341F: CCTACGGGNGGCWGCAG; and 806R: GGACTACHVGGGTATCTAAT. Using the AxyPrep DNA Gel Extraction Kit from Axygen Biosciences, amplicons were extracted from 2% agarose gels, purified, and quantified using QuantiFluor-ST (Promega, U.S.). Amplicons were paired-end sequenced (2 × 250) on an Illumina MiSeq platform.

Quantitative Insights Into Microbial Ecology was used to demultiplex, quality-filter, and evaluate the raw 16S rRNA gene sequence data (QIIME). Using UPARSE, the sequences with a similarity of less than 97% were grouped into operational taxonomic units (OTUs). By using the RPD classifier (version 2.2), which is based on the SILVA Database, the representative sequences were divided into organisms. The β diversity and relative abundance of microorganisms were compared using the Kruskal–Wallis ANOVA test. All data were presented in the text as the means ± SE, and $p < 0.05$ was considered to be a significant difference.

2.10. *SCFAs Content in Feces*

The concentrations of SCFAs in feces were determined with a gas chromatograph (Shimadzu Corporation, GC-2014C, Kyoto, Japan)as reported previously [39]. The standards of SCFAs (Aladdin Bio-Chem Technology Co., Ltd., Shanghai, China) were propionate (P110445), butyrate (B11se0438), isobutyrate (I103524), and isovalerate (I108280). Approximately 200 mg of the fecal content sample was homogenized with 1 mL of distilled water;

then, 0.15 mL of 50% H2SO4 (*w*/*w*) and 1.6 mL of diethyl ether were added. After the samples were incubated at 4 °C for 20 min, they were centrifuged at 7155 g for 5 min, and then collected and filtered the organic phase. The conditions were used for GC: a temperature of 50 °C was started, kept for 3 min, then increased to 130 °C at 10 °C/min, 170 °C at 5 °C/min, 220 °C at 15 °C/min, and held at 220 °C for 3 min. The injector and detector had respective temperatures of 250 °C and 270 °C.

2.11. Statistical Analysis

Except for 16S rRNA sequencing data, other data were reported as mean ± SE, and the significant differences between the means were determined by one-way ANOVA followed by the Newman–Keuls multiple comparison post-test method using GraphPad Prism 6.0 software (GraphPad Software Inc., San Diego, CA, USA). The Kruskal–Wallis H test, a nonparametric test method, was used to evaluate the 16S rRNA sequencing data in order to find the significant differences between the several groups. Means were considered to be statistically significant if $p < 0.05$.

3. Results

3.1. Composition of PSPc

The composition of extracted and purified peanut skin procyanidins was analyzed by UPLC-QTOF-MS/MS based on the different cleavage modes of procyanidins, combined with information of [M-H]-(m/z) and corresponding fragment ions. The total ion flow diagram of the PSPc is shown in Figure 1, and the secondary mass spectrometry of each compound is shown in Figure S1. The specific analysis of the main substances according to the secondary mass spectrometry is shown in Table 2. PSPc are mainly characterized by A-type procyanidins dimer, Protocatechualdehyde and Catechins; it also includes very little A-type procyanidins trimer, A-type procyanidins tetramer, Protocatechuic acid, and B-type procyanidins.

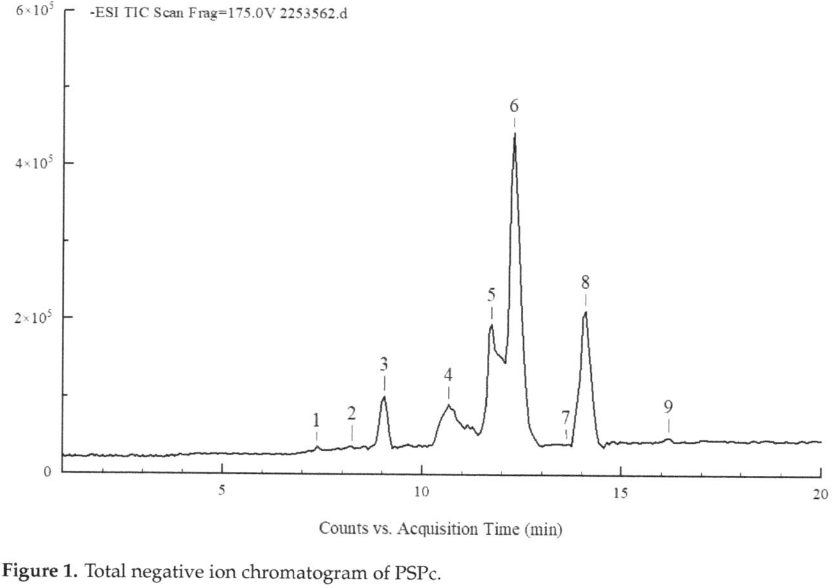

Figure 1. Total negative ion chromatogram of PSPc.

Table 2. Mass spectrometry information of the main compounds in PSPc.

Compound	Matter	Rt (min)	MS[M-H] (m/z)	MS/MS (m/z)
1	Protocatechuic acid	7.388	153	109.03
2	B-type procyanidins	8.286	577	125.02, 287.05, 289.07, 407.07, 425.08, 451.10
3	Protocatechualdehyde	9.052	137	119.01, 108.02
4	A-type procyanidins dimer	10.665	575	285.04, 289.07, 407.07, 449.08
5	Catechins	11.741	289	125.02, 137.02, 165.02, 179.03, 205.05, 245.08
6	A-type procyanidins dimer	12.279	575	285.04, 289.07, 407.07, 449.08
7	A-type procyanidins trimer	13.648	863	287.05, 289.07, 411.07, 451.07, 575.12, 711.13
8	A-type procyanidins dimer	14.110	575	285.04, 289.07, 407.07, 423.07, 49.08
9	A-type procyanidins tetramer	16.126	1149	573.10, 575.12, 997.17

3.2. Effects of PSPE and PSPc Supplementation on Pathological Changes in DSS-Induced Colitis Mice

The mice were treated with PSPE and PSPc for 17 days (Figure 2A). Compared with the control group, the body weight loss was more severe in the DSS models on the end day of DSS treatment ($p < 0.01$) (Figure 2B,C). Although there was no significant difference in the DSS models (Figure 2C), the DAI score was significantly elevated in the DSS-treated mice compared with the control group (Figure 2D). Compared with the DSS group mice, DAI values were reduced in both DSS + PSPE and DSS + PSPc groups (23.0% and 25.3%, respectively) (Figure 2E). Thus, the PSPE and PSPc supplementation exhibited a beneficial effect on alleviating UC development in mice.

3.3. Effect of PSPE and PSPc Supplementation on the Inflammatory Factor Expressions and Biomarker of Oxidative Stress in DSS-Induced Colitis Mice

The cytokines IL-1β, TNF-α, and IL-6 play essential roles in intestinal epithelial cells and inflammatory cells, and the overexpression of COX-2 and iNOS may be involved in the occurrence and development of UC [40]. The protein levels of IL-1β and TNF-α in colon tissue were evaluated by ELISA assays. The level of the RNA expressions was quantified using RT-PCR. Compared with the control group, the IL-1β, TNF-α, and IL-6 protein levels were significantly upregulated in the DSS group. However, compared with the DSS group, the protein expressions of IL-1β, TNF-α and IL-6 were significantly down-regulated in the DSS + PSPE and DSS + PSPc groups (Figure 3A–C). Compared with the control group, the mRNA expressions of the inflammatory mediators, including COX-2, and iNOS, were significantly elevated in the DSS group. However, the mRNA expressions of COX-2 and iNOS were significantly lower in the PSPE and PSPc treatment in DSS-treated mice. (Figure 3D,E). The level of MDA in colon tissues, as a biomarker of oxidative stress, has also been detected [41]. The PSPE and PSPc treatment significantly down-regulated the DSS-increased levels of MDA (Figure 3F). These results indicated that PSPE and PSPc supplementation reduced inflammatory factor expressions and suppressed oxidative stress in DSS-induced colitis mice.

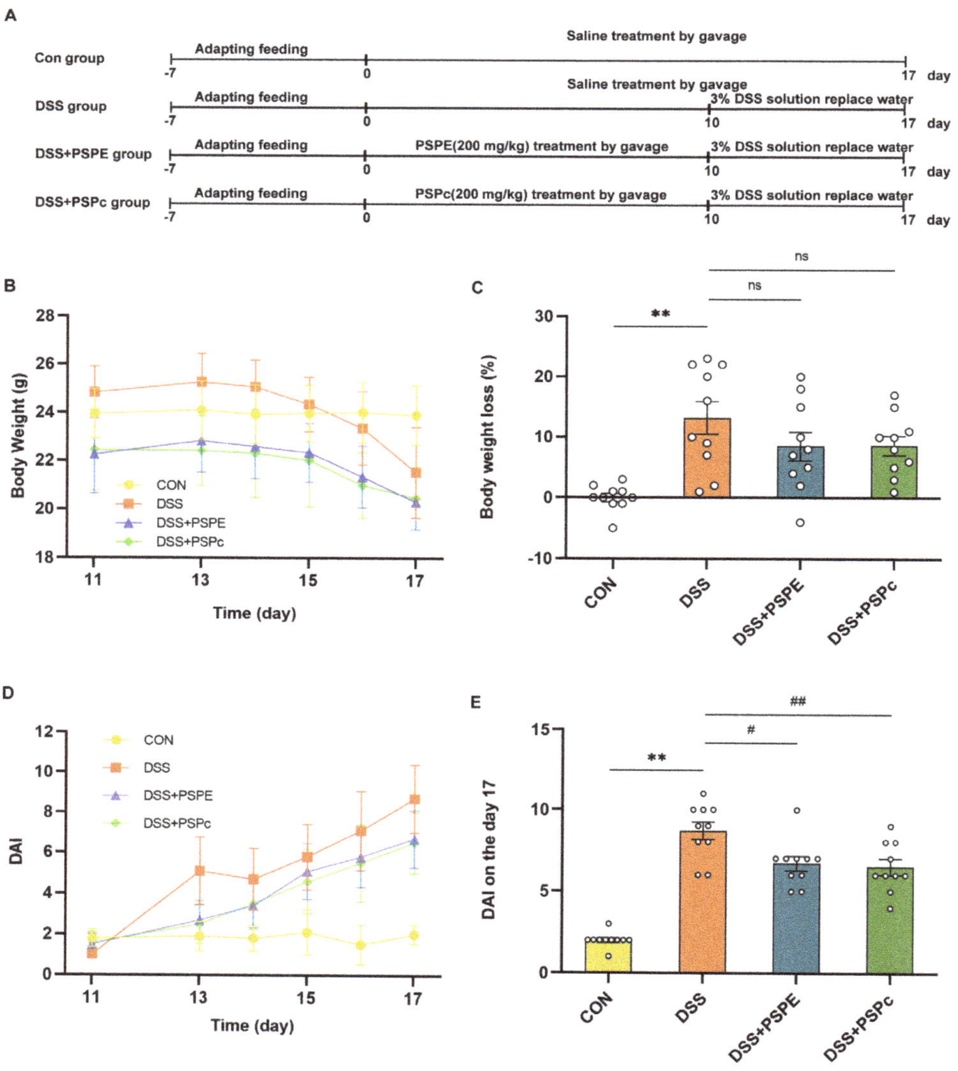

Figure 2. Effect of PSPE and PSPc on the symptoms in the colon tissue of DSS−induced colitis mice. (**A**) Con group: control group (n = 10 mice); DSS group: DSS alone group (n = 10 mice); DSS + PSPE group: PSPE 200 mg/kg + DSS group (n = 10 mice); DSS + PSPc group: PSPc 200 mg/kg + DSS group (n = 10 mice). Schematic representation of the experimental design; (**B**) body weight during DSS treatment (11th–17th day); (**C**) body weight loss during the DSS treatment ([body weight on 11th day-body weight on 17th day]/body weight on 11th day); (**D**) DAI score during DSS treatment; (**E**) DAI on the seventh day after DSS-induced UC. Data were expressed as mean ± SE (n = 3 mice). ** $p < 0.01$, compared with the Con group, # $p < 0.05$, ## $p < 0.01$, compared with the DSS group, ns = not significant.

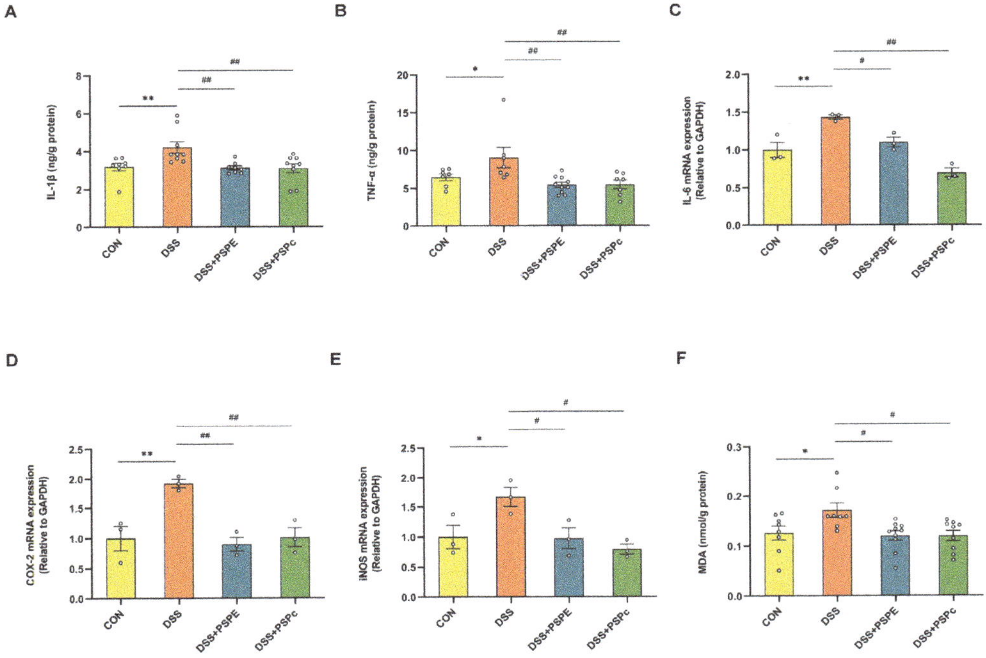

Figure 3. Effect of PSPE and PSPc on the proinflammatory cytokines and mRNA expression in colonic tissue. (**A**,**B**) Proinflammatory cytokines (IL-1β and TNF-α) in colonic tissue. Data were expressed as mean ± SE (n = 7–10 mice). (**C**–**E**) qRT-PCR expression (IL-6, COX-2, and iNOS). Data were expressed as mean ± SE (n = 3 mice). (**F**) MDA in colonic tissue. Data were expressed as mean ± SE (n = 7–10 mice). * $p < 0.05$, ** $p < 0.01$, compared with the Con group, # $p < 0.05$, ## $p < 0.01$, compared with the DSS group, ns = not significant.

3.4. Effects of PSPE and PSPc Supplementation on the Histopathological Changes in the Colon of the DSS-Induced Colitis Mice

The H&E staining showed that, compared with the control group, the DSS treatment caused severe colon injury, crypt structure damage, goblet cell loss, massive inflammatory cell infiltration, muscularis thinning, and cortical destruction (Figure 4A). These were typical symptoms of UC development. However, the colon tissue damage induced by DSS was significantly reversed in the DSS + PSPE and DSS + PSPc groups ($p < 0.05$) (Figure 4B). Moreover, Alcian blue staining was employed to detect the number of goblet cells (Figure 4A). The loss of goblet cells was significantly attenuated in the DSS + PSPE and DSS + PSPc groups (Figure 4C). Compared with the DSS group, the DSS-induced mucosal damage was significantly suppressed, and the down-regulated MUC-2 mRNA expression was increased in the DSS + PSPc group ($p < 0.01$) (Figure 4D).

Claudin-1 protein, concentrated in epithelial tissue, plays an essential role in forming impermeable barriers [42]. The IHC analysis shows that reducing Claudin-1 expression in the DSS-treated mice was prevented in the DSS + PSPE and DSS + PSPc groups (Figure 4A,E). Meanwhile, compared with the DSS group, the mRNA expressions of tight junction proteins, including Claudin-1 and occludin were significantly increased in PSPE and PSPc-treated colitis mice ($p < 0.05$) (Figure 4F,G). These results indicated that PSPE and PSPc treatment effectively alleviated the inflammation, epithelial layer damage, goblet cell loss, and Claudin-1 loss in the colon of the colitis mice.

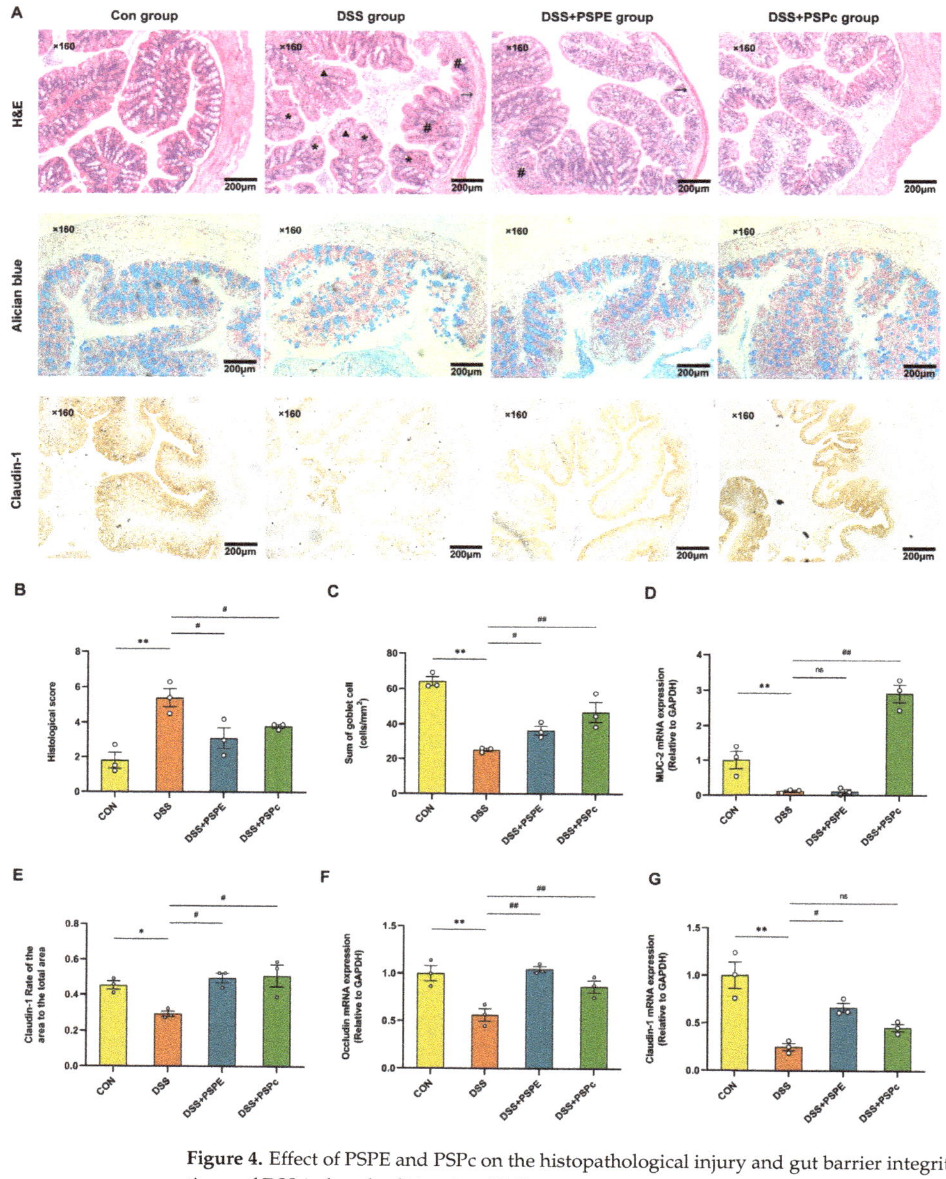

Figure 4. Effect of PSPE and PSPc on the histopathological injury and gut barrier integrity in colon tissue of DSS-induced colitis mice. (**A**) Representative images of H&E staining, Alcian blue staining, and claudin-1 protein IHC staining of the colon for each group. * damaged crypts structure, # loss of goblet cells, ▲ massive infiltrating inflammatory cells, → muscularis thinning. Scale bar: 200 μm (×160); (**B**) histopathology scores of colonic tissues; (**C**) numbers of goblet cells in the colon; (**D**) qRT-PCR expression (MUC-2); (**E**) quantitative analysis of claudin-1 protein content in colonic tissue; (**F**,**G**) qRT-PCR expression (Claudin-1 and Occludin). Data were expressed as mean ± SE (n = 3 mice). * $p < 0.05$, ** $p < 0.01$, compared with the Con group, # $p < 0.05$, ## $p < 0.01$, compared with the DSS group, ns = not significant.

3.5. Effects of PSPE and PSPc Supplementation on the Gut Microbiome Diversity in DSS-Induced Colitis Mice

In order to investigate the effects of PSPE and PSPc treatment on the gut microbiome composition, the 16S rDNA sequencing analysis was used in this study. As shown in the principal co-ordinates analysis (PCoA), a significant separation was observed among the Con, DSS, DSS + PSPE, and DSS + PSPc groups (Figure 5A). The linear discriminant analysis (LDA) revealed that *Olsenella* was enriched in the DSS group. Meanwhile, the microbes including *Turicibacter, Pseudomonadales, Moraxellaceae, and Acinetobacter* were enriched in the DSS + PSPE group, and the microbes including *Bacteroides, Bacteroidaceae, Gammaproteobacteria, Cupriavidus, Burkholderiaceae, Clostridium_XVIII* were enriched in the DSS + PSPc group (Figure 5B).

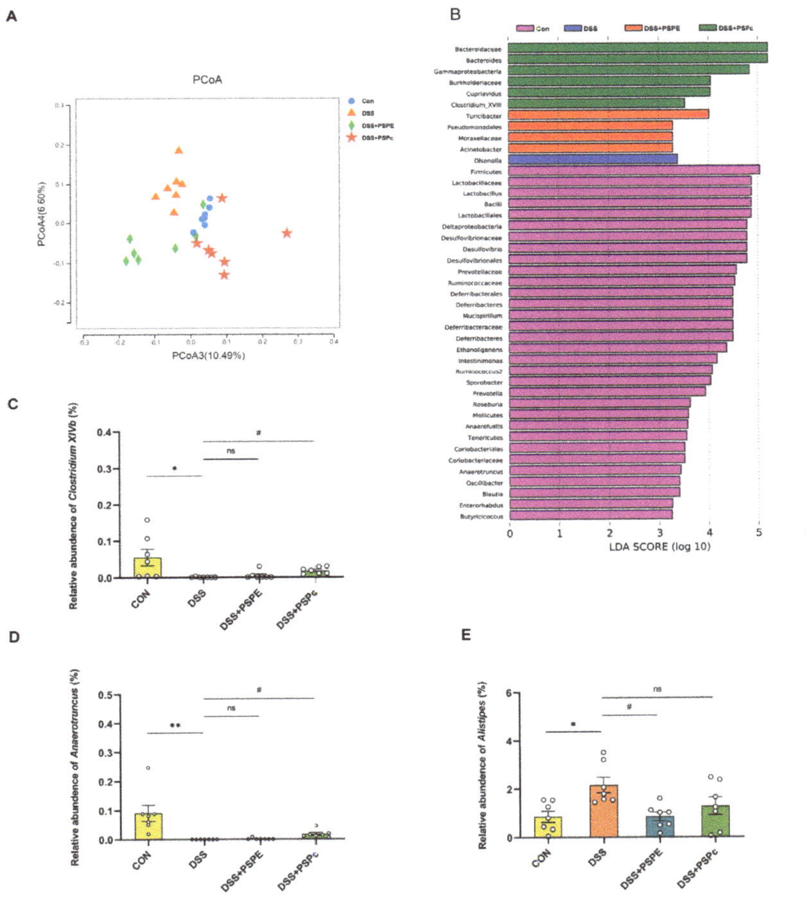

Figure 5. Effect of PSPE and PSPc on the gut microbiome compositions in DSS-induced colitis mice. (**A**) Principal components analysis (PCoA); (**B**) linear discriminant snalysis (LDA) score. (**C**) Relative abundance (%) of *Clostridium XIVb*; (**D**) relative abundance (%) of *Anaerotruncus;* (**E**) relative abundance (%) of *Alistipes;* data are presented as mean ± SE, and statistical significance was determined by Kruskal–Wallis test (H test) with, $n = 7$ mice. * $p < 0.05$, ** $p < 0.01$, compared with the Con group, # $p < 0.05$, compared with the DSS group, ns = not significant.

The relative abundances of *Clostridium XlVb* and *Anaerotruncus* increased in the DSS + PSPc group compared with the DSS group ($p < 0.05$). Compared to the control group, the relative abundances of *Alistipes* in the DSS groups increased significantly ($p < 0.05$), whereas the relative abundances decreased in the DSS + PSPE group (Figure 5C–E).

3.6. Effect of PSPE and PSPc Supplementation on the SCFAs Generation in DSS-Induced Colitis Mice

Many studies have shown that SCFAs, especially butyrate, have therapeutic effects in patients with IBD [43]. Compared with the DSS group, butyrate, isobutyrate, and isovalerate contents were significantly promoted in the DSS + PSPE and DSS + PSPc groups (Figure 6A–C). The DSS group significantly increased propionate formation in feces compared with the CON group. However, propionate content significantly decreased in the DSS + PSPE and DSS + PSPc groups (Figure 6D). PSPE and PSPc treatment altered SCFAs production in the colon.

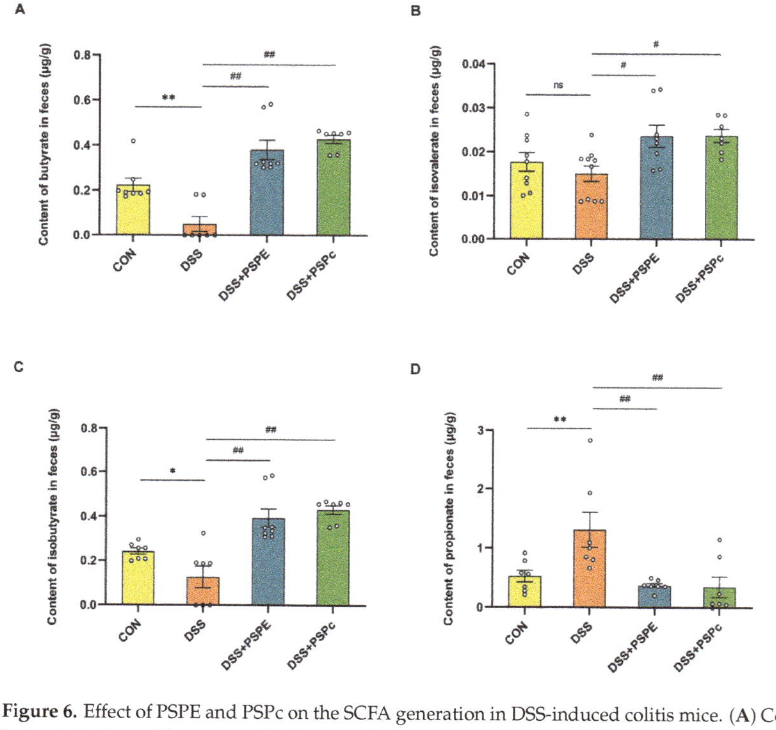

Figure 6. Effect of PSPE and PSPc on the SCFA generation in DSS-induced colitis mice. (**A**) Content of butyrate in feces; (**B**) content of isobutyrate in feces; (**C**) content of isovalerate in feces; (**D**) content of propionate in feces; data were expressed as mean ± SE (n = 7–10 mice). * $p < 0.05$, ** $p < 0.01$, compared with the Con group, # $p < 0.05$, ## $p < 0.01$, compared with the DSS group, ns = not significant.

3.7. Correlation Analysis among the DAI, Fecal SCFAs Levels, Colon IL-1β, TNF-α, MDA Level, and Gut Microbiota

To clarify the correlation between gut microbiota, fecal SCFAs level, and DAI score, Spearman's correlation analysis was implemented based on these experimental data (Figure 7). The body weight loss was positively correlated with the DAI on the 17 (r = 0.74, $p < 0.01$) but negatively correlated with the fecal isobutyrate (r = −0.58, $p < 0.05$), isovalerate (r = −0.68, $p < 0.05$). The DAI on the 17 was positively correlated with the MDA levels (r = 0.58, $p < 0.05$) but negatively correlated with the fecal propionate levels (r = −0.66,

$p < 0.05$). The colonic IL-1β, TNF-α, and MDA levels were decreased in DSS + PSPE and DSS + PSPc groups. The colonic IL-1β levels showed a negative correlation with the fecal butyrate ($r = -0.55$, $p < 0.05$), but showed a positive correlation with the colonic TNF-α ($r = 0.89$, $p < 0.01$) and MDA levels ($r = 0.76$, $p < 0.01$). The colonic TNF-α levels showed a negative correlation with the fecal butyrate ($r = -0.72$, $p < 0.01$), but showed a positive correlation with the colonic MDA levels ($r = 0.67$, $p < 0.05$). The fecal propionate levels were positively correlated relative abundance of *Anaerotruncus* ($r = 0.63$, $p < 0.05$). The fecal butyrate levels were positively correlated relative the fecal isobutyrate ($r = 0.96$, $p < 0.01$), isovalerate ($r = 0.79$, $p < 0.01$). The fecal isobutyrate levels were positively correlated the fecal isovalerate ($r = 0.79$, $p < 0.01$). These results suggested that the beneficial effects of PSPE and PSPc on UC development were related to the balance of gut microbiota compositions and the formation of microbial metabolites.

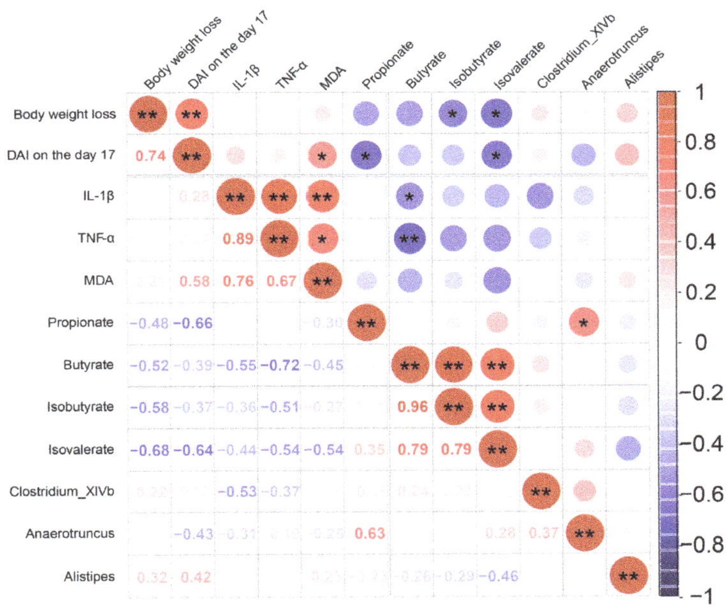

Figure 7. Correlation analysis among the DAI, fecal SCFAs levels, colon IL-1β, TNF-α, and MDA levels, and gut microbiota. Spearman's correlation analysis was performed. In the upper right corner, you could find the size and color of the circle used to represent the relevant indicator level, with red indicating positive and blue indicating negative. The corresponding value of the corresponding index could be found in the lower left corner. Correlation analysis among indicators,* $p < 0.05$, ** $p < 0.01$.

4. Discussion

In the present study, we found that the major components of PSPc are A-type procyanidin dimers, catechins, protocatechualdehyde, etc. Moreover, it has been found that PSPE and PSPc supplementation can reduce DSS-induced body weight loss and DAI value in the colitis mice. PSPE and PSPc supplementation could also reduce intestinal inflammation, epithelial layer damage, goblet cell loss, and elevate the tight junction protein Claudin-1 expression. The PSPE and PSPc treatment significantly down-regulated the levels of inflammatory mediators and the biomarker expression of oxidative stress. Furthermore, the PSPE and PSPc treatments reshaped the gut microbiome compositions and enhanced the generation of microbial metabolite SCFAs in the feces of colitis mice. The correlation analysis demonstrated that the inflammatory responses were highly correlated with the gut microbiota compositions and microbial metabolite formation.

Peanut skin procyanidin extract (PSPE) has been reported to possess various biological activities [22,44–51]. Peanut skin contains a variety of dietary polyphenols, especially procyanidins, which may be major active components of peanut skin [29] and are composed of flavan-3-alcohol structural units, and the monomers, mainly including catechin, epicatechin, catechin gallate, epicatechin gallate, and epicatechin gallate, etc. [52]. The procyanidins can be divided into type A and type B procyanidins according to the different positions of chemical bonds between monomers. Type B procyanidins are formed by C4-C8 or C4-C6 phase linkage of catechin or/with epicatechin monomer and are widely distributed in plants [53]. Type A procyanidins are linked by C2-O-C7 or C2-O-C5 in addition to C4-C8 or C4-C6, which are mainly found in peanuts and cranberries. It has been reported that type A procyanidins which are more stable with higher bioavailability were the main ingredient in peanut red skins [54]. In addition, according to the degree of polymerization between monomers, they can be divided into a dimer, trimer, tetramer, etc. [55]. However, the bioactivity of procyanidins is inversely proportional to the degree of polymerization—the lower the degree of polymerization, the higher the bioactivity of Procyanidins [56]. In this study, PSPc were obtained after purification of PSPE, it was found that PSPc were mainly composed of A-type procyanidins dimer, A-type proanthocyanidin trimer, A-type proanthocyanidin tetramer, catechins, and protocatechualdehyde, etc. These active components suggested that PSPc had better biological activities.

UC is usually accompanied by weight loss, inflammation, oxidative stress, intestinal barrier disruption, and other phenomena. We found that, compared with the control group, DSS-treated mice showed significantly severe body weight loss, significantly elevated the DAI score, and significantly increased the protein expressions of IL-1β and TNF-α, the mRNA expressions of IL-6, COX-2, and iNOS, and the protein expression of MDA in colonic tissue, which showed that the DSS-treated mice had typical symptoms of UC. The intake of polyphenols has been shown to be effective in relieving the symptoms of UC [7–11]. A-type procyanidins could improve the intestinal barrier effect, reduce the expression levels of inflammatory cytokines (TNF-α, IL-β, IL-6, and IL-10) and oxidative stress (MDA, T-SOD, NO, and iNOS) in DSS-induced UC mice [57]. Catechins could significantly inhibit excessive oxidative stress through direct or indirect antioxidant effects, hence promoting activation of the anti-oxidative substances and reducing oxidative damage to the colon [58]. Protocatechualdehyde could play a significant anti-inflammatory role by regulating the activation or inactivation of inflammatory and oxidative stress-related cell signaling pathways, such as NF-κB, MAPKs, and STAT1/3 pathways [59,60]. In the current study, we found that, compared with the DSS group, both PSPE and PSPc supplementation could reduce the DAI values, decrease the levels of TNF-α and IL-1β, decrease the mRNA expressions of IL-6, COX-2, and iNOS, down-regulated the levels of MDA, and increase the levels of tight junction protein, MUC-2 and goblet cells in colitis mice. These results indicated that PSPE and PSPc could improve the symptoms of colitis mice. In addition, PSPc is mainly composed of A-type procyanidins, catechins, protocatechualdehyde, which may play a major role in alleviating ulcerative colitis.

Dysbiosis of the gut microbiome is increasingly considered to be causatively related to UC [30]. Consistent with previous studies, the gut microbiota was dysregulated, with Firmicutes reduced and Bacteroidetes increased in the DSS treatment group. It was accompanied by a loss of diversity in the microbiota [61]. PSPE-treated and PSPc-treated mice had a good protective effect on the intestinal flora of colitis mice. The results showed that *Olsenella*, which observed a significant increase in patients with UC [62], was enriched in the DSS group. The microbes including *Turicibacter, Pseudomonadales, and Moraxellaceae* enriched in the PSPE group have been reported negatively correlated with inflammatory factors and significantly associated with great immune recovery [63–65]. Both *Bacteroides* and *Clostridium_XVIII*, enriched in the DSS + PSPc group, were reported to have anti-inflammatory action. Moreover, the abundance of *Clostridium XlVb* and *Anaerotruncus* was significantly lower in the DSS group than in the control group. PSPc treatment significantly increased the level of *Clostridium XlVb* and *Anaerotruncus* in colitis mice.

Previous studies have found that SCFAs are the most important metabolites of gut microbiota and play a very important role in colitis, the fecal SCFAs levels were reduced in active UC [37,38]. In our study, the results showed that the SCFAs, such as butyrate, isobutyrate, and isovalerate, were significantly decreased in the DSS group compared with the control group. The PSPE and PSPc treatment significantly increased the levels of butyrate, isobutyrate, and isovalerate in DSS-treated mice. We conducted a correlation study and found that the fecal butyrate levels were positively correlated with the fecal isobutyrate and isovalerate levels, but negatively correlated with the IL-1β and TNF-α levels. Meanwhile, it has been reported that butyrate is a key metabolite of UC, which regulates immune responses by inhibiting the release of proinflammatory cytokines and enhances intestinal barrier function by upregulating MUC2 expression in UC [43,66].

Previous studies have detected that the genus of *Anaerotruncus*, commonly found in the intestinal tracts of healthy humans and animals, can produce acetic and butyric acids through sugar metabolism pathways [67]. PSPE treatment significantly increased the level of *Anaerotruncus*, it suggested that the peanut skin procyanidins may attenuate DSS-induced ulcerative colitis by increasing SCFAs content, such as butyrate, isobutyrate, and isovalerate. Despite the multiple beneficial effects of SCFAs on host gastrointestinal activity, excessive quantities of propionic acid have been reported in gingival inflammation, acne, and irritable bowel syndrome [68]. In this study, we found that propionic acid increased in the DSS group compared with the control group, and supplementation of PSPc significantly decreased propionate content in DSS-induced colitis mice. Meanwhile, It has been found that the abundance of *Alistipes*, one of the mainly propionate producers, highly related to dysbiosis and chronic intestinal inflammation [69,70], increased in DSS-induced colitis mice and decreased in the DSS + PSPE group. Based on these results, we suggest that PSPE and PSPc had beneficial effects on alleviating UC development by reconstructing the microbiota compositions and changing the generation of SCFA in a DSS-induced colitis mouse model. In the future, we need to further confirm the effects of PSPE and PSPc on gut microbial metabolism changes and the underlying mechanism involved.

5. Conclusions

Taken together, the present results demonstrated that both PSPE and PSPc supplementation significantly ameliorated DSS-induced colitis by alleviating colonic inflammation, inhibiting colonic injury, and maintaining the integrity of the intestinal barrier. In addition, PSPc, mainly A-type procyanidins, could regulate the metabolism by maintaining gut microbiota. It suggests that purified PSPc have potential as dietary supplements or for the production of fortified foods as a promising nutritional strategy for reducing UC.

Supplementary Materials: The following supporting information can be downloaded at: https://www.mdpi.com/article/10.3390/antiox11112098/s1, Figure S1: Composition of PSPc.

Author Contributions: Conceptualization, N.W., C.X., and G.Z.; methodology, N.W., W.C., C.C., and Y.Z.; investigation, N.W., W.C., Q.Y., H.R., and Z.L.; resources, N.W., W.C., C.C., Y.Z., Q.Y., H.R., and Z.L.; data curation, N.W., W.C., C.C., C.X., and Z.L.; writing—original draft preparation, N.W., W.C., Z.L., and C.X.; writing—review and editing, N.W., W.C., Z.L., C.X., and G.Z.; visualization, W.C., Q.Y., Z.L., and C.X.; supervision, N.W., Z.L., C.X., and G.Z.; project administration, N.W. and C.X.; funding acquisition, N.W., Q.Y., and G.Z. All authors have read and agreed to the published version of the manuscript.

Funding: This work was funded by the National Key Research and Development Program of China (Grant No. 2018YFC1602902), the Key Scientific Research Project of Henan Province (Grant No. 22A550014), the Scientific and Technological Breakthroughs Project of Henan Province (Grant No. 222102110150) and the Opening Foundation of Key Laboratory of Nutrition and Healthy Food of Zhengzhou (KF20200515).

Institutional Review Board Statement: The animal study protocol was approved by The Animal Ethics Committee of Northwest A&F University (81803231) and the BGI Institutional Review Board on Bioethics and Biosafety (BGI-IRB).

Informed Consent Statement: Not applicable.

Data Availability Statement: Data is contained within the article and supplementary material.

Acknowledgments: The authors would like to thank the instrument-shared platform of the college of food science & engineering of NWAFU, for the assistance in the GC analysis (Yuan Zhou).

Conflicts of Interest: The authors declare no conflict of interest.

References

1. Hodson, R. Inflammatory Bowel Disease. *Nature* **2016**, *540*, S97. [CrossRef] [PubMed]
2. Baumgart, D.C.; Sandborn, W.J. Inflammatory Bowel Disease: Clinical Aspects and Established and Evolving Therapies. *Lancet* **2007**, *369*, 1641–1657. [CrossRef]
3. Ananthakrishnan, A.N. Epidemiology and Risk Factors for IBD. *Nat. Rev. Gastroenterol. Hepatol.* **2015**, *12*, 205–217. [CrossRef] [PubMed]
4. Feuerstein, J.D.; Moss, A.C.; Farraye, F.A. Ulcerative Colitis. *Mayo Clin. Proc.* **2019**, *94*, 1357–1373. [CrossRef]
5. Zhao, B.; Wu, J.; Li, J.; Bai, Y.; Luo, Y.; Ji, B.; Xia, B.; Liu, Z.; Tan, X.; Lv, J.; et al. Lycopene Alleviates DSS-Induced Colitis and Behavioral Disorders via Mediating Microbes-Gut–Brain Axis Balance. *J. Agric. Food Chem.* **2020**, *68*, 3963–3975. [CrossRef]
6. Chiu, H.-F.; Venkatakrishnan, K.; Golovinskaia, O.; Wang, C.-K. Gastroprotective Effects of Polyphenols against Various Gastro-Intestinal Disorders: A Mini-Review with Special Focus on Clinical Evidence. *Molecules* **2021**, *26*, 2090. [CrossRef]
7. Oz, H.S.; Chen, T.; de Villiers, W.J.S. Green Tea Polyphenols and Sulfasalazine Have Parallel Anti-Inflammatory Properties in Colitis Models. *Front. Immunol.* **2013**, *4*, 132. [CrossRef]
8. Morimoto, M.; Watanabe, T.; Yamori, M.; Takebe, M.; Wakatsuki, Y. Isoflavones Regulate Innate Immunity and Inhibit Experimental Colitis. *J. Gastroenterol. Hepatol.* **2009**, *24*, 1123–1129. [CrossRef]
9. Andújar, I.; Recio, M.C.; Giner, R.M.; Cienfuegos-Jovellanos, E.; Laghi, S.; Muguerza, B.; Ríos, J.L. Inhibition of Ulcerative Colitis in Mice after Oral Administration of a Polyphenol-Enriched Cocoa Extract Is Mediated by the Inhibition of STAT1 and STAT3 Phosphorylation in Colon Cells. *J. Agric. Food Chem.* **2011**, *59*, 6474–6483. [CrossRef]
10. Aldini, R.; Budriesi, R.; Roda, G.; Micucci, M.; Ioan, P.; D'Errico-Grigioni, A.; Sartini, A.; Guidetti, E.; Marocchi, M.; Cevenini, M.; et al. Curcuma Longa Extract Exerts a Myorelaxant Effect on the Ileum and Colon in a Mouse Experimental Colitis Model, Independent of the Anti-Inflammatory Effect. *PLoS One* **2012**, *7*, e44650. [CrossRef]
11. Larrosa, M.; González-Sarrías, A.; Yáñez-Gascón, M.J.; Selma, M.V.; Azorín-Ortuño, M.; Toti, S.; Tomás-Barberán, F.; Dolara, P.; Espín, J.C. Anti-Inflammatory Properties of a Pomegranate Extract and Its Metabolite Urolithin-A in a Colitis Rat Model and the Effect of Colon Inflammation on Phenolic Metabolism. *J. Nutr. Biochem.* **2010**, *21*, 717–725. [CrossRef] [PubMed]
12. Ma, Y.; Kerr, W.L.; Swanson, R.B.; Hargrove, J.L.; Pegg, R.B. Peanut Skins-Fortified Peanut Butters: Effect of Processing on the Phenolics Content, Fibre Content and Antioxidant Activity. *Food Chem.* **2014**, *145*, 883–891. [CrossRef] [PubMed]
13. Constanza, K.E.; White, B.L.; Davis, J.P.; Sanders, T.H.; Dean, L.L. Value-Added Processing of Peanut Skins: Antioxidant Capacity, Total Phenolics, and Procyanidin Content of Spray-Dried Extracts. *J. Agric. Food Chem.* **2012**, *60*, 10776–10783. [CrossRef] [PubMed]
14. Sobolev, V.S.; Cole, R.J. Note on Utilisation of Peanut Seed Testa. *J. Sci. Food Agric.* **2004**, *84*, 105–111. [CrossRef]
15. Yang, K.; Hashemi, Z.; Han, W.; Jin, A.; Yang, H.; Ozga, J.; Li, L.; Chan, C.B. Hydrolysis Enhances Bioavailability of Proanthocyanidin-Derived Metabolites and Improves β-Cell Function in Glucose Intolerant Rats. *J. Nutr. Biochem.* **2015**, *26*, 850–859. [CrossRef]
16. Appeldoorn, M.M.; Vincken, J.-P.; Sanders, M.; Hollman, P.C.H.; Gruppen, H. Combined Normal-Phase and Reversed-Phase Liquid Chromatography/ESI-MS as a Tool to Determine the Molecular Diversity of A-Type Procyanidins in Peanut Skins. *J. Agric. Food Chem.* **2009**, *57*, 6007–6013. [CrossRef]
17. Chang, M.; Sun, X.; Guo, X.; Bai, H.; Liu, R.; Jin, Q.; Wang, X. Composition and Antioxidant Study of Procyanidins from Peanut Skins. *J. Food Meas. Charact.* **2020**, *14*, 2781–2789. [CrossRef]
18. Cos, P.; De Bruyne, T.; Hermans, N.; Apers, S.; Berghe, D.V.; Vlietinck, A.J. Proanthocyanidins in Health Care: Current and New Trends. *Curr. Med. Chem.* **2004**, *11*, 1345–1359. [CrossRef]
19. Pan, C.; Wang, C.; Zhang, L.; Song, L.; Chen, Y.; Liu, B.; Liu, W.-T.; Hu, L.; Pan, Y. Procyanidins Attenuate Neuropathic Pain by Suppressing Matrix Metalloproteinase-9/2. *J. Neuroinflammation* **2018**, *15*, 187. [CrossRef]
20. Jia, Z.; Song, Z.; Zhao, Y.; Wang, X.; Liu, P. Grape Seed Proanthocyanidin Extract Protects Human Lens Epithelial Cells from Oxidative Stress via Reducing NF-kB and MAPK Protein Expression. *Mol. Vis.* **2011**, *17*, 210–217.
21. Kramer, K.; Yeboah-Awudzi, M.; Magazine, N.; King, J.M.; Xu, Z.; Losso, J.N. Procyanidin B2 Rich Cocoa Extracts Inhibit Inflammation in Caco-2 Cell Model of in Vitro Celiac Disease by down-Regulating Interferon-Gamma-or Gliadin Peptide 31-43-Induced Transglutaminase-2 and Interleukin-15. *J. Funct. Foods* **2019**, *57*, 112–120. [CrossRef]
22. Bansode, R.R.; Randolph, P.; Ahmedna, M.; Hurley, S.; Hanner, T.; Baxter, S.A.S.; Johnston, T.A.; Su, M.; Holmes, B.M.; Yu, J.; et al. Bioavailability of Polyphenols from Peanut Skin Extract Associated with Plasma Lipid Lowering Function. *Food Chem.* **2014**, *148*, 24–29. [CrossRef] [PubMed]

23. Lu, M.; Xu, L.; Li, B.; Zhang, W.; Zhang, C.; Feng, H.; Cui, X.; Gao, H. Protective Effects of Grape Seed Proanthocyanidin Extracts on Cerebral Cortex of Streptozotocin-Induced Diabetic Rats through Modulating AGEs/RAGE/NF-KappaB Pathway. *J. Nutr. Sci. Vitaminol.* **2010**, *56*, 87–97. [CrossRef] [PubMed]
24. Jiang, X.; Liu, J.; Lin, Q.; Mao, K.; Tian, F.; Jing, C.; Wang, C.; Ding, L.; Pang, C. Proanthocyanidin Prevents Lipopolysaccharide-Induced Depressive-like Behavior in Mice via Neuroinflammatory Pathway. *Brain Res. Bull.* **2017**, *135*, 40–46. [CrossRef]
25. Ungaro, R.; Mehandru, S.; Allen, P.B.; Peyrin-Biroulet, L.; Colombel, J.-F. Ulcerative Colitis. *Lancet* **2017**, *389*, 1756–1770. [CrossRef]
26. Larrosa, M.; Yañéz-Gascón, M.J.; Selma, M.V.; González-Sarrías, A.; Toti, S.; Cerón, J.J.; Tomás-Barberán, F.; Dolara, P.; Espín, J.C. Effect of a Low Dose of Dietary Resveratrol on Colon Microbiota, Inflammation and Tissue Damage in a DSS-Induced Colitis Rat Model. *J. Agric. Food Chem.* **2009**, *57*, 2211–2220. [CrossRef]
27. Liu, M.; Huang, B.; Wang, L.; Lu, Q.; Liu, R. Peanut Skin Procyanidins Ameliorate Insulin Resistance via Modulation of Gut Microbiota and Gut Barrier in Type 2 Diabetic Mice. *J. Sci. Food Agric.* **2022**. [CrossRef]
28. Lewis, W.E.; Harris, G.K.; Sanders, T.H.; White, B.L.; Dean, L.L. Antioxidant and Anti-Inflammatory Effects of Peanut Skin Extracts. *Sci. Res.* **2013**, *4*, 35240. [CrossRef]
29. Xu, M.; Lv, C.; Wang, H.; Lu, Q.; Ye, M.; Zhu, X.; Liu, R. Peanut Skin Extract Ameliorates High-Fat Diet-Induced Atherosclerosis by Regulating Lipid Metabolism, Inflammation Reaction and Gut Microbiota in ApoE−/− Mice. *Food Res. Int.* **2022**, *154*, 111014. [CrossRef]
30. Shen, Z.-H.; Zhu, C.-X.; Quan, Y.-S.; Yang, Z.-Y.; Wu, S.; Luo, W.-W.; Tan, B.; Wang, X.-Y. Relationship between Intestinal Microbiota and Ulcerative Colitis: Mechanisms and Clinical Application of Probiotics and Fecal Microbiota Transplantation. *World J. Gastroenterol.* **2018**, *24*, 5–14. [CrossRef]
31. Liu, Z.; Dai, X.; Zhang, H.; Shi, R.; Hui, Y.; Jin, X.; Zhang, W.; Wang, L.; Wang, Q.; Wang, D.; et al. Gut Microbiota Mediates Intermittent-Fasting Alleviation of Diabetes-Induced Cognitive Impairment. *Nat. Commun.* **2020**, *11*, 855. [CrossRef] [PubMed]
32. Smith, P.M.; Howitt, M.R.; Panikov, N.; Michaud, M.; Gallini, C.A.; Bohlooly-Y, M.; Glickman, J.N.; Garrett, W.S. The Microbial Metabolites, Short-Chain Fatty Acids, Regulate Colonic Treg Cell Homeostasis. *Science* **2013**, *341*, 569–573. [CrossRef] [PubMed]
33. Peng, Y.; Yan, Y.; Wan, P.; Chen, D.; Ding, Y.; Ran, L.; Mi, J.; Lu, L.; Zhang, Z.; Li, X.; et al. Gut Microbiota Modulation and Anti-Inflammatory Properties of Anthocyanins from the Fruits of Lycium Ruthenicum Murray in Dextran Sodium Sulfate-Induced Colitis in Mice. *Free Radic. Biol. Med.* **2019**, *136*, 96–108. [CrossRef] [PubMed]
34. Mo, J.; Ni, J.; Zhang, M.; Xu, Y.; Li, Y.; Karim, N.; Chen, W. Mulberry Anthocyanins Ameliorate DSS-Induced Ulcerative Colitis by Improving Intestinal Barrier Function and Modulating Gut Microbiota. *Antioxidants* **2022**, *11*, 1674. [CrossRef] [PubMed]
35. Rangan, P.; Choi, I.; Wei, M.; Navarrete, G.; Guen, E.; Brandhorst, S.; Enyati, N.; Pasia, G.; Maesincee, D.; Ocon, V.; et al. Fasting-Mimicking Diet Modulates Microbiota and Promotes Intestinal Regeneration to Reduce Inflammatory Bowel Disease Pathology. *Cell Rep.* **2019**, *26*, 2704–2719.e6. [CrossRef]
36. Alipour, M.; Zaidi, D.; Valcheva, R.; Jovel, J.; Martínez, I.; Sergi, C.; Walter, J.; Mason, A.L.; Wong, G.K.-S.; Dieleman, L.A.; et al. Mucosal Barrier Depletion and Loss of Bacterial Diversity Are Primary Abnormalities in Paediatric Ulcerative Colitis. *J. Crohn's Colitis* **2015**, *10*, 462–471. [CrossRef]
37. Zou, Q.; Zhang, X.; Liu, X.; Li, Y.; Tan, Q.; Dan, Q.; Yuan, T.; Liu, X.; Liu, R.H.; Liu, Z. Ficus Carica Polysaccharide Attenuates DSS-Induced Ulcerative Colitis in C57BL/6 Mice. *Food Funct.* **2020**, *11*, 6666–6679. [CrossRef]
38. Zhang, X.; Zou, Q.; Zhao, B.; Zhang, J.; Zhao, W.; Li, Y.; Liu, R.; Liu, X.; Liu, Z. Effects of Alternate-Day Fasting, Time-Restricted Fasting and Intermittent Energy Restriction DSS-Induced on Colitis and Behavioral Disorders. *Redox Biol.* **2020**, *32*, 101535. [CrossRef]
39. Yan, S.; Shi, R.; Li, L.; Ma, S.; Zhang, H.; Ye, J.; Wang, J.; Pan, J.; Wang, Q.; Jin, X.; et al. Mannan Oligosaccharide Suppresses Lipid Accumulation and Appetite in Western-Diet-Induced Obese Mice Via Reshaping Gut Microbiome and Enhancing Short-Chain Fatty Acids Production. *Mol. Nutr. Food Res.* **2019**, *63*, e1900521. [CrossRef]
40. Yao, D.; Dong, M.; Dai, C.; Wu, S. Inflammation and Inflammatory Cytokine Contribute to the Initiation and Development of Ulcerative Colitis and Its Associated Cancer. *Inflamm. Bowel Dis.* **2019**, *25*, 1595–1602. [CrossRef]
41. Samarghandian, S.; Azimi-Nezhad, M.; Farkhondeh, T.; Samini, F. Anti-Oxidative Effects of Curcumin on Immobilization-Induced Oxidative Stress in Rat Brain, Liver and Kidney. *Biomed. Pharmacother.* **2017**, *87*, 223–229. [CrossRef] [PubMed]
42. Zhu, L.; Han, J.; Li, L.; Wang, Y.; Li, Y.; Zhang, S. Claudin Family Participates in the Pathogenesis of Inflammatory Bowel Diseases and Colitis-Associated Colorectal Cancer. *Front. Immunol.* **2019**, *10*, 1441. [CrossRef] [PubMed]
43. Segain, J.P.; Raingeard de la Blétière, D.; Bourreille, A.; Leray, V.; Gervois, N.; Rosales, C.; Ferrier, L.; Bonnet, C.; Blottière, H.M.; Galmiche, J.P. Butyrate Inhibits Inflammatory Responses through NFkappaB Inhibition: Implications for Crohn's Disease. *Gut* **2000**, *47*, 397–403. [CrossRef]
44. Zhao, L.; Zhu, X.; Yu, Y.; He, L.; Li, Y.; Zhang, L.; Liu, R. Comprehensive Analysis of the Anti-Glycation Effect of Peanut Skin Extract. *Food Chem.* **2021**, *362*, 130169. [CrossRef]
45. Fernandes, A.C.F.; Vieira, N.C.; de Santana, Á.L.; de Pádua Gandra, R.L.; Rubia, C.; Castro-Gamboa, I.; Macedo, J.A.; Macedo, G.A. Peanut Skin Polyphenols Inhibit Toxicity Induced by Advanced Glycation End-Products in RAW264.7 Macrophages. *Food Chem. Toxicol.* **2020**, *145*, 111619. [CrossRef] [PubMed]
46. Xiang, L.; Wu, Q.; Osada, H.; Yoshida, M.; Pan, W.; Qi, J. Peanut Skin Extract Ameliorates the Symptoms of Type 2 Diabetes Mellitus in Mice by Alleviating Inflammation and Maintaining Gut Microbiota Homeostasis. *Aging* **2020**, *12*, 13991–14018. [CrossRef] [PubMed]

47. Gam, D.H.; Hong, J.W.; Kim, J.H.; Kim, J.W. Skin-Whitening and Anti-Wrinkle Effects of Bioactive Compounds Isolated from Peanut Shell Using Ultrasound-Assisted Extraction. *Molecules* **2021**, *26*, 1231. [CrossRef] [PubMed]
48. Li, K.; Zhang, M.; Chen, H.; Peng, J.; Jiang, F.; Shi, X.; Bai, Y.; Jian, M.; Jia, Y. Anthocyanins from Black Peanut Skin Protect against UV-B Induced Keratinocyte Cell and Skin Oxidative Damage through Activating Nrf 2 Signaling. *Food Funct.* **2019**, *10*, 6815–6828. [CrossRef]
49. Chen, L.; Yan, F.; Chen, W.; Zhao, L.; Zhang, J.; Lu, Q.; Liu, R. Procyanidin from Peanut Skin Induces Antiproliferative Effect in Human Prostate Carcinoma Cells DU145. *Chem. Biol. Interact.* **2018**, *288*, 12–23. [CrossRef]
50. Xiang, L.; Wu, Q.; Cheng, L.; Sun, K.; Li, J.; Yoshida, M.; Qi, J. Leptin and Adiponectin Signaling Pathways Are Involved in the Antiobesity Effects of Peanut Skin Extract. *Oxid. Med. Cell. Longev.* **2019**, *2019*, 2935315. [CrossRef]
51. Bansode, R.R.; Plundrich, N.J.; Randolph, P.D.; Lila, M.A.; Williams, L.L. Peanut Flour Aggregation with Polyphenolic Extracts Derived from Peanut Skin Inhibits IgE Binding Capacity and Attenuates RBL-2H3 Cells Degranulation via MAPK Signaling Pathway. *Food Chem.* **2018**, *263*, 307–314. [CrossRef] [PubMed]
52. Liu, X.; Chen, T.; Wang, Q.; Liu, J.; Lu, Y.; Shi, Y. Structure Analysis and Study of Biological Activities of Condensed Tannins from Bruguiera Gymnorhiza (L.) Lam and Their Effect on Fresh-Cut Lotus Roots. *Molecules* **2021**, *26*, 1369. [CrossRef] [PubMed]
53. Hellström, J.K.; Mattila, P.H. HPLC Determination of Extractable and Unextractable Proanthocyanidins in Plant Materials. *J. Agric. Food Chem.* **2008**, *56*, 7617–7624. [CrossRef]
54. Longo, E.; Rossetti, F.; Merkyte, V.; Boselli, E. Disambiguation of Isomeric Procyanidins with Cyclic B-Type and Non-Cyclic A-Type Structures from Wine and Peanut Skin with HPLC-HDX-HRMS/MS. *J. Am. Soc. Mass Spectrom.* **2018**, *29*, 2268–2277. [CrossRef]
55. Donovan, J.L.; Lee, A.; Manach, C.; Rios, L.; Morand, C.; Scalbert, A.; Rémésy, C. Procyanidins Are Not Bioavailable in Rats Fed a Single Meal Containing a Grapeseed Extract or the Procyanidin Dimer B3. *Br. J. Nutr.* **2002**, *87*, 299–306. [CrossRef]
56. Yang, H.; Tuo, X.; Wang, L.; Tundis, R.; Portillo, M.P.; Simal-Gandara, J.; Yu, Y.; Zou, L.; Xiao, J.; Deng, J. Bioactive Procyanidins from Dietary Sources: The Relationship between Bioactivity and Polymerization Degree. *Trends Food Sci. Technol.* **2021**, *111*, 114–127. [CrossRef]
57. Huang, B.; Wang, L.; Liu, M.; Wu, X.; Lu, Q.; Liu, R. The Underlying Mechanism of A-Type Procyanidins from Peanut Skin on DSS-Induced Ulcerative Colitis Mice by Regulating Gut Microbiota and Metabolism. *J. Food Biochem.* **2022**. [CrossRef]
58. Fan, F.Y.; Sang, L.X.; Jiang, M.; McPhee, D.J. Catechins and Their Therapeutic Benefits to Inflammatory Bowel Disease. *Molecules* **2017**, *22*, 484. [CrossRef]
59. Fraga, C.G.; Oteiza, P.I. Dietary Flavonoids: Role of (-)-Epicatechin and Related Procyanidins in Cell Signaling. *Free Radic. Biol. Med.* **2011**, *51*, 813–823. [CrossRef]
60. Dryden, G.W.; Lam, A.; Beatty, K.; Qazzaz, H.H.; McClain, C.J. A Pilot Study to Evaluate the Safety and Efficacy of an Oral Dose of (-)-Epigallocatechin-3-Gallate-Rich Polyphenon E in Patients with Mild to Moderate Ulcerative Colitis. *Inflamm. Bowel Dis.* **2013**, *19*, 1904–1912. [CrossRef]
61. Tilg, H.; Zmora, N.; Adolph, T.E.; Elinav, E. The Intestinal Microbiota Fuelling Metabolic Inflammation. *Nat. Rev. Immunol.* **2020**, *20*, 40–54. [CrossRef] [PubMed]
62. Gryaznova, M.V.; Solodskikh, S.A.; Panevina, A.V.; Syromyatnikov, M.Y.; Dvoretskaya, Y.D.; Sviridova, T.N.; Popov, E.S.; Popov, V.N. Study of Microbiome Changes in Patients with Ulcerative Colitis in the Central European Part of Russia. *Heliyon* **2021**, *7*, e06432. [CrossRef] [PubMed]
63. Liu, P.; Gao, M.; Liu, Z.; Zhang, Y.; Tu, H.; Lei, L.; Wu, P.; Zhang, A.; Yang, C.; Li, G.; et al. Gut Microbiome Composition Linked to Inflammatory Factors and Cognitive Functions in First-Episode, Drug-Naive Major Depressive Disorder Patients. *Front. Neurosci.* **2021**, *15*, 800764. [CrossRef] [PubMed]
64. Guo, N.; Zhou, L.-X.; Meng, N.; Shi, Y.-P. Associations of Oral and Intestinal Florae and Serum Inflammatory Factors with Pathogenesis of Oral Cancer. *Eur. Rev. Med. Pharmacol. Sci.* **2020**, *24*, 11090–11095. [CrossRef]
65. Serrano-Villar, S.; Sanchez-Carrillo, S.; Talavera-Rodríguez, A.; Lelouvier, B.; Gutiérrez, C.; Vallejo, A.; Servant, F.; Bernadino, J.I.; Estrada, V.; Madrid, N.; et al. Blood Bacterial Profiles Associated With Human Immunodeficiency Virus Infection and Immune Recovery. *J. Infect. Dis.* **2021**, *223*, 471–481. [CrossRef]
66. Mowat, A.M.; Agace, W.W. Regional Specialization within the Intestinal Immune System. *Nat. Rev. Immunol.* **2014**, *14*, 667–685. [CrossRef]
67. Zhao, S.; Peng, X.; Zhou, Q.; Huang, Y.; Rao, X.; Tu, J.; Xiao, H.; Liu, D. Bacillus Coagulans 13002 and Fructo-Oligosaccharides Improve the Immunity of Mice with Immunosuppression Induced by Cyclophosphamide through Modulating Intestinal-Derived and Fecal Microbiota. *Food Res. Int.* **2021**, *140*, 109793. [CrossRef]
68. Nankova, B.B.; Agarwal, R.; MacFabe, D.F.; La Gamma, E.F. Enteric Bacterial Metabolites Propionic and Butyric Acid Modulate Gene Expression, Including CREB-Dependent Catecholaminergic Neurotransmission, in PC12 Cells–Possible Relevance to Autism Spectrum Disorders. *PLoS One* **2014**, *9*, e103740. [CrossRef]
69. Parker, B.J.; Wearsch, P.A.; Veloo, A.C.M.; Rodriguez-Palacios, A. The Genus Alistipes: Gut Bacteria With Emerging Implications to Inflammation, Cancer, and Mental Health. *Front. Immunol.* **2020**, *11*, 906. [CrossRef]
70. Polansky, O.; Sekelova, Z.; Faldynova, M.; Sebkova, A.; Sisak, F.; Rychlik, I. Important Metabolic Pathways and Biological Processes Expressed by Chicken Cecal Microbiota. *Appl. Environ. Microbiol.* **2015**, *82*, 1569–1576. [CrossRef]

antioxidants

MDPI

Article

Quantitative Chemical Composition, Anti-Oxidant Activity, and Inhibition of TNF Release by THP-1 Cells Induced by Extracts of *Echinodorus macrophyllus* and *Echinodorus grandiflorus*

Marina Pereira Rocha [1], Lyandra Maciel Cabral da Silva [1], Laura Paulino Maia Silva [1], José Hugo de Sousa Gomes [1], Rodrigo Maia de Pádua [1], João Aguiar Nogueira Batista [2], Marcelo Martins Sena [3,4], Priscilla Rodrigues Valadares Campana [1] and Fernão Castro Braga [1,*]

[1] Departament of Pharmaceutical Products, Faculty of Pharmacy, Universidade Federal de Minas Gerais, Campus Pampulha, Avenida Antônio Carlos 6627, Belo Horizonte 31270-901, MG, Brazil; procha.marina@gmail.com (M.P.R.); yandramcabral@gmail.com (L.M.C.d.S.); lpmaias@hotmail.com (L.P.M.S.); josehugoufmg@gmail.com (J.H.d.S.G.); rodrigomaiapadua@farmacia.ufmg.br (R.M.d.P.); prvcampana@gmail.com (P.R.V.C.)

[2] Departament of Botany and Molecular Biology, Universidade Federal de Minas Gerais, Campus Pampulha, Avenida Antônio Carlos 6627, Belo Horizonte 31270-901, MG, Brazil; janb@icb.ufmg.br

[3] Chemistry Department, Universidade Federal de Minas Gerais, Av. Antônio Carlos 6627, Belo Horizonte 31270-901, MG, Brazil; marcsen@ufmg.br

[4] Instituto Nacional de Ciência e Tecnologia em Bioanalítica (INCT-Bio), Campinas 13083-970, SP, Brazil

* Correspondence: fernaobraga@farmacia.ufmg.br; Tel.: +55-31-3409-6951; Fax: +55-31-3409-6935

Citation: Rocha, M.P.; da Silva, L.M.C.; Silva, L.P.M.; Gomes, J.H.d.S.; Pádua, R.M.d.; Batista, J.A.N.; Sena, M.M.; Campana, P.R.V.; Braga, F.C. Quantitative Chemical Composition, Anti-Oxidant Activity, and Inhibition of TNF Release by THP-1 Cells Induced by Extracts of *Echinodorus macrophyllus* and *Echinodorus grandiflorus*. *Antioxidants* **2023**, *12*, 1365. https://doi.org/10.3390/antiox12071365

Academic Editors: Dimitrios Kouretas, Jicheng Zhan, Zhigang Liu and Hui-Min David Wang

Received: 17 May 2023
Revised: 6 June 2023
Accepted: 14 June 2023
Published: 29 June 2023

Abstract: This study investigated the similarities between *Echinodorus macrophyllus* and *Echinodorus grandiflorus*, plant species that are traditionally used in Brazil to treat rheumatism and arthritis, whose anti-inflammatory effects are supported by scientific evidence. The contents of *cis*- and *trans*-aconitic acid, homoorientin, chicoric acid, swertisin, caffeoyl-feruloyl-tartaric acid, and di-feruloyl-tartaric acid were quantified by UPLC-DAD in various hydroethanolic extracts from the leaves, whereas their anti-oxidant activity and their effect on TNF release by LPS-stimulated THP-1 cells were assessed to evaluate potential anti-inflammatory effects. The 50% and 70% ethanol extracts showed higher concentrations of the analyzed markers in two commercial samples and a cultivated specimen of *E. macrophyllus*, as well as in a commercial lot of *E. grandiflorus*. However, distinguishing between the species based on marker concentrations was not feasible. The 50% and 70% ethanol extracts also exhibited higher biological activity, yet they did not allow differentiation between the species, indicating similar chemical composition and biological effects. Principal component analysis highlighted comparable chemical composition and biological activity among the commercial samples of *E. macrophyllus*, while successfully distinguishing the cultivated specimen from the commercial lots. In summary, no differences were observed between the two species in terms of the evaluated chemical markers and biological activities.

Keywords: *Echinodorus grandiflorus*; *Echinodorus macrophyllus*; chemical markers; molecular analysis; anti-oxidant activity; inhibition of TNF release; principal component analysis

1. Introduction

Echinodorus macrophyllus (Kunth) Micheli (syn. *Aquarius macrophyllus* (Kunth) Christenh. & Byng., *Alisma macrophyllum* Kunth) and *Echinodorus grandiflorus* (Cham & Schltdl) Micheli (syn. *Aquarius grandiflorus* (Cham. & Schltdl.) Christenh. & Byng, *Alisma grandiflorum* Cham. & Schltdl.) (Alismataceae) are plant species popularly known as *chapéu-de-couro* in Brazil. The aerial parts of these species are traditionally used to treat rheumatism and arthritis in the country, among other uses [1,2]. Both species occur in tropical regions of South America [3], being predominantly found in the northeast and central-west regions of Brazil, mainly in floodplains, swamps, and coastal areas [4].

The first and second editions of the *Brazilian Pharmacopeia* contain a monograph of *E. macrophyllus* [5,6], while a monograph of *E. grandiflorus* is found in the fifth and sixth editions [7,8]. These two species are utilized for similar medicinal purposes in various regions of Brazil as teas or herbal preparations [2]. The interchangeable use of *E. macrophyllus* and *E. grandiflorus* may be attributed to difficulties in distinguishing them based on morphological characteristics. Both species are aquatic plants with large leathery leaves, which have led to their popular name *chapéu-de-couro* (leather hat) [9]. *E. macrophyllus* and *E. grandiflorus* share inflorescences with delicate white flowers, which are also used for decorative purposes, and their leaf and floral anatomy do not exhibit noticeable differences to non-specialists. The presence of phenotypic plasticity and the striking similarity in morphological structures [10] makes it challenging to differentiate these species unequivocally.

The chemical composition of leaves from *E. macrophyllus* and *E. grandiflorus* has been extensively investigated, revealing similarities that include the presence of diterpenes, flavone C-glycosides, hydroxycinnamoyl tartaric acid derivatives, alkaloids, saponins, and phenolic acids [11–14]. Moreover, the observed biological activities of these species have been linked to their chemical composition. For instance, a flavonoid-rich fraction from *E. macrophyllus* exhibited a more potent anti-inflammatory effect than the extract in the air pouch model in mice, reducing leukotriene B_4 release and neutrophil migration in vitro [11]. The anti-arthritic activity of *E. grandiflorus* leaves has been attributed to flavone C-glycosides, as evidenced by the potent effect observed when administering a fraction enriched in these compounds in an antigen-induced arthritis model in mice [13]. Additionally, the flavonoid-rich fraction elicited the most potent inhibition of tumor necrosis factor (TNF) release by lipopolysaccharide (LPS)-stimulated THP-1 cells, along with its constituents isovitexin and isoorientin [12]. Other classes of metabolites present in these species, such as diterpenes, hydroxycinnamoyl tartaric acid derivatives, and *cis*- and *trans*-aconitic acid, have also been reported for their anti-inflammatory activity by inhibiting TNF release in LPS-stimulated THP-1 cells [12].

Although the leaves of *E. macrophyllus* and *E. grandiflorus* are consumed locally for medicinal purposes, there are no clear-cut differences based on morphological characteristics or chemical composition that can be used to unambiguously differentiate them. Therefore, the aim of this study was to undertake molecular, chemical, and biological investigations of different extracts prepared from *E. macrophyllus* and *E. grandiflorus* leaves in order to highlight similarities and differences between these species. The extracts were analyzed by UPLC-DAD (ultra-efficiency chromatography coupled to diode array detector) for the quantification of chemical markers (*cis*- and *trans*-aconitic acid, homoorientin, chicoric acid, swertisin, caffeoyl-feruloyl-tartaric acid, and di-feruloyl-tartaric acid), using a method developed and validated in the present work. The extracts had their anti-oxidant activity assayed in vitro, along with their effect on the release of TNF by LPS-stimulated THP-1 cells, to access their potential anti-inflammatory effect. Finally, a multi-variate exploratory principal component analysis (PCA) model [15,16] was built, aiming to investigate the possible existence of correlations between the quantitative chemical composition of the extracts and their biological effects.

2. Material and Methods

2.1. Plant Drug Samples

The leaves of *E. macrophyllus* and *E. grandiflorus* were obtained from Indústria Farmacêutica Catedral (Belo Horizonte, Brazil). Two lots of the first species were purchased and labeled as DV1 (April 2014) and DV2 (March 2019), along with a lot of the second species identified as EG (November 2019). These plant drugs were provided dried and underwent quality control checks as indicated in the supplier's reports. Upon receipt, the leaves were pulverized using a knife mill (Marconi, São Paulo, Brazil) and stored in glass flasks for further use. A third sample of *E. macrophyllus* aerial parts was collected at the Natural History Museum of the Universidade Federal de Minas Gerais (UFMG) and a voucher was incorporated into the BHCB herbarium at UFMG, under the number BHCB 28557. After drying in a ventilated oven at 40 °C, this plant (drug-coded DVMus) was pulverized in a knife mill and stored in a glass flask.

2.2. Chemicals and Reagents

All chemicals, reagents, and reference compounds used were of analytical or LC-MS (liquid chromatography coupled to mass spectrometer)/UPLC grade. The water was filtered through a Milli-Q water purification system (Millipore, Bedford, MA, USA) before use. *trans*-aconitic acid (purity > 99.0%), chicoric acid (purity > 98%), and homoorientin (purity > 99%) were purchased from Extrasynthese, France; *cis*-aconitic acid (purity > 98%), 1,1-diphenyl-2-picryl-hydrazyl-hydrate (DPPH), pyrogallol (purity > 98%), quercetin (purity > 95%), β-carotene (purity > 93%), lipopolysaccharide from *Escherichia coli* (Sigma-Aldrich, St. Louis, MO, USA), *o*-phenylenediamine (OPD), dexamethasone (purity > 98%), and 3-(4,5-dimethyl-thiazol-2-yl)-2,5-diphenyltetrazolium bromide were obtained from Sigma-Aldrich (St. Louis, MO, USA). Swertisin (purity > 95%) was kindly donated by Prof. Eloir Paulo Schenkel (Universidade Federal de Santa Catarina, Brazil). Human TNF-α DY210 Duo Set was acquired from R&D Systems (Minneapolis, MN, USA).

2.3. DNA Isolation, Amplification, Sequencing, Alignment, and Pairwise Distance Calculation

Total genomic DNA was extracted from fresh leaves using a modified version of the 2 × CTAB protocol of Ref. [17]. For PCR amplifications, we used primers described in the Supplementary Table S1 and the parameters previously described [10,18], with minor modifications. The amplified and purified products were sequenced on both DNA strands using the same PCR primers by Macrogen Inc. (Seoul, Korea). The DNA sequence electropherograms were edited and a consensus generated through the Staden program [19]. The sequences were aligned using the MUSCLE program [20] and thereafter manually adjusted using the MEGA7 program [21], according to the procedures described by Ref. [22]. Pairwise distances of the combined dataset (47 terminals and 3182 base pairs) between the major groups recovered in the phylogenetic analyses were calculated using the *p*-distance method in MEGA7 [21]. Gaps and missing data were treated with the pairwise deletion option.

2.4. Molecular Markers, Taxon Sampling, and Molecular Phylogenetic Analyses

Nucleotide sequences from three nuclear genome regions, the nrITS consisting of ITS1, ITS2 and the intervening 5.8S gene, the 5S non-transcribed region (5S-NTS) and the second intron of *LEAFY*, and two plastid regions (*matK-trnK* and *psbA-trnH*) were used in the analysis. Marker selection was based on previous molecular phylogenetic analyses of *Echinodorus* [10,18]. The taxon sampling was based on a selection of taxa available in GenBank selected to cover the taxonomic diversity of *Echinodorus* and included 38 taxa and 47 terminals. Voucher information, geographic origins, and GenBank accession numbers are provided in Supplementary Table S2. To root the trees, we used *Echinodorus berteroi*, which was recovered as sister to the other species of the genus according to the results of Lehtonen and [10,18]. Most species of *Echinodorus*, except *E. berteroi* (the type species of the genus), have recently been transferred to the genus *Aquarius* Christenh. & Byng [23]. An evaluation of this proposal was beyond the objectives of this study and we followed the previous classification here, keeping all species in *Echinodorus*.

To compare the outcomes of different methodologies in phylogenetic analysis, we employed both maximum parsimony (MP) and Bayesian inference (BI) techniques. Our investigation focused on evaluating a model-free nucleotide substitution approach (MP) in comparison with a model-based approach (BI). We conducted an extensive examination of individual markers and a merged dataset to detect any potential inconsistencies. For the maximum parsimony analysis, we utilized PAUP* version 4 [24] and implemented Fitch parsimony with equal weights for unordered characters (Fitch, 1971) as the criterion for determining optimality. Each search involved 2000 replicates, with random taxon additions; branch swapping was carried out using the tree-bisection and reconnection (TBR) algorithm.

To mitigate excessive swapping on suboptimal islands, we stored a maximum of 10 trees per replicate. To evaluate the internal support, we employed character bootstrapping [25], with 2000 replicates using simple addition, and TBR branch swapping. Throughout the bootstrapping process, we retained up to 15 trees per replicate. The assessment of bootstrap support levels relied on bootstrap percentages (BP), where values falling between 50% and 70% were classified as weak, 71% and 85% as moderate, and values exceeding 85% as strong [26].

Bayesian analysis using MrBayes 3.2.7a [27] was implemented in the Cyberinfrastructure for Phylogenetic Research (CIPRES) Portal 2.0 (San Diego Supercomputer Center, La Jolla, CA, USA [28], treating each DNA region as a separate partition. An evolutionary model for each DNA region was selected in MrModeltest 2 [29] using the hierarchical likelihood ratio tests (hLRTs). To estimate model parameters for each partition separately, we utilized the unlink command. Our analysis involved conducting two independent runs, each with four chains, spanning a total of 10,000,000 generations. During the Markov chain Monte Carlo (MCMC) sampling, we sampled one tree every 1000 generations, employing a temperature parameter of 0.2. We assessed convergence between the runs using multiple criteria. The average standard deviation of split frequencies was evaluated and confirmed to be less than 0.01, indicating convergence. Additionally, we calculated the potential scale reduction factor (PSRF), which yielded a value of 1.0, further validating convergence. Convergence was achieved after 1,970,000 generations. We discarded the initial 2500 trees (25%) as burn-in and utilized the remaining trees to evaluate the topology and to calculate posterior probabilities (PP) through a 50% majority-rule consensus. It is worth noting that, in Bayesian analysis, posterior probabilities cannot be directly compared with bootstrap percentages (BP), as PP values tend to be higher [30]. To evaluate support, we applied criteria similar to a standard statistical test. Groups with a PP value exceeding 0.95 were considered strongly supported, while PP values ranging from 0.90 to 0.95 were categorized as moderately supported. Groups with PP values below 0.90 were considered weakly supported.

2.5. *Preparation of Extracts*

Portions of 1.0 g of the dried leaves of *E. macrophyllus* (DV1, DV2, DVMus) and *E. grandiflorus* were sonicated with 150 mL of 96 °GL EtOH or with hydroethanolic solutions at 90% (90% EtOH *v/v*), 70% (70% EtOH *v/v*), and 50% (50% EtOH *v/v*) in 3 cycles of 10 min each. The ethanol content of the hydroethanolic solutions was adjusted with the aid of an alcoholmeter. The obtained extracts were filtered, pooled, and concentrated in a rotary evaporator, under reduced pressure, at an average temperature of 55 °C. Subsequently, the hydroethanolic extracts were lyophilized for complete elimination of residual water. The extracts were transferred to glass flasks and kept in a desiccator until constant weight for yield calculation.

2.6. *Quantification of Chemical Markers by UPLC-DAD*

The contents of the chemical markers *cis*- and *trans*-aconitic acid, homoorientin, chicoric acid, swertisin, caffeoyl-feruloyl-tartaric acid, and di-feruloyl-tartaric acid were quantified in the extracts described in Section 2.5, using an UPLC-DAD method developed and validated by us.

2.6.1. Sample Preparation and Chromatographic and Analytical Conditions

For the analysis, portions (2.00 mg) of the samples were dissolved with 1.0 mL of MeOH/H_2O (1:1 *v/v*) for 70% EtOH and 50% EtOH extracts or solubilized in methanol for 96 °GL EtOH and 90% EtOH extracts. The extracts were sonicated in an ultrasound bath for 10 min, at room temperature. Thereafter, the extracts were centrifuged at 8400× *g* for 10 min and filtered through a 0.22 μm membrane of PTFE; the obtained supernatants (5 μL) were injected onto the UPLC system.

A *Waters Acquity UPLC* H-Class Bio System (2015 *model*) system composed of a quaternary pump, an autosampler, a photodiode array detector (DAD) 2996, and a Waters Empower pro data handling system was employed (Waters Corporation, Milford, MA, USA). The analyses were performed on a BEH Shield RP-18 column (100 × 2.1 mm i.d., 1.7 μm; Waters) in combination with a BEH Shield RP-18 guard column (4 × 4 mm i.d., 1.7 μm; Waters) at 40 °C. Water (A) and acetonitrile (B) were used as eluents, both containing 0.1% (*v*/*v*) of formic acid, at a flow rate of 0.3 mL/min. A segmented gradient was employed for elution, comprising concave (type 7) and linear gradient steps, along with isocratic elution steps, as follows: 0 min, 4% B; 1.5 min, 15% B (type 7 gradient); 2.5 min, 16.2% B (linear gradient); 3.5 min, 16.2% B; 4.0 min, 17.4% B (type 7 gradient); 5.0 min, 17.4% B (isocratic); 8.5 min, 20.6% B (type 7 gradient); 9.5 min, 20.6% B (isocratic); 10.5 min, 21.3% B (linear gradient); 11.5 min, 21.3% B (isocratic); 30.5 min, 90% B (type 7 gradient); 32.5 min, 95% B (linear gradient); 33.5 min, 95% B (isocratic). A reverse gradient was adopted to return to the initial elution conditions (4% B) in 1 min; a 5-min equilibrium time was adopted between injections. UV spectra from 190 to 400 nm were recorded online; chromatograms were extracted at 350 nm to quantify flavonoids and tartaric acid derivatives and at 220 nm for aconitic acids quantitation.

2.6.2. Identification of Chemical Markers by LC-MS Analysis

The identity of the chemical markers found in the extracts was confirmed by ultra-performance liquid chromatography with electrospray ionization and *triple quadrupole* mass spectrometry (UPLC-ESI-MS) analysis, carried out in a Waters *Acquity* UPLC system (Waters, Milford, MA, USA), composed of a binary pump, an auto sampler, an in-line degasser, a photodiode array detector (Waters), and a mass spectrometer Xeco™ Triple Quadrupole MS (Waters). Data were processed using MassLynx4.1 (Waters). The chromatographic conditions were those described for the UPLC-DAD method (Section 2.6.1). The mass spectrometer was operated in the exploratory mode (scan), both at the negative and positive ionization modes using the following conditions: capillary voltage of 3.54 kV; 10–70 V ramp cone voltage; source temperature 120 °C; desolvation temperature 450 °C. Mass/charge ratios (*m*/*z*) from 100 to 900 were evaluated. The samples were injected at 1 mg/mL, obtained by solubilizing 1 mg of each extract in 1.0 mL of methanol HPLC grade. Extracts were sonicated in an ultrasound bath for 10 min and centrifuged at 8400× *g* for 10 min; the supernatants were filtered through a PVDF membrane (0.22 μm) and automatically injected into the UPLC-ESI-MS system.

2.6.3. Validation of the UPLC-DAD Method

The method was validated according to the guidelines of Refs. [31–33]. The 70% EtOH extract was selected for validation studies because of its higher complexity compared with other extracts.

Selectivity and system suitability. Purity of peaks from the analytes was evaluated by UV spectra recorded by DAD at different points, in chromatograms of sample solutions, and analyzed by the relationship between the purity angle and the threshold. System suitability was assessed by the outcomes for resolution, retention factor, tailing, number of theoretical plates, and RSD (relative standard deviation) of the retention time of peaks, calculated by Empower 3 software (Waters, Milford, MA, USA).

Linearity, precision, limit of quantification (LOQ), and limit of detection (LOD). Standard solutions were prepared for each compound using a 1:1 mixture of *cis*- and *trans*-aconitic acid (6.25, 12.5, 25.0, 50.0, and 100.0 μg/mL), homoorientin (1.25, 2.5, 5.0, 10.0, and 20.0 μg/mL), chicoric acid (9.37, 18.75, 37.5, 75.0, 150.0, and 300 μg/mL), and swertisin (30.0, 45.0, 67.0, 100.0, and 150.0 μg/mL). The solvents used for preparing the standard solutions were either MeOH or ultrapure water, depending on the solubility of each substance. Triplicate aliquots (5 μL) of these prepared solutions were injected onto the UPLC-DAD system using the developed method. The calibration curves were constructed by analyzing the injected mass of each compound and quantifying the corresponding peak areas.

The analysis was performed on two independent days and linear regression analysis was employed to determine the calibration curves using Prism 6.01 software (GraphPad, San Diego, CA, USA). The calibration curve of chicoric acid was also used to quantify the contents of caffeoyl-feruloyl-tartaric acid and di-feruloyl-tartaric acid. The intra-day precision was evaluated by calculating the relative standard deviation (RSD) values of sample solutions analyzed on the same day ($n = 6$) at 100% of the analyte concentration. Similarly, the inter-day precision was assessed by analyzing the sample solutions on two different days, performed by two different analysts ($n = 12$). The limits of quantification (LOQ) and detection (LOD) were determined by injecting standard solutions at progressively lower concentrations, in quintuplicate. The LOQ was determined as the concentration at which the peak area exhibited an RSD below 3%, while the LOD was defined as the concentration that yielded a signal-to-noise ratio (S/N) of 3.

Recovery and robustness. Accuracy of the method was accessed by recovery studies.

Recovery was assessed by adding known aliquots of reference compound solutions to the extracts at three concentration levels falling within the calibration curve. To evaluate robustness, six sample solutions were prepared and analyzed using both the established conditions and by altering the following parameters: column temperature (38 and 42 °C), wavelength (352 and 348 nm to flavonoids and phenolic acids; 218 and 222 nm to aconitic acids), and formic acid content in the mobile phase (0.08 and 0.12% *v*/*v*). The concentrations of the analytes were compared by ANOVA, followed by Tukey's test ($p < 0.05$).

2.7. Effect on the Release of TNF In Vitro

The effect of the extracts on the release of TNF was evaluated in vitro in THP-1 cell culture (monocyte lineage derived from human acute monocytic leukemia: ATCC TIB-202) stimulated by LPS. Cell viability was evaluated by the MTT (3-4,5-dimethyl-thiazol-2-yl-2,5-diphenyltetrazolium bromide) assay, as previously described, with modifications [34]. A cell suspension was prepared at 1.0×10^6 cells/mL, obtained by centrifuging a culture flask and sequentially resuspending the pellet with RPMI 10% FBS and 2 µL PMA (1 mg/mL). Then, 100 µL aliquots of this suspension were transferred to the wells of a microplate, which were incubated at 37 °C in a 5% CO_2 atmosphere for 6 h to allow cell differentiation and adhesion. After this period, the cells were starved with RPMI medium, supplemented with 2% FBS, and incubated again under the same conditions for 12 h. The cells were transferred to a 96-well plate at 1×10^6 cells/well and incubated for 24 h in an oven at 37 °C and a humidified atmosphere of 5% CO_2. Subsequently, 100 µL of complete medium with 2% of FBS containing different concentrations of the extracts (60 µg/mL) were added in triplicate and the plate was incubated for 3 h under the conditions described above. Then, 20 µL of LPS (1 ng/mL) were added to the wells and the plate was incubated again for 24 h under the same conditions. After incubation, the microplate was centrifuged (1800× *g* for 5 min) and 100 µL of the supernatant was collected to quantify cytokines by the ELISA method, according to the protocol described in the sequence. The cell pellet was used to assess cell viability by MTT and 28 µL/well of tetrazolium salt (MTT) was added at a concentration of 2 mg/mL in phosphate buffer (PBS). After 90 min of incubation with the MTT, the entire contents were aspirated by a Pasteur pipette, followed by the addition of DMSO to each well (100 µL/well) to solubilize the formed formazan crystals. The microplates were read in a spectrophotometer at 510 nm. Cell viability was calculated according to the equation $(A - B/C - B) \times 100$, where A, B, and C are the absorbance of the samples, blank, and negative control, respectively.

ELISA Protocols for In Vitro Assays

The assay was conducted using an ELISA kit from R&D Systems (DY210-KIT TNF), following the manufacturer's instructions and guidelines. An amount of 100 µL/well of TNF capture antibody (4 µg/mL), solubilized in PBS, was added to a 96-well microplate and the plate was kept at 4 °C for 12 h. After incubation, the plate was washed (PBS/Tween 20, 0.1% *v*/*v*) and completely dried. Thereafter, 200 µL of the blocking solution (1% bovine serum albumin in PBS) was added. After 2 h, the plate was washed and dried as described above. Then, 100 μL of samples or TNF standard solution (1000, 500, 250, 125, 62.5, 31.25, and 15.6, pg/mL) was added. The microplate was incubated for 18 h at 4 °C and then the plates were washed and dried. In the sequence, 100 μL of TNF detection antibody solution (500 ng/mL) was added. After 2 h, the plates were washed and dried and then 100 μL of a solution containing streptavidin bound to peroxidase (R&D system) was added to the microplate wells. The plate was incubated for 30 min and washed and dried as described above. Subsequently, 100 μL of *o*-phenylenediamine (OPD) in citrate buffer (0.4 mg/mL) and 2.4 μL of hydrogen peroxide solution (35% *v*/*v*) were added to the wells. The reaction was stopped after 30 min by adding 50 μL of 1 mol/L hydrochloric acid. The OPD oxidation product was detected in a microplate reader at 490 nm. The concentration of TNF was calculated using the calibration curve.

2.8. *In Vitro Anti-Oxidant Activity*

2.8.1. DPPH Radical Scavenger Activity

The assessment of anti-oxidant activity by the 2,2-diphenyl-1-picrylhydrazyl radical (DPPH) assay was carried out as previously described [35], with adaptations. All extracts were initially tested at 200 μg/mL; subsequently, the active samples were evaluated at 1–200 μg/mL and concentration–response curves were constructed. To carry out the experiments, 250 μL of each sample, methanol (negative control), and pyrogallol (50 μg/mL, positive control) were added to a 96-well plate. Then, 100 μL of methanol (blank wells) and DPPH (120 μg/mL, reaction wells) were added to the microplate. The measurements were carried out using a microplate reader (EL808IU-Biotek, Winooski, VT, USA) at 515 nm, with readings taken every 5 min for a total of 25 min. The percentage of radical scavenging activity (% RSA) was determined using the following formula: % RSA = (AC − AS)/AC) × 100, where AC represents the absorbance of the control and AS represents the absorbance of the samples taken at 35 min. The results were expressed as the EC_{50} value, which was the effective concentration at which 50% of DPPH radicals were scavenged. The EC_{50} was determined through non-linear regression analysis using GraphPad Prism, version 6.0.

2.8.2. β-Carotene/Linoleic Acid Co-Oxidation Assay

The assays were carried out as previously described [36], with minor adaptations. The extracts were solubilized in methanol (2.2 mg/mL) and subsequently diluted from 2.47 to 200 μg/mL (dilution factor 1:3). For the experiments, 25 μL aliquots of the solutions were added to a 96-well plate. A mixture of 25 mg of linoleic acid and 100 mg Tween 20 (Sigma, St. Louis, MO, USA) was added to a round-bottom flask containing 1 mL of a β-carotene solution in chloroform (1 mg/mL, Sigma) and transferred to a rotary evaporator for complete evaporation of the solvent. Then, 50 mL of aerated water was added to the flask, affording the β-carotene emulsion. A blank emulsion was prepared similarly, except for the addition of the β-carotene solution. Aliquots of 250 μL from both emulsions were added to the even and odd columns of the microplate. Methanol was used as a negative control and quercetin (20 μg/mL, Sigma) was used as a positive control. Readings were taken immediately at a wavelength of 470 nm and subsequently at 15 min intervals up to 120 min. Throughout the reading period, the microplate was maintained at an incubation temperature of 45 °C. The anti-oxidant activity was expressed as % of inhibition of lipid peroxidation (%I), using the equation I% = control absorbance (Ac) (initial abs − final abs) − sample absorbance drops (Aam) (final abs − initial abs)/Ac × 100). Assays were performed in triplicate and IC_{50} values were determined by non-linear regression using GraphPad Prism, version 6.0 (GraphPad Software, San Diego, CA, USA).

2.8.3. ROS Activity in THP-1 Cells

The effect of the extracts on the production of reactive oxygen species (ROS) by THP-1 cells was performed according to [37] with adaptations. THP-1 cells were cultured in RPMI medium supplemented with 10% fetal bovine serum (FBS). A cell suspension was prepared at a density of 6×10^5 cells/mL, obtained by centrifuging a culture flask and sequentially resuspending the pellet with RPMI 10% FBS and 4 µL PMA (100 nM), following incubation for 72 h in an oven at 37 °C and 5% CO_2. Thereafter, PMA was washed with RPMI 10% FBS, with subsequent incubation of the cells without adding PMA and 2% of FBS in an oven at 37 °C and 5% CO_2 for 24 h. Subsequently, 1×10^6 cells per well were placed (250 µL) in a 24-well plate and incubated for 24 h in an oven at 37 °C and 5% CO_2. After incubation, the cells were treated with the extracts (60 µg/mL in RPMI with 0.2% DMSO) and the plate was incubated for another 3 h in the previous conditions. Then, the cells were treated with 20 µL of a LPS solution (10 µg/mL) at 15 min intervals up to 75 min. The following controls were added to the plate: wells containing only RPMI medium without LPS, wells containing RPMI medium with LPS, and wells containing non-differentiated THP-1 cells. Quercetin (100 µmol/L in methanol) was used as a positive control. Following LPS stimulation, the cells were incubated with 10 µM H_2DCFDA (2′,7′-dichlorodihydrofluorescein diacetate) for 15 min at 37 °C and 5% CO_2. The supernatants were collected and fluorescence was read in a microplate reader (Varioskan™ LUX, Thermo Scientific, Waltham, MA, USA) at 493 nm for excitation and at 522 nm for emission.

2.9. Statistical Analysis

The concentrations of the analytes were compared using analysis of variance (ANOVA), followed by Tukey's test for post hoc multiple comparisons, with a significance level set at $p < 0.05$. Statistical analysis was performed using GraphPad Prism software, version 6.0 (GraphPad Software Inc.). The results were expressed as mean ± standard deviation (SD) and differences were considered statistically significant when $p < 0.05$.

2.10. PCA Model

Principal component analysis (PCA) was performed employing MATLAB R2010b software (The MathWorks, Natick, MA, USA) jointly with PLS_Toolbox 6.5 (Eigenvector Technologies, Manson, IA, USA).

3. Results and Discussion

3.1. Molecular and Phylogenetic Analysis of E. macrophyllus

The BOLD (Barcode of Life Data System) database of *Echinodorus* is completely based on GenBank sequences and therefore we used BLASTN (nucleotide blast) on the NCBI BLAST homepage [38] to search in the nucleotide collection database and to identify the sequences/taxa with higher identity with the sequences we generated from *E. macrophyllus*. The best match varied between markers (Table 1). For *psbA-trnH*, *matK-trnK*, and Leafy, the best match was with *E. paniculatus*, while, for ITS, the best match was with *E. osiris*, but the differences between the additional hits were very small.

Since an unequivocal match was not possible, we performed phylogenetic analysis to verify which taxa were most related to the sequenced *E. macrophyllus* sample using the ITS, 5S-NTS, Leafy, *matK-trnK*, and *psbA-trnH* markers. Initial analysis with each marker individually did not detect any case of strongly supported incongruence, so a combined matrix was used for the final analysis. Table 2 presents the general features of the datasets and parsimony statistics, along with a summary of the models implemented for each partition. The strict consensus tree from the parsimony analysis and the Bayesian majority-rule consensus tree were for the most part congruent and, as the latter was more fully resolved and had stronger overall support, it was chosen for presentation and discussion (Figure 1). Our analysis recovered three main clades with high support within *Echinodorus*: clade A (PP 1.00, BP 100%), clade B (PP 1.00, BP 98%), and clade C (PP 1.00, BP 62%) (Figure 1). Clade C in turn was divided into three main clades: a clade formed by *E. grisebachii* and *E. heikobleheri* (clade D: PP 1.00, BP 100%), another formed by *E. trialatus*, *E. emersus*, and *E. scaber* (clade E: PP 1.00, BP 100%), and clade F (PP 1.00, BP 100%), which included our sample of *E. macrophyllus* and 20 other taxa. Relationships within clade F were poorly resolved and poorly supported. Our sample of *E. macrophyllus* was recovered as sister to a clade containing *E. osiris* and six other taxa, but with low support (PP 0.85, BP < 50%). By using neither analysis, our *E. macrophyllus* sample was recovered as closest to the *E. macrophyllus* sample available in GenBank. This result was probably due to a combination of factors. First, sequence divergence between *Echinodorus* species was very low. Mean pairwise distances within and between the major clades identified in the molecular phylogenetic analysis (Figure 1) showed low values, particularly within the clades (Table 3). In clade F, for example, which included the *E. macrophyllus* samples, the mean pairwise distance among the sequences of all markers was only 0.007; for some markers, such as *matK-trnK*, these values were even lower (Table 4). These results suggested a recent origin and diversification of some species groups of the genus and indicated that the combination *matK* and *rbcL*, for example, (proposed as the standard for plant DNA barcoding [39]) in most cases would not be sufficient for the discrimination of closely related species of *Echinodorus*. Second, the sampling of *Echinodorus* in nucleotide collection databases was still small relative to the breadth of geographic distribution and morphological variability of the species. For *E. macrophyllus*, for example, distributed throughout Brazil and also in Bolivia [40], there was only one sample available in GenBank. Finally, it was also not possible to rule out the possibility of conflicting identifications between our sample and those of the GenBank.

Table 1. Best matches between the sequences of *Echinodorus macrophyllus* (MHNJB) and the GenBank database using blastn.

Marker	Best Matches	% Identity	Identities bp	Query Cover (%)	Accession No.
ITS	*Echinodorus osiris*	99.56	684/687	100	KT437651
	Echinodorus longiscapus	99.42	683/687	100	KT437653
	Echinodorus portoalegrensis	99.42	683/687	100	KT437648
	Echinodorus barthii	99.27	682/687	100	KT437652
	Echinodorus opacus	99.13	681/687	100	KT437650
psbA-trnH	*Echinodorus paniculatus*	100.00	463/463	100	HM367285
	Echinodorus grandiflorus	99.78	462/463	100	HM367294
	Echinodorus longiscapus	99.78	462/463	100	HM367292
	Echinodorus muricatus	99.57	462/464	100	DQ786514
	Echinodorus cordifolius	99.35	460/463	100	HM367302
matK-trnK	*Echinodorus paniculatus*	99.85	1309/1311	100	OK587808
	Echinodorus cordifolius	99.85	1309/1311	100	OK587807
	Echinodorus paniculatus	99.85	1309/1311	100	EF088097
	Echinodorus longiscapus	99.77	1308/1311	100	EF088112
	Echinodorus grandiflorus	99.69	1307/1311	100	EF088113
Leafy	*Echinodorus cordifolius*	99.07	212/214	100	HM367210
	Echinodorus paniculatus	99.07	212/214	100	EF088144
	Echinodorus uruguayensis	98.60	211/214	100	EF088159
	Echinodorus cordifolius	98.60	211/214	100	EF088172
	Echinodorus bracteatus	98.60	212/215	100	EF088170

Table 2. Characterization of the markers used in the phylogenetic analyses. VI—number of variable informative characters; CI—consistency index; RI—retention index; hLRTs—hierarchical likelihood ratio tests.

Marker	Terminals	Characters	VI (%)	CI	RI	Model (hLRTs)
ITS	37	725	100 (13.8%)	0.71	0.8	GTR+G
5S-NTS	28	266	98 (36.8%)	0.68	0.7	HKY+G
LEAFY	43	306	33 (10.8%)	0.93	0.95	K80+G
matK-trnK	35	1311	164 (12.5%)	0.86	0.92	GTR+G
psbA-trnH	46	574	108 (18.8%)	0.81	0.91	GTR+G
Combined	47	3182	503 (15.8%)	0.73	0.82	

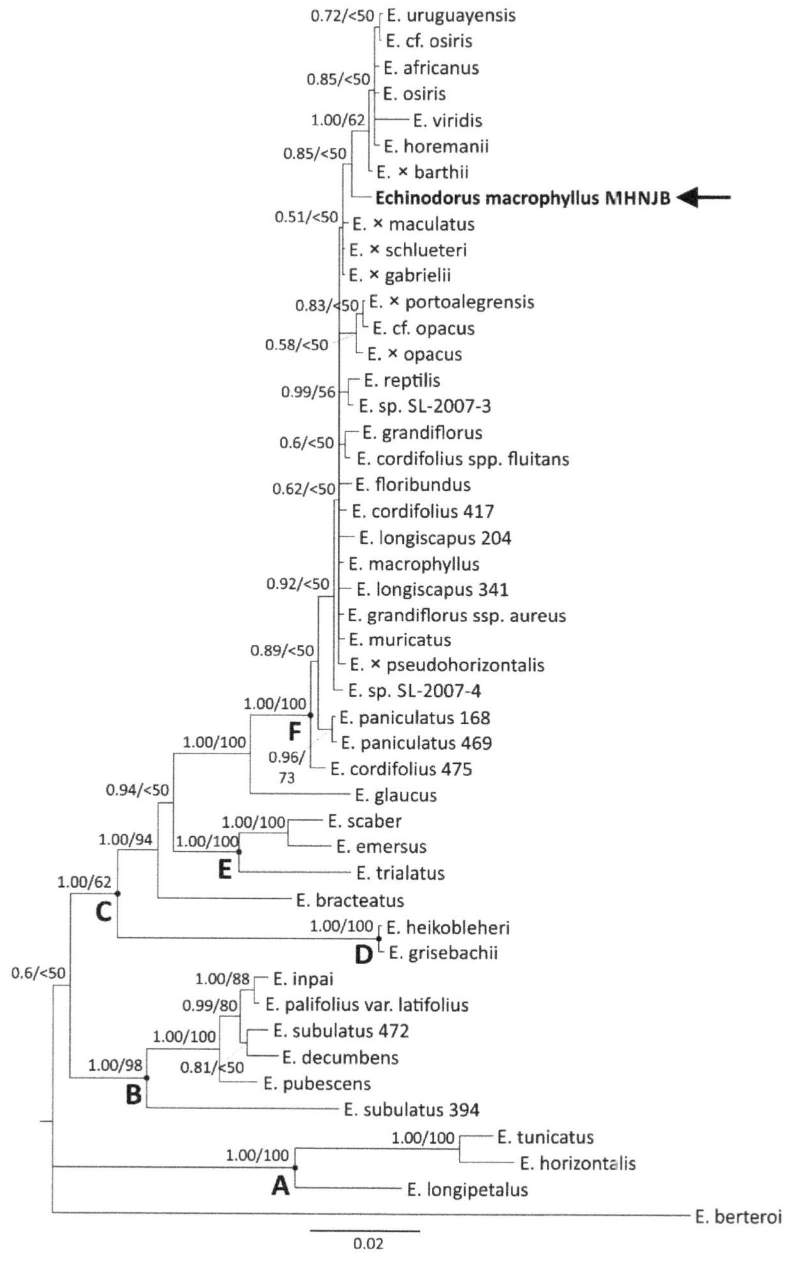

Figure 1. Bayesian 50% majority-rule consensus tree of the combined ITS, 5S-NTS, LEAFY, *matK-trnK*, and *psbA-trnH* datasets. Numbers above the lines, next to the nodes, represent the posterior probabilities (PP) from the Bayesian analysis and bootstrap percentages (BP) from parsimony analysis. Major clades discussed in the text are indicated by the letters A to F. The sequenced sample from *E. macrophyllus* is highlighted in bold and by an arrow. The generic names for all *Echinodorus* species are abbreviated.

Table 3. Mean pairwise distances within and between the major clades recovered in the phylogenetic analysis. Mean distances within groups are highlighted in bold. N indicates the number of terminals in each group. n/c denotes cases in which it was not possible to estimate evolutionary distance.

	N	1	2	3	4	5	6	7	8
1. Clade F	30	**0.007**							
2. *E. glaucus*	1	0.029	**n/c**						
3. Clade E	3	0.051	0.052	**0.026**					
4. *E. bracteatus*	1	0.042	0.050	0.046	**n/c**				
5. Clade D	2	0.071	0.073	0.073	0.068	**0.001**			
6. Clade B	6	0.065	0.067	0.065	0.059	0.078	**0.022**		
7. Clade A	3	0.094	0.089	0.090	0.090	0.095	0.092	**0.037**	
8. *E. berteroi*	1	0.109	0.112	0.109	0.107	0.121	0.111	0.134	**n/c**

Table 4. Mean pairwise distances for each marker within the major clades recovered in the phylogenetic analysis.

	ITS	5S-NTS	LEAFY	*matK-trnK*	*psbA-trnH*	Combined
Clade F	0.008	0.062	0.015	0.003	0.004	0.007
Clade E	0.033	0.067	0.047	0.012	0.025	0.026
Clade D	0.003	0.004	0.000	0.000	0.000	0.001
Clade B	0.043	0.167	0.019	0.008	0.015	0.022
Clade A	0.019	0.127	0.029	0.025	0.058	0.037

In general, the identification of related and morphologically similar species of *Echinodorus* is difficult and requires specialized taxonomic expertise. The use of DNA sequences can represent a great differential in this sense, particularly when it comes from the certification of dried samples processed for the industry, when the identification based on morphological characters is not possible. However, this will require increased population and geographic sampling of species and the use of more polymorphic and more informative markers, such as the nuclear markers 5S-NTS and Leafy, which proved to be the most diversified among those tested (Table 4) and thus may be good candidates for an initial examination.

3.2. Development and Validation of an UPLC-DAD Method for the Analysis of Chemical Markers

Aiming to investigate the chemical composition of *E. macrophyllus* and *E. grandiflorus*, different extracts were prepared by sonicating the dried leaves with distinct solvents. The extracts prepared with 50% EtOH and 96 °GL EtOH showed the highest and lowest yields, respectively (Table 5). These findings indicate the predominance of hydrophilic compounds in both species. UPLC-DAD profiles were initially recorded for all extracts using an exploratory run and pointed out the 70% EtOH extract as the one with the most complex matrix. Therefore, this extract was employed for development and validation of the analytical method to quantify chemical markers.

Table 5. Results of system suitability tests for chromatographic determination of compounds **1–6** from *E. macrophyllus*.

Compound	Retention Factor (*k*)	USP Tailing Factor	Number of Theoretical Plates (N)	RSD Values of Retention Time (%)	Resolution (Rs)
cis- and *trans*-aconitic acid (**1**)	0.49	1.07	5656	0.12	1.87
homoorientin (**2**)	4.49	1.29	7480	0.07	3.41
swertisin (**3**)	7.20	1.02	6441	0.40	1.50
chicoric acid (**4**)	9.40	1.19	6756	0.08	1.55
caffeoyl-feruloyl-tartaric acid (**5**)	13.38	1.13	2147	0.22	2.65
di-feruloyl-tartaric acid (**6**)	15.42	1.05	6857	0.11	2.00

A segmented gradient (concave and linear) of water and acetonitrile, intercalated with isocratic elution steps, was used for UPLC analysis. The elution conditions were set and optimized to allow the adequate separation of compounds **1–6** in extracts of leaves from *E. grandiflorus* and *E. macrophyllus*. Two different wavelengths were used for quantifying the analytes: at 220 nm for the analysis of *cis*- and *trans*-aconitic acid (Figure 2A) and at 350 nm to quantify flavonoids and tartaric acid (Figure 2B).

The identification of the chemical markers in the extracts was performed by the analysis of reference compounds (*cis*- and *trans*-aconitic acid, homoorientin, chicoric acid, and swertisin) in the established chromatographic conditions and comparison of their retention times and UV data recorded by DAD with those of peaks registered in the chromatograms. The identity of the chemical markers was further confirmed by UPLC-ESI-MS analysis. Two other constituents of the extracts were assigned by UPLC-ESI-MS analysis: caffeoyl-feruloyl-tartaric acid (**5**) and di-feruloyl-tartaric acid (**6**). Compound **5** produced the deprotonated anion $[M-H]^-$ at m/z 487 Da in the negative ionization mode, along with fragment ions at m/z 149 Da $[M-338]^-$, ascribed to tartaric acid, at m/z 193 Da $[M-294]^-$, credited to the feruloyl unit, and at m/z 311 Da $[M-176]^-$, resulting from the loss of the caffeoyl unit. The identification of this compound by LC-MS analysis in a fraction from the ethanolic extract of *E. macrophyllus* was reported by [11]. In its turn, compound **6** produced the molecular ions at m/z 501 $[M-H]^-$ in the negative ionization mode, which, along with the fragment ions observed at m/z 149 $[M-352]^-$, was ascribed to tartaric acid, and at m/z 325 $[M-176]^-$, credited to the loss of one of the caffeoyl units, thus allowing its assignation. To the best of our knowledge, compounds **5** and **6** have not been isolated so far from *E. macrophyllus*. On the other hand, several caffeoyl-feruloyl-tartaric acid derivatives along with cinnamoyl-tartaric acid derivatives were obtained from *E. grandiflorus* [41].

Among the compounds identified in the extracts of *E. macrophyllus*, only homoorientin [42], swertisin [43], and caffeoyl-feruloyl-tartaric acid [11] have been previously reported; the occurrence of all other constituents is herein described for the first time. On the other hand, all of them have been previously reported as constituents of *E. grandiflorus* [12,13,41], corroborating the similar chemical composition of these species.

In the sequence, the developed UPLC-DAD method was validated for the quantitation of the six chemical markers identified in the extracts (Figure 2). The selectivity was assessed by spectral homogeneity of the peaks, evaluated by DAD. The ascending, upper, and descending regions of the spectrum of each peak matched exactly, indicating that other compounds were not co-eluted, i.e., the threshold angle was higher than the purity angle. The reliability of the chromatographic conditions was assured by the results of system suitability tests, as depicted in Table 5. Relative standard deviation (RSD) values of retention time ($\leq$1%), number of theoretical plates (N > 2000), retention factor (k > 0.5), USP tailing factor ($\leq$2.0), and resolution (Rs > 1.5) obtained for all analytes fell within the limits recommended by [31,44] for quantitative analysis.

The calibration curves constructed for the reference compounds were linear along the established ranges, with r^2 values above 0.99 (Table 6). Curves obtained in two consecutive days were statistically similar (p > 0.05). Furthermore, statistical analysis showed that calibration data presented normal distribution, with randomly distributed residuals. The method showed good precision with RSD values below 4.3% for all analytes (Table 6). Additionally, the mean concentrations determined for the inter-day precision experiment did not differ statistically (p > 0.05) for all analytes. LOQ values between 0.04 and 1.92 µg and LOD between 0.01 and 0.58 µg (Table 6) demonstrated the high sensitivity of the method. The method demonstrated robustness (p > 0.05) under all tested conditions, as the introduced variations did not significantly alter the concentrations of the analytes when compared with the results obtained using the established conditions. The method was also accurate, as indicated by the recovery rates ranging from 95.8% to 110.8% for all tested compounds. The results of the method validation are summarized in Table 6.

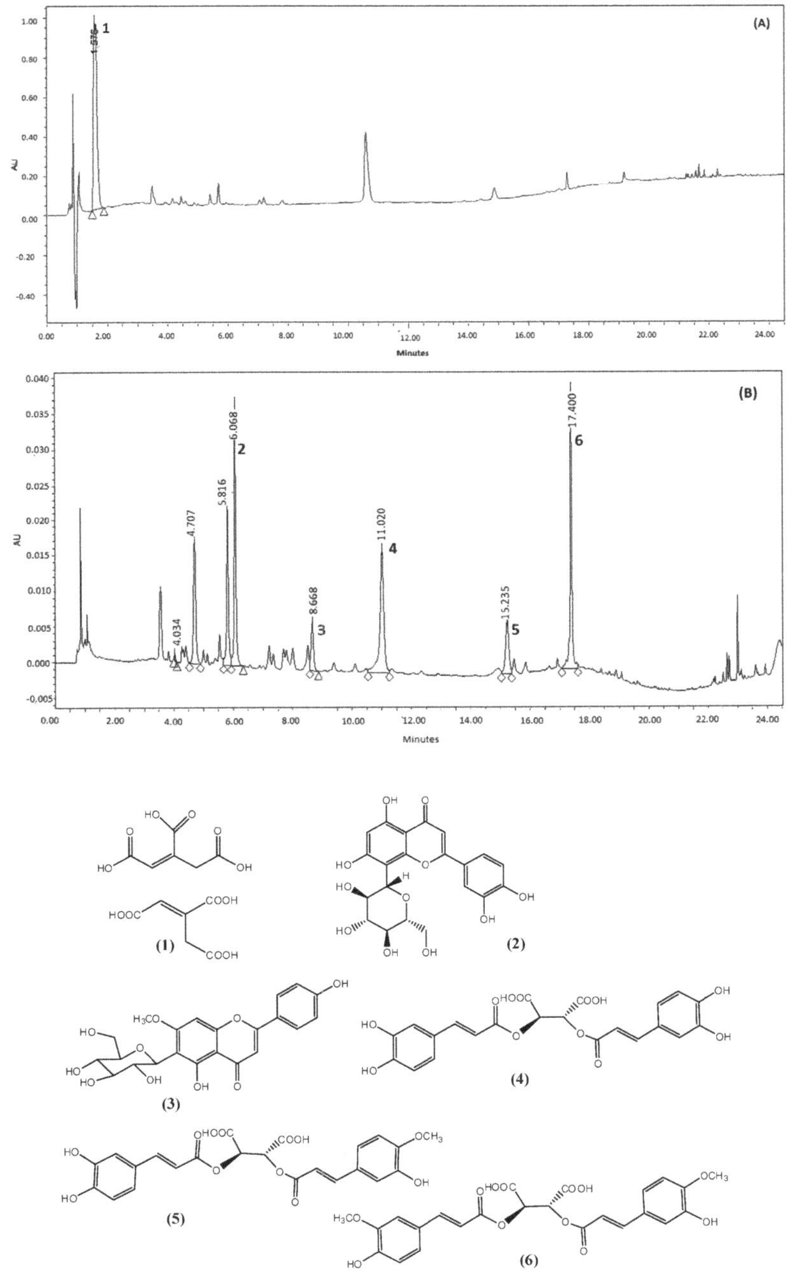

Figure 2. Typical UPLC-DAD chromatograms obtained for 70% EtOH extract of *E. macrophyllus* leaves. The chemical structures and corresponding peaks are (**1**) *cis*- and *trans*-aconitic acid, (**2**) homoorientin, (**3**) swertisin, (**4**) chicoric acid, (**5**) caffeoyl-feruloyl-tartaric acid, and (**6**) di-feruloyl-tartaric acid. Chromatograms registered at 220 nm (**A**) and 350 nm (**B**). Chromatographic conditions: see Materials and Methods (Section 2.6.1).

Table 6. Calibration results, LOQ, LOD, precision of the intra-day ($n = 6$) and inter-day ($n = 12$) assays, and recovery measurements of constituents quantified by UPLC-DAD in *E. grandiflorus* and *E. macrophyllus*.

Compound	Regression Equation	Linear Range (µg)	r^2	LOQ (µg)	LOD (µg)	Intra-Day RSD	Inter-Day RSD	Recovery (% ± SD)
1	$y = 43{,}603x - 19{,}041$	6.25–100	0.9999	0.990	0.300	2.65	2.60	96.0 ± 3.9 110.8 ± 4.8 100.5 ± 3.8
2	$y = 38{,}042x - 14{,}758$	1.25–20	0.9998	0.216	0.065	0.23	1.55	110.2 ± 0.8 95.8 ± 4.4 98.8 ± 2.1
3	$y = 9441x + 23{,}371$	3.0–150	0.9969	1.928	0.584	4.90	4.27	99.5 ± 1.9 107.9 ± 1.1 100.4 ± 2.9
4	$y = 47{,}326x - 53{,}912$	9.37–300	0.9993	0.048	0.014	0.84	1.79	99.9 ± 0.7 100.0 ± 1.1 98.9 ± 2.1

3.3. Quantitation of Chemical Markers in Extracts of *E. macrophyllus* and *E. grandiflorus*

The developed UPLC-DAD method was applied to the analysis of 12 extracts (96 °GL EtOH, 90% EtOH, 70% EtOH, and 50% EtOH) prepared from leaves of three different samples of *E. macrophyllus* (commercial lots DV1 and DV2; cultivated sample DVMus), along with four extracts of *E. grandiflorus* (96 °GL EtOH, 90% EtOH, 70% EtOH, and 50% EtOH) prepared from a single commercial lot. The quantitative results expressed as percentage mean (w/w) ± SD of triplicates are depicted in Table 7. As expected, all extracts prepared with 50% EtOH and 70% EtOH showed high concentrations of the polar compounds *cis*- and *trans*-aconitic acid, analyzed jointly (**1**), with variable contents in the samples. The 70% EtOH extracts obtained from DV1 (8.09 ± 0.02% w/w) and DV-Mus (7.47 ± 0.03%) showed similar contents of **1**, whereas the highest concentration was measured in the 50% EtOH of *E. grandiflorus* (20.12 ± 0.16%) (Table 7).

Homoorientin (**2**) was found in all extracts prepared with 70% EtOH and 90% EtOH, with variable contents. The extracts obtained from DVMus (70% EtOH, 0.02 ± 0.00%; 90% EtOH, 0.04 ± 0.01%) presented the lowest concentration of **2**, whereas the 50% EtOH from *E. grandiflorus* (0.73 ± 0.01%) exhibited the highest concentration. Swertisin (**3**) was found in all extracts from both species, its highest content being measured in extracts of *E. grandiflorus* (70% EtOH, 3.70 ± 0.02%) and the lowest concentration in extracts prepared with DV2 of *E. macrophyllus* (70% EtOH, 0.08 ± 0.00%).

Chicoric acid (**4**) was quantified in all extracts prepared with 50% EtOH and 90% EtOH from DV1 of *E. macrophyllus* (50% EtOH, 3.50 ± 0.06%) and from *E. grandiflorus* (50% EtOH, 6.76 ± 0.12%); the last one showing a significantly higher concentration ($p < 0.05$), whereas the 50% EtOH extract from DvMus (0.57 ± 0.01%) showed the lowest content. Caffeoyl-feruloyl-tartaric acid (**5**) and di-feruloyl-tartaric acid (**6**) were found in all extracts, except **5** in the 96°GL EtOH extract of DVMus. The 50% EtOH extract of *E. grandiflorus* showed the highest contents of **5** (1.89 ± 0.04%) and **6** (1.13 ± 0.02%).

The quantification of chemical markers in the extracts of *E. macrophyllus* and *E. grandiflorus* did not provide sufficient differentiation between these species.

Up until now, the chemical composition of *E. grandiflorus* has been investigated more than *E. macrophyllus*; previous reports have included the quantitation of homoorientin by HPLC-DAD in 70% EtOH extracts [45], along with the determination of *cis*- and *trans*-aconitic acid and compounds **2**, **3**, and **4** in different extracts using a distinct HPLC-DAD method [12,13]. The *Brazilian Pharmacopoeia* [8] establishes caffeic acid, isoorientin, and swertiajaponin as chemical markers of *E. grandiflorus*. Among these markers, homoorientin was also described and quantified for *E. macrophyllus* and *E. grandiflorus* in this work. To the best of our knowledge, the quantitative chemical composition of *E. macrophyllus* extracts has not been reported so far, although some of the compounds herein quantified (**4**, **5**, and **6**) have been detected in this species, along with several other flavone C-glycosides and hydroxycinnamoyl tartaric acid derivatives [11,43].

Table 7. Extractive yields and concentrations (mean ± SD of triplicates) of chemical markers in extracts of *E. macrophyllus* and *E. grandiflorus*.

Plant Material	Extract	Extractive Yield (% *w/w*)	Concentration (% *w/w*)					
			cis- and *trans*- Aconitic Acid (1)	Homoorientin (2)	Swertisin (3)	Chicoric Acid (4)	Caffeoyl-Feruloyl-Tartaric Acid (5)	Di-Feruloyl-Tartaric Acid (6)
DV1	96 °GL EtOH	8.51	0.491 ± 0.012 ^{a}	0.052 ± 0.001 ^{a}	0.171 ± 0.002 ^{a}	0.750 ± 0.013 ^{a}	0.636 ± 0.003 ^{a}	0.630 ± 0.005 ^{a}
	90% EtOH	9.98	2.992 ± 0.036 ^{b}	0.095 ± 0.002 ^{b}	0.260 ± 0.013 ^{b}	1.531 ± 0.012 ^{b}	0.794 ± 0.002 ^{b}	0.751 ± 0.006 ^{b}
	70% EtOH	21.31	8.094 ± 0.024 ^{c}	0.164 ± 0.007 ^{b}	0.382 ± 0.009 ^{c}	2.853 ± 0.042 ^{c}	0.991 ± 0.007 ^{c}	0.862 ± 0.008 ^{c}
	50% EtOH	25.34	10.333 ± 0.082 ^{d}	0.190 ± 0.006 ^{c}	0.480 ± 0.011 ^{d}	3.505 ± 0.059 ^{d}	1.081 ± 0.008 ^{c}	0.913 ± 0.003 ^{c}
DV2	96 °GL EtOH	11.14	ND	0.140 ± 0.002 ^{b}	0.164 ± 0.004 ^{a}	0.577 ± 0.002 ^{e}	0.570 ± 0.001 ^{a}	0.595 ± 0.001 ^{a}
	90% EtOH	14.40	2.311 ± 0.106 ^{e}	0.261 ± 0.003 ^{d}	0.743 ± 0.015 ^{e}	2.408 ± 0.060 ^{c}	0.798 ± 0.007 ^{b}	0.944 ± 0.005 ^{c}
	70% EtOH	19.13	1.263 ± 0.036 ^{f}	0.141 ± 0.001 ^{b}	0.082 ± 0.004 ^{f}	0.740 ± 0.005 ^{a}	0.992 ± 0.003 ^{c}	0.743 ± 0.004 ^{b}
	50% EtOH	22.58	1.244 ± 0.020 ^{f}	0.083 ± 0.009 ^{a}	0.172 ± 0.004 ^{a}	0.572 ± 0.001 ^{e}	1.089 ± 0.009 ^{c}	0.690 ±0.004 ^{b}
DVMus	96 °GL EtOH	4.07	ND	ND	0.181 ± 0.004 ^{a}	ND	ND	0.578 ± 0.010 ^{a}
	90% EtOH	19.65	0.762 ± 0.030 ^{g}	0.046 ± 0.008 ^{a}	0.312 ± 0.015 ^{c}	0.612 ± 0.005 ^{e}	0.581 ± 0.018 ^{a}	0.607 ± 0.008 ^{a}
	70% EtOH	21.51	7.473 ± 0.027 ^{h}	0.020 ± 0.004 ^{a}	0.235 ± 0.005 ^{b}	ND	0.574 ± 0.004 ^{a}	0.585 ± 0.012 ^{a}
	50% EtOH	22.21	7.595 ± 0.080 ^{h}	ND	0.244 ± 0.007 ^{b}	0.576 ± 0.001 ^{e}	0.574 ± 0.001 ^{a}	0.581 ± 0,013 ^{a}
EG	96 °GL EtOH	2.97	ND	0.130 ± 0.001 ^{a}	1.432 ± 0.063 ^{g}	0.718 ± 0.005 ^{a}	0.620 ± 0.002 ^{a}	0.630 ± 0.034 ^{a}
	90% EtOH	11.37	ND	0.262 ± 0.005 ^{d}	2.742 ± 0.130 ^{h}	1.320 ± 0.050 ^{b}	0.852 ± 0.017 ^{b}	0.831 ± 0.026 ^{c}
	70% EtOH	22.13	1.731 ± 0.063 ^{i}	0.352 ± 0.006 ^{e}	3.701 ± 0.022 ^{i}	4.471 ± 0.087 ^{f}	1.262 ± 0.008 ^{c}	0.852 ± 0.014 ^{c}
	50% EtOH	22.76	20.123 ± 0.161 ^{j}	0.734 ± 0.007 ^{f}	0.452 ± 0.017 ^{d}	6.762 ± 0.118 ^{g}	1.893± 0.005 ^{c}	1.134 ± 0.017 ^{d}

Legend: SD—standard deviation; DV1—*E. macrophyllus* commercial sample 1; DV2—*E. macrophyllus* commercial sample 2; DVMus—*E. macrophyllus* UFMG sample; EG—*E. grandiflorus* commercial sample; EtOH—96 °GL ethanol extract; 90% EtOH—90% hydroethanolic extract; 70% EtOH—70% hydroethanolic extract; 50% EtOH—50% hydroethanolic extract. Differences between the mean concentrations of the extracts were analyzed by ANOVA, followed by Tukey test ($p < 0.05$). For the contents of chemical markers, different letters within the columns indicate significant differences. ND—not determined (bellow the LOQ).

3.4. In Vitro Inhibition of TNF Release

The potential anti-inflammatory activity of extracts from *E. macrophyllus* and *E. grandiflorus* leaves was assayed in vitro, by measuring their inhibitory effect on the release of TNF by LPS-stimulated THP-1 cells. The extracts were not cytotoxic for THP-1 cells at 60 μg/mL, showing cell viability above 90%, according to results from SRB and MTT assays. Furthermore, none of the extracts or dexamethasone (Dx) (used as a positive control) induced the release of TNF in the absence of the inflammatory stimulus (LPS), indicating an absence of a pro-inflammatory effect.

The extracts prepared from the three samples of *E. macrophyllus* induced a similar inhibition of TNF release by LPS-stimulated THP-1 cells (Figure 3). None of the extracts obtained with 90% EtOH induced a significant reduction of this cytokine. Among the 96 °GL ethanol extracts, only those prepared from DV2 exhibited a significant reduction in TNF levels (44.1 ± 2.1%). All extracts prepared from *E. macrophyllus*, with either 70% EtOH or 50% EtOH, reduced the release of TNF significantly, reaching 61.9 ± 3.0% (70% EtOH extract) and 46.6 ± 10.2% (50% EtOH extract) inhibition for the extracts obtained from DV1. Furthermore, 70% EtOH and 50% EtOH extracts of DV2 induced TNF inhibition rates similar to the extracts prepared with DVMus ($p > 0.05$), whereas significant differences were obtained for those 70% EtOH and 50% EtOH extracts prepared from DV1 ($p < 0.05$). Taken together, the obtained results indicated that extracts prepared with 50% EtOH and 70% EtOH induced a more potent inhibition of TNF release for the three plant drugs of *E. macrophyllus*. It is worth noting that the 70% EtOH and 50% EtOH extracts of DV1 and DVMus were the most active extracts, showing similar activity without significant differences among them.

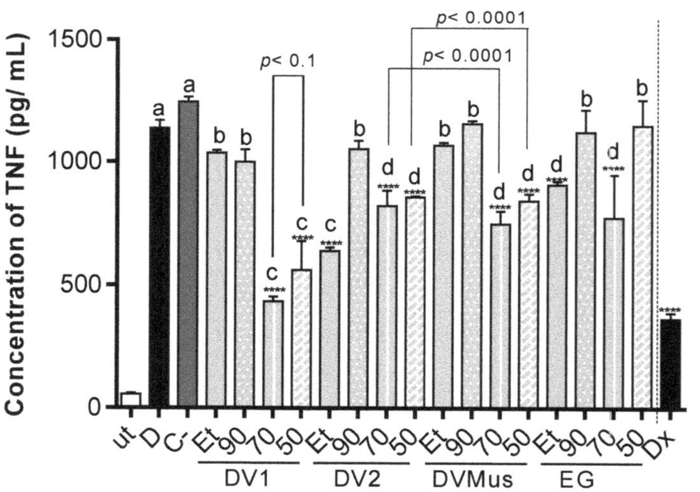

Figure 3. Effect on TNF release by LPS-stimulated THP-1 cells induced by extracts of leaves from *E. macrophyllus* prepared from different plant samples (DV1, DV2, and DVMus) and from a sample of *E. grandiflorus* (EG). The extracts were tested at 60 μg/mL. Data represent the mean ± SD of biological duplicates, with three experimental replicates each. **** $p < 0.0001$ compared with the challenged group. The significance level of the statistical test was set at $p < 0.05$. Ut—untreated cells; D—cells + LPS; C—DMSO solvent (control); Dx—dexamethasone (0.2 μM); Et—96 °GL ethanol extract; 90—90% hydroethanolic extract; 70—70% hydroethanolic extract; 50—50% hydroethanolic extract. Different letters in the bars indicate significant differences.

As for *E. grandiflorus*, the 96 °GL, 50%, and 70% EtOH extracts promoted significant reductions of TNF release by LPS-stimulated THP-1 cells (44.1 ± 2.1%, 20.4 ± 0.5%, and

32.2 ± 6.2%, respectively), whereas the 90% EtOH extract did show a less significant reduction of TNF.

We have previously investigated the effect of *E. grandiflorus* extracts on the release of TNF by LPS-stimulated THP-1 cells, non-differentiated into macrophages. In the previous study [12,13], it was observed that the 50% and 70% EtOH extracts exhibited significantly higher inhibition rates of TNF release (respectively, 100 ± 0.0% and 66.7 ± 5.0%) than those herein reported. Such differences may be related to the lower contents of flavone C-glycosides in the extracts here evaluated, since the authors ascribed the anti-inflammatory activity mainly to these compounds. Moreover, they did not employ cells differentiated into macrophages, as herein reported.

In general, the extracts showing high inhibitory activity of TNF release (70% EtOH and 50% EtOH) were those with the highest contents of aconitic acids and hydroxycinnamoyl tartaric acid derivatives. The data presented in this study supported the findings reported by [11], who observed a decrease in TNF and IL-1β levels in exudates when mice were treated with a fraction enriched in tartaric acid derivatives from *E. macrophyllus*, using an air pouch model of inflammation.

Data reported here for the in vitro inhibition of THF release by LPS-stimulated THP-1 cells treated with different extracts of *E. macrophyllus* and *E. grandiflorus* showed no significant differences in the responses induced by these species. Therefore, it was not possible to differentiate them based on the assayed biological activity, as similarly observed for the chemical markers quantified in the extracts.

3.5. In Vitro Anti-Oxidant Activity and Effect on ROS Production

The anti-oxidant activity of the extracts was assayed in vitro by the DPPH and β-carotene/linoleic acid co-oxidation radical scavenging methods. The extracts derived from *E. macrophyllus* DV1 exhibited significantly higher activity in the DPPH assay compared with those obtained from DV2 and DVMus. The EC_{50} values of the DV1 extracts were below 200 μg/mL, with the 50% EtOH extract demonstrating the highest activity (EC_{50} = 109.6 ± 1.7 μg/mL), as shown in Table 8. As for *E. grandiflorus*, only the 70% EtOH extract presented an EC_{50} value < 200 μg/mL (135.2 ± 2.0). The active extracts showed high contents of hydroxycinnamoyl tartaric acids (Table 7), suggesting the participation of these compounds in the observed anti-oxidant effect. In the same direction, *E. macrophyllus* fractions containing tartaric acid derivatives have been reported to promote significant reduction of DPPH radical by approximately 70% [11].

Regarding the β-carotene/linoleic acid co-oxidation assay, the most active extracts from *E. macrophyllus* were those prepared with 50% EtOH, either for DV1 (IC_{50} 40.1 ± 1.6 μg/mL), DV2 (IC_{50} 23.9 ± 1.8 μg/mL), or DVMus (IC_{50} 17.7 ± 1.5 μg/mL). In turn, all *E. grandiflorus* extracts were active, the 50% EtOH extract being the most active one (IC_{50} 6.8 ± 1.5 μg/mL) (Table 8). As far as we know, this is the first report on the anti-oxidant activity of extracts from *E. macrophyllus* and *E. grandiflorus* in the β-carotene/linoleic acid co-oxidation assay. Notably, among all extracts prepared with different plant drugs of both species, the 50% EtOH extracts showed the highest anti-oxidant activity, probably due to their higher contents of flavone-C-glycosides and hydroxycinnamoyl tartaric acid derivatives. This supposition was supported by several reports on the anti-oxidant action of different phenolic derivatives [46,47].

The results from the β-carotene/linoleic acid co-oxidation assay pointed out the 50% EtOH extracts as the most active ones and, for this reason, their effect on the ROS production by LPS-stimulated THP-1 cells was assayed. None of the assayed extracts inhibited ROS formation at high rates, with inhibition values below 20% (Table 8).

The anti-oxidant activity of a compound or extract can be more accurately estimated by using three experimental models that associate different oxidation mechanisms, including: (i) the formation, chain reactions, and action of free radicals; (ii) lipid peroxidation and degradation of defense cells systems; (iii) the formation of ROS [48]. In line with this recommendation, we utilized these different models to demonstrate the anti-oxidant potential of *E. macrophyllus* and *E. grandiflorus* extracts in our study.

Table 8. Anti-oxidant activity of extracts of *E. macrophyllus* (DV1, DV2, and DVMus) and *E. grandiflorus* assayed in vitro by the DPPH and β-carotene/linoleic acid co-oxidation models and effect of selected extracts on ROS produced by LPS-stimulated THP-1 cells.

Plant Drug	Extract	DPPH (EC_{50} μg/mL ± SD)	β-Carotene (IC_{50} μg/mL ± SD)	ROS (% Reduction)
DV1	96 °GL EtOH	>200 [a]	>200 [a]	NA
	90% EtOH	140.2 ± 1.5 [b]	44.4 ± 1.6 [b]	NA
	70% EtOH	139.2 ± 1.8 [b]	41.7 ± 1.5 [b]	NA
	50% EtOH	109.6 ± 1.7 [c]	40.1 ± 1.6 [b]	9.1 ± 5.5 [a]
DV2	96 °GL EtOH	>200 [a]	>200 [a]	NA
	90% EtOH	>200 [a]	>200 [a]	NA
	70% EtOH	>200 [a]	>200 [a]	NA
	50% EtOH	>200 [a]	23.9 ± 1.8 [c]	14.8 ± 3.4 [b]
DVMus	96 °GL EtOH	180.8 ± 1.5 [d]	22.6 ±1.6 [c]	NA
	90% EtOH	>200 [a]	36.0 ± 1.9 [b]	NA
	70% EtOH	>200 [a]	32.8 ± 1.5 [b]	NA
	50% EtOH	>200 [a]	17.7 ± 1.5 [d]	16.3 ± 2.2 [c]
EG	96 °GL EtOH	>200 [a]	8.7 ±1.5 [e]	NA
	90% EtOH	>200 [a]	11.9 ± 1.4 [f]	NA
	70% EtOH	135.2 ± 1.9 [b]	8.3 ± 1.1 [e]	NA
	50% EtOH	>200 [a]	6.8 ± 1.5 [g]	19.0 ± 6.2 [d]

Legend: DPPH—2,2-diphenyl-1-picrylhydrazyl radical; NA—not analyzed; IC_{50}—50% inhibitory concentration; EC_{50}—50% effective concentration; SD—standard deviation; DV1—*E. macrophyllus* commercial sample 1; DV2—*E. macrophyllus* commercial sample 2; DVMus—*E. macrophyllus* UFMG sample; EG—*E. grandiflorus* commercial sample; EtOH—96 °GL ethanol extract; 90% EtOH—90% hydroethanolic extract; 70% EtOH—70% hydroethanolic extract; 50% EtOH—50% hydroethanolic extract. Differences between the mean concentrations of the extracts were analyzed by ANOVA, followed by Tukey test ($p < 0.05$). Different letters within the columns indicate significant differences.

Phenolic compounds are known to inhibit ROS due to the presence of hydroxyl groups, capable of scavenging free radicals by electron donation [49,50]. ROS inhibition can also result from the increase of cell anti-oxidant defenses [51]. Extracts of *Echinodorus* prepared with 50% EtOH promoted high DPPH radical capture, decreased lipid peroxidation, reduced ROS production, and diminished the release of TNF. These extracts possessed high contents of hydroxycinnamoyl tartaric acids and flavonoids, which may explain their anti-oxidant and anti-inflammatory action. These results demonstrated the potential of these extracts for further investigation as potential agents for the treatment of inflammatory disorders [52].

3.6. PCA Model

An exploratory unsupervised classification PCA model was built to investigate the similarity between *E. macrophyllus* and *E. grandiflorus*, based on the contents of chemical markers and biological activities of the assayed extracts. The extracts were allocated at the rows and the chemical markers (*cis-* and *trans*-aconitic acids, CTAA; homoorientin, HOMO; swertisin, SWER; chicoric acid, CHA; caffeoyl-feruloyl-tartaric acid, CATA; diferuloyl-tartaric acid, DFTA) were put in the columns, along with data from the biological assays (lipid peroxidation in the β-carotene assay, Bcar; DPPH radical scavenging, DPPH; inhibition of TNF release, TNF), generating a matrix of dimensions 144 × 11. Data were preprocessed by autoscaling to provide equal weights for all investigated variables in both methods [15].

The obtained results for PCA model are depicted in Figure 4 as a biplot of PC1 × PC2. A biplot is a graphical representation simultaneously showing the relations among scores (describing samples) and loadings (describing variables). A model with two PCs accounted for 67.83% of the total variance data. In this model, none of the samples presented Q residues above the limit at a 95% confidence level, thus indicating the absence of outliers in the dataset. PC1 accounted for 39.86% of variance and discriminated DVMus (positive scores) from DV1/DV2 (negative scores) samples (these two clusters are indicated by red rectangles in Figure 4). EG samples were not clearly discriminated by this PCA model, even when PC3 was considered (10.8% of the variance), and it was not possible to ascribe its similarity to any of the evaluated samples of *E. macrophyllus*. Based on the PC1 of this model, we could conclude that the variation within different samples of the same species (DV, *E. macrophyllus*) was greater than the variation between the two species. In other words, the differences observed within *E. macrophyllus* samples were more pronounced than the differences between *E. macrophyllus* and *E. grandiflorus*. In addition, the anti-inflammatory (TNF) and anti-oxidant (DPPH/Bcar) activities were captured by PC2 (27.97%), respectively, with positive and negative loadings (green circles).

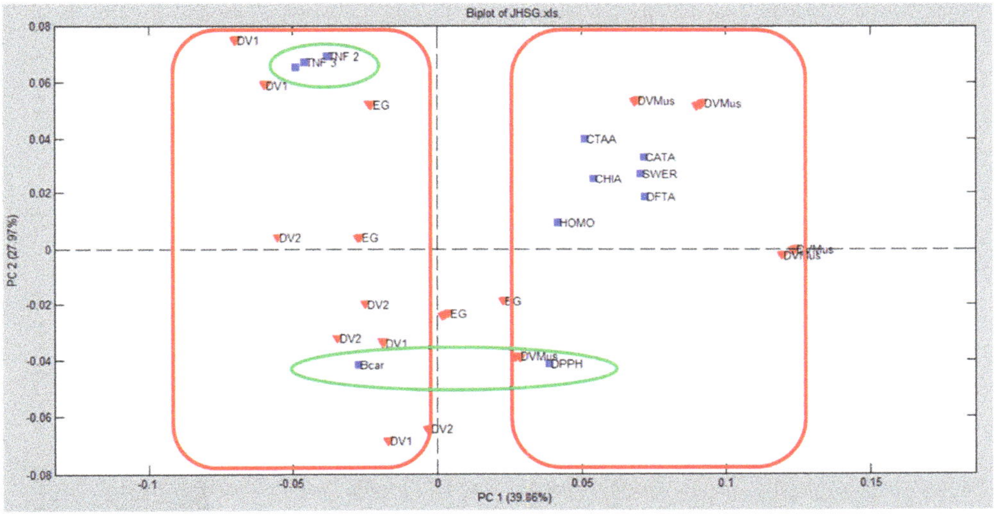

Figure 4. Biplot provided by a PCA model built for *E. macrophyllus* and *E. grandifloras* samples based on the quantitative chemical composition and in vitro biological activity of extracts. Red down triangles represent scores/samples and blue squares represent loadings/variables.

In Figure 4, samples were clustered as a function of the scores/samples in PC1. Aiming to better visualize interrelations, Figure 5A shows the same PCA biplot of Figure 4, but with marked clusters as a function of the loadings/variables. As depicted in Figure 5, the inhibition of TNF release was related to DV1, suggesting a positive correlation between these variables (see the black ellipse). Moreover, the anti-oxidant activity Bcar was associated to DV2 (red ellipse), while the DPPH activity was weakly correlated to EG (green ellipse). Chemical markers (CTAA, HOMO, SWER, CHA, CATA, and DFTA) were positively correlated to DVMus (grey circle in Figure 5A), which was consistent with the higher concentrations of these assayed constituents in this sample.

A heatmap (correlation map) was also generated to characterize the relationship between the analyzed variables (Figure 5B). The inhibition of TNF release showed weak correlations with CTAA ($r_{TNF\text{-}1}$ = 0.0198; $r_{TNF\text{-}2}$ = 0.1543; $r_{TNF\text{-}3}$ = 0.0732), which, in turn, correlated moderately with DPPH activity (r = 0.2149). DPPH activity showed weak/moderate correlations to all assayed chemical markers: HOMO (r = 0.1545), SWER (r = 0.1654), CHIA (r = 0.303), CATA (r = 0.2755), and DFTA (r = 0.2507). Conversely, these chemical markers were inversely correlated to Bcar anti-oxidant activity: CTAA, r = −0.2709; HOMO, r = −0.2344; SWER, r = −0.4801; CHIA, r = −0.3293; CATA, r = −0.2847; DFTA, r = −0.3321. The apparent contradictory results observed in the anti-oxidant assays could be attributed to the differences in the methods employed. The DPPH assay was based on the measurement of radical scavenging activity, while the β-carotene test evaluated the inhibition of lipid peroxidation. These assays targeted different mechanisms of anti-oxidant activity and may have elicited distinct responses from the constituents present in the extracts [53].

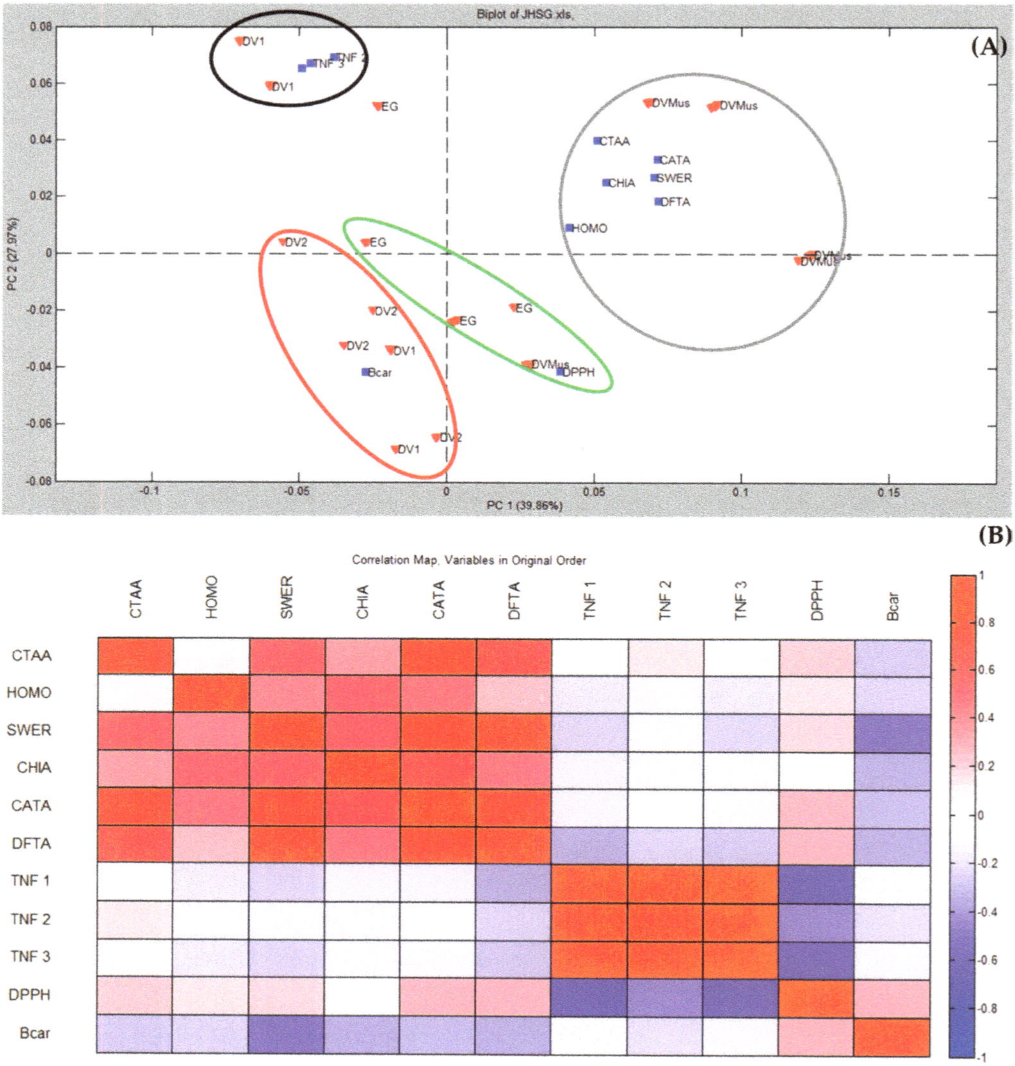

Figure 5. (**A**) The same PCA biplot of Figure 4 showing variable clusters: red down triangles represent scores and blue squares represent loadings. (**B**) Heatmap of chemical markers' contents and biological activity. CATA—*cis* and *trans*-aconitic acids; HOMO—homoorientin; SWER—swertisin; CHIA—chicoric acid; CATA—caffeoyl-feruloyl-tartaric acid; TNF 1, 2, and 3 (inhibition of TNF release); Bcar—lipid peroxidation in β-carotene assay; DPPH—DPPH radical scavenging assay.

The inhibition of lipid peroxidation is regarded as a more complex and selective method for identifying anti-oxidant compounds than DPPH, although both methods are important to characterize the anti-oxidant response [54]. The DPPH assay has been known to exhibit nonspecific responses to various compounds and may yield misleading results when used in isolation. To ensure a more accurate assessment of anti-oxidant activity, it is recommended to analyze the DPPH assay in conjunction with other complementary anti-oxidant tests [55]. The literature data have indicated low correlation between results from DPPH test and other anti-oxidant assays [56], as herein observed for the low positive correlation with the β-carotene assay ($r = 0.2690$).

PCA results clearly demonstrated the similarities in the quantitative chemical composition and biological activities of *E. macrophyllus* and *E. grandiflorus* extracts, which was in accordance with previous reports [57]. However, PCA was not able to discriminate *E. grandiflorus* from the analyzed samples of *E. macrophyllus*. The analyzed chemical markers showed a stronger correlation with DVMus, allowing for the differentiation between *E. macrophyllus* samples (DVMus vs. DV1 and DV2). However, these same chemical markers, which had positive loadings on PC1, did not show a significant correlation with *E. grandiflorus* extracts. Therefore, based on the quantification of these chemical markers and the assessed biological activities in this study, it was suggested that *E. macrophyllus* and *E. grandiflorus* could not be distinguished from each other.

Furthermore, our findings indicated that the two commercial lots of *E. macrophyllus* (DV1 and DV2) exhibited comparable contents of the analyzed chemical markers and presented similar biological activities. However, according to the PCA model, DVMus displayed distinct chemical marker contents compared with DV1 and DV2. This divergence could be attributed to the fact that DVMus was sourced from a single cultivated specimen, while DV1 and DV2 were composed of multiple specimens grown in different environments. These varying growth conditions could impact the chemical profiles of the plants [58].

4. Conclusions

Our study explored, for the first time, molecular analysis, quantitative chemical composition, and biological activity of various extracts from *E. macrophyllus* and *E. grandiflorus* leaves aiming to highlight the similarities and differences between these species. The molecular analysis of *E. macrophyllus* did not allow its genetic differentiation from other *Echinodorus* species, including *E. grandiflorus*, suggesting that there are only a few distinct genetic characteristics separating these species. The quantitation of six chemical markers in extracts prepared from commercial samples (DV1 and DV2) and a cultivated specimen (DVMus) of *E. macrophyllus*, along with a commercial lot of *E. grandiflorus* (EG), revealed higher contents of these markers in the 50% and 70% ethanol extracts. However, we were unable to differentiate between the species based on the concentrations of these markers. Moreover, the 50% and 70% ethanol extracts also elicited higher biological activity—inhibition of TNF release by LPS-stimulated THP-1 cells and anti-oxidant activity—in comparison with the 90% and 96°GL ethanol extracts. This difference in activity was attributed to the higher contents of chemical markers in the former extracts. Nevertheless, the results from the biological assays did not allow us to discriminate between the two species. Overall, our data demonstrated that *E. macrophyllus* and *E. grandiflorus* have similar chemical composition and biological activity. Additionally, the PCA analysis revealed a comparable chemical composition and biological activity among the commercial samples of *E. macrophyllus*. However, the analysis successfully distinguished the chemical composition of the cultivated specimen from the commercial lots. Interestingly, the cultivated and commercial samples of *E. macrophyllus* exhibited a higher level of intra-specific variance compared with the inter-specific variance observed with *E. grandiflorus*. It is worth noting, though, that the quantified chemical markers and biological activities assessed in this study did not reveal any discernible differences between the two *Echinodorus* species.

As part of our future perspectives, we plan to conduct in vivo studies to investigate the anti-inflammatory activity of the 50% and 70% ethanol extracts from *E. macrophyllus*.

These studies will involve the use of proof-of-concept models as well as disease models, including acute and chronic antigen-induced arthritis. By evaluating the efficacy of these extracts in vivo, we aim to contribute to the development of an herbal preparation from *E. macrophyllus* that can be potentially utilized for managing arthritis and other inflammatory diseases. These studies will provide valuable insights into the therapeutic potential of *E. macrophyllus* and its application in the field of inflammatory disorders.

Supplementary Materials: The following supporting information can be downloaded at: https://www.mdpi.com/article/10.3390/antiox12071365/s1, Table S1: Primers used in PCR and sequencing; Table S2: GenBank accessions and voucher data of the terminals used in the phylogenetic analyses of Echinodorus, Supplementary References are [59–64].

Author Contributions: Conceptualization, P.R.V.C. and F.C.B.; Formal Analysis, M.P.R., J.H.d.S.G. and M.M.S.; Funding Acquisition, F.C.B.; Investigation, M.P.R., L.M.C.d.S., L.P.M.S. and J.A.N.B.; Methodology, M.P.R., L.M.C.d.S., L.P.M.S., J.A.N.B., J.H.d.S.G. and M.M.S.; Project Administration, R.M.d.P., P.R.V.C. and F.C.B.; Resources, M.M.S., R.M.d.P., P.R.V.C. and F.C.B.; Supervision, M.M.S., R.M.d.P., P.R.V.C. and F.C.B.; Validation, M.M.S., P.R.V.C. and F.C.B.; Visualization, M.P.R., J.H.d.S.G. and M.M.S.; Writing—Original Draft Preparation, M.P.R. and J.H.d.S.G.; Writing—Review and Editing, M.M.S., P.R.V.C. and F.C.B. All authors have read and agreed to the published version of the manuscript.

Funding: This research was funded by Conselho Nacional de Desenvolvimento Científico e Tecnológico, Brazil (CNPq/MS/SCTIE/Decit N° 19/2018—Fitoterápicos), grant number 405889/2018-6.

Institutional Review Board Statement: Not applicable.

Informed Consent Statement: Not applicable.

Data Availability Statement: Data is contained within the article or supplementary material.

Acknowledgments: We thank CNPq/Brazil for the DTI fellowship awarded to M.P.R., as well as the undergraduate fellowships granted to L.M.C.d.S. and L.P.M.S. We also extend our thanks to CNPq for the research fellowship provided to F.C.B. Fundação Ezequiel Dias (Funed/Brazil) and Vera Lúcia de Almeida are acknowledged for providing the facilities to conduct the β-carotene and DPPH radical scavenging assays.

Conflicts of Interest: The sponsors had no role in the design, execution, interpretation, or writing of the study.

References

1. Corrêa, M.P. *Dictionary of Useful Plants in Brazil and Cultivated Exotics*; National Press: Rio de Janeiro, Brazil, 1984.
2. Lorenzi, H.; Matos, F.J.A. *Plantas Medicinais No Brasil: Nativas e Exóticas*; Instituto Plantarum: Nova Odessa, Brazil, 2000; p. 43.
3. Tabosa, A.B.; Matias, L.Q.; Martins, F.R. Live fast and die young: The aquatic macrophyte dynamics in a temporary pool in the Brazilian semiarid region. *Aquat. Bot.* **2012**, *102*, 71–78. [CrossRef]
4. Lorenzi, H. *Weeds of Brazil: Terrestrial, Aquatic, Parasitic and Toxic*, 6th ed.; Nova Odessa's Edition; Editora Vozes: Petrópolis, Brazil, 1998; pp. 110–111.
5. Brazil Ministério da Saúde. *Agência Nacional de Vigilância Sanitária. Farmacopeia Brasileira*, 1st ed.; Agência Nacional de Vigilância Sanitária: São Paulo, Brazil, 1926; 1149p.
6. Brazil Ministério da Saúde. *Agência Nacional de Vigilância Sanitária. Farmacopeia Brasileira*, 2nd ed.; Agência Nacional de Vigilância Sanitária: São Paulo, Brazil, 1959; 208p.
7. Brazil Ministério da Saúde. *Agência Nacional de Vigilância Sanitária. Farmacopeia Brasileira*, 5th ed.; Agência Nacional de Vigilância Sanitária: São Paulo, Brazil, 2010; Volume 2, 808p. Available online: https://www2.fcfar.unesp.br/Home/Instituicao/Departamentos/principiosativosnaturaisetoxicologianovo/farmacognosia/5-edicao{-}{-}-volume-2.pdf (accessed on 14 April 2023).
8. Brazil Ministério da Saúde. *Agência Nacional de Vigilância Sanitária. Farmacopeia Brasileira*, 6th ed.; Agência Nacional de Vigilância Sanitária: São Paulo, Brazil, 2019; Volume 2, 746p. Available online: https://www.gov.br/anvisa/pt-br/assuntos/farmacopeia/farmacopeia-brasileira/6a-edicao-volume-2 (accessed on 14 April 2023).
9. Marques, A.M.; Provance, D.W.; Kaplan, M.A.C.; Figueiredo, M.R. *Echinodorus grandiflorus*: Ethnobotanical, phytochemical and pharmacological overview of a medicinal plant used in Brazil. *Food Chem. Toxicol.* **2017**, *109*, 1032–1047. [CrossRef] [PubMed]
10. Lehtone, S.L.; Myllys, L. Cladistic analysis of *Echinodorus* (Alismataceae): Simultaneous analysis of molecular and morphological data. *Cladistics* **2008**, *24*, 218–239. [CrossRef]

11. Silva, G.P.; Fernandes, D.C.; Vigliano, M.V.; Pinto, F.A.; Fonseca, E.N.; Santos, S.V.M.; Marques, P.R.; Gayer, C.R.M.; Velozo, L.S.M.; Lima, C.K.F.; et al. *Echinodorus macrophyllus*: Hydroxycinnamoyl derivatives reduce neutrophil migration through modulation of cytokines, chemokines, and prostaglandin in the air-pouch model. *J. Ethnopharmacol.* **2022**, *284*, 114757. [CrossRef]
12. Garcia, E.F.; Oliveira, M.A.; Dourado, L.P.A.; Souza, D.G.; Teixeira, M.M.; Braga, F.C. In vitro TNF-α inhibition elicited by extracts from *Echinodorus grandiflorus* leaves and correlation with their phytochemical composition. *Planta Med.* **2016**, *82*, 337–343. [CrossRef]
13. Garcia, E.; Oliveira, M.A.; Candido, L.C.M.; Coelho, F.M.; Costa, V.V.; Queiroz-Junior, C.M.; Boff, D.; Amaral, F.A.; Souza, D.G.; Teixeira, M.M.; et al. Effect of the hydroethanolic extract from *Echinodorus grandiflorus* leaves and a fraction enriched in flavone-C-glycosides on antigen-induced arthritis in mice. *Planta Med.* **2016**, *82*, 407–413. [CrossRef]
14. Kobayashi, J.; Sekiguchi, M.; Shigemori, H.; Ohsaki, A.; Chapecoderins, A.-C. New labdane derived diterpenoids from *Echinodorus macrophyllus*. *J. Nat. Prod.* **2000**, *63*, 375–377. [CrossRef]
15. Bro, R.; Smilde, A.K. Principal component analysis. *Anal. Methods* **2014**, *6*, 2812–2831. [CrossRef]
16. Pereira, A.B.D.; Gomes, J.H.S.; Pereira, A.C.; de Pádua, R.M.; Côrtes, S.F.; Sena, M.M.; Braga, F.C. Definition of chemical markers for *Hancornia speciosa* Gomes by chemometric analysis based on the chemical composition of extracts, their vasorelaxant effect and α-glucosidase inhibition. *J. Ethnopharmacol.* **2022**, *299*, 115692. [CrossRef]
17. Doyle, J.L.; Doyle, D.J. A rapid DNA isolation procedure for small quantities of fresh leaf tissue. *Phytochem. Bull.* **1987**, *19*, 11–15.
18. Lehtonen, S.; Falck, D. Watery varieties: Aquarium plant diversity from aesthetic, commercial, and systematic perspectives. In *Ornamental Plants: Types, Cultivation and Nutrition*; Joshua, C.A., Ed.; Nova Science Publishers: New York, NY, USA, 2011; ISBN 978-1-61761-736-2.
19. Staden, R.; Beal, K.F.; Bonfield, J.K. The Staden Package, 1998. *Methods Mol. Biol.* **2000**, *132*, 115–130. [CrossRef]
20. Edgar, R.; Muscle, C. Multiple sequence alignment with high accuracy and high throughput. *Nucleic Acids Res.* **2004**, *32*, 1792–1797. [CrossRef] [PubMed]
21. Kumar, S.; Stecher, G.; Tamura, K. MEGA7: Molecular Evolutionary Genetics Analysis version 7.0 for bigger datasets. *Mol. Biol. Evol.* **2016**, *33*, 1870–1874. [CrossRef] [PubMed]
22. Kelchner, S.A. The Evolution of Non-Coding Chloroplast DNA and Its Application in Plant Systematics. *Ann. Mo. Bot. Gard.* **2000**, *87*, 482–498. [CrossRef]
23. Christenhusz, M.J.M.; Fay, M.F.; Byng, J.W. *GLOVAP Nomenclature Part 1. The Global Flora: A Practical Flora to Vascular Plant Species of the World*; Special Edition; Plant Gateway: Bradford, UK, 2018; Volume 4, pp. 1–155.
24. Swofford, D.L. *PAUP (Phylogenetic Analysis Using PAUP)*; Sinauer Associates: Sunderland, MA, USA, 2018; p. 4.0a159.
25. Felsenstein, J. Confidence limits on phylogenies: An approach using the bootstrap. *Evolution* **1985**, *39*, 783–791. [CrossRef]
26. Kress, W.J.; Prince, L.M.; Williams, K.J. The phylogeny and a new classification of the gingers (Zingiberaceae): Evidence from molecular data. *Am. J. Bot.* **2002**, *89*, 1682–1696. [CrossRef]
27. Ronquist, F.; Teslenko, M.; Van Der Mark, P.; Ayres, D.L.; Darling, A.; Höhna, S.; Larget, B.; Liu, L.; Suchard, M.A.; Huelsenbeck, J.P. MRBAYES 3.2: Efficient Bayesian phylogenetic inference and model selection across a large model space. *Syst. Biol.* **2012**, *61*, 539–542. [CrossRef]
28. Miller, M.A.; Pfeiffer, W.; Schwartz, T. Creating the CIPRES Science Gateway for inference of large phylogenetic trees. In Proceedings of the Gateway Computing Environments Workshop (GCE), Gateway Computing, New Orleans, LA, USA, 14 November 2010; pp. 1–8.
29. Nylander, J.A.A. *MrModeltest v2. Program Distributed by the Author*; Evolutionary Biology Centre, Uppsala University: Uppsala, Sweden, 2004.
30. Erixon, P.; Svennblad, B.; Britton, T.; Oxelman, B. Reliability of Bayesian posterior probabilities and bootstrap frequencies in phylogenetics. *Syst. Biol.* **2003**, *52*, 665–673. [CrossRef]
31. Food and Drug Administration (FDA). Guidance for Industry: Analytical Procedures and Methods Validation. 2015. Available online: https://www.fda.gov/files/drugs/published/Analytical-Procedures-and-Methods-Validation-for-Drugs-and-Biologics.pdf (accessed on 13 March 2023).
32. ICH—International Conference on Harmonisation of Technical Requirements for Registration of Pharmaceutical for Human Use. ICH Harmonised Tripartite Guideline. In *Validation of Analytical Procedures: Methodology, Q2B (EMA/CHMP/ICH/82072/2006)*; European Agency for the Evaluation of Medicinal Products: London, UK, 2006.
33. Agência Nacional de Vigilância Sanitária—Brazil. RDC n° 166 de 24 of July of 2017. Provides for the Validation of Analytical Methods and Other Measures. Available online: http://antigo.anvisa.gov.br/documents/10181/2721567/RDC_166_2017_COMP.pdf/d5fb92b3-6c6b-4130-8670-4e3263763401#:~:text=QU%C3%8DMICAS%20DE%20REFER%C3%8ANCIA-,Art.,comp%C3%AAndios%20oficialmente%20reconhecidos%20pela%20Anvisa (accessed on 13 March 2023).
34. Mosmann, T. Rapid colorimetric assay for cellular growth and survival: Application to proliferation and cytotoxicity assays. *J. Immunol. Methods* **1983**, *65*, 55–63. [CrossRef] [PubMed]
35. Mensor, L.L.; Menezes, F.S.; Leitão, G.G.; Reis, A.S.; Dos Santos, T.C.; Coube, C.S.; Leitão, S.G. Screening of Brazilian plant extracts for antioxidant activity by the use of DPPH free radical method. *Phytother. Res.* **2001**, *15*, 127–130. [CrossRef] [PubMed]
36. Duarte-Almeida, J.M.; dos Santos, R.J.; Genovese, M.I.; Lajolo, F.M. Avaliação da atividade antioxidante utilizando sistema beta-caroteno/ácido linoléico e método de sequestro de radicais DPPH•. *Food Sci. Technol.* **2006**, *26*, 446–452. [CrossRef]

37. Ryan, K.A.; Smith, M.F., Jr.; Sanders, M.K.; Ernst, P.B. Reactive oxygen and nitrogen species differentially regulate Toll-like receptor 4-mediated activation of NF-kappa B and interleukin-8 expression. *Infect. Immun.* **2004**, *72*, 2123–2130. [CrossRef] [PubMed]
38. Johnson, M.; Zaretskaya, I.; Raytselis, Y.; Merezhuk, Y.; Mcginnis, S.; Madden, T.L. NCBI BLAST: A better web interface. *Nucleic Acids Res.* **2008**, *1*, 36. [CrossRef] [PubMed]
39. CBOL Plant Working Group. A DNA barcode for land plants. *Proc. Natl. Acad. Sci. USA* **2009**, *106*, 12794–12797. [CrossRef] [PubMed]
40. Govaerts, R. *Echinodorus*. Plants of the World Online. Facilitated by the Royal Botanic Gardens, Kew. Published on the Internet. Available online: http://www.plantsoftheworldonline.org/ (accessed on 26 March 2023).
41. Schnitzler, M.; Petereit, F.; Nahrstedt, A. Trans-Aconitic acid, glucosylflavones and hydroxycinnamoyltartaric acids from the leaves of *Echinodorus grandiflorus* ssp. *aureus*, a Brazilian medicinal plant. *Rev. Bras. Farmacogn.* **2007**, *17*, 149–154. [CrossRef]
42. Portella, V.G.; Cosenza, G.P.; Diniz, L.R.; Pacheco, L.F.; Cassali, G.D.; Caliari, M.V.; Brandão, M.D.; Vieira, M.A. Nephroprotective Effect of *Echinodorus macrophyllus* Micheli on Gentamicin-Induced Nephrotoxicity in Rats. *Nephron Extra* **2012**, *2*, 177–183. [CrossRef]
43. Fernandes, D.C.; Martins, B.P.; Silva, G.P.D.; Fonseca, E.N.D.; Santos, S.V.M.; Velozo, L.S.M.; Gayer, C.R.M.; Sabino, K.C.C.; Coelho, M.G.P. *Echinodorus macrophyllus* fraction with a high level of flavonoid inhibits peripheral and central mechanisms of nociception. *J. Tradit. Complement. Med.* **2021**, *12*, 123–130. [CrossRef]
44. Snyder, L.R.; Kirkland, J.J.; Glajch, J.L. *Practical HPLC Method Development*, 2nd ed.; Wiley-Interscience: New York, NY, USA, 1997; 765p.
45. Strada, C.L.; Da Lima, K.C.; Da Silva, V.C.; Ribeiro, R.V.; Dores, E.F.G.C.; Dall'oglio, E.; Schmeda-Hirschmann, G.; Carollo, C.C.; Martins, D.T.O.; Junior, P.T.S. Isovitexin as marker and bioactive compound in the antinociceptive activity of the Brazilian crude drug extracts of *Echinodorus scaber* and *E. grandiflorus*. *Rev. Bras. Farmacogn.* **2017**, *27*, 619–626. [CrossRef]
46. Rani, R.; Arora, S.; Kaur, J.; Manhas, R.K. Phenolic compounds as antioxidants and chemopreventive drugs from *Streptomyces cellulosae* strain TES17 isolated from rhizosphere of *Camellia sinensis*. *BMC Complement. Altern. Med.* **2018**, *18*, 82. [CrossRef]
47. Gasparotto, F.M.; Palozi, R.A.C.; Da Silva, C.H.F.; Pauli, K.B.; Donadel, G.; Lourenço, B.H.L.B.; Nunes, B.C.; Lívero, F.A.D.R.; De Souza, L.M.; Lourenço, E.L.B.; et al. Antiatherosclerotic Properties of *Echinodorus grandiflorus* (Cham. & Schltdl.) Micheli: From Antioxidant and Lipid-Lowering Effects to an Anti-Inflammatory Role. *J. Med. Food* **2019**, *22*, 919–927. [CrossRef]
48. Munteanu, I.G.; Apetrei, C. Analytical Methods Used in Determining Antioxidant Activity: A Review. *Int. J. Mol. Sci.* **2021**, *22*, 338. [CrossRef]
49. Conforti, F.; Menichini, F. Phenolic compounds from plants as nitric oxide production inhibitors. *Curr. Med. Chem.* **2011**, *18*, 1137–1145. [CrossRef]
50. Zielinska-Blizniewska, H.; Sitarek, P.; Merecz-Sadowska, A.; Malinowska, K.; Zajdel, K.; Jablonska, M.; Zajdel, R. Plant Extracts and Reactive Oxygen Species as Two Counteracting Agents with Anti- and Pro-Obesity Properties. *Int. J. Mol. Sci.* **2019**, *20*, 4556. [CrossRef] [PubMed]
51. Snezhkina, A.V.; Kudryavtseva, A.V.; Kardymon, O.L.; Savvateeva, M.V.; Melnikova, N.V.; Krasnov, G.S.; Dmitriev, A.A. ROS Generation and Antioxidant Defense Systems in Normal and Malignant Cells. *Oxidative Med. Cell. Longev.* **2019**, *2019*, 6175804. [CrossRef]
52. Arulselvan, P.; Fard, M.T.; Tan, W.S.; Gothai, S.; Fakurazi, S.; Norhaizan, M.E.; Kumar, S.S. Role of Antioxidants and Natural Products in Inflammation. *Oxidative Med. Cell. Longev.* **2016**, *2016*, 5276130. [CrossRef]
53. Amorati, R.; Valgimigli, L. Advantages and limitations of common testing methods for antioxidants. *Free. Radic. Res.* **2015**, *49*, 633–649. [CrossRef]
54. Prieto, M.A.; Rodríguez-Amado, I.; Vázquez, J.A.; Murado, M.A. β-Carotene assay revisited. application to characterize and quantify antioxidant and prooxidant activities in a microplate. *J. Agric. Food Chem.* **2012**, *60*, 8983–8993. [CrossRef]
55. Bibi-Sadeer, N.; Montesano, D.; Albrizio, S.; Zengin, G.; Mahomoodally, M.F. The Versatility of Antioxidant Assays in Food Science and Safety-Chemistry, Applications, Strengths, and Limitations. *Antioxidants* **2020**, *9*, 709. [CrossRef] [PubMed]
56. Koleva, I.I.; Van Beek, T.A.; Linssen, J.P.; De Groot, A.; Evstatieva, L.N. Screening of plant extracts for antioxidant activity: A comparative study on three testing methods. *Phytochem. Anal.* **2002**, *13*, 8–17. [CrossRef] [PubMed]
57. Gilbert, B.; Alves, L.F.; Favoreto, R.F. Echinodorus grandiflorus, *E. macrophyllus*. In *Monografias de Plantas Medicinais Brasileiras e Aclimatadas: Volume II [Online]*; Abifisa; Editora FIOCRUZ: Rio de Janeiro, Brazil, 2022; pp. 97–120. ISBN 978-65-5708-177-8. [CrossRef]
58. Pang, Z.; Chen, J.; Wang, T.; Gao, C.; Li, Z.; Guo, L.; Xu, J.; Cheng, Y. Linking Plant Secondary Metabolites and Plant Microbiomes: A Review. *Front. Plant Sci.* **2021**, *2*, 12. [CrossRef]
59. Cox, A.V.; Bennet, M.D.; Dyer, T.A. Use of the polymerase chain reaction to detect spacer size heterogeneity in plant 5S-rRNA gene clusters and to locate such clusters in wheat (*Triticum aestivum* L.). *Theor. Appl. Genet.* **1992**, *83*, 684–690. [CrossRef] [PubMed]
60. Grob, G.B.J.; Gravendeel, B.; Eurlings, M.C.M. Potential phylogenetic utility of the nuclear FLORICAULA/LEAFY second intron: Comparison with three cloroplast DNA regions in *Amorphophallus* (Araceae). *Mol. Phylogenet. Evol.* **2004**, *30*, 13–23. [CrossRef] [PubMed]
61. Liang, H.; Hilu, K.W. Application of matK gene sequences to grass systematics. *Can. J. Bot.* **1996**, *74*, 125–134. [CrossRef]
62. Sang, T.; Crawford, D.J.; Stuessy, T.F. Chloroplast DNA phylogeny, reticulate evolution, and biogeography of *Paeonia* (Paeoniaceae). *Am. J. Bot.* **1997**, *84*, 1120–1136. [CrossRef]

63. Tate, J.A.; Simpson, B.B. Paraphyly of *Tarasa* (Malvaceae) and diverse origins of the polyploid species. *Syst. Bot.* **2003**, *28*, 723–737. [CrossRef]
64. White, T.J.; Bruns, T.; Lee, S.; Taylor, J. Amplification and direct sequencing of fungal ribosomal RNA genes for phylogenetics. In *PCR Protocols: A Guide to Methods and Applications*; Innis, M.A., Gelfund, D.H., Sninsky, J.J., White, T.J., Eds.; Academic Press: San Diego, CA, USA, 1990; pp. 315–332.

antioxidants MDPI

Article

Cyanidin-3-*O*-glucoside Regulates the Expression of *Ucp1* in Brown Adipose Tissue by Activating Prdm16 Gene

Suping Han [1,†], Yafan Yang [1,†], Yanan Lu [1,2,†], Jielong Guo [1], Xue Han [1,3], Yunxiao Gao [1], Weidong Huang [1], Yilin You [1,*] and Jicheng Zhan [1,*]

[1] Beijing Key Laboratory of Viticulture and Enology, College of Food Science and Nutritional Engineering, China Agricultural University, Tsinghua East Road 17, Beijing 100083, China; hansuping@cau.edu.cn (S.H.); S20213061075@cau.edu.cn (Y.Y.); luyanan@bcu.edu.cn (Y.L.); b20183060515@cau.edu.cn (J.G.); hanxuehx313@bjmu.edu.cn (X.H.); s20193060991@cau.edu.cn (Y.G.); weidonghuang@cau.edu.cn (W.H.)

[2] School of Biomedicine, Beijing City University, Beijing 100094, China

[3] Department of Physiology and Pathophysiology, School of Basic Medical Sciences, Peking University Health Science Center, Beijing 100191, China

* Correspondence: yilinyou@cau.edu.cn (Y.Y.); zhanjicheng@cau.edu.cn (J.Z.); Tel.: +86-10-62737535 (Y.Y.); +86-10-62737553 (J.Z.)

† These authors contributed equally to this work.

Abstract: (1) Background: Brown adipose tissue (BAT) burns energy to produce heat. Cyanidin-3-*O*-glucoside (C3G) can then enhance the thermogenic ability of BAT in vivo. However, the mechanism by which C3G regulates Ucp1 protein expression remains unclear. (2) Methods: In this study, C3H10T12 brown adipose cells and db/db mice and mice with high-fat, high-fructose, diet-induced obesity were used as the model to explore the effect of C3G on the expression of the *Ucp1* gene. Furthermore, the 293T cell line was used for an in vitro cell transgene, a double luciferase reporting system, and yeast single hybridization to explore the mechanism of C3G in regulating Ucp1 protein. (3) Results: we identified that, under the influence of C3G, Prdm16 directly binds to the −500 to −150 bp promoter region of *Ucp1* to activate its transcription and, thus, facilitate BAT programming. (4) Conclusions: This study clarified the mechanism by which C3G regulates the expression of the *Ucp1* gene of brown fat to a certain extent.

Keywords: brown adipose tissue; Cyanidin-3-*O*-glucoside; PR domain containing 16; *Ucp1* promoter; peroxisomes proliferator-activated receptors

Citation: Han, S.; Yang, Y.; Lu, Y.; Guo, J.; Han, X.; Gao, Y.; Huang, W.; You, Y.; Zhan, J. Cyanidin-3-*O*-glucoside Regulates the Expression of *Ucp1* in Brown Adipose Tissue by Activating Prdm16 Gene. *Antioxidants* **2021**, *10*, 1986. https://doi.org/10.3390/antiox10121986

Academic Editor: Marco G. Alves

Received: 1 December 2021
Accepted: 8 December 2021
Published: 14 December 2021

Publisher's Note: MDPI stays neutral with regard to jurisdictional claims in published maps and institutional affiliations.

1. Introduction

Anthocyanins are polyphenolic compounds that are abundant in dark-colored fruits, vegetables, and pigmented cereals (such as berries, cherries, grapes, purple onions, black beans, purple cabbage, black rice, red sorghum, and purple maize (In addition, anthocyanins are commonly processed as colorants in beverages, fruit fillings, snacks, and dairy products, which account for a considerable amount of the anthocyanins consumed in the average diet [1]. Anthocyanin is a common component in functional foods for preventing cardiovascular diseases and inflammatory diseases (including diabetes and metabolic syndrome), mainly due to its excellent antioxidant activity [2]. In anthocyanin-related structures, the carboxypyranoside -3-*O*-glucoside, in particular, exhibits a good inhibition effect on active oxygen generation and high structural stability [3]. A large number of studies have proven that anthocyanins offer a variety of therapeutic effects, including that of increased energy expenditure and of limiting weight gain provided by Cyanidin-3-*O*-glucoside (C3G) [4].

The rising prevalence of excess body mass and obesity in nearly all countries around the world has been described as a global pandemic [5]. Currently available anti-obesity strategies are largely dependent on limiting energy uptake and/or absorption, but these tend to have little effect and often cause unwanted side effects [6,7]. An alternative strategy is urgently required to increase the energy expenditure in key metabolic organs, such as brown adipose tissue (BAT) [8]. In rodents, increasing BAT activity can effectively enhance metabolic rate and protect the animal from obesity [9].

BAT plays an essential role in non-shivering thermogenesis in mammals. Its cells are rich in mitochondria, the inner membranes of which are, in turn, rich in uncoupling protein 1 (Ucp1). BAT can specifically overexpress Ucp1, thus uncoupling transmembrane electron transfer and oxidative phosphorylation in the mitochondrial respiratory chain, thereby releasing a large amount of energy in the form of heat [10]. The known anti-obesity effect of functional BAT has drawn research and attention to the activation and regulation mechanism of brown fat functioning, and the specific expression of Ucp1 by BAT has become a hot topic for scientists.

The expression of the *Ucp1* gene is regulated by a variety of transcription factors, such as peroxisome proliferator-activated receptors (PPARs), PR domain containing 16 (Prdm16), peroxisome proliferator-activated receptor γ coactivator-1(PGC-1), and mitochondrial transcription factor A(TFAM). PPARα participates in the regulation of fatty acid oxidation mainly by inducing the expression of genes related to fatty acid oxidation [11]. Prdm16 is a zinc finger structure transcription factor containing PR domain, which is highly expressed in mouse BAT. It is also a key factor in determining whether adipose precursor cells develop into skeletal muscle cells or brown adipose cells. The Prdm16 protein is greatly enriched in BAT and causes increased expression of mitochondrial genes and greater density of mitochondria [12]. The loss of Prdm16 can hinder the differentiation of brown adipocytes, while overexpression can significantly increase their numbers. Animals lacking Prdm16 in BAT have a dramatically reduced capacity to produce heat [13]. Prdm16 regulates the thermogenic function of brown fat, mainly by binding to PPARγ and coactivating its transcriptional function [14].

The anti-obesity and anti-diabetic effects of anthocyanins are well-accepted experimentally [15,16]. Anthocyanins C3G and C3R have been reported to increase the mitochondrial copy number in vitro and to promote the metabolism of carbohydrates and fat [17]. C3G can enhance whole-body metabolism by upregulating the mitochondrial function of BATT and beige formation in subcutaneous WAT [4]. However, its molecular targets have yet to be elucidated.

Here, via quantitative real-time PCR, to quantify the expression of BAT-related genes *Ucp1* and *Prdm16* following the treatment of $C_3H_{10}T_{1/2}$ cells with C3G during brown adipogenesis, we speculate that *Prdm16* may be a novel transcriptional activator of *Ucp1*. The results of a luciferase assay and yeast one-hybrid screening ascertained that C3G activates the thermogenic gene *Ucp1* by regulating the expressions of *Prdm16*. Thus, Prdm16 can bind directly to the *Ucp1* promoter region, from −500 to −150 bp to active transcription. Our results, thus, suggest that Prdm16 is indeed a novel transcriptional activator of Ucp1 and that C3G can increase energy metabolism through a previously unrecognized molecular mechanism. Consequently, these findings may support an innovative approach to the treatment of obesity and its related diseases.

2. Materials and Methods

2.1. Materials and Reagents

C3G (purity 95.7%) was purchased from Chengdu Manchester Co., Ltd. (Chengdu, China), A ready-to-use PCR kit 3.0 was purchased from Beijing Tianenze Biotechnology Co., Ltd. (Beijing, China), which is used for PCR identification of bacterial liquid, and the method is carried out according to the instruction manual. Dimethyl sulfoxide (DMSO), a double luciferase reporter gene detection kit, and Lipo8000TM transfection reagent were purchased from Beyotime Biotechnology Co., Ltd. (Shanghai, China), while a HiFi Script cDNA Synthesis Kit, Ultra SYBR Mixture, and 200 bp Ladder DNA Marker were purchased from Kangwei Century Biology Science and Technology Co., Ltd. (Beijing, China), and a mass extraction and purification kit for plasmid was purchased from Weigela Biology

Technology Co., Ltd .(Chengdu, China). Furthermore, recombinant human PPARA (P0645) was purchased from Wuhan Fine Biotechnology Co., Ltd. (Wuhan, China).

2.2. Cell Culture

The mouse mesenchymal stem cell line, known as $C_3H_{10}T_{1/2}$ mouse embryonic fibroblasts, was purchased from the cell bank of the Chinese Academy of Sciences (Shanghai, China), and was induced to differentiate into brown adipocytes. The $C_3H_{10}T_{1/2}$ cells used in this experiment were cell samples frozen in liquid nitrogen in the laboratory.

The cryovials were taken out of the liquid nitrogen and immediately melted in a 37 °C water bath, centrifuged at 500× *g* for 3 min, and cultured in a 37 °C, 5% CO_2 incubator for passage. When the cells grew to a density of 80–90%, the medium was changed to brown adipose differentiation medium (DMEM HIGH + 10% FBS + 0.02 μM INSULIN + 1 × 10^{-3} μM T3) and an appropriate concentration of C3G was added. Two days later (recorded as day 0), the medium was changed to brown fat induction medium (DMEM HIGH + 10% FBS + 0.02 μM INSULIN + 1 × 10^{-3} μM T3 + 0.125 mM INDOMETHACIN + 0.5 mM IBMX + 2 μg/mL DEX) and an appropriate concentration of C3G was added. Two days later (recorded as day 2), the medium was changed to the brown adipose differentiation medium described above, and an appropriate concentration of C3G was added. The liquid was changed again on the 4th day. On the sixth day, the cells were observed under an inverted microscope. There were obvious fat droplets in the cytoplasm, i.e., the differentiation was completed.

2.3. Animals Studies

Animal experiments were conducted according to the methods reported by You et al. [4,18]. Animal experiments were conducted in two stages. Firstly, 20 three-week-old obese C57BLKS/J-Leprdb/Leprdb (db/db) male mice (mouse model of hereditary obesity) were purchased from the Model Animal Research Center of Nanjing University, China. After an acclimation of one week, the mice were randomly divided into two groups (n = 10 per group) with overall equal body weights. The mice received C3G (1 mg/mL) dissolved in drinking water for 12 weeks. DMSO was used as the vehicle for different treatments. In addition, a further 36 three-week-old male C57BL/6J mice (obesity model mice induced by high fat and high glucose diet) were purchased from Vital River Laboratory Animal Technology Co., Ltd., China. After one week of acclimation, these mice were randomly divided into three groups (n = 12/group) with overall equal body weights. They were assigned to one of three dietary treatments for 12 weeks: (1) normal chow diet group (CHOW, 3.2 kcal/g, 4.5% fat, w/w); (2) diet-induced obesity group (DIO, 4.7 kcal/g, 25% fructose and 25% lard); and (3) C3G group (DIO + C3G, 4.7 kcal/g, 25% fructose and 25% lard). C3G-group mice received C3G dissolved in drinking water (1 mg/mL), while drinking water also served as the vehicle for the CHOW and DIO treatments. Three mice were housed per cage in an Office of Laboratory Animal Welfare-certified animal facility, with a 12-h light/12-h dark cycle. Principles of laboratory animal care were followed, with all procedures conducted according to the guidelines established by the National Institute of Health, and every effort was made to minimize suffering.

2.4. Isolation of Total RNA Extraction and Analysis by qPCR

The total RNA of cells and BAT were extracted using RNAiso Plus reagent. Isolated RNA was quantified by measuring optical density (OD) at 260 and 280 nm with a NanoDrop 2000 spectrophotometer (Thermo Scientific, Waltham, MA, USA). The cDNA was synthesized by means of a PrimeScript RT reagent kit (Takara Biotechnology Co., Ltd., Dalian, China) and the assay was performed using real-time PCR with a SYBR Premix Ex Taq^{TM} I and 7500 Real-Time PCR System (Applied Biosystems, Foster City, CA, USA). Data were normalized to the internal control actin and analyzed using the ΔΔCt method. Primer sequences for the *Ucp1* promoter and *Prdm16* are listed in Table 1.

Table 1. Primer sequences used for qRT-PCR.

Gene	Primer Sequence (5′→3′)
Ucp1	F:GGCAAAAACAGAAGGATTGC R:TAAGCCGGCTGAGATCTTGT
Prdm16	F:GAAGTCACAGGAGGACACGGR: CTCGCTCCTCAACACACCTC

2.5. Construction of Plasmid Vector

The primers were designed and cloned according to the selected gene fragments, and the target gene was recombined with the vector and transformed into *E. coli*. Thereafter, PCR and double enzyme digestion were performed, followed by sequencing for identification, and the detected correct plasmid vector was used for subsequent experiments. The cleavage sites and vector plasmids of the different gene fragments involved in this experiment are shown in Table 2.

Table 2. The cleavage sites and vector plasmids of the different gene fragments.

Vector	Gene	Primer Sequence (5′→3′)
Double luciferase reporter gene vector of *Ucp1* promoter	*Ucp1-KPN I-F* *Ucp1-MluI-R*	GGGGTACCcattcctctaagaccatagctt CGACGCGTacttctgcgccctgacct
Eukaryotic vector for overexpression of *Prdm16*	*Prdm16-EcoRI-F* *Prdm16-XbaI-R*	tccaagcttctgcaggaattcA TGCGA TCCAAGGCGAGG accgggcccactagttctagaTCA TTGCA TA TGCCTCCGGG
Double luciferase reporter gene vector of *Ucp1* promoter with different fragment size	*Ucp1(-150bp)-Kpn I-F* *Ucp1(-150bp)-Mlu I-R* *Ucp1(-500bp)-Kpn I-F* *Ucp1(-500bp)-Mlu I-R* *Ucp1(-1000bp)-Kpn I-F* *Ucp1(-1000bp)-Mlu I-R* *Ucp1(-1500bp)-Kpn I-F* *Ucp1(-1500bp)-Mlu I-R*	atttctctatcgataggtaccGAGTGACGCGCGGCTGGG cgagcccgggctagcacgcgtCTGCGCCCTGACCTGGGA atttctctatcgataggtaccTCCAGTCACCCAAA TCTGAAGG cgagcccgggctagcacgcgtCTGCGCCCTGACCTGGGA atttctctatcgataggtaccAGCAGAACCTGGCCAACCA cgagcccgggctagcacgcgtCTGCGCCCTGACCTGGGA atttctctatcgataggtaccTTATTATACTGTTGTTGCTGCTGCT cgagcccgggctagcacgcgtCTGCGCCCTGACCTGGGA
pLacZi-Ucp1	*Ucp1-EcoR I-F* *Ucp1-Xho I-R* *Prdm16-EcoR I-F* *Prdm16-Xho I-R*	ctttgatattggatcgaattcCATTCCTCTAAGACCATAGCTTGGT atacagagcacatgcctcgagACTTCTGCGCCCTGACCTG gattatgcctctcccgaattcTACGCTAGGTTCCGCTCCC agaagtccaaagcttctcgagTAGTAACGTA TACGGAGGCCCA T
PJG-Prdm16	*Prdm16-EcoR I-F* *Prdm16-Xho I-R*	gattatgcctctcccgaattcTACGCTAGGTTCCGCTCCC agaagtccaaagcttctcgagTAGTAACGTA TACGGAGGCCCA T

The primer sequence was designed by CE Design, via the official website of Nanjing Vazyme Biotechnology Co., Ltd. (Nanjing, China), (www.vazyme.com (accessed on 31 May 2020)), according to the gene sequence provided by NCBI, and the primer was synthesized by Shanghai Sangon Biological Engineering Technology Co., Ltd. (Shanghai, China).

2.6. Cell Transfection and Luciferase Activity Detection

Cell transfection was performed using the method described in the Lipo8000 Transfection Reagent, Beyotime Biotechnology, Inc., with some modification. The transfection reagent was configured to 50 μL of Opti-MEM® (Thermo Scientific, Waltham, MA, USA) Medium, 1 μg of plasmid DNA, and 1.6 μL of Lipo8000™ (Beyotime Biotechnology, Shanghai, China) transfection reagent. After 24 h transfection, the cells were lysed, and firefly luciferase detection reagent and renal luciferase detection working solution were added, respectively, to measure RLU on a multifunctional microplate reader.

2.7. Yeast One-Hybrid Assay

In order to verify the interaction between *Prdm16* and the *Ucp1* promoter, in this experiment, the *p*LacZi-Ucp1 reporter vector and *p*JG-Prdm16 fusion expression vector were constructed and transfected into yeast to screen for blue leukoplakia.

Respectively, the *Prdm16* was connected with the PJG plasmid after enzyme digestion, the *Ucp1* promoter was connected with the pLacZi plasmid after enzyme digestion, and co-transfection was performed into EGY48 competent cells. The cells transfected with Placezi+ and PJG+ were used as the positive control, *p*LacZi-Ucp1 and PJG empty vectors were used as the self-activated control, and the Placezi empty vector and *p*JG-Prdm16 were used as the negative control. The cells were cultured in a SD-trp/ura-deficient medium for 48 h, after which single colonies were picked and cultured in X-α-gal medium for 48 h, and the results were observed.

2.8. shRNA-Mediated Gene Silencing of Prdm16

According to the gene sequence number information provided by NCBI, a primer sequence was designed through the GPP Web Portal website: 5′-CCGGGACGGTGACGTTGT AAATAATCTCGAGATTATTTACAACGTCACCGTCTTTTTG-3′. The pLKO.1 vector was enzymatically cleaved with AgeI (NEB, #R0552S) and EcoRI (NEB, #R0101S) for ligation and transformation. The collected plasmids were transferred into $C_3H_{10}T_{1/2}$ cells and inoculated onto a 6-well cell culture plate. After further differentiation for 6 days, according to the brown adipose differentiation method, the mRNA expressions of prdm16 and Ucp1 mRNA the cells were detected by RT-PCR.

2.9. Microscale Thermophoresis (MST) Study

The MST micro-thermophoroscope (Monolith NT.115) was derived from Northrop. First, the PPARα protein was fluorescently labeled to make it become a reporter molecule, and the gradient-diluted C3G molecule was mixed with the PPARα protein of the same concentration in an equal amount and loaded into the capillary. The affinity of molecular interactions was quantitatively analyzed by detecting the fluorescence changes caused by thermophoresis in capillaries with different concentrations under the temperature gradient field. The experimental results were automatically analyzed by MO Affinity Analysis software, and the Kd value was calculated accurately.

2.10. Analysis of Experimental Data

All data were expressed as means ± SD. Statistical analysis was conducted using GraphPad Prism (version 7.0). The statistical differences between the groups were evaluated using one-way ANOVA with a least significant difference (LSD) test, and a *p*-value less than 0.05 was considered statistically significant.

3. Results

3.1. C3G Upregulates the Expression of Ucp1 Gene Both In Vitro and In Vivo

As mentioned above, it has previously been reported that C3G treatment specifically upregulates the thermogenic gene expressions in BAT-cMyc cells [17]. In this study, we treated $C_3H_{10}T_{1/2}$ cells with various concentrations of C3G to explore the effect of C3G on the expression of the *Ucp1* gene. However, this proved to have no effect on BAT-cMyc cell viability or proliferation during brown adipogenesis. In comparison with the PBS medium, C3G at concentrations of 10 μM, 20 μM, and 40 μM was found to significantly upregulate the relative expression level of *Ucp1* mRNA in the $C_3H_{10}T_{1/2}$ cells ($p < 0.05$) and, at 20 μM, the C3G could significantly upregulate the expression of Ucp1 (Figure 1A) ($p < 0.01$). Similarly, through the animal experiments, it was found that C3G treatment could significantly increase the expression of the *Ucp1* gene of BAT in db/db mice and mice with DIO. The expression of *Ucp1* in the C3G-treated mice was 3.82 times and 3.06 times higher than those in the two control groups (Figure 1B,C). These results are consistent with our previous research results [4,17], further demonstrating C3G-induced upregulation of *Ucp1* mRNA expression, both in vitro and in vivo.

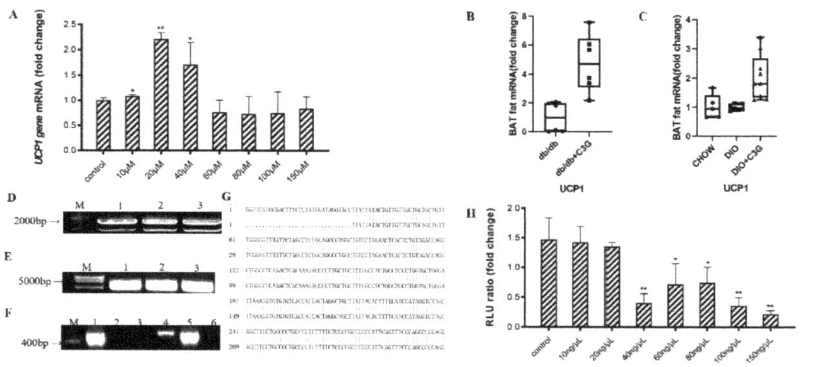

Figure 1. C3G was able to up-regulate the expression of *Ucp1* gene both in vitro and in vivo. (**A**) Effects of C3G at different concentrations on *Ucp1* gene expression in $C_3H_{10}T_{1/2}$ cells. (**B**,**C**) C3G up-regulated the expression of *Ucp1* gene both in db/db mice model (**B**) and mice with DIO model (**C**) (n = 6–8). (**D**) Electrophoretic map of cloned Ucp1, M: Kang was 200Bpladder; Lane 1–3: Ucp1 amplification products (**E**) Electrophoresis diagram of pGL3-Basic enzyme digestion product, M: DL15000 marker; Lane 1–3: pGL3-Basic double enzyme digestion product.(**F**) PCR electrophoresis diagram of pGL3-Ucp1 bacterial solution; M: Kang Wei 200Bpladder; Lane 1–6: PCR products from the picked six single-colony bacterial solutions.(**G**) Comparative diagram of sequencing results of PGL3-Ucp1 (**H**) Activation of *Ucp1* promoter by C3G with different concentrations. Values represent means ± SD. Error bars represent SD; mean values with asterisk or different letters are significantly different (* means $p < 0.05$; ** means $p < 0.01$).

All figures and tables should be cited in the main text as Figure 1, Table 1, etc.

Next, we successfully constructed the luciferase reporter gene vector (Figure 1D–G), to examine the role of C3G in the activation of *Ucp1*. The luciferase assay showed that, compared with the PBS-supplemented medium, treatments with various concentrations of C3G and transfection with the −2.0 kb *Ucp1* promoter–luciferase did not activate the *Ucp1* promoter (Figure 1H) ($p < 0.01$). Thus, these results indicate that, while C3G significantly accelerates *Ucp1* expression, it does not directly activate its promoter.

3.2. C3G Upregulates the Expression of Prdm16 Gene Both In Vitro and In Vivo

As the expression level of the BAT-specific gene *Ucp1* was increased by C3G treatment and this process was not dependent on the activation of the *Ucp1* promoter, we speculated the existence of other regulators as target genes of C3G, such as transcription factors. *Prdm16* determines the brown fat-like program and thermogenesis in both brown and white adipose tissues [19–22], so we consequently sought to determine whether this thermogenesis-related gene participates in this process, and found that treatment with C3G (20 μM) resulted in a 4.42-fold increase in *Prdm16* mRNA expression in $C_3H_{10}T_{1/2}$ cells, compared to the control group. Interestingly, the expression levels of *Prdm16* followed the same trend as that of *Ucp1* after treatment with various concentrations of C3G (Figure 2A) ($p < 0.01$). Moreover, the promotional effect of C3G on *Prdm16* was consistent with that in previous reports [4,17]. Similarly, in light of the animal experiments in this study, it was ascertained that C3G treatment can significantly increase the expression of *Prdm16* and *PPARα* genes in the BAT of mice. In the db/db mouse model and the DIO model mice, the expressions of *Prdm16* and *PPARα* genes in C3G-treated mice increased by 7.21 times, 10.16 times, 3.91 times, and 4.58 times ($p < 0.05$), respectively, compared with the control group (Figure 2B,C). These results are consistent with our previous research results [4,17], indicating clearly that C3G treatment increased the expression of *Prdm16* gene, both in vitro and in vivo.

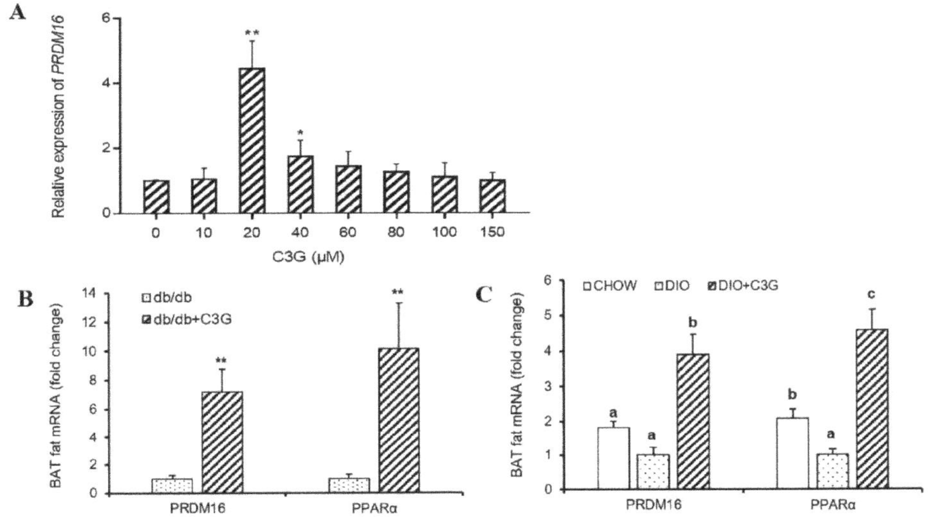

Figure 2. C3G was able to up-regulate the expression of *Prdm16* gene both in vitro and vivo. (**A**) Effect of C3G at different concentrations on *Prdm16* gene expression in C3H10T12 cells. (**B**,**C**) Gene expression profile in BAT, real-time PCR analysis of thermogenic-related genes, including *Prdm16*, PPARα in db/db mice model (**B**) and mice with DIO model (**C**) (n = 6–8). Values represent means SD. Error bars represent SD; mean values with asterisk or different letters are significantly different (* means $p < 0.05$; ** means $p < 0.01$).

3.3. *Prdm16* Regulates the Expression of *Ucp1* Gene in Brown Adipocytes

We hypothesized accordingly that C3G may first upregulate the expression of gene *Prdm16* and then activate Ucp1. To explore and examine the effect of transcription factor *Prdm16* on *Ucp1* expression, we constructed a *p*CMV-Prdm16 vector to transfect $C_3H_{10}T_{1/2}$ cells and further ascertained the level of *Ucp1* mRNA via quantitative real-time PCR after the completion of adipocyte differentiation (Figure 3A–C). As expected, *Prdm16* did regulate *Ucp1* expression at the level of transcription. In the treatment group in which *Prdm16* was overexpressed, the expression of *Ucp1* increased over four times more than that in the control group, with a significant up-regulation effect (Figure 3D) ($p < 0.01$). Further, to validate the loss-of-function studies, we constructed shPRDM16 to transfect $C_3H_{10}T_{1/2}$ cells and further determine *Ucp1* mRNA levels by quantitative real-time polymerase chain reaction after adipocyte differentiation. As shown in Figure 3E, shPRDM16 reduced Ucp1 expression at the transcriptional level by more than 0.5 times that of the control group. Taken together, these results indicate that C3G did, at least partly, activate *Ucp1* by regulating the expression of the transcription factor *Prdm16*.

3.4. *Prdm16* Directly Binds and Activates *Ucp1* Promoter

As a transcription factor, Prdm16 often regulates gene expression by binding to the promoter region of the target gene. To determine the mechanism by which Prdm16 regulates *Ucp1*, we performed a double luciferase reporter experiment to quantitate promoter activation. Compared to the empty vector control, co-transfection of *p*CMV-Prdm16 with the −2.0 kb *Ucp1* promoter–luciferase resulted in an obvious activation of the *Ucp1* promoter (Figure 4A) ($p < 0.01$).

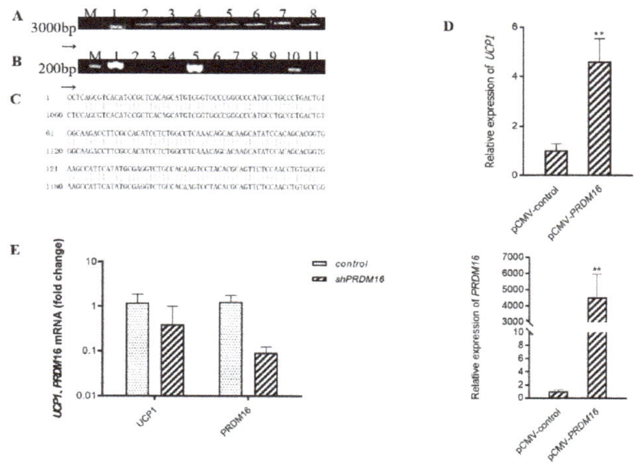

Figure 3. Overexpression of *Prdm16* upregulates *Ucp1* gene expression in brown adipose cells (**A**) Clone *Prdm16* electrophoretic map, M: Kang Wei 200B PLA DD; Lane 1–8: *Prdm16* amplification product. (**B**) PCR electrophoresis diagram of pCMV-Prdm16 bacterial solution; M: Kang Wei 200B PLA DD; Lane 1–11: PCR products from the picked bacterial solutions of 11 single colonies. (**C**) Comparative diagram of pCMV-Prdm16 sequencing results. (**D**) Effect of overexpression of *Prdm16* on *Ucp1* gene expression C3H10T12 cells. (**E**) Effects of shPRDM16 on Ucp1 gene expression during $C_3H_{10}T_{1/2}$ cell differentiation. Compared with control, ** means $p < 0.01$.

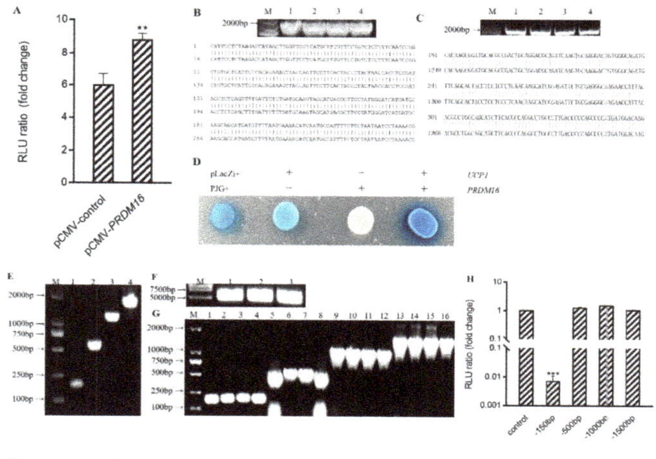

Figure 4. Prdm16 capable of activating *Ucp1* promoter (**A**) Effect of Prdm16 on *Ucp1* promoter (double luciferase report gene method) (**B**) PCR electrophoresis diagram and sequencing comparison diagram of pLacZi-Ucp1 bacterial solution. M: Kang Wei 200B PLA DD;Lane 1–4: PCR products from the picked four single-colony bacterial solutions. (**C**) PCR electrophoresis diagram and sequencing comparison diagram of pJG-Prdm16 bacterial solution. M: Kang Wei 200B PLA DD; Lane 1–4: PCR products from the picked four single-colony bacterial solutions. (**D**) Effect of Prdm16 on *Ucp1* promoter (yeast one-hybrid assay) (**E**) Clone electrophoretic images of Ucp1 promoter with different lengths. (**F**) PGL3-Basic enzyme digestion electrophoresis diagram (**G**) PCR electrophoresis map of bacterial solution after ligation of *Ucp1* promoter with different lengths to pGL3-Basic. (**H**) Effects of Prdm16 on promoters of different fragments *Ucp1*. Compared with control, ** means $p < 0.01$; *** means $p < 0.001$.

This interaction of *Prdm16* with the *Ucp1* promoter was subsequently reverified. A *p*LacZi-Ucp1 reporter vector and *p*JG-Prdm16 fusion expression vector were constructed and transfected into yeast for blue-streak screening (Figure 4B–D). Similar to the positive control group of Plazzi+ and PJG+, treatment groups transfected with *p*LacZi-*Ucp1* and *p*JG-Prdm16 were blue and dark in color. The results verified by the single yeast hybrid assay showed that *Prdm16* could directly bind to the *Ucp1* promoter.

To start defining how Prdm16 activates the *Ucp1* promoter, 5′deletions of the *Ucp1* promoters driving a luciferase reporter were generated and co-transfected along with *p*CMV-Prdm16 into 293 cells (Figure 4E–G). All *Ucp1* promoter constructs deleted down to −500 bp, thus showing remarkable activation upon *Prdm16* co-transfection. However, *Ucp1* promoter activation by Prdm16 was lost when the promoter was deleted to −150 bp, indicating that Prdm16 functions through the sequence from −500 to −150 bp (Figure 4H).

3.5. PPARα Binds to Ucp1 Promoter to Activate Transcription

The above experiments showed that Prdm16, regulated by C3G, plays a novel role in upregulating the expression of *Ucp1*; however, it has been previously reported that the expression of *Ucp1* is regulated by many factors. We found that *PPARα*, a gene associated with fatty acid oxidation, could also activate *Ucp1* promotion in the same way as Prdm16 (Figure 5A–E) ($p < 0.01$). In addition, the results of the luciferase activity assay showed that the activity of the −150 bp *Ucp1* promoter was significantly reduced (Figure 5F) ($p < 0.01$). Analysis, combined with the results of this assay, showed that the region where the *Ucp1* promoter binds to *PPARα* is between −500 and −150 bp upstream of its initiation site.

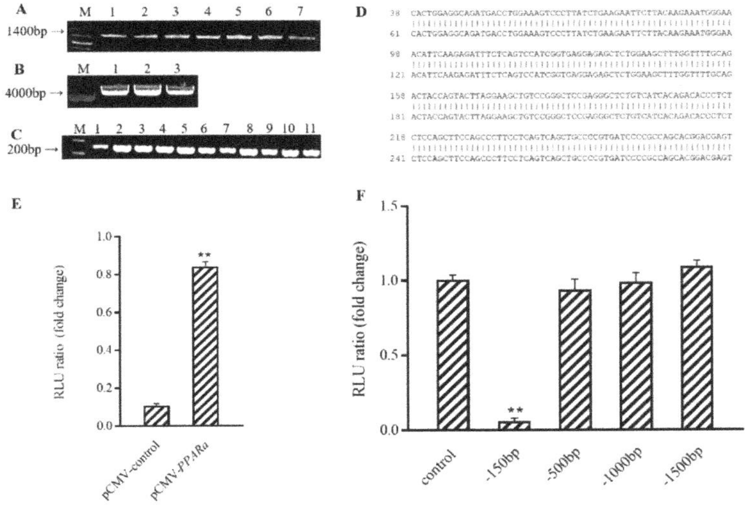

Figure 5. PPARα binds to *Ucp1* promoter to activate transcription. (**A**) Cloned PPARα electrophoresis M: Kang Wei 200B PLA DD; Lane 1–7: PPARα amplification product. (**B**) Electrophoresis map of pCMV-N-Flag enzyme digestion product, M: Kang Wei 200B PLA DD; Lane 1–3: pCMV-N-Flag double enzyme digestion product. (**C**) PCR electrophoresis diagram of pCMV-PPARα bacterial solution;M: Kang Wei 200B PLA DD; Lane 1–11: PCR products from picked 11 single-colony bacterial solutions (**D**) Comparison of pCMV-PPARα sequencing results. (**E**) Effect of PPARα on the intact *Ucp1* promoter (−2000 bp). (**F**) Effect of PPARα on promoter of *Ucp1* with different fragments. ** mesns $p < 0.01$.

3.6. Conformational Investigation and Interaction Study of PPARα–C3G

Molecular modeling of the PyMOL program was used to improve understanding of the PPARα–C3G interaction. The optimal energy sequencing results for PPARα and

C3G binding modes are shown in Figure 6A. As the thermophoresis law of molecules in solution is closely related to their molecular size, charge, and hydration layer properties, the change in any property will cause a change in its movement law, MST technology has extremely high sensitivity and wide application range. MST was utilized to further prove the interaction results by fluorescence spectroscopy. As shown in Figure 6B, the fitting curve of PPARα and C3G binding was well-shaped, with the Kd value of 1.12 μm. The Kdre value indicated the degree of dissociation of the complex in the equilibrium state. The smaller Kd value indicated that the degree of dissociation was lower, and the affinity between PPARα–C3G was stronger.

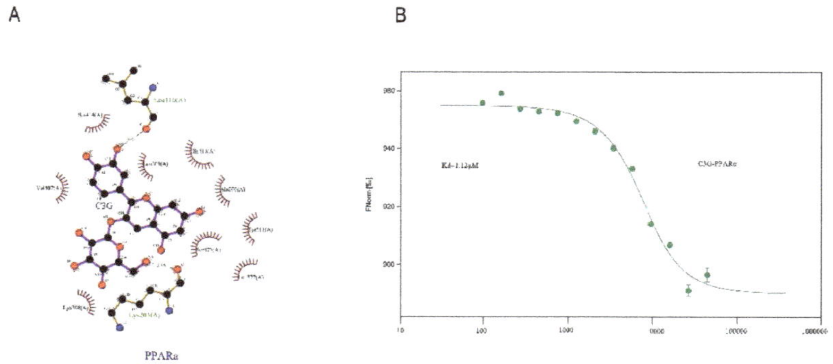

Figure 6. Conformational investigation and interaction study of PPARα–C3G (**A**).The molecular visualization system predicted the binding conformation of pparα to C3G. (**B**) MST technique to quantitate the affinity of pparα for C3G molecular interactions. The final C3G was added at varying concentrations (ranging from 5 μM to 23,000 μM).

4. Discussion

This study proved, for the first time, that C3G does not upregulate the expression of Ucp1 by activating the *Ucp1* promoter. Another novel finding herein is that Prdm16 can bind to the region −500 to −150 bp upstream of the transcription start site of *Ucp1* gene, thereby activating the promoter activity. It was also proposed for the first time that C3G participates in the synthesis and thermogenesis of brown adipocyte Ucp1 by regulating *Prdm16*.

Thermogenesis is one of the most important functions of BAT, a tissue that has recently been shown to play a significant role in energy and glucose metabolism in adults, making it an interesting potential target for strategies aimed at augmenting the energy metabolic rate [23]. In short, increasing BAT activity may be a highly effective therapeutic approach to combating obesity and its related metabolic diseases.

Uncoupling proteins (UCPs) are a class of heat-producing proteins distributed on the inner membrane of mitochondria. Their activity is closely related to heat loss, regulated by purine nucleosides (ADP, ATP), guanosine (GDP, GTP), and free fatty acids (FFA), and they, therefore, play an important role in mammalian heat production and energy metabolism regulation [24]. There are five known main types of UCPs, namely Ucp1, UCP2, UCP3, UCP4, and UCP5. Ucp1 activity exists in mammalian BAT, while UCP2 is widely distributed in whole body tissues, such as white adipose tissue (WAT), heart, liver, kidney, spleen, thyroid, and gastrointestinal tract [25]. UCP3 is mainly distributed in skeletal muscle and UCP4 is mainly distributed in brain [26,27], while UCP5 is mainly found in brain and neural tissues [28]. Ucp1 is the only expression of uncoupling protein found in BAT.

Brown fat cells are characterized by densely packed mitochondria that contain Ucp1 in their inner mitochondrial membrane [29]. BAT thermogenesis is dependent on Ucp1 as it is a proton transporter that allows protons to leak across the mitochondrial inner membrane, thereby dissipating the electrochemical gradient normally used for ATP synthesis [29]. The regulation of levels of the *Ucp1* gene, thus, plays a decisive role in the control of Ucp1 content and is at the center of the physiological regulation of BAT heat production [30]. Revealing the involved signal pathway and transcription regulation of Ucp1 in BAT's energy metabolic process not only helps us to better understand the important role of Ucp1 in BAT energy metabolism control but also provides a theoretical foundation for obesity treatment based on BAT.

Our previous research showed that dietary supplementation with the functional ingredients of natural plants exhibited a positive effect on BAT functionality in mice. Such plant-based ingredients include mulberry polyphenol extract, such as C3G and rutin [4,18,31]. Moreover, some phenolic acids, such as vanillic acid and chlorogenic acid, which are components of mulberry fruit polyphones, also induce the thermogenesis of BAT and could, therefore, be applied to the prevention of dietary-induced obesity and insulin resistance [31,32]. We previously reported that mulberry and mulberry wine extract (ME and MWE) can not only improve the expression of BAT specific gene *Ucp1*, but also improve the mitochondrial copy number [33]. As the most abundant anthocyanin in ME and MWE, C3G has also been proven to increase the quantity of mitochondria in subsequent experiments [17]. In order to further explore the heat production mechanism of BAT promoted by C3G, we studied the mitochondrial biogenesis and function. The results show that C3G significantly upregulated the expression of mitochondrial synthesis-related genes *Tfam, NRF1* and *NRF2*, in both the BAT of both the db/db and mice with DIO, and increased the number of mitochondria and the expression levels of mitochondrial oxidative phosphorylation-related proteins ATP5A, UQCR2, and NDUFB8 [4,18].

This strongly suggested the promotion of BAT heat production by C3G, thus warranting further exploration of the mechanism by which C3G regulates the action of *Ucp1* and its related heat production gene. In this study, we first verified the promotional effect of C3G on *Ucp1* gene expression through animal models. Thereafter, we found that C3G can upregulate the BAT specific gene *Ucp1* expression during brown adipocyte differentiation, thereby effectively increasing the mitochondrial respiration uncoupling of brown adipocytes, which is consistent with the previously reported conclusion [34]. However, our study found that there is no concentration dependence, and also that the activation of *Ucp1* by C3G is not achieved by activating the promoter, thus suggesting the existence of other regulatory factors in the process.

Numerous studies have shown that the activity of Ucp1 is inhibited by nucleotides and activated by non-lipidated fatty acids [33,35]. Norepinephrine (NE) released by sympathetic nerves or hypothermia are the main physiological signals that activate Ucp1 synthesis in BAT cells [36]. Furthermore, the synthesis of Ucp1 is strongly regulated by the transcription level; cAMP is the main activator of *Ucp1* gene transcription [37–40]; and β3-AR-specific agonists L-755507, CGP12177, CL316243, and BRL35135 can significantly increase the levels of *Ucp1* mRNA in mammalian BAT cells [38–40]. In addition, several transcription factors and coregulators found in both WAT and BAT, as well as in other tissues, including PPARs, PGC1α, and ATF2, have been implicated in the transcriptional activation of *Ucp1* [12,41]. Our data show that C3G can activate *Ucp1* by upregulating the expression of the transcription factor *Prdm16*. Since Prdm16 is related to heat production [42,43], it is even clearer that C3G does indeed play a role in weight loss by activating heat production.

Prdm16 can form transcription complexes with other transcription factors, such as C/EBPβ, PGC1α, PPARα, and PPARγ, to regulate the expression of thermogenic genes [14,44]. Furthermore, it was found in our previous research that C3G can upregulate PPARα [4]. Consequently, on the basis of the above discussion, it would be reasonable to believe that PPARα is also involved in *Ucp1* transcriptional regulation. Through double luciferase experiments, we ascertained that PPARα can indeed activate the transcriptional expression of *Ucp1* in a similar manner to that of Prdm16. Unlike Prdm16, however, PPARα is a fatty acid oxidation-related gene that is also expressed in WAT, which suggests that the transcriptional activation of C3G to *Ucp1* is not limited to BAT. C3G is expected to achieve weight loss and lipid reduction by increasing WAT heat production, which would be consistent with previous conclusions [18].

On the one hand, we proved that C3G could enhance the thermogenesis of BAT. On the other hand, anthocyanins were reported to have anti-cancer and anti-inflammatory effects [45], which were closely related to their antioxidant activities. Therefore, the role of antioxidant activity of C3G in thermogenesis could be further explored. However, there are a variety of metabolites of C3G, and 41 metabolites have been identified. Among them, protocatechuic acid and vanilloid are metabolites that exert the main physiological functions of C3G in the body [46]. For example, PCA and PGA, the degradation products of C3G, inhibit the proliferation of cancer cells, which is closely related to their antioxidant activity [47]. The effect of C3G on the thermogenesis of brown fat suggests that its metabolites may have the same effect or even more obvious effects. Therefore, the above C3G metabolites can be further studied in future studies to further clarify the effective activators that activate *Ucp1*.

5. Conclusions

Taken together, the findings of our study indicate that C3G can upregulate the expression level of Prdm16, and then activate the transcriptional expression of *Ucp1* by directly binding to the −500 to −150 bp region of the *Ucp1* promoter.

In future research, it would be valuable to determine whether Prdm16 interacts with PPARα or other activating factors to activate *Ucp1* and to determine the molecular mechanism by which C3G regulates Prdm16 expression. The results of this study provide an effective approach for the development of novel methods to improve energy metabolism with the activation of Ucp1 as the target, and also help to establish an innovative theoretical basis for the in-depth utilization and development of C3G-rich fruit and vegetable resources.

Author Contributions: Conceptualization, Y.Y. (Yilin You) and J.Z.; methodology, W.H.; software, J.G.; validation, Y.Y. (Yafan Yang), S.H. and Y.L.; formal analysis, X.H.; investigation, Y.G.; resources, Y.Y. (Yilin You); data curation, Y.Y. (Yilin You); writing—original draft preparation, Y.Y. (Yafan Yang) and S.H.; writing—S.H. and Y.L.; visualization, S.H.; supervision, Y.Y. (Yilin You); project administration, J.Z.; funding acquisition, Y.Y. (Yilin You). All authors have read and agreed to the published version of the manuscript.

Funding: This research was funded by the National Natural Science Foundation of China Youth Fund (81900777) to Yilin You and the National "Thirteenth Five-Year" Plan for Science and Technology Support (2016YFD0400500) to Weidong Huang.

Institutional Review Board Statement: This study was approved by the Animal Experiment Committee of the College of Food Science and Nutritional Engineering, China Agricultural University (authorization reference number CFSNE20190038).

Informed Consent Statement: Not applicable.

Data Availability Statement: The data presented in this study are available in this manuscript.

Conflicts of Interest: The authors declare no conflict of interest.

References

1. You, Y.; Zhou, F.; Huang, D. Eating the Right Color: Dietary Anthocyanins and Obesity Control. *Food Beverage Asia* **2018**, *12*, 57–59.
2. Duchowicz, P.R.; Szewczuk, N.A.; Pomilio, A.B. QSAR Studies of the Antioxidant Activity of Anthocyanins. *J. Food Sci. Technol. -Mysore* **2019**, *56*, 5518–5530. [CrossRef] [PubMed]
3. Correia, P.; Araújo, P.; Ribeiro, C.; Oliveira, H.; Pereira, A.R.; Mateus, N.; de Freitas, V.; Brás, N.F.; Gameiro, P.; Coelho, P.; et al. Anthocyanin-Related Pigments: Natural Allies for Skin Health Maintenance and Protection. *Antioxidants* **2021**, *10*, 1038. [CrossRef] [PubMed]
4. You, Y.; Yuan, X.; Liu, X.; Liang, C.; Zhan, J. Front Cover: Cyanidin-3-glucoside Increases Whole Body Energy Metabolism by Upregulating Brown Adipose Tissue Mitochondrial Function. *Mol. Nutr. Food Res.* **2017**, *61*, 1770111. [CrossRef]

5. Ng, T. Global, Regional, and National Prevalence of Overweight and Obesity in Children and Adults during 1980–2013: A Systematic Analysis for the Global Burden of Disease Study 2013. *Lancet* **2014**, *384*, 766–781. [CrossRef]
6. Al-Tahami, B.A.M.; Ismail, A.A.A.-S.; Bee, Y.T.G.; Awang, S.A.; Rani, W.R.S.W.A.; Sanip, Z.; Rasool, A.H.G. The effects of anti-obesity intervention with orlistat and sibutramine on microvascular endothelial function. *Clin. Hemorheol. Microcirc.* **2015**, *59*, 323–334. [CrossRef]
7. Siebenhofer, A.; Jeitler, K.; Horvath, K.; Berghold, A.; Semlitsch, T. Long-Term Effects of Weight-Reducing Drugs in People with Hypertension. *Cochrane Database Syst. Rev.* **2016**, *3*, CD007654. [CrossRef]
8. Cypess, A.M.; Chen, Y.C.; Sze, C.; Wang, K.; English, J.; Chan, O.; Holman, A.R.; Tal, M.; Palmer, M.R.; Kolodny, G.M. Cold but Not Sympathomimetics Activates Human Brown Adipose Tissue in Vivo. *Proc. Natl. Acad. Sci. USA* **2012**, *109*, 10001–10005. [CrossRef]
9. van Marken Lichtenbelt, W.D.; Vanhommerig, J.W.; Smulders, N.M.; Drossaerts, J.M.; Kemerink, G.J.; Bouvy, N.D.; Teule, G.J. None Cold-Activated Brown Adipose Tissue in Healthy Men. *N. Engl. J. Med.* **2009**, *360*, 1917. [CrossRef]
10. Villarroya, F.; Peyrou, M.; Giralt, M. Transcriptional Regulation of the Uncoupling Protein-1 Gene. *Biochimie* **2017**, *134*, 86–92. [CrossRef]
11. Escher, P.; Braissant, O.; Basu-Modak, S.; Michalik, L.; Wahli, W.; Desvergne, B. Wahli Rat PPARs: Quantitative Analysis in Adult Rat Tissues and Regulation in Fasting and Refeeding. *Endocrinology* **2001**, *142*, 4195–4202. [CrossRef]
12. Collins, S.; Yehuda-Shnaidman, E.; Wang, H. Positive and Negative Control of Ucp1 Gene Transcription and the Role of β-Adrenergic Signaling Networks. *Int. J. Obes.* **2010**, *34*, S28–S33. [CrossRef]
13. Harms, M.J.; Ishibashi, J.; Wang, W.; Lim, H.W.; Seale, P. Prdm16 Is Required for the Maintenance of Brown Adipocyte Identity and Function in Adult Mice. *Cell Metab.* **2014**, *19*, 593–604. [CrossRef] [PubMed]
14. Kajimura, S.; Seale, P.; Kubota, K.; Lunsford, E.; Frangioni, J.V.; Gygi, S.P.; Spiegelman, B.M. Initiation of Myoblast to Brown Fat Switch by a Prdm16–C/EBP-β Transcriptional Complex. *Nature* **2009**, *460*, 1154–1158. [CrossRef]
15. Matsukawa, T.; Inaguma, T.; Han, J.; Villareal, M.O.; Isoda, H. Cyanidin-3-Glucoside Derived from Black Soybeans Ameliorate Type 2 Diabetes through the Induction of Differentiation of Preadipocytes into Smaller and Insulin-Sensitive Adipocytes. *J. Nutr. Biochem.* **2015**, *26*, 860–867. [CrossRef] [PubMed]
16. Wu, T.; Yu, Z.; Tang, Q.; Song, H.; Gao, Z.; Chen, W.; Zheng, X. Honeysuckle Anthocyanin Supplementation Prevents Diet-Induced Obesity in C57BL/6 Mice. *Food Funct.* **2013**, *4*, 1654–1661. [CrossRef]
17. You, Y.; Liang, C.; Han, X.; Guo, J.; Ren, C.; Liu, G.; Huang, W.; Zhan, J. Mulberry Anthocyanins, Cyanidin 3-Glucoside and Cyanidin 3-Rutinoside, Increase the Quantity of Mitochondria during Brown Adipogenesis. *J. Funct. Foods* **2017**, *36*, 348–356. [CrossRef]
18. You, Y.; Han, X.; Guo, J.; Guo, Y.; Yin, M.; Liu, G.; Zhan, J. Cyanidin-3-Glucoside Attenuates High-Fat and High-Fructose Diet-Induced Obesity by Promoting the Thermogenic Capacity of Brown Adipose Tissue. *J. Funct. Foods* **2018**, *41*, 62–71. [CrossRef]
19. Becerril, S.; Gómez-Ambrosi, J.; Martín, M.; Moncada, R.; Frühbeck, G. Role of Prdm16 in the Activation of Brown Fat Programming. Relevance to the Development of Obesity. *Histol. Histopathol.* **2013**, *28*, 28. Available online: http://hdl.handle.net/10201/61619 (accessed on 18 May 2021).
20. Hondares, E.; Rosell, M.; Diaz-Delfin, J.; Olmos, Y.; Monsalve, M.; Iglesias, R.; Villarroya, F.; Giralt, M. Peroxisome Proliferator-Activated Receptor α (PPARα) Induces PPARγ Coactivator 1α (PGC-1α) Gene Expression and Contributes to Thermogenic Activation of Brown Fat: Involvement of Prdm16. *J. Biol. Chem.* **2011**, *286*, 43112. [CrossRef]
21. Ohno, H.; Shinoda, K.; Ohyama, K.; Sharp, L.Z.; Kajimura, S. EHMT1 Controls Brown Adipose Cell Fate and Thermogenesis through the Prdm16 Complex. *Nature* **2013**, *504*, 163–167. [CrossRef]
22. Seale, P.; Kajimura, S.; Yang, W.; Chin, S.; Rohas, L.; Uldry, M.; Tavernier, G.; Langin, D.; Spiegelman, B. Transcriptional Control of Brown Fat Determination by Prdm16. *Cell Metab.* **2007**, *6*, 38–54. [CrossRef]
23. Cypess, A.M.; Lehman, S.; Williams, G.; Tal, I.; Rodman, D.; Goldfine, A.B.; Kuo, F.C.; Palmer, E.L.; Tseng, Y.H.; Doria, A. Identification and Importance of Brown Adipose Tissue in Adult Humans. *Obstet. Gynecol. Surv.* **2009**, *64*, 519–520. [CrossRef]
24. Himms-Hagen, J. Brown Adipose Tissue Thermogenesis: Interdisciplinary Studies. *Faseb. J.* **1990**, *4*, 2890–2898. [CrossRef] [PubMed]
25. Hidaka, S.; Kakuma, T.; Yoshimatsu, H.; Yasunaga, S.; Kurokawa, M.; Sakata, T. Molecular Cloning of Rat Uncoupling Protein 2 CDNA and Its Expression in Genetically Obese Zucker Fatty (Fa/Fa) Rats. *Biochim. Biophys. Acta* **1998**, *1389*, 178–186. [CrossRef]
26. Mao, W.; Yu, X.X.; Zhong, A.; Li, W.; Brush, J.; Sherwood, S.W.; Adams, S.H.; Pan, G. UCP4, a Novel Brain-Specific Mitochondrial Protein That Reduces Membrane Potential in Mammalian Cells. *FEBS Lett.* **1999**, *443*, 326–330. [CrossRef]
27. Solanes, G.; Vidal-Puig, A.; Grujic, D.; Flier, J.S.; Lowell, B.B. The Human Uncoupling Protein-3 Gene: Genomic structure, chromosomal localization, and genetic basis for short and long form transcripts *. *J. Biol. Chem.* **1997**, *272*, 25433–25436. [CrossRef]
28. Yu, X.X.; Mao, W.; Zhong, A.; Schow, P.; Brush, J.; Sherwood, S.W.; Adams, S.H.; Pan, G. Characterization of Novel UCP5/BMCP1 Isoforms and Differential Regulation of UCP4 and UCP5 Expression through Dietary or Temperature Manipulation. *Faseb. J. Off. Publ. Fed. Am. Soc. Exp. Biol.* **2000**, *14*, 1611. [CrossRef]
29. Rosen, E.; Spiegelman, B. What We Talk About When We Talk About Fat. *Cell* **2014**, *156*, 20–44. [CrossRef]
30. Rehnmark, S.; Bianco, A.C.; Kieffer, J.D.; Silva, J.E. Transcriptional and Posttranscriptional Mechanisms in Uncoupling Protein MRNA Response to Cold. *Am. J. Physiol.* **1992**, *262*, 58–67. [CrossRef] [PubMed]

31. You, Y.; Yuan, X.; Lee, H.J.; Huang, W.; Jin, W.; Zhan, J. Zhan Mulberry and Mulberry Wine Extract Increase the Number of Mitochondria during Brown Adipogenesis. *Food Funct.* **2015**, *6*, 401–408. [CrossRef]
32. Han, X.; Guo, J.; You, Y.; Yin, M.; Liang, J.; Ren, C.; Huang, W. Juan Vanillic Acid Activates Thermogenesis in Brown and White Adipose Tissue. *Food Funct.* **2018**, *9*, 4366–4375. [CrossRef]
33. Han, X.; Zhang, Y.; Guo, J.; You, Y.; Huang, W. Chlorogenic Acid Stimulates the Thermogenesis of Brown Adipocytes by Promoting the Uptake of Glucose and the Function of Mitochondria. *J. Food Sci.* **2019**, *84*, 3815–3824. [CrossRef]
34. Echtay, K. Mitochondrial Uncoupling Proteins—What Is Their Physiological Role? *Free Radic. Biol. Med.* **2007**, *43*, 1351–1371. [CrossRef]
35. Jastroch, M.; Withers, K.; Klingenspor, M. Klingenspor Uncoupling Protein 2 and 3 in Marsupials: Identification, Phylogeny, and Gene Expression in Response to Cold and Fasting in Antechinus Flavipes. *Physiol. Genom.* **2004**, *17*, 130–139. [CrossRef] [PubMed]
36. Collins, S.; Daniel, K.W.; Petro, A.E.; Surwit, R.S. Strain-Specific Response to Beta 3-Adrenergic Receptor Agonist Treatment of Diet-Induced Obesity in Mice. *Endocrinology* **1997**, *138*, 405–413. [CrossRef] [PubMed]
37. Kumar, M. Differential Effects of Retinoic Acid on Uncoupling Protein-1 and Leptin Gene Expression. *J. Endocrinol.* **1998**, *157*, 237–243. [CrossRef] [PubMed]
38. Rabelo, R.; Camirand, A.; Silva, J.E. Silva 3′,5′-Cyclic Adenosine Monophosphate-Response Sequences of the Uncoupling Protein Gene Are Sequentially Recruited during Darglitazone-Induced Brown Adipocyte Differentiation. *Endocrinology* **1997**, *138*, 5325–5332. [CrossRef]
39. Sasaki, N.; Uchida, E.; Niiyama, M.; Yoshida, T.; Saito, M. Anti-Obesity Effects of Selective Agonists to the Beta 3-Adrenergic Receptor in Dogs. II. Recruitment of Thermogenic Brown Adipocytes and Reduction of Adiposity after Chronic Treatment with a Beta 3-Adrenergic Agonist. *J. Vet. Med. Sci.* **1998**, *60*, 465. [CrossRef]
40. Savontaus, E.; Rouru, J.; Boss, O.; Huupponen, R.; Koulu, M. Differential Regulation of Uncoupling Proteins by Chronic Treatments with Beta 3-Adrenergic Agonist BRL 35135 and Metformin in Obese Fa/Fa Zucker Rats. *Biochem. Biophys. Res. Commun.* **1998**, *246*, 899–904. [CrossRef] [PubMed]
41. Kang, S.; Bajnok, L.; Longo, K.A.; Petersen, R.K.; Hansen, J.B.; Kristiansen, K.; Macdougald, O.A. Effects of Wnt Signaling on Brown Adipocyte Differentiation and Metabolism Mediated by PGC-1α. *Mol. Cell. Biol.* **2005**, *25*, 1272–1282. [CrossRef]
42. Seale, P.; Conroe, H.M.; Estall, J.; Kajimura, S.; Frontini, A.; Ishibashi, J.; Cohen, P.; Cinti, S.; Spiegelman, B.M. Prdm16 Determines the Thermogenic Program of Subcutaneous White Adipose Tissue in Mice. *J. Clin. Investig.* **2011**, *121*, 96–105. [CrossRef]
43. Ricquier, D.; Casteilla, L.; Bouillaud, F. Molecular Studies of the Uncoupling Protein. *FASEB J.* **1991**, *5*, 2237–2242. [CrossRef] [PubMed]
44. Seale, P.; Bjork, B.; Yang, W.; Kajimura, S.; Chin, S.; Kuang, S.; Scimè, A.; Devarakonda, S.; Conroe, H.M.; Erdjument-Bromage, H.; et al. Prdm16 Controls a Brown Fat/Skeletal Muscle Switch. *Nature* **2008**, *454*, 961–967. [CrossRef] [PubMed]
45. Wang, L.; Zhan, J.; Huang, W. Grape Seed Proanthocyanidins Induce Apoptosis and Cell Cycle Arrest of HepG2 Cells Accompanied by Induction of the MAPK Pathway and NAG-1. *Antioxidants* **2020**, *9*, 1200. [CrossRef] [PubMed]
46. Kay, C.D.; Kroon, P.A.; Cassidy, A. The Bioactivity of Dietary Anthocyanins Is Likely to Be Mediated by Their Degradation Products. *Mol. Nutr. Food Res.* **2010**, *53*, 92–101. [CrossRef]
47. Pace, E.; Jiang, Y.; Clemens, A.; Crossman, T.; Rupasinghe, H.V. Impact of Thermal Degradation of Cyanidin-3-*O*-glucoside of Haskap Berry on Cytotoxicity of Hepatocellular Carcinoma HepG2 and Breast Cancer MDA-MB-231 Cells. *Antioxidants* **2018**, *7*, 24. [CrossRef]

antioxidants

MDPI

Article

Isoorientin Attenuated the Pyroptotic Hepatocyte Damage Induced by Benzo[a]pyrene via ROS/NF-κB/NLRP3/Caspase-1 Signaling Pathway

Hao Li, Li Yuan *, Xueyi Li, Ying Luo, Zhong Zhang and Jianke Li

Engineering Research Center of High Value Utilization of Western China Fruit Resources, College of Food Engineering and Nutritional Science, Shaanxi Normal University, Xi'an 710119, China; lihr13@snnu.edu.cn (H.L.); lixueyi@snnu.edu.cn (X.L.); luoying@snnu.edu.cn (Y.L.); zzhang@snnu.edu.cn (Z.Z.); jiankel@snnu.edu.cn (J.L.)

* Correspondence: yuanli112086@snnu.edu.cn

Citation: Li, H.; Yuan, L.; Li, X.; Luo, Y.; Zhang, Z.; Li, J. Isoorientin Attenuated the Pyroptotic Hepatocyte Damage Induced by Benzo[a]pyrene via ROS/NF-κB/NLRP3/Caspase-1 Signaling Pathway. *Antioxidants* **2021**, *10*, 1275. https://doi.org/10.3390/antiox10081275

Academic Editors: Zhigang Liu, Jicheng Zhan and Hui-Min David Wang

Received: 20 July 2021
Accepted: 9 August 2021
Published: 11 August 2021

Abstract: Isoorientin (Iso), a natural bioactive flavonoid, possesses significant anti-tumor and antioxidant activities. Benzo[a]pyrene (BaP) is a food processing injurant with carcinogenicity, teratogenicity, and genotoxicity. Our preliminary study demonstrates that Iso attenuated the pyroptotic hepatocyte damage induced by BaP; however, the molecular mechanism remains unknown. The present study showed that Iso reduced the increase caused by BaP in the overflow of LDH, NO, and the electrical conductivity and the protein expressions of GSDMD-N, IL-18, and IL-1β, further showing that Iso could reduced the pyroptotic damage in HL-7702 cells induced by BaP. Caspase-1 inhibitor (Z-VAD-FMK) inhibited the characteristic pyroptosis protein expressions of Caspase-1, GSDMD-N, IL-18, and IL-1β, showing that the classic pyroptosis pathway depending on Caspase-1 was caused by BaP in HL-7702 cells. Consistent with the effects of the NLRP3 inhibitor (MCC950), NF-κB inhibitor (PDTC), ROS, and mtROS inhibitor (NAC and Mito-TEMPO), Iso weakened the stimulatory effects of BaP on the levels of ROS, the nuclear localization of NF-κB, and the activation of NLRP3 inflammasome and the characteristic indices of pyroptosis, demonstrating that Iso could alleviate the BaP-induced pyroptotic hepatocytes injury through inhibiting the ROS/NF-κB/NLRP3/Caspase-1 signaling pathway, which provides a new perspective and strategy to prevent liver injury induced by BaP.

Keywords: isoorientin; benzo[a]pyrene; pyroptosis; ROS; NF-κB; NLRP3

1. Introduction

Isoorientin (3′,4′, 5, 7-tetrahydroxy-6-C-glucopyranosy flavone, Iso, Figure 1A), a natural flavonoid compound, is abundant in food such as hawthorn [1], cucumber [2], and pueraria lobata [3], with anti-oxidation, anti-inflammation, and anti-tumor activity [4,5]. It is able to prevent liver injury by reducing lipid metabolism, enhancing the ability of antioxidation, regulating the secretion of pro-inflammatory cytokines [6], and suppressing the enzyme activities of the respiratory chain complexes and the phase II [7].

Benzo[a]pyrene (BaP, Figure 1B), is a group I carcinogen and possesses strong carcinogenicity, teratogenicity, and genotoxicity [8–10]. It is widespread in the high temperature heat processed foods, such as smoked fish, sausage, fried chicken, vegetable oil, and cereals [11]. Pyroptosis is an inflammatory programmed cells death, accompanied by the formation of pores on the cell membrane and the release of contents, which in turn triggers an inflammatory response [12]. It has been reported that Gossypol [13], Arsenic [14], and Cadmium [15] induced the pyroptotic damage.

NLRP3 (NLR family pyrin domain-containing 3) inflammasome is a multiprotein complex, formed by NLRP3, apoptosis associated speck-like protein containing a CARD (ASC), and Pro-Caspase-1. It could be expressed in large quantities, when inflammation occurs in hepatocytes [16]. Existing studies have proven that NLRP3 inflammasome plays a critical role in the process of pyroptosis [17,18], it induces pyroptosis by promoting the cleavage and activation of Caspase-1, and resulting in the maturation and secretion of pro-inflammatory cytokines [19]. NLRP3 inflammasome could be activated by the extra stimuli, the mitochondrial dysfunction, reactive oxygen species (ROS), lysosome damage, or disorder of intracellular ion homeostasis [20–22]. Besides, the activation of NLRP3 inflammasome requires the stimulation of nuclear factor-κb (NF-κB) [23]. NF-κB is the key to the biological processes of inflammation, proliferation, differentiation, and it controls the transcription of pro-inflammatory genes that regulate the synthesis of chemokines, cytokines, and adhesion molecules [24].

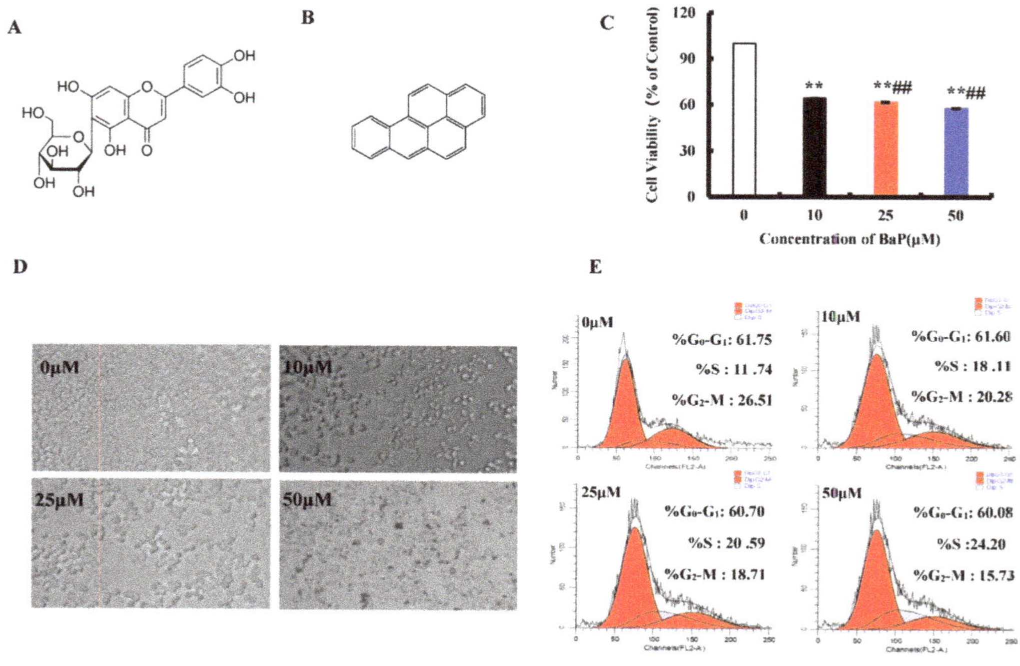

Figure 1. Effects of BaP on the proliferation of HL-7702 cells. Chemical structure of Iso (**A**) and BaP (**B**). Cells were incubated with BaP at different concentrations for 24 h and then processed for MTT assay. ** $p < 0.01$ compared with control, ## $p < 0.01$ compared with 10 μM BaP (**C**). Effect of BaP on cell morphology, which was observed with an optic microscope (**D**). Representative histograms of DNA content and cell cycle phases (G_0-G_1, S and G_2-M) in BaP treated cells (**E**).

Our preliminary study demonstrates that BaP induced pyroptotic liver injury through improving the electrical conductivity, stimulating the release of LDH and the production of NO, up-regulating the protein expression of pyroptosis characteristic indices, such as Caspase-1, COX-2, IL-1β, and IL-18 in vitro and vivo [25], and Iso could significantly attenuate the pyroptotic hepatocytes damage induced by BaP [26]; however, the underlying mechanism remains unknown.

This study aimed to explore whether or not the natural flavonoid Iso attenuated BaP-induced pyroptotic hepatocyte damage through the ROS/NF-κB/NLRP3/Caspase-1 signaling pathway in HL-7702 liver cells by detecting ROS levels, the nuclear localization of NF-κB, the activation of NLRP3 inflammasome, and the characteristic protein expressions of pyroptosis.

2. Materials and Methods

2.1. Chemicals and Reagents

Iso (purity ≥ 98%) was obtained from Pharmaceutical Technology Co., Ltd. (Jiangsu, China) and BaP was purchased from Sigma (St. Louis, MO, USA). Caspase-1 activity assay kit and 2,7-dichlorodihydrofluorescein diacetate (H_2DCFDA) were obtained from Beyotime biotechnology Co., Ltd. (Shanghai, China). RPMI-1640 medium, BCA protein kit, and penicillin–streptomycin solution were from Thermo Fisher (Shanghai, China).

TritonX-100, DAPI-Fluoromount-G, Bovine Serum Albumin (BSA), and CoraLite488-conjugated Affinipure Goat Anti-Rabbit IgG(H+L) were from Proteintech Group, Inc (Rosemont, IL, USA).

Polyclonal antibodies specific to NLRP3 (GTX106313), ASC (10500-1-AP), NF-κB (10745-1-AP), GSDMD (20770-1-AP), IL-1β (16806-1-AP) and IL-18 (SC-6179) were purchased from Santa Cruz Biotechnology (Dallas, Texas, United States). GAPDH (BA2913) was obtained from Bioworld Technology, Inc. (Louis Park, MN, USA). Caspase-1 (2225) was purchased from Cell Signaling Technology (Shanghai, China).

Specific inhibitors MCC950, N-acetylcysteine (NAC), and Z-YVAD-FMK were obtained from Beyotime biotechnology Co., Ltd. (Shanghai, China). Ammonium pyrrolidinedithiocarbamate (PDTC) and Mito-TEMPO were obtained from Sigma (St. Louis, MO, USA). All other reagents were dissolved in water or DMSO, and then diluted with fresh RPMI-1640 medium.

2.2. Cell Culture

HL-7702 liver cells were from the Shanghai Zhong Qiao Xin Zhou Biotechnology Co., Ltd. (Shanghai, China). Cells were cultured in RPMI-1640 medium with 10% fetal bovine serum and 1% penicillin–streptomycin, in a humidified incubator (5% CO_2, 95% air) at 37 °C.

2.3. Cytotoxicity Assay

Cell viability was detected by the MTT assay. HL-7702 cells with the density of 1×10^6 cells/mL were seeded into plates. After treatment, the MTT (0.5 mg/mL) was added and incubated for 4 h. Then, MTT was removed and 100 μL DMSO was added to each well. The absorbance at 490 nm was detected using a microplate reader (Thermo Fisher, Ltd., USA). Cell viability of HL-7702 cells was expressed as the ratio of the absorbance of the treated group cells to the control group cells.

2.4. Cell Cycle Analysis

Cell cycle was examined by Flow cytometry. After treatment for 24 h, trypsin was added to digest and blow the cells to prepare a cell suspension. Then, they were centrifuged and resuspended several times. Finally, RNase A and PI were added to the resuspended cells which were then cultured for 30 min at room temperature in the dark, and analyzed by the Flow cytometry (Partec GmbH, Münster, Germany).

2.5. Measurement of Intracellular ROS

ROS was detected using Flow cytometer and Fluorescence microscopy was used to detect ROS levels. Cells were harvested and incubated with 10 μM H_2DCFDA, which was added to the treated cells for 20 min at 37 °C. After washing with cold PBS, the fluorescence of cells was observed under Fluorescence microscope (Olympus Optical, Tokyo, Japan) or Flow cytometer (Partec GmbH, Münster, Germany).

2.6. Western Blot Analysis

HL-7702 cells were lysed with a lysis solution containing RIPA buffer, and then the protein concentration of the supernatant after centrifugation was measured with BCA kits, and finally, the protein concentration of all treatment groups were balanced with a loading buffer, and denatured in a 95 °C water bath for 10 min. Subsequently, the treated proteins were subjected to such steps as loading, electrophoresis, and transfer to a polyvinylidene fluoride membrane, followed by washing, blocking with skimmed milk powder for 2 h, and incubation with the primary antibody for 4 °C overnight and the secondary antibody for 2 h. Finally, the ECL kit was used for coloration and proteins were visualized on the Chemical XRS Imaging System (UVP, Ltd., USA).

2.7. Release of LDH

The contents of Lactate dehydrogenase (LDH) in the cell-free supernatant reflect the cellular permeability. The absorbance value of the cell-free supernatant at 450 nm was determined by using an LDH assay kit (Nanjing Jiancheng Bioengineering Institute, Nanjing, China).

2.8. NO Assay

Griess reagent (including 0.1% (*w*/*v*) N-(1-naphathyl)-ethylenediamine and 1% (*w*/*v*) sulfanilamide) was used to determine the content of NO in the culture supernatant. After treatment, the cell-free supernatant was collected and reacted with Griess reagent at 25 °C for 15 min. The absorbance at 540 nm was detected by the microplate reader (Thermo Fisher, Ltd.,USA). The value of NO was calculated using $NaNO_2$ standard curve.

2.9. Relative Electrical Conductivity Assay

The relative electrical conductivity of the cell-free supernatant was measured by the DDSJ-308A conductivity meter (Shanghai precision scientific instrument co., Ltd., Shanghai, China).

2.10. Scanning Electron Microscopy

Cells were fixed with 2.5% glutaraldehyde, and were sequentially dehydrated in graded ethanol (30, 50, 70, 90, 100% (*v*/*v*)), ethanol:isoamyl acetate (3:1, 1:1, 1:3 (*v*/*v*)), and isopropyl acetate for 15 min at room temperature. Finally, cells were observed under a scanning electron microscope (HITACHI, SU8220, Tokyo, Japan) at a voltage of 5 kV.

2.11. Caspase-1 Enzyme Measurements

The activity of Caspase-1 enzyme in cells was detected by a Caspase-1 enzyme kit (Beyotime Biotechnology, Shanghai, China). Briefly, cells lysates containing 10–30 mg of protein reacted with the reaction buffer containing Ac-YVAD-ρNA at 37 °C. After 30 min, the absorbance at 405 nm was detected using a microplate reader (Thermo Fisher, Ltd., USA.) and the Caspase-1 activity was normalized for total proteins of cell lysates.

2.12. NF-κB Activation Assay

Firstly, HL-7702 cells were added to the cell coverslips for 24 h. After fixing (ice-cold methanol), permeabilizing (0.25% Trition X-100), and sealing, cells were incubated with NF-κB antibody (1:100) and FITC-conjugated antibody (1:200) for 2 h and 1 h, respectively. Finally, cells were stained with DAPI for one minute, and then observed with an Olympus FV1200 confocal microscope.

2.13. Statistical Analysis

All experimental results are expressed as mean ± standard deviation (SD) and were analyzed by ANOVA and Duncan test (SPSS v19.0); p value < 0.05 was considered as statistically significant.

3. Results

3.1. Effects of BaP on the Proliferation of HL-7702 Cells

MTT assay was used to measure cell viability. BaP induced the death of HL-7702 cells in a dose-dependent manner at three concentrations of 10 μM, 25 μM, and 50 μM ($p < 0.01$). With the increasing concentration of BaP, the cell viability decreased by 35.73%,

38.24%, and 42.15%, respectively (Figure 1C). There were significant differences among these groups ($p < 0.01$). With microscope observation, we also found that the degree of cell damage was positively correlated with the concentration of BaP (Figure 1D). In order to explore whether the inhibitive effects of BaP on cell proliferation is related to blocking the cell cycle, the flow cytometry was calculated. As shown in Figure 1E, compared with the control, BaP significantly ($p < 0.01$) prolonged the S-phase by 12.46% and shortened the G_2-M-phase by 10.78%. These results indicated that BaP not only could damage HL-7702 cells, but also suppress the proliferation of HL-7702 cells by blocking the S-phase.

3.2. BaP Induced Pyroptotic Damage in HL-7702 Cells

Pyroptosis is a form of the programmed cell death which accompanies inflammation. When pyroptosis occurs, the GSDMD-N is activated to oligomerize on the cell membrane and form pores, which leads to the change of cell membrane permeability. As shown in Figure 2E, there were few pores in the control group, and the number of pores increased with the increase of the concentration of BaP. Furthermore, compared with 0 µM BaP treatment, 50 µM BaP notably ($p < 0.01$) raised the overflow of LDH (Figure 2A) and NO (Figure 2B), and the electrical conductivity (Figure 2C) by 57.21%, 88.86%, and 87.17%, respectively, and resulted in the significant increasing trend in the protein expression of GSDMD-N (Figure 2D). These results indicated that BaP could induce the pyroptotic hepatocyte damage in HL-7702 cells.

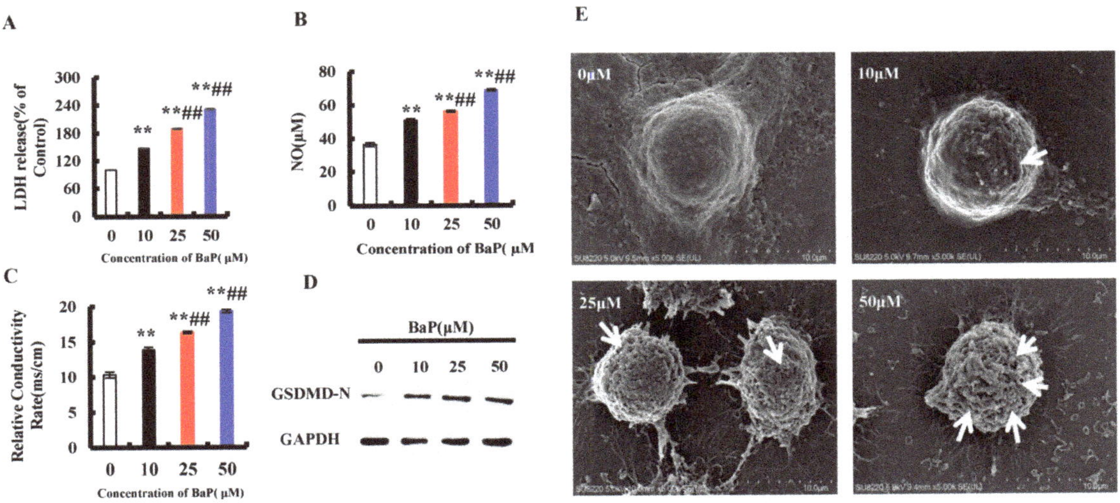

Figure 2. BaP induced pyroptotic hepatocyte damage in HL-7702 cells. LDH (**A**), NO (**B**), and electrical conductivity (**C**) were detected ** $p < 0.01$ compared with control, ## $p < 0.01$ compared with 10 µM BaP. Western blot image of protein expression acquired by Chemical XRS Imaging System (**D**). Scanning electron microscope observation of membrane pores; white arrows indicate the membrane pore (**E**).

3.3. BaP Induced Caspase-1-Dependent Pyroptotic Hepatocytes Damage

The classic pathway of pyroptosis is dominated by Caspase-1. Z-VAD-FMK is an irreversible Pan-Caspase inhibitor that can inhibit the activity of the Caspase family. We chose Z-VAD-FMK to verify whether or not the classic pathway of pyroptosis has been induced by BaP in HL-7702 cells. As shown in Figure 3A, in comparison with the control group, 25 µM BaP obviously ($p < 0.01$) decreased the cell viability by 39.40%, while the viability of cells which were exposed by Z-VAD-FMK and BaP only decreased by 7.3%, comparing with the control group. The same trends were found in the overflow of LDH (Figure 3B), NO (Figure 3C), and the electrical conductivity (Figure 3D). Similarly, Z-VAD-

FMK weakened the enhancing effects of BaP on the Caspase-1 enzyme activity (Figure 3E), the protein expressions of Caspase-1, GSDMD-N, and inflammatory factors (IL-18, IL-1β) (Figure 3F), showing that BaP induced Caspase-1-dependent pyroptotic hepatocyte damage in HL-7702 liver cells.

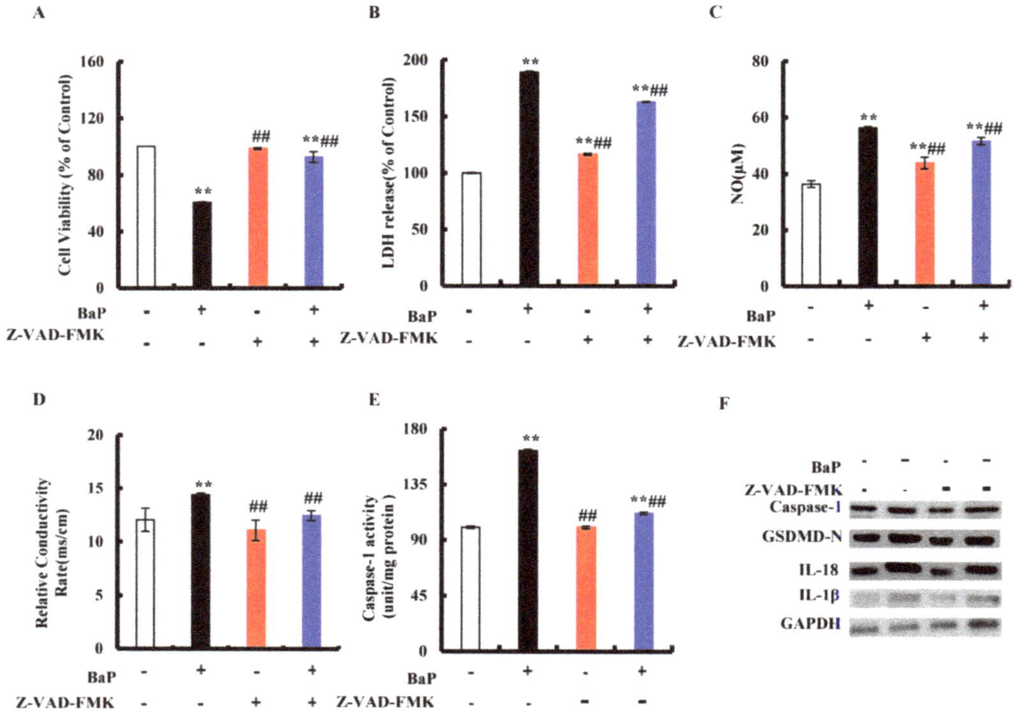

Figure 3. Effects of Caspase-1 inhibitor (Z-VAD-FMK, 20 μM) on pyroptotic hepatocyte damage. Cell viability (**A**), LDH (**B**), NO (**C**), electrical conductivity (**D**), and Caspase-1 enzyme activity (**E**) were detected. ** $p < 0.01$ compared with control, ## $p < 0.01$ compared with 25 μM BaP. Western blot image of protein expression acquired by Chemical XRS Imaging System (**F**).

3.4. Effects of NLRP3 Inflammasome on BaP-Induced Pyroptotic Hepatocyte Damage

We assessed whether or not the activation of NLRP3 inflammasome is related to the occurrence of pyroptosis. MCC950 is an effective and selective inhibitor of NLRP3. It was obvious from Figure 4A that the addition of MCC950 could greatly ($p < 0.01$) reduce the degree of cell damage and improve cell viability. In the co-treatment with MCC950 and BaP group, LDH release (Figure 4B), NO (Figure 4C), and the electrical conductivity (Figure 4D) were significantly ($p < 0.01$) lower than in the BaP group. Moreover, the characteristic protein expressions of pyroptosis (Figure 4F) were also reduced by MCC950, as well as the Caspase-1 enzyme activity (Figure 4E), contrary to the result of the BaP group. These results fully proved that BaP induced the pyroptotic hepatocyte damage through activating NLRP3 inflammasome in HL-7702 liver cells.

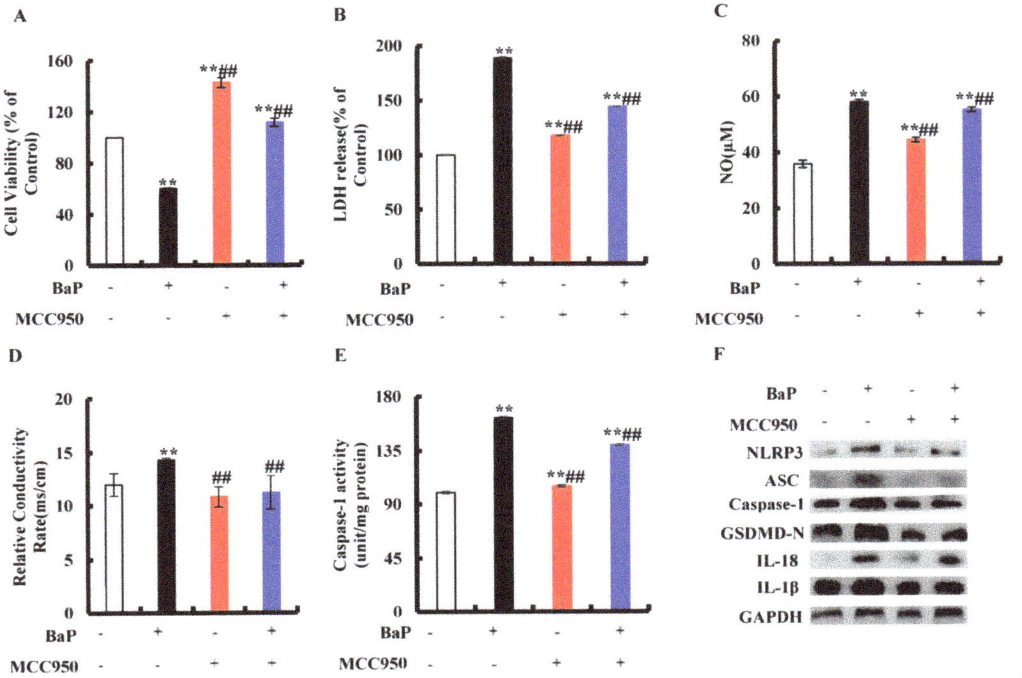

Figure 4. Effects of NLRP3 inhibitor (MCC950, 10μM) on pyroptotic hepatocytes damage induced by BaP (25μM). Cell viability (**A**), LDH (**B**), NO (**C**), electrical conductivity (**D**), and Caspase-1 enzyme activity (**E**) were detected. ** $p < 0.01$ compared with control, ## $p < 0.01$ compared with 25 μM BaP. Western blot image of protein expression acquired by Chemical XRS Imaging System (**F**).

3.5. Effects of NF-κB on NLRP3 and Pyroptotic Hepatocytes Damage

Inhibition of NF-κB can effectively reduce the level of pro-inflammatory cytokines. PDTC is an inhibitor of NF-κB activation that can inhibit the nuclear localization of NF-κB in cells. To explore whether or not BaP induced Caspase-1-dependent pyroptosis through activating NF-κB, then stimulating NLRP3 inflammasome, cells were exposed to PDTC. As shown in Figure 5A, PDTC could significantly ($p < 0.01$) alleviate the decline of cell viability caused by BaP. In comparison with the BaP treatment, 50 μM PDTC would significantly ($p < 0.01$) weaken the increase of LDH release (Figure 5B), NO (Figure 5C), and the electrical conductivity (Figure 5D) caused by BaP. The protein expressions of NF-κB, NLRP3, ASC, Caspase-1, GSDMD-N, IL-18, and IL-1β in the BaP and PDTC co-treatment group were also lower than those of the BaP group (Figure 5E). Meanwhile, BaP resulted in an obvious nuclear localization of NF-κB (p65), while PDTC could suppress the nuclear localization of NF-κB (p65) (Figure 5F). From the above, these results indicated that BaP induced Caspase-1-dependent pyroptosis through activating NF-κB, and then stimulating NLRP3.

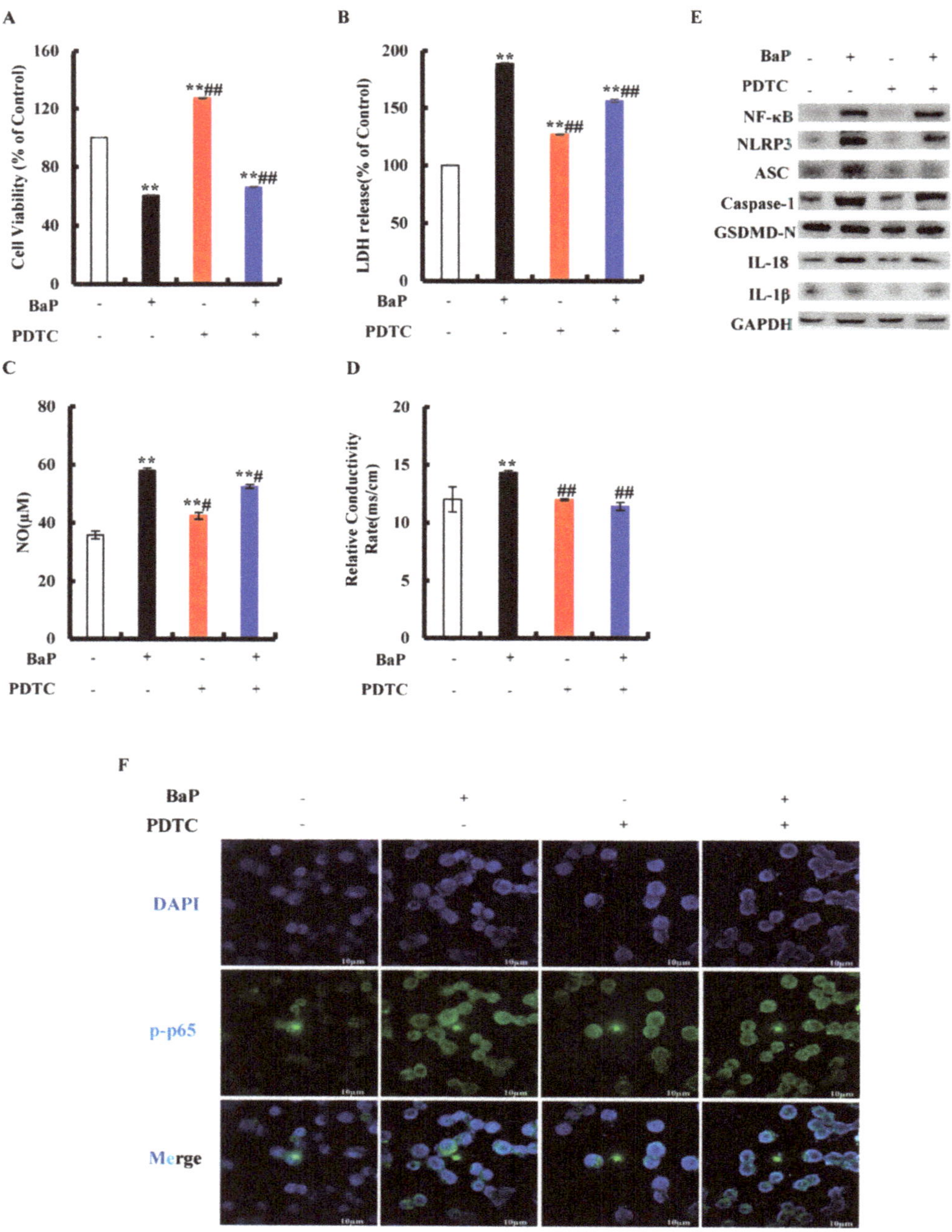

Figure 5. Effects of NF-κB inhibitor (PDTC, 50 μM) on pyroptotic hepatocyte damage induced by BaP (25 μM). Cell viability (**A**), LDH (**B**), NO (**C**), and electrical conductivity (**D**) were detected. ** $p < 0.01$ compared with control, ## $p < 0.01$ compared with 25 μM BaP. Western blot image of protein expression acquired by Chemical XRS Imaging System (**E**). Nuclear translocation of phosphorylated p65 was observed by immunofluorescent labeling with DAPI (blue) and anti-phosphorylated p65 (green) (**F**).

3.6. Effects of ROS on NF-κB, NLRP3, and Pyroptotic Hepatocyte Damage

BaP exposure would generate a large amount of ROS, and ROS could activate the nuclear localization of NF-κB and NLRP3 inflammasome. NAC is a common antioxidant, which can remove ROS in cells. Mito-TEMPO was proven to specifically remove ROS in mitochondria. In order to understand what role ROS plays in the pyroptotic hepatocyte damage caused by BaP, NAC and Mito-TEMPO were used to treat the HL-7720 cells. It was shown that NAC or Mito-TEMPO significantly ($p < 0.01$) improved the viability of cells after BaP exposure (Figure 6A) and obviously slowed down the production of ROS (Figure 6B). Our results also showed that the addition of NAC or Mito-TEMPO would greatly ($p < 0.01$) decrease the overflow of LDH (Figure 6C), NO (Figure 6D), and the electrical conductivity (Figure 6E). In addition, the protein expressions of NF-κB, NLRP3, ASC, Caspase-1, GSDMD-N, IL-18, and IL-1β in the BaP and NAC or Mito-TEMPO co-treatment group were lower than those of the BaP group (Figure 6F). According to the previous results, it was concluded that BaP induced the pyroptotic hepatocyte damage in HL-7702 cells by activating the ROS/NF-κB/NLRP3/Caspase-1 signaling pathway.

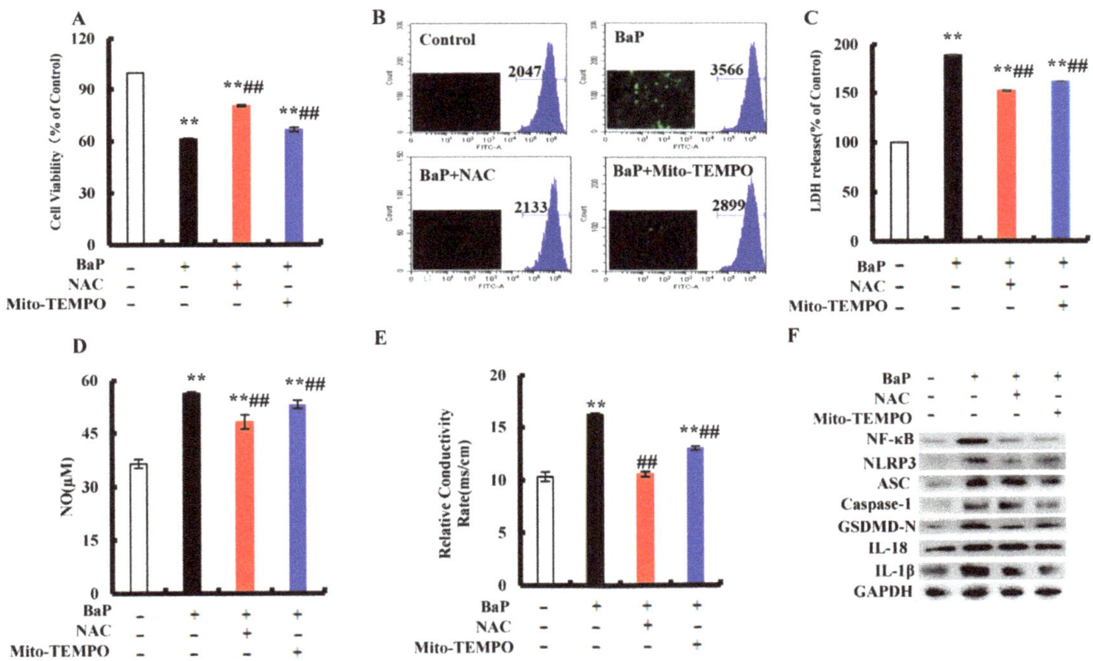

Figure 6. Effects of ROS inhibitor (NAC, 2 mM) and mtROS inhibitor (Mito-TEMPO, 50 μM) on BaP-induced pyroptotic hepatocyte damage. Cell viability (**A**), LDH (**C**), NO (**D**), and electrical conductivity (**E**) were detected. ** $p < 0.01$ compared with control, ## $p < 0.01$ compared with 25 μM BaP. ROS were examined by Flow cytometer and Fluorescence microscope (**B**). Western blot image of protein expression acquired by Chemical XRS Imaging System (**F**).

3.7. Effects of Iso on the BaP-Induced Pyroptotic Hepatocyte Damage

Our previous work proved that Iso inhibited the pyroptosis induced by BaP; this study further verified whether or not Iso inhibited the pyroptotic hepatocytes damage induced by BaP in HL-7702 cells through the ROS/NF-κB/NLRP3/Caspase-1 signaling pathway. Firstly, we carried out the assay of MTT and cell cycle to evaluate the effects of Iso on HL-7720 cells proliferation. The viability of cells decreased in the BaP group, but increased in the Iso and BaP co-treatment group (Figure 7A). Compared with the BaP group, the percentage of S-phase in the cell cycle was decreased by 3.88% after Iso and BaP co-exposure (Figure 7B). At the same time, Iso decreased the levels of ROS (Figure 7C) and the Caspase-1 enzyme activity (Figure 7D), as well as the characteristic protein expressions of pyroptosis (Figure 7F). Besides, in comparison with the BaP treatment, 5 μM Iso reduced the number of cell membrane pores (Figure 7E) and significantly suppressed the nuclear localization of NF-κB (p65) (Figure 7G). All the results allowed to conclude that Iso had an alleviative effect on the BaP-induced pyroptotic hepatocyte damage in HL-7702 cells through the inactivation of the ROS/NF-κB/NLRP3/Caspase-1 signaling pathway.

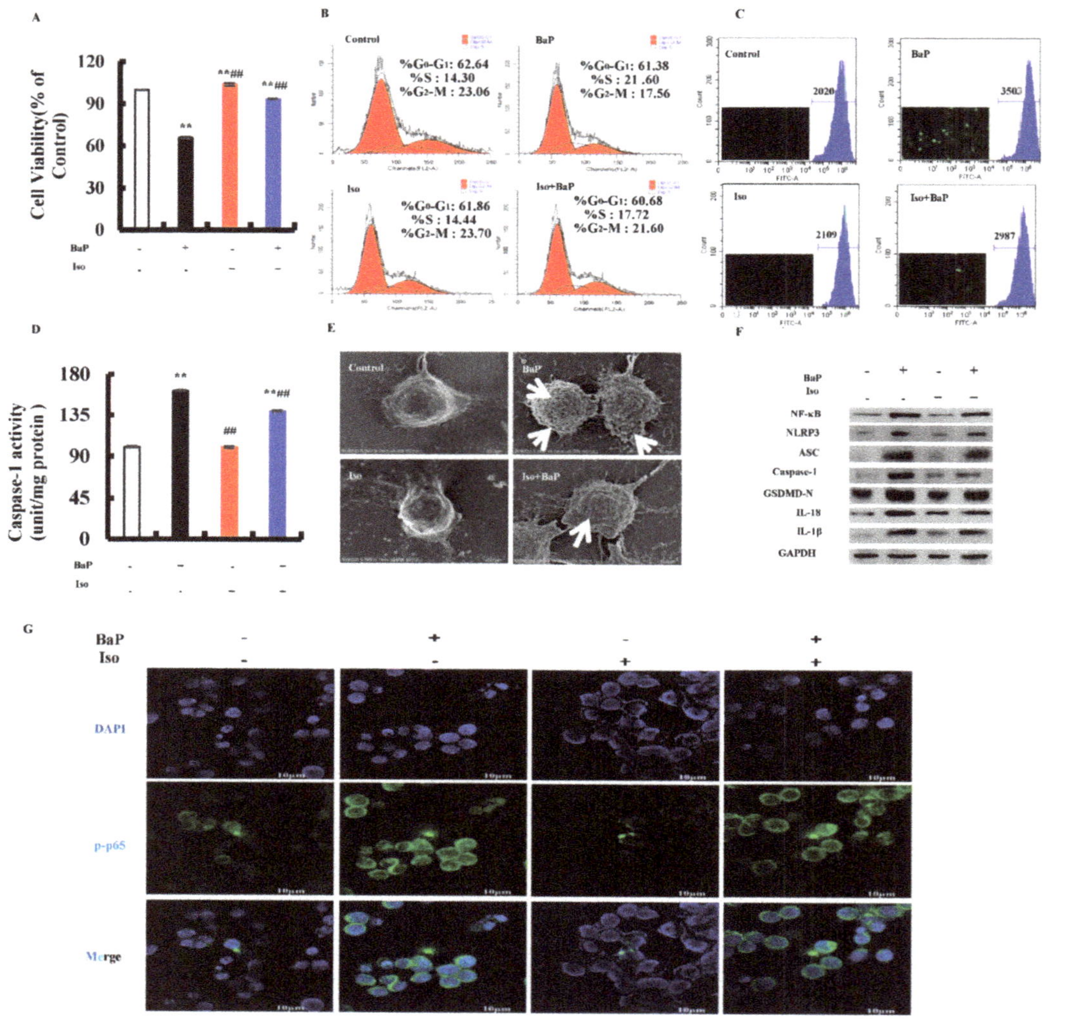

Figure 7. Effects of Iso (5 μM) on the pyroptotic hepatocyte damage induced by BaP (25 μM). Cell viability (**A**), cell cycle (**B**), ROS levels (**C**), and Caspase-1 enzyme activity (**D**) were detected. ** $p < 0.01$ compared with control, ## $p < 0.01$ compared with 25 μM BaP. Scanning electron microscope observation of membrane pores; white arrows indicate the membrane pore (**E**). Western blot image of protein expression acquired by Chemical XRS Imaging System (**F**). Nuclear translocation of phosphorylated p65 was observed by immunofluorescent labeling with DAPI (blue) and anti-phosphorylated p65 (green) (**G**).

4. Discussion

Pyroptosis is a programmed cell death mode which depends on the inflammatory Caspase family and is accompanied by an inflammatory reaction. Long-term and excessive pyroptosis will lead to cell death, tissue damage, organ failure, septic shock, and other pathological conditions [27]. It has been reported that Gossypol, Arsenic, and Cadmium induced the pyroptotic damage [13–15]. Our preliminary study also demonstrates that BaP caused hepatocytes damage via inducing pyroptosis [25,28]. Iso is a natural flavonoid with a luteolin structure that is widespread in some foods and edible plants [29,30]. It could improve cell viability, reduce the electrical conductivity, NO release, and the protein expression of Caspase-1 to alleviate BaP-induced pyroptotic liver injury [26]. In this study, we further verified that Iso could reduce the membrane pores, the nuclear localization of the p65 subunit of NF-κB, the activation of NLRP3 inflammasome, the increase of Caspase-1 enzyme activity, and the characteristic protein expression of pyroptosis (Caspase-1, GSDMD-N, IL-1β, and IL-18), ultimately ameliorating BaP-induced pyroptotic damage by inhibiting the ROS/NF-κB/NLRP3/Caspase-1 signaling pathway in HL-7702 cells (Figure 8).

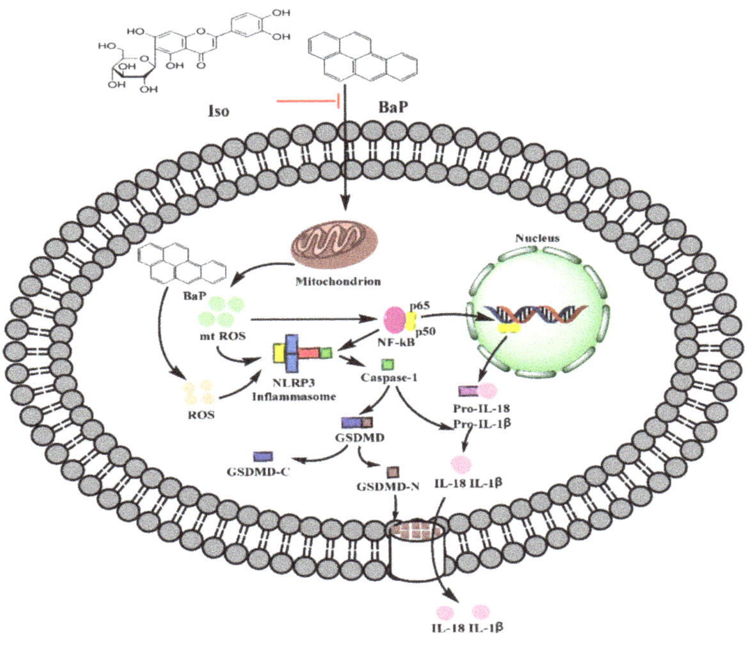

Figure 8. Possible molecular mechanism of Iso on the pyroptotic hepatocyte damage induced by BaP in HL-7702 liver cells.

Gasdermin-D (GSDMD), a member of the Gasdermin family, is the common substrate of all inflammatory Caspase (Caspase-1/4/5/11) enzymes and is also the direct executor of pyroptosis in human and murine animals [31]. When pyroptosis occurs, GSDMD is cleaved by activated Caspase-1/4/5/11 into the N-terminal domain (GSDMD-N) with lipophilicity and pore-forming activity, and C-terminal domain (GSDMD-C) with hydrophilicity. GSDMD-N then selectively combined with the inner membrane of the cell membrane and oligomerized to form membrane pores with a diameter ranging from 10 to 20 nm, which caused the cell membrane permeability disorder and eventually led to cell swelling and rupture, resulting in pyroptotic death. At the same time, GSDMD-N could also promote the spillage of inflammatory factors IL-1β and IL-18 in large quantities, and stimulate a strong inflammatory reaction [32,33]. This work showed that Iso notably weakened the stimulatory effect of BaP on the oligomerization of GSDMD-N, the formation of cells membrane pores, the increase of the permeability of cell membrane, and the electrical conductivity, as well as the level of IL-1β and IL-18 (Figures 2 and 7), indicating that Iso could suppress the BaP-induced pyroptotic hepatocytes damage.

According to different inflammatory Caspase and external stimuli, pyroptosis can be divided into the classic pathway depending on Caspase-1 and the non-canonical pathway depending on Caspase-4/5/11 [34]. The classic pyroptosis pathway is that Caspase-1 specifically cleaves and activates GSDMD to form GSDMD-N and GSDMD-C, and GSDMD-N further forms membrane pores on the cell membrane through oligomerization, which eventually leads to cell death. Meanwhile, Caspase-1 shears ProIL-1β and Pro-IL-18 into IL-1β and IL-18 overflowed through membrane pores. It has been reported that arsenic trioxide (As_2O_3) resulted in the nonalcoholic fatty liver disease/nonalcoholic steatohepatitis by inducing the classic pyroptosis pathway in HepG2 cells, accompanied by up-regulating the protein expressions of Caspase-1 and IL-1β [14]. As a Caspase inhibitor, Z-VAD-FMK alleviated HUVECs cells pyroptosis induced by $CdCl_2$ [15]. To prove the genre of pyroptosis induced by BaP, Z-VAD-FMK was used to treat cells together with BaP. The results revealed that BaP resulted in the Caspase-1-dependent pyroptotic hepatocyte damage in HL-7702 cells (Figure 3). Consistent with the effect of Z-VAD-FMK, Iso decreased the Caspase-1 enzyme activity and the protein expressions of Caspase-1, GSDMD-N, IL-1β, and IL-18, demonstrating that Iso could inhibit the Caspase-1-dependent pyroptotic hepatocyte damage caused by BaP in HL-7702 cells (Figure 7).

Inflammasome is a type of polyprotein complex, including NLRP3, AIM2, Pyrin, NLRC4, and NLRP1. After the formation of the NLRP3 inflammasome with NLRP3, ASC and Pro-Caspase-1, the Pro-Caspase-1 will self-shear to form active Caspase-1, subsequently cleave and activate GSDMD to form GSDMD-N, and finally initiate pyroptosis [35,36]. MCC950 is a specific inhibitor of NLRP3. It was found that the addition of MCC950 could alleviate the pyroptosis induced by Mesoporous silica nanoparticles [37]. In this study, we also found that the MCC950 or Iso co-treatment with BaP could obviously weakened the activity of the NLRP3 inflammasome and the characteristic protein expressions of pyroptosis induced by BaP (Figures 4 and 7), indicating that Iso could constrain the activation of the NLRP3 inflammatory response and the degree of damage of pyroptotic hepatocytes.

NF-κB is a nuclear transcription factor in the cells. It participates in the inflammatory response and regulates the expressions of inflammasomes [38]. It was found that the inhibition of NF-κB and NLRP3 inflammasome in macrophages could ameliorate colitis of mice [39]. The inhibitor of NF-κB (PDTC) was used to analyze whether or not NF-κB was involved in the activation of the NLRP3 inflammasome and pyroptosis reaction with cells. Iso and PDTC had similar effects, and both inhibited the nuclear localization of the p65 subunit of NF-κB, the activation of the NLRP3 inflammasome, and then initiating the Caspase-1-dependent pyroptotic hepatocyte damage induced by BaP (Figures 5 and 7). Moreover, NF-κB did promote the activation of the NLRP3 inflammasome, and subsequently initiated pyroptotic hepatocyte damage.

BaP exposure would generate a large amount of ROS [40]. ROS is a trigger for the activation of the NLRP3 inflammasome, [21] and mtROS also activates the NLRP3 inflammasome through the NF-κB pathway [20,41,42]. ROS stimulated by Nicotine [43] and chronic ethanol [44] could induce the activation of the NLRP3 inflammasome and Caspase-1, and ultimately result in cells pyroptotic death. Chen's research proved that pre-treatment with Mito-TEMPO (mtROS scavenger) or NAC (total ROS scavenger) suppressed Cd-induced activation of NLRP3 and pyroptotic cell death [15]. In the same way, our results revealed that ROS was a main contributor to BaP-induced pyroptosis and the activation of NF-κB and NLRP3 was confirmed by the fact that both ROS scavenger (NAC and Mito-TEMPO) and Iso could strikingly decrease the level of ROS and mtROS, and ameliorate the BaP-induced pyroptotic hepatocyte damage in HL-7702 cells through the inhibition of the ROS/NF-κB/NLRP3/Caspase-1 signaling pathway (Figures 6–8).

5. Conclusions

In summary, Iso was proven to ameliorate BaP-induced pyroptotic hepatocytes damage in HL-7702 cells by suppressing the ROS/NF-κB/NLRP3/Caspase-1 signaling pathway, which will contribute to a theoretical basis for the hepatoprotective effect of Iso and the full revealment of the toxic effect and the control of BaP.

Author Contributions: H.L., L.Y. and X.L. designed research; H.L., L.Y. and X.L. performed research; H.L., L.Y. and X.L. analyzed data; Y.L., Z.Z. and J.L. contributed to discussion; H.L., L.Y. and X.L. wrote the paper; all authors approved the final version of the paper. All authors have read and agreed to the published version of the manuscript.

Funding: This work was financially supported by the National Natural Science Foundation of China (31972183), the Supporting Program for Youth Talent (095920201311), the Fundamental Research Funds for the Central Universities (GK202002006), and the Fund of the Engineering Research Center of High Value Utilization of Western China Fruit Resources (XGZX2021-G10). The funders had no role in the design and conduct of this study, collection and interpretation of the data, or preparation and approval of the manuscript.

Institutional Review Board Statement: Not applicable.

Informed Consent Statement: Not applicable.

Data Availability Statement: Data is contained within the article.

Conflicts of Interest: The authors declare no conflict of interest.

Abbreviations

Iso	Isoorientin
BaP	Benzo[a]pyrene
HL-7702 cells	HL-7702 human normal liver cells
NO	nitric oxide
LDH	lactate dehydrogenase
NAC	N-acetylcysteine
ASC	apoptosis associated speck-like protein containing a CARD
NLRP3	NLR family pyrin domain-containing 3
mtROS	mitochondria-derived reactive oxygen species
PDTC	pyrrolidinedithiocarbamate
NF-κB	Nuclear factor-κb
GSDMD	Gasdermin-D

References

1. Prinz, S.; Ringl, A.; Huefner, A.; Pemp, E.; Kopp, B. 4‴-Acetylvitexin -2″-O-rhamnoside, isoorientin, orientin, and 8-methoxykaempferol-3-O-glucoside as markers for the differentiation of Crataegus monogyna and Crataegus pentagyna from Crataegus laevigata (Rosaceae). *Chem. Biodivers.* **2007**, *4*, 2920–2931. [CrossRef]
2. Zou, J.H.; Yang, J.; Zhou, L. Acylated flavone C-glycosides from Trollius ledebouri. *J. Nat. Prod.* **2004**, *67*, 664. [CrossRef] [PubMed]
3. Anilkumar, K.; Reddy, G.V.; Azad, R.; Yarla, N.S.; Dharmapuri, G.; Srivastava, A.; Kamal, M.A.; Pallu, R. Evaluation of Anti-Inflammatory Properties of Isoorientin Isolated from Tubers of Pueraria tuberosa. *Oxid. Med. Cell. Longev.* **2017**, *2017*, 5498054. [CrossRef] [PubMed]
4. Lee, W.; Ku, S.-K.; Bae, J.-S. Vascular barrier protective effects of orientin and isoorientin in LPS-induced inflammation in vitro and in vivo. *Vasc. Pharmacol.* **2014**, *62*, 3–14. [CrossRef] [PubMed]
5. Yang, J.H.; Choi, M.-H.; Yang, S.H.; Cho, S.S.; Park, S.J.; Shin, H.-J.; Ki, S.H. Potent Anti-Inflammatory and Antiadipogenic Properties of Bamboo (Sasa coreana Nakai) Leaves Extract and Its Major Constituent Flavonoids. *J. Agric. Food Chem.* **2017**, *65*, 6665–6673. [CrossRef]
6. Yuan, L.; Han, X.; Li, W.; Ren, D.; Yang, X. Isoorientin Prevents Hyperlipidemia and Liver Injury by Regulating Lipid Metabolism, Antioxidant Capability, and Inflammatory Cytokine Release in High-Fructose-Fed Mice. *J. Agric. Food Chem.* **2016**, *64*, 2682–2689. [CrossRef]
7. Yuan, L.; Wang, J.; Wu, W.; Liu, Q.; Liu, X. Effect of isoorientin on intracellular antioxidant defence mechanisms in hepatoma and liver cell lines. *Biomed. Pharmacother.* **2016**, *81*, 356–362. [CrossRef] [PubMed]

8. Sun, D.; Chen, Q.; Zhu, B.; Lan, Y.; Duan, S. Long-Term Exposure to Benzo[a]Pyrene Affects Sexual Differentiation and Embryos Toxicity in Three Generations of Marine Medaka (Oryzias Melastigma). *Int. J. Environ. Res. Public Health* **2020**, *17*, 970. [CrossRef]
9. Reizer, E.; Csizmadia, I.G.; Palotás, B.; Viskolcz, B.; Fiser, B. Formation Mechanism of Benzo(a)pyrene: One of the Most Carcinogenic Polycyclic Aromatic Hydrocarbons (PAH). *Molecules* **2019**, *24*, 1040. [CrossRef]
10. Filippov, S.; Yarushkin, A.A.; Kalinina, T.; Ovchinnikov, V.; Knyazev, R.A.; Gulyaeva, L.F. Effect of Benzo(a)pyrene on the Expression of miR-483-3p in Hepatocyte Primary Culture and Rat Liver. *Biochemistry* **2019**, *84*, 1197–1203. [CrossRef] [PubMed]
11. Jan, A.; Diane, B.; Andrew, C. Polycyclic aromatic hydrocarbons in food- scientific opinion of the panel on contaminants in the food chain. *EFSA J.* **2008**, *6*, 724.
12. Kroemer, G.; Galluzzi, L.; Vandenabeele, P.; Abrams, J.; Alnemri, E.S.; Baehrecke, E.H.; Blagosklonny, M.V.; El-Deiry, W.S.; Golstein, P.; Green, D.; et al. Classification of cell death: Recommendations of the Nomenclature Committee on Cell Death 2009. *Cell Death Differ.* **2008**, *16*, 3–11. [CrossRef] [PubMed]
13. Lin, Q.-R.; Li, C.-G.; Zha, Q.-B.; Xu, L.-H.; Pan, H.; Zhao, G.-X.; Ouyang, D.-Y.; He, X.-H. Gossypol induces pyroptosis in mouse macrophages via a non-canonical inflammasome pathway. *Toxicol. Appl. Pharmacol.* **2016**, *292*, 56–64. [CrossRef] [PubMed]
14. Qiu, T.; Pei, P.; Yao, X.; Jiang, L.; Wei, S.; Wang, Z.; Bai, J.; Yang, G.; Gao, N.; Yang, L.; et al. Taurine attenuates arsenic-induced pyroptosis and nonalcoholic steatohepatitis by inhibiting the autophagic-inflammasomal pathway. *Cell Death Dis.* **2018**, *9*, 946. [CrossRef]
15. Chen, H.; Lu, Y.; Cao, Z.; Ma, Q.; Pi, H.; Fang, Y.; Yu, Z.; Hu, H.; Zhou, Z. Cadmium induces NLRP3 inflammasome-dependent pyroptosis in vascular endothelial cells. *Toxicol. Lett.* **2016**, *246*, 7–16. [CrossRef] [PubMed]
16. Vernon, P.J.; Tang, D. Eat-me: Autophagy, phagocytosis and reactive oxygen species signaling. *Antioxid. Redox Signal* **2013**, *18*, 677–691. [CrossRef] [PubMed]
17. Bergsbaken, T.; Fink, S.; Cookson, B.T. Pyroptosis: Host cell death and inflammation. *Nat. Rev. Genet.* **2009**, *7*, 99–109. [CrossRef]
18. Wree, A.; Eguchi, A.; McGeough, M.D.; Pena, C.A.; Johnson, C.D.; Canbay, A.; Hoffman, H.M.; Feldstein, A.E. NLRP3 inflammasome activation results in hepatocyte pyroptosis, liver inflammation, and fibrosis in mice. *Hepatology* **2013**, *59*, 898–910. [CrossRef] [PubMed]
19. Gao, J.; Cui, J.Z.; To, E.; Cao, S.; Matsubara, J.A. Evidence for the activation of pyroptotic and apoptotic pathways in RPE cells associated with NLRP3 inflammasome in the rodent eye. *J. Neuroinflamm.* **2018**, *15*, 15. [CrossRef]
20. Chen, W.; Zhang, X.; Fan, J.; Zai, W.; Luan, J.; Li, Y.; Wang, S.; Chen, Q.; Wang, Y.; Liang, Y.; et al. Tethering Interleukin-22 to Apolipoprotein A-I Ameliorates Mice from Acetaminophen-induced Liver Injury. *Theranostics* **2017**, *7*, 4135–4148. [CrossRef] [PubMed]
21. Zhou, R.; Yazdi, A.S.; Menu, P.; Tschopp, J. A role for mitochondria in NLRP3 inflammasome activation. *Nat. Cell Biol.* **2010**, *469*, 221–225. [CrossRef] [PubMed]
22. Takahashi, M. NLRP3 Inflammasome as a Novel Player in Myocardial Infarction. *Int. Heart J.* **2014**, *55*, 101–105. [CrossRef] [PubMed]
23. Bauernfeind, F.G.; Horvath, G.; Stutz, A.; Alnemri, E.S.; MacDonald, K.; Speert, D.; Fernandes-Alnemri, T.; Wu, J.; Monks, B.G.; Fitzgerald, K.A.; et al. Cutting edge: NF-kappaB activating pattern recognition and cytokine receptors license NLRP3 inflflammasome activation by regulating NLRP3 expression. *J. Immunol.* **2009**, *183*, 787–791. [CrossRef]
24. Xiang, H.C.; Lin, L.X.; Hu, X.F.; Zhu, H.; Li, H.P.; Zhang, R.Y.; Hu, L.; Liu, W.T.; Zhao, Y.L.; Shu, Y.; et al. AMPK activation attenuates inflammatory pain through inhibiting NF-kappaB activation and IL-1beta expression. *J. Neuroinflamm.* **2019**, *16*, 34. [CrossRef] [PubMed]
25. Yuan, L.; Liu, J.; Deng, H.; Gao, C. Benzo[a]pyrene Induces Autophagic and Pyroptotic Death Simultaneously in HL-7702 Human Normal Liver Cells. *J. Agric. Food Chem.* **2017**, *65*, 9763–9773. [CrossRef] [PubMed]
26. Li, X.; He, S.; Gao, C.; Deng, H.; Liu, Y.; Li, C.; Yuan, L.; Luo, Y. Isoorientin attenuates benzo[a]pyrene-induced liver injury by inhibiting autophagy and pyroptosis in vitro and vivo. *Food Agric. Immunol.* **2019**, *30*, 841–861. [CrossRef]
27. Songane, M.; Khair, M.; Saleh, M. An updated view on the functions of caspases in inflammation and immunity. *Semin. Cell Dev. Biol.* **2018**, *82*, 137–149. [CrossRef]
28. Li, Q.; Gao, C.; Deng, H.; Song, Q.; Yuan, L. Benzo[a]pyrene induces pyroptotic and autophagic death through inhibiting PI3K/Akt signaling pathway in HL-7702 human normal liver cells. *J. Toxicol. Sci.* **2019**, *44*, 121–131. [CrossRef] [PubMed]
29. Watanabe, M. An Anthocyanin Compound in Buckwheat Sprouts and Its Contribution to Antioxidant Capacity. *Biosci. Biotechnol. Biochem.* **2007**, *71*, 579–582. [CrossRef] [PubMed]
30. Toda, K.; Hitoe, S.; Takeda, S.; Shimizu, N.; Shimoda, H. Passionflower extract induces high-amplitude phythms without phase shifts in the expression of several circadian clock genes in vitro and in vivo. *Int. J. Biomed. Sci.* **2017**, *13*, 84–92. [PubMed]
31. Shi, J.; Gao, W.; Shao, F. Pyroptosis: Gasdermin-Mediated Programmed Necrotic Cell Death. *Trends Biochem. Sci.* **2017**, *42*, 245–254. [CrossRef] [PubMed]
32. Liu, X.; Zhang, Z.; Ruan, J.; Pan, Y.; Magupalli, V.G.; Wu, H.; Lieberman, J. Inflammasome-activated gasdermin D causes pyroptosis by forming membrane pores. *Nat. Cell Biol.* **2016**, *535*, 153–158. [CrossRef] [PubMed]
33. Mulvihill, E.; Sborgi, L.; Mari, S.A.; Pfreundschuh, M.; Hiller, S.; Müller, D.J. Mechanism of membrane pore formation by human gasdermin-D. *EMBO J.* **2018**, *37*, e98321. [CrossRef]
34. Gong, W.; Shi, Y.; Ren, J. Research progresses of molecular mechanism of pyroptosis and its related diseases. *Immunobiology* **2019**, *225*, 151884. [CrossRef] [PubMed]

35. Man, S.M.; Hopkins, L.J.; Nugent, E.; Cox, S.; Glück, I.M.; Tourlomousis, P.; Wright, J.A.; Cicuta, P.; Monie, T.P.; Bryant, C.E. Infiammasome activation causes dual recruiment of NLRC4 and NLRP3 to the same macromolecular complex. *Proc. Natl. Acad. Sci. USA* **2014**, *111*, 7403–7408. [CrossRef] [PubMed]
36. Van, O.N.; Gurung, P.; Vande, W.L.; Fossoul, A.; Kanneganti, T.D.; Lamkanfi, M. Activation of the NLRP1b infiammasome independently of ASC-mediated caspase-1 autoproteolysis and speck formation. *Nat. Commun.* **2014**, *5*, 3209.
37. Zhang, X.; Luan, J.; Chen, W.; Fan, J.; Nan, Y.; Wang, Y.; Liang, Y.; Meng, G.; Ju, D. Mesoporous silica nanoparticles induced hepatotoxicity via NLRP3 inflammasome activation and caspase-1-dependent pyroptosis. *Nanoscale* **2018**, *10*, 9141–9152. [CrossRef] [PubMed]
38. Li, D.; Zhou, J.; Yang, B.; Yu, Y. microRNA-340-5p inhibits hypoxia/reoxygenation-induced apoptosis and oxidative stress in cardiomyocytes by regulating the Act1/NF-κB pathway. *J. Cell Biochem.* **2019**, *120*, 14618–14627. [CrossRef] [PubMed]
39. Wu, X.F.; Ouyang, Z.J.; Feng, L.L.; Chen, G.; Guo, W.J.; Shen, Y.; Wu, X.D.; Sun, Y.; Xu, Q. Suppression of NF-κB signaling and NLRP3 inflammasome activation in macrophages is responsible for the amelioration of experimental murine colitis by the natural compound fraxinellone. *Toxicol. Appl. Pharmacol.* **2014**, *281*, 146–156. [CrossRef]
40. Cui, Q.; Chen, F.Y.; Chen, H.Y.; Peng, H.; Wang, K.J. Benzo[a]pyrene (BaP) exposure generates persistent reactive oxygen species (ROS) to inhibit the NF-κB pathway in medaka (Oryzias melastigma). *Environ. Pollut.* **2019**, *251*, 502–509. [CrossRef] [PubMed]
41. Sun, B.; Wang, X.; Ji, Z.; Wang, M.; Liao, Y.-P.; Chang, C.H.; Li, R.; Zhang, H.; Nel, A.E.; Xiang, W. NADPH Oxidase-Dependent NLRP3 Inflammasome Activation and its Important Role in Lung Fibrosis by Multiwalled Carbon Nanotubes. *Small* **2015**, *11*, 2087–2097. [CrossRef] [PubMed]
42. Guo, C.; Wang, J.; Jing, L.; Ma, R.; Liu, X.; Gao, L.; Cao, L.; Duan, J.; Zhou, X.; Li, Y.; et al. Mitochondrial dysfunction, perturbations of mitochondrial dynamics and biogenesis involved in endothelial injury induced by silica nanoparticles. *Environ. Pollut.* **2018**, *236*, 926–936. [CrossRef]
43. Wu, X.; Zhang, H.; Qi, W.; Zhang, Y.; Li, J.; Li, Z.; Lin, Y.; Bai, X.; Liu, X.; Chen, X.; et al. Nicotine promotes atherosclerosis via ROS-NLRP3-mediated endothelial cell pyroptosis. *Cell Death Dis.* **2018**, *9*, 171. [CrossRef] [PubMed]
44. Hoyt, L.; Randall, M.; Ather, J.; DePuccio, D.; Landry, C.C.; Qian, X.; Janssen-Heininger, Y.M.; van der Vliet, A.; Dixon, A.E.; Amiel, E.; et al. Mitochondrial ROS induced by chronic ethanol exposure promote hyper-activation of the NLRP3 inflammasome. *Redox Biol.* **2017**, *12*, 883–896. [CrossRef] [PubMed]

antioxidants

MDPI

Article

Styrylpyrones from *Phellinus linteus* Mycelia Alleviate Non-Alcoholic Fatty Liver by Modulating Lipid and Glucose Metabolic Homeostasis in High-Fat and High-Fructose Diet-Fed Mice

Chun-Hung Chiu [1,2,†], Chun-Chao Chang [3,4,†], Jia-Jing Lin [1], Chin-Chu Chen [5], Charng-Cherng Chyau [1,*] and Robert Y. Peng [1,6,*]

1 Research Institute of Biotechnology, Hungkuang University, Shalu District, Taichung City 43302, Taiwan; chchiu@hk.edu.tw (C.-H.C.); n404user@gmail.com (J.-J.L.)
2 Department of Program in Animal Healthcare, Hungkuang University, Shalu District, Taichung City 43302, Taiwan
3 Division of Gastroenterology and Hepatology, Department of Internal Medicine, Taipei Medical University Hospital, Taipei 11031, Taiwan; chunchao@tmu.edu.tw
4 Division of Gastroenterology and Hepatology, Department of Internal Medicine, School of Medicine, College of Medicine, Taipei Medical University, Taipei 11031, Taiwan
5 Biotech Research Institute, GrapeKing Bio Ltd., Taoyuan 32542, Taiwan; gkbioeng@grapeking.com.tw
6 Graduate Institute of Clinical Medicine, College of Medicine, Taipei Medical University, Taipei 11031, Taiwan
* Correspondence: ccchyau@hk.edu.tw (C.-C.C.); ypeng@sunrise.hk.edu.tw (R.Y.P.); Tel.: +886-4-26318652 (C.-C.C.); Fax: +886-4-26525386 (C.-C.C.)
† These authors contributed equally to this work.

Citation: Chiu, C.-H.; Chang, C.-C.; Lin, J.-J.; Chen, C.-C.; Chyau, C.-C.; Peng, R.Y. Styrylpyrones from *Phellinus linteus* Mycelia Alleviate Non-Alcoholic Fatty Liver by Modulating Lipid and Glucose Metabolic Homeostasis in High-Fat and High-Fructose Diet-Fed Mice. *Antioxidants* **2022**, *11*, 898. https://doi.org/10.3390/antiox11050898

Academic Editor: Stanley Omaye

Received: 19 April 2022
Accepted: 28 April 2022
Published: 30 April 2022

Abstract: *Phellinus linteus* (PL), an edible and medicinal mushroom containing a diversity of styrylpyrone-type polyphenols, has been shown to have a broad spectrum of bioactivities. In this study, the submerged liquid culture in a 1600-L working volume of fermentor was used for the large-scale production of PL mycelia. Whether PL mycelia extract is effective against nonalcoholic fatty liver disease (NAFLD) is still unclear. In the high fat/high fructose diet (HFD)-induced NAFLD C57BL/6 mice study, the dietary supplementation of ethyl acetate fraction from PL mycelia (PL-EA) for four weeks significantly attenuated an increase in body weight, hepatic lipid accumulation and fasting glucose levels. Mechanistically, PL-EA markedly upregulated the *pgc-1α*, *sirt1* genes and *adiponectin*, downregulated *gck* and *srebp-1c*; upregulated proteins PPARγ, pAMPK, and PGC-1α, and downregulated SREBP-1 and NF-κB in the liver of HFD-fed mice. Furthermore, the major purified compounds of hispidin and hypholomine B in PL-EA significantly reduced the level of oleic and palmitic acids (O/P)-induced lipid accumulation through the inhibition of up-regulated lipogenesis and the energy-metabolism related genes, *ampk* and *pgc-1α*, in the HepG2 cells. Consequently, these findings suggest that the application of PL-EA is deserving of further investigation for treating NAFLD.

Keywords: NAFLD; *Phellinus linteus*; styrylpyrone polyphenolics; hispidin; hypholomine B; centrifugal partition chromatography (CPC); hepatoprotection; dyslipidemia; mice

1. Introduction

Nonalcoholic fatty liver disease (NAFLD) is one of the most salient causes of liver disease worldwide that will likely emerge as the leading cause of end-stage liver disease in the coming decades [1]. The global prevalence of NAFLD is 25.24%, with highest prevalence in the Middle East and South America and lowest in Africa [2], in contrast to 24.13% in the USA [1,2], and 11.5% in Taiwan [3]. In NAFLD, dyslipidemia manifests as an increase in serum triglyceride and low-density lipoprotein cholesterol levels and decreased high-density lipoprotein cholesterol levels [4]. NAFLD without histological changes is associated with complications from dyslipidemia and type 2 diabetes (T2DM), while NAFLD with histological changes is classified as NASH [4].

The majority of the population with NAFLD have isolated steatosis (non-alcoholic fatty liver, NAFL) and a smaller proportion develop non-alcoholic steatohepatitis (NASH), with an increase in hepatic fibrosis leading progressively to cirrhosis, liver cancer, end-stage liver disease and death [5–7]. Overall, cardiovascular disease (CVD) may become the leading cause of death in patients with NAFLD [5].

Phellinus linteus (PL), known as 'Sanghuang' mushroom in China and Korea, and "meshimakobu" in Japan, is one of the most important groups of medicinal macrofungi which have long been used in clinical settings for the past two centuries in many Asian countries [8,9]. A representative group of medicinal fungi, including *P. linteus*, *P. igniarius*, *P. ribis*, *Inonotus obliquus* and *I. xeranticus* was shown to produce a large and diverse spectrum of styrylpyrone-type polyphenols [10]. In a previous study, two predominant active substances were isolated and identified from the culture broth of *P. linteus*. Their chemical structures were identified as hispidin and hypholomine B and helped to elucidate the neuraminidase inhibitory activity, which plays an important role in viral proliferation for the prevention of the spread of influenza infection [11]. The styrylpyrone-type polyphenols of PL, i.e., hypholomine B and hispidin, have been reported to have a significant scavenging activity against radical species in a concentration-dependent manner. In ABTS$^{\bullet+}$ scavenging capabilities, hypholomine B was found to be four times greater than that of Trolox, and superior to that of hispidin [12]. In addition to the antioxidant and anti-neuraminidase activities, PL also possesses diverse bioactivities, including anti-viral [13], anti-diabetic [14], and anti-dementia properties [15]; more importantly, another traditional major indication is hypoglycemic and hypolipidemic effects for preventing type 2 diabetes [16].

Treatments of NAFLD with multimodal interventions such as weight loss, life style modifications and possible medication have been considered as available options [17]. Although there is a growing body of literature demonstrating the hepatoprotective [18] and anti-diabetic effects of Pl [14], it is unclear whether PL from submerged liquid culture is effective in the inhibition of NAFLD in vivo. Few studies have been performed on the production of Pl and its bioactive compounds on the large industrial cultivation of *P. linteus* (>100 L fermenter). Due to the difficulty of process handling, whether the industrial scale could produce similar active components is still unclear. To delineate the protective and therapeutic effects of *P. linteus* on NAFLD, a certain amount of the active compound that was obtained experimentally was isolated from the 2000 L-airlift fermenter products by centrifugal partition chromatography (CPC), which was used in vivo and in an in vitro model to examine whether the cited risk factors of NAFLD could be alleviated.

2. Materials and Methods

2.1. Chemicals and Equipments

Minimum Essential Medium Alpha Medium, MEM NEAA (100×), trypsin (0.25%), Fetal bovine serum, penicillin-streptomycin (10,000 units/mL penicillin and 10,000 µg/mL streptomycin) were provided by Gibco (Grand Island, NE, USA). 2′,7′-Dichlorodihydrofluorescein diacetate (DCFH-DA), 2,2′-azinobis (3-ethyl-benzothiazoline-6-sulfonic acid) (ABTS$^{\bullet+}$) and 2,2-diphenyl-1-picrylhydrazyl (DPPH) free radicals and Oil Red O solution 0.5% in isopropanol were purchased from Sigma-Aldrich (St. Louis, MO, USA). The protein Assay was a product of Bio-Rad (Hercules, CA, USA). Other chemicals not mentioned were provided by Merck (Darmstadt, Germany). The antibodies used included anti-AMPK alpha 1/2 (Abcam, ab131512, 62 kDa), anti-PPAR gamma antibody- ChIP grade (Abcam, ab45036, 57 kDa), anti-PGC1 alpha (Abcam, ab54481, 92 kDa), AMPKα (Cell signaling, 2532s, 62 kDa), SREBP-1(2A4) (Novus, NB600-582), SIRT-1 poly antibody (Proteintech, 13161-1-AP, 82 kDa), NF-κB p65 (C22B4) (Cell signaling, 4764s, 65 kDa), β-actin (Taiclone, tab913655, 42 kDa). High performance liquid chromatography (HPLC) was performed with a Hitachi HPLC system including an L-2130, L-2200 autosampler and an L-2400 diode array detector, and was operated using the D-2000 Elite system software. A column block heater (Jones Chromatography, Hengoed, Wales, UK) was used and controlled at 35 °C. LC-ESI-MS data were obtained with a 6420 triple quadruple LC/MS system (Agilent, Santa Clara, CA, USA) including a 1260 Infinity HPLC system (Agilent Technologies, Santa Clara, CA, USA), a MassHunter Workstation was used for data acquisition, a degasser (model G1379B), a binary gradient pump (model G1312B), an autosampler (model G1329B), a column oven (model G1316A, maintained at 35 °C) and a photodiode array detection (PDA) system (model G1315D, 210–400 nm) were also used. The analytical column used was the Waters Symmetry C18 analysis column (150 × 2 mm i.d.; 100 Å, 3.5 μm,) which was linked with a precolumn (SceurityGuard C18 (ODS), 4 × 3.0 mm i.d., Phenomenex Inc., Torrance, CA, USA). The column oven was maintained at 35 °C. Centrifugal partition chromatography (CPC) experiments were performed on an Armen fully integrated Spot Prep instrument (Armen instrument, Saint-Avé, France). The CPC instrument equipped with one 250 mL capacity column was placed on two separated rotors. Each column is composed of eight stacked disks engraved with a total of 576 twin chambers (250 mL capacity). The CPC columns were coupled with a Spot-Prep I (Armen Instrument) integrated preparative HPLC instrument equipped with a built-in two-headed quaternary gradient HPLC pump, an injector loop (10 mL), a Flash 10 DAD 600 detector (Ecom, Prague, Czech Republic), an automatic fraction collector, and was operated using Armen Glider software.

2.2. Cultivation of *P. linteus* Mycelia

The cultivation of *Phellinus linteus* ATCC 26710 was conducted at the Grape King Biotech Research Institute (LongTan, Taoyuan City, Taiwan) as in the previous report [19]. In brief, the mycelia of *P. linteus* were inoculated into a 2L-broth containing 1% glucose and soybean milk (Brix 2.5) (pH 4.5) and incubated at 28 °C for 8 days with continuous aeration at 1 *vvm* and agitation at 100 rpm. Then, they were transferred into a pilot scale 500 L fermenter (temperature 28 °C; aeration rate 0.5 *vvm*; agitation speed 60 rpm) and fermented for 8 days, and finally into a 2000 L fermenter (working volume 1600 L; temperature 28 °C; aeration rate 0.5 *vvm*; agitation speed 60 rpm), in which they were fermented for 12 days. The broth was centrifuged to collect the mycelia, which were lyophilized and stored at −20 °C for use.

2.3. Preparation of *P. linteus* Mycelial Extract

The lyophilized *P. linteus* mycelia powder was extracted with pure water, with 25, 50, 75 or 100%% methanol, respectively, at a solid to solvent ratio 1:10 (*w*/*v*), were then ultrasonicated at ambient temperature for 30 min and suction filtered through a 0.45 μm PTFE membrane to collect the filtrate. The residue was similarly re-extracted twice. The filtrates were combined and evaporated under reduced pressure until dry. The reclaimed percentage was calculated, and the desiccated powder was stored at −30 °C for use. The best yield containing the highest total polyphenol content (TPC) from the extraction was then selected as the sample for the following solvent partition (hereafter denoted as PL for this extract).

2.4. Solvent Partition of PL

The PL was dissolved in 20 mL deionized water and ultrasonicated for 10 min. The aqueous PL solution was sequentially partitioned with n-hexane, ethyl acetate, and n-butanol, thrice by each. Each partition was separately combined and evaporated under vacuum. The desiccated powder was taken for its weight and stored at −20 °C. The ethyl acetate fraction was used for animal experiments and denoted as PL-EA hereafter.

2.5. Purification of PL-EA with Centrifugal Partition Chromatography

Based on our previous report [20] concerning CPC purification methods, the obtained PL-EA products were further purified using CPC using the two-phase solvent system. In brief, petroleum ether-ethyl acetate-methanol-water solvent systems in different proportions (Supplementary Table S3) were used in the study for the purification of hispidin and hypholomine B with one-step separation.

2.6. HPLC and HPLC/ESI-MS-MS Analyses of PL-EA Extract

The sample injection volume was 10 μL, and the flow rate of the mobile phase was 0.3 mL/min. The mobile phase consisted of solvent A (H_2O containing 0.1% formic acid) and solvent B (acetonitrile containing 0.1% formic acid). The gradient elution was programmed as (time in min, A:B): at 0 min, A:B = 90:10; at 3 min, A:B = 90:10; at 10 min, A:B = 65:35; at 15 min, A:B = 50:50; at 30 min, A:B = 5:95; at 40 min, A:B = 5:95; at 45 min, A:B = 90:10; and at 60 min, A:B = 90:10. After the compounds were eluted and separated, they were further identified with a triple quadrupole mass spectrometer. Nitrogen was used as the drying gas whose flow rate was 9 L/min and temperature was 300 °C. The nebulizing gas was operated at 35 psi. The other parameters were as follows: the potential, 3500 V; the fragmentor voltage, 90 V; and the collision voltage, 15 V. The quadrupole 1 filtered the calculated the m/z of compound of interest, while the quadrupole 2 scanned the ions produced from nitrogen collision with these ionized compounds in the range 100–800 m/z within a scan time of 200 ms/cycle. A multiple reaction monitoring (MRM) mode was used for the MS data acquisition of PL-EA extract. The detection parameters of target compounds and internal standard (IS, quercetin) are summarized in Supplementary Table S1 and their MRM chromatograms are shown in Supplementary Figure S1. A quantitative analysis was performed using the IS method acquired in the MRM mode which was employed for quantifying each target from the obtained peak area. Data were collected and analyzed with the Agilent MassHunter Workstation B.01.04 Software.

2.7. Animal Experiment

To validate the inhibition of PL-EA on NAFLD in vivo, biochemical and histological changes regarding mice liver tissues were investigated.

Male C57BL/6 mice were purchased at 5 weeks of age from BioLASCO Taiwan Co., Ltd. (Yi-Lan, Taiwan). These mice were authorized for admission to the University animal room by The University Ethic Committee of Animals Care and Protection (license code: 10419). In the first week, the animals were caged in the animal room and maintained at a light control of 12 h light/12 h dark cycle, the animal room was maintained at 22 ± 2 °C, in relative humidity (RH) 65 ± 5%. Mice were provided ad libitum access to water from a reverse osmosis system. During a 14-week period study, the mice were grouped into four experimental groups, the normal control (C), the high fat high fructose diet (HFD) control, the HFD + high dose PL-EA (PL-EA-H), and the HFD + low dose PL-EA (PL-EA-L). Group C was fed with regular chow, while the other three groups were given HFD. The NAFLD model was induced in mice with the HFD chow containing fat 40%, fructose 22%, and cholesterol 2% (Research diet, New Brunswick, NJ, USA) [21]. The oral glucose tolerance tests (OGTT) were carried out, respectively (1 g of glucose/kg bw) at week 9 (after induction) and week 14 (after treated). The PL-EA-H (70 mg/kg) and PL-EA-L (35 mg/kg) were administered daily by oral gavage after week 10. The whole experiment lasted for 14 weeks. The animals were kept fasted overnight, CO_2-euthanized, and the organs were dissected immediately, rinsed with sterile saline and stored at −20 °C for further use.

2.8. Oral Glucose Tolerance Test (OGTT)

The oral glucose tolerance test (OGTT) was carried out at week 9 after induction for 8 weeks, and a second time OGTT was conducted at week 14 after the administration of PL-EA for 4 weeks. In brief, after having been fasted for 12 h, the experimental mice were tail-vein bled, the blood sugar level was measured to serve the zero point reference. The mice were then fed glucose solution 0.2 mL by oral gavage, at a dose of 3 g/kg. After tube fed, the blood sugar level was tested at 30, 60, 90, and 120 min, to establish the plasma glucose concentration–time course curve, from which the result of OGTT was determined.

At week 14, the experimental mice were fasted for 12 h and CO_2-euthanized. After assured unconsciousness, blood was withdrawn from the heart using a 1 mL syringe with a 26 G needle and transferred into the heparin coated gel separator green tubes. The blood was centrifuged with 3500× *g* at 4 °C. The supernatant plasma was separated and subjected to biochemical tests within 24 h. Livers, kidneys, and spleens were dissected and rinsed twice with sterile saline. The adhering water was adsorbed with heavy tissues. The organs were weighed, wrapped with aluminum foil, and frozen with liquid nitrogen. The liver was divided into five sections, the left lobe was divided into two pieces, one immersed in formalin, and the other and the remaining parts were separately classified and wrapped with aluminum foil and frozen at −80 °C for use.

2.9. Biochemical Measurements for Blood

The plasma was analyzed with the Fuji automatic biochemical analyzer (FUJI DRI-CHEM 3500s) for total cholesterol (T-CHO), low density lipoprotein-cholesterol (LDL-C), high density lipoprotein-cholesterol (HDL-C), triglycerides (TG), uric acid (UA), glutamic oxaloacetic transaminase (GOT; or aspartate transaminase, AST), and glutamic pyruvic transaminase (GPT; or alanine aminotransferase, ALT).

2.10. Histological Examination

The organs dissected from the mice were immersed in 10-fold volume formalin (10%) and agitated overnight to fix. The fixed tissues were rinsed under flowing tap water for 30 min, immersed in sequentially increasing concentrations of ethanol, starting from 70%, 80%, 85%, 90%, and 95%, each for 1 h, and then in absolute alcohol for 1.5 h and this was repeated thrice before finally being immersed twice in xylene. The treated tissues were immersed twice in liquid paraffin at 57 °C, each time for 2 h, to prepare the paraffin-embedded tissues and slices.

2.11. Hematoxylin-Eosin Staining

Liver tissues were formalin-fixed, embedded in paraffin, sectioned into 2 μm, and subjected to H&E using the conventional protocol, and the images were photographed according to the previous report [21].

2.12. Oil-Red O Staining

We adapted the method of Cui et al. [22] with a slight modification, in which the paraffin-embedded tissues were sliced with a microtome and stained with 0.3% Oil Red O solution for 1 h at ambient temperature, rinsed with deionized water and mounted onto the microscope to inspect the oil drop distribution profile.

2.13. Protein Extraction and Western Blot Analysis

A previous report on the expression analysis of proteins, including PGC-1α, AMPK, p-AMPK, SREBP-1, NF-κB and PPARγ in liver tissues, was followed, although a slight modification was introduced [21]. In brief, the liver tissues were homogenized in the RIPA buffer containing protease inhibitors. An total of 30–40 μg protein was loaded and separated in a 10% SDS-PAGE and electro-blotted to the nitrocellulose membranes. After blocking with TBS buffer (20 mM Tris–HCl, 150 mM NaCl, pH 7.4) containing 5% non-fat milk, the membrane was incubated overnight at 4 °C with various specific antibodies including PGC-1α (1:1000; #ab5448), AMPK (1:1000; #ab1315120), p-AMPK (1:1000; #ab23875), PPARγ (1:500; #ab45036), NF-κB (1:1000, #ab16502) and SREBP-1 (1:5000; #ab26481) from Abcam (Cambridge, UK), and β-actin (1:3000; #MAB1501; Millipore, Billerica, MA, USA), followed by treatment with horseradish peroxidase-conjugated anti-mouse IgG. The results were visualized with the ECL chemiluminescent detection kit (PerkinElmer, Waltham, MA, USA) and quantified using the Image J gel analysis software.

2.14. RNA Isolation and Quantitative Real-Time PCR (qPCR)

RNA from hepatic tissues was isolated to quantify gene expression with RT-qPCR. Total RNA was extracted by using TRIzol® reagent (ThermoFisher Scientific, Waltham, MA, USA), and 1.5 μg of total mRNA was reverse-transcribed using a Takara PrimeScript RT Reagent Kit (Takara Bio, Mountain View, CA, USA), following the instructions provided by the manufacturer. Amplification and detection were performed with the StepOnePlus™ Real-Time PCR System (Applied Biosystems, Foster City, CA, USA). The DNA fragments were amplified for 40 cycles (enzyme activation: 20 sec at 95 °C, hold; denaturation: 3 s at 95 °C; annealing: 40 s at 60 °C). The gene expression of β-actin was determined as the internal control and the relative expression level was calculated by using the standard $2^{-\Delta\Delta Ct}$ method. Primers sequences are listed in Supplementary Table S2.

2.15. HepG2 Cells Experiments

The HepG2 human hepatocellular carcinoma cell line (ATCC CRL-11997) was purchased from the Bioresources Collection and Research Center (Shin-Chu, Taiwan). HepG2 cells were cultured in a minimum essential medium (MEM) containing 10% fetal bovine serum, 1% penicillin-streptomycin, 1% sodium pyruvate, 1% non-essential amino acids and maintained in humidified 5% CO_2/95% air at 37 °C.

2.16. Fatty Acid Induced Mimic Hepatosteatosis in HepG2 Cells and the Treatment of Purified Compounds

Oleic (O) and palmitic (P) acids, both fatty acids, were applied at a molar ratio of 2:1 to induce lipid deposition of HepG2 cells [23]. In brief, after reaching 80% confluence, the HepG2 cells were cultured with serum-free medium containing 1% fat-free bovine serum albumin (BSA) and exposed to 400 μM of fatty acids O/P (2:1) and incubated for 24 h to induce a mimic steatosis. Later, the supplementation of hispidin or hypholomine B at 10 or 50 μM, respectively, was added to the wells and plates incubated for another 24 h.

To detect the lipid accumulation, the control and hispidin- or hypholomine B-treated HepG2 cells were fixed with 10% formalin for 30 min, and then stained with Oil Red O solution for 10 min. The cells were washed three times with physiological saline (PBS) and observed under an inverted microscope. To quantify lipid accumulation in the HepG2 cells, isopropanol was added to dissolve the Oil Red O reagent and the absorbance was measured at 500 nm.

2.17. Analysis for Gene Expression in HepG2 Cells

In our previous report [24] on the extraction of RNA, reverse transcription of RNA to cDNA and the quantification of gene expression using real-time PCR was followed. In brief, each of the 1.5 μg RNA isolated from the O/P-induced and hispidin or hypholomine B-treated HepG2 cell were used to synthesize cDNA. A real-time polymerase chain reaction was conducted according to the manufacturer's instructions with the KAPA SYBR® Fast one step ABI Prism® (Sigma-Aldrich). The $2^{-\Delta\Delta CT}$ value for each sample was analyzed with the StepOnePlus TM Real Time PCR System (Applied Biosystems, Thermo Fisher). Primer sequences are listed in Table S2.

2.18. Statistical Analysis

The data are presented as a mean ± SD result and further analyzed using the GraphPad Prism program (GraphPad, San Diego, CA, USA). The comparison within groups was evaluated using a one-way analysis of variance (ANOVA). Tukey's post hoc test was further used for an analysis of the significance of differences among the means. A confidence level of $p < 0.05$ was considered to be statistically significant.

3. Results and Discussion

3.1. Yield of Extraction from Different Solvent and the Solvent Partition

The following five extraction solvents were used: pure water, 25, 50, 75 and 100% methanol. The results are given in Table 1. Although the yields from *P. linteus* mycelia using the 50% methanol extraction were slightly higher than when using the 75% methanol extract, the best results for the total polyphenol contents (TPC) were obtained using the 75% methanol extract (contents were 1.34 fold as high as those obtained with 50% methanol). The 75% methanol extract (PL) was then selected to obtain the PL extracts, and was used for the liquid–liquid solvent partition in the preparation of the polyphenol-enriched sample. As a result, the yield from the fraction of ethyl acetate (PL-EA) partition was 20.61 ± 1.16 mg/g d.w.

Table 1. The yield and total polyphenol content of *Phellinus linteus* mycelium freeze-dried powder from different extraction methods.

Extract	Yield (%)	Total Polyphenols (mg/g Freeze-Dried Mycelium)
Water	18.71 ± 0.31^{b}	23.70 ± 6.09^{c}
25% MeOH	23.68 ± 0.89^{a}	21.22 ± 3.13^{c}
50% MeOH	24.55 ± 1.64^{a}	52.23 ± 4.58^{b}
75% MeOH	21.93 ± 1.46^{ab}	70.13 ± 5.90^{a}
100% MeOH	19.83 ± 1.29^{b}	47.42 ± 2.02^{b}

Each value represents the mean ± SD of triplicate experiments. Values with different letters within the same column are significantly different ($p < 0.05$).

3.2. Evaluation of the Distribution Coefficient (K_d) for CPC

A suitable K value should be between 0.5 and 3.0 [25], from which the optimum solvent system can be selected for the purification of targets by using the centrifugal partition chromatography (CPC). The two-solvent system consisting of petroleum ether-ethyl acetate-methanol-water (1.6:2.3:1.0:2.1, *v*/*v*/*v*/*v*) was determined for the isolation and purification of hispidin and hapholomine B (Supplementary Table S3).

3.3. Identification and Quantification of Constituents of PL-EA by HPLC-ESI-MS/MS

The 75% methanol extract (yield of 21.93%, *w*/*w*) of *P. linteus* mycelia (PL) showed an HPLC profile containing hispidin (peak 1), hypholomine B (peak 2), and hypholomine B isomer (peak 3) and many other unidentified compounds (Figure 1A), which after liquid–liquid extraction with n-hexane to remove the lipids, followed by extraction with ethyl acetate (PL-EA), yielded a fraction that was abundant in antioxidants (Figure 1B) with a yield of 2.06% (*w*/*w*). These styrylpyrones were further confirmed and quantified using a HPLC-ESI(−)-MRM analysis, as shown in Supplementary Figure S1. Finally, isolation with CPC yielded purified hispidin (purity > 95%, Figure 1C), and hypholomine B and hypholomine B isomer (PL-HB) (purity > 90%, Figure 1D). As can be seen, CPC efficiently improved their purity and the majority of non-phenolic compounds were efficiently removed.

Previously, in an analysis of the ethyl acetate fraction of the 70% methanolic extract of the fruiting bodies of *Phellinus linteus*, Min et al. [26] showed the occurrence of styrylpyrone-class compounds, davallialactone, hispidin, hypholomine B, and caffeic acid [26], while Lee and Yun isolated nine compounds with ethyl acetate-soluble fractionation of *P. linteus* fruiting bodies and identified new compounds, such as protocatechuic acid, protocatechualdehyde, ellagic acid, interfungin A, and inoscavin A [10]. More recently, the same group further isolated new compounds from the culture broth of *P. linteus*, which included inotilone, 4-(3,4-dihydroxyphenyl)-3-buten-2-on, phellilane H, (2E,4E)-(+)-4′-hydroxy-γ-ionylideneacetic acid, and (2E,4E)-γ-ionylideneacetic acid [27].

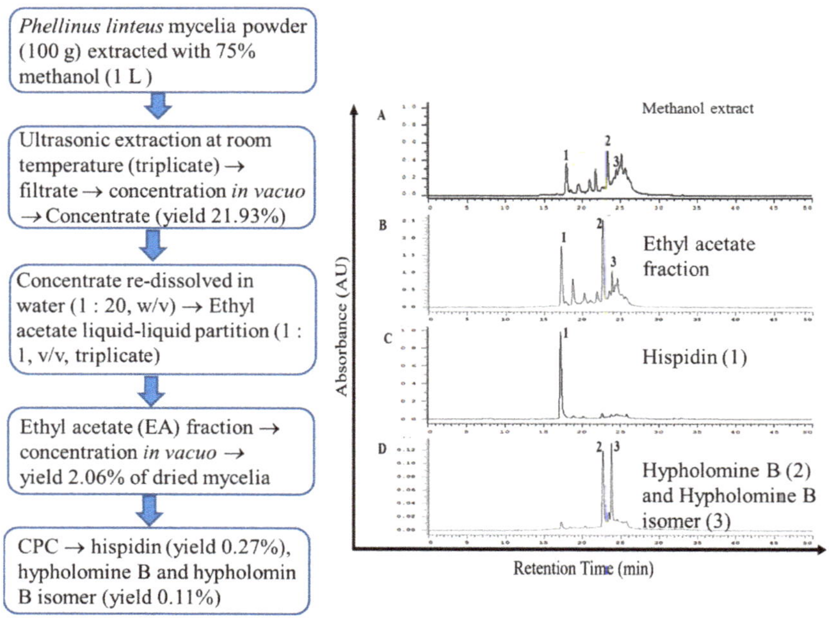

Figure 1. Experimental procedures of solvent extraction, partition and purification of extract prepared using the *Phelinus linteus* mycelia extract and HPLC analysis of extracts obtained in each procedure. (**A**) 75% methanol. (**B**) ethyl acetate. (**C**) hispidin. (**D**) hypholomine B and hypholomine B isomer. Peaks 1: hispidin; 2: hypholomine B, and 3: hypholomine B isomer.

3.4. *The Antioxidative Capability*

3.4.1. DPPH Radical Scavenging Capability

The DPPH radical scavenging capability of each partition (or isolated compound) was found to increase in a dose-dependent manner. The strongest was PL-HB (hypholomine B and its isomers) which was comparable to Trolox (Figure 2a). The order of antioxidative capabilities at 125 μg/mL was as follows: Trolox = PL-HB = PL-H > PL-EA > PL (Figure 2a), consistent with [12], the capability of hispidin oligomers to scavenge DPPH free radical was in the following order: hypholomine B > 1,1-distyrylpyrylethane > 3,14′-bihispidinyl > hispidin [12]. Chang et al. indicated that the ethyl acetate fraction exhibited strong DPPH radical-scavenging activity as well as antioxidant activities (IC_{50} = 0.66 ± 0.01 mg/mL) [28].

Hispidin exhibited quenching effects against DPPH radicals, superoxide radicals, and hydrogen peroxide in a dose-dependent manner [29–31], and at 1.0 mM it inhibited 85.5% of the DPPH radicals [31].

Previously, Jeon et al. isolated 10 antioxidants from the fruiting bodies of *P. linteus* including hispidin, davalliallactone, interfungins A, and hypholomine B, etc., and demonstrated that davalliallactone and interfungins A exhibited the strongest inhibitory effect against the DPPH radicals [32], indicating that more powerful and therapeutically useful antioxidants can be achieved if the whole fungi are utilized.

A wide spectrum of the literature demonstrated that many species of *Phellinus* including *P. linteus*, *P. igniarius* and *P. durissimus* all showed significant DPPH scavenging capability [33–35], which, as suggested, can be attributed to various antioxidant compounds contained in genus Phellinus, such as caffeic acid, davalliallactone, ellagic acid, hispidin, hypholomine B, inoscavin A, interfungins A, methyldavallialactone, protocatechualdehyde and protocatechuic acid [36], which is in agreement with our findings (Figure 1B,C).

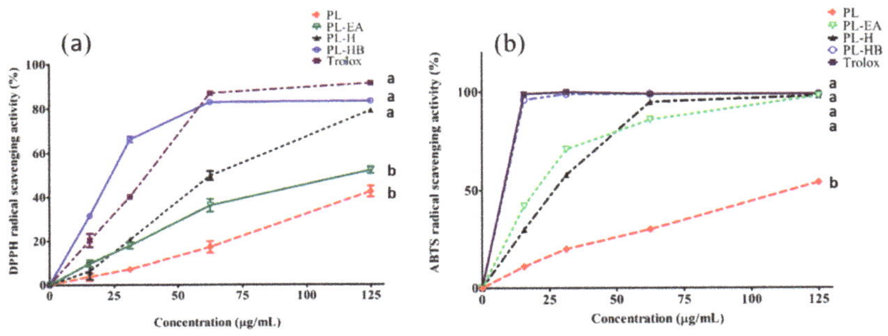

Figure 2. Radical scavenging capabilities of different preparations fractionated from *P. linteus mycelia*. (**a**) for DPPH radicals. (**b**) for $ABTS^+$ radicals. PL: 75% methanol crude extract. PL-EA: ethyl acetate fraction. PL-H: CPC isolated hispidin. PL-HB: CPC isolated hypholomine B and hypholomine B isomer. Data are expressed as mean ± SD from triplicate experiments. Different letters in lower case on each curve indicate significantly different from each other ($p < 0.05$).

3.4.2. $ABTS^+$ Radical Scavenging Capability

Similar to the results found for the scavenging capability of DPPH, the anti-$ABTS^+$ radical capability of different extracts from lyophilized *P. linteus* mycelia increased in a dose dependent fashion, and the order (in decreasing tendency) was as follows: Trolox = PL-H = PL-EA = PL-HB > PL at dose 125 μg/mL (Figure 2b) ($p < 0.05$). With regard to the anti-$ABTS^+$ capability, it was positively correlated with the total phenolic contents (Figure 2b), which is in good agreement with Chang et al. [28]. In particular, caffeic acid, inotilone, and 4-(3,4-dihydroxyphenyl)-3-buten-2-one were potent $ABTS^+$ scavengers which showed IC_{50} values of 0.52 ± 0.10, 1.10 ± 0.10, and 1.69 ± 0.11 μM, respectively [24]. The *P. linteus* ethanolic extract showed a powerful capacity in scavenging DPPH and $ABTS^+$ radicals, of 2–10 fold stronger than any other species of mushrooms [26].

3.5. Body Weight Variation of Experimental Mice

A mouse model of obesity was successfully established by feeding them a high-fat high-fructose diet (HFD) for 10 weeks (control: 32.36 ± 1.71 g; HFD: 40.00 ± 1.62 g) (Table 2). The body weight of all experimental mice increased steadily during the whole period of the experiment within 14 weeks, while the HFD mice group consistently showed the largest body gain (Figure 3). PL-EA-H and PL-EA-L treatments did not affect the mice's slow weight gain trend, indicating the nontoxic nature of the PL-EA fraction. More importantly, the PL-EA extract seemed to exhibit a rather promising sliming effect (Figure 3). A similar experiment conducted by Noh et al. demonstrated davalliallactone to be the active compound responsible for reducing the body weight gain [37].

Table 2. Effects of ethyl acetate fraction from 75% methanol extracts of *Phellinus linteus* mycelium freeze-dried powder (PL-EA) on body weight; liver weight and liver to body weight ratio in a high-fat/high-fructose diet (HFD)-fed mouse model.

	Control	HFD	PL-EA-L	PL-EA-H
Body weight(g)				
Initial	21.32 ± 0.49	21.40 ± 0.81	21.10 ± 0.68	21.37 ± 0.94
Final	32.36 ± 1.71	40.00 ± 1.62 ***	35.83 ± 3.16 **##	35.50 ± 2.15 *###
Liver weight (g)	1.89 ± 0.25	3.22 ± 0.48 ***	2.49 ± 0.44 **##	2.31 ± 0.04 ###
Liver weight/ Body weight (%)	5.83 ± 0.62	8.06 ± 1.19 ***	7.02 ± 0.70 *#	6.48 ± 0.75 ##

HFD: high-fat/high-fructose diet, PL-EA-L: high-fat/high-fructose diet + low dose PL-EA (35 mg/kg b.w.), PL-EA-H: high-fat/high-fructose diet + high dose PL-EA (70 mg/kg b.w.). Data are expressed as means ± SE ($n = 9$). # $p < 0.05$; ## $p < 0.01$; ### $p < 0.001$ vs. the HFD group; * $p < 0.05$; ** $p < 0.01$; *** $p < 0.001$ vs. the control.

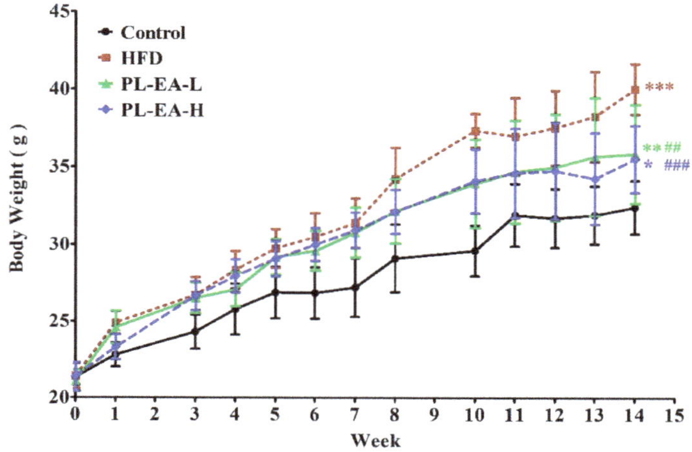

Figure 3. PL-EA inhibits high-fat high-fructose (HFD)-induced obese in mice. HFD: high-fat/high-fructose diet. PL-EA-H: HDF + high dose PL-EA (70 mg/kg). PL-EA-L: HDF + low dose PL-EA (35 mg/kg). Data are expressed as mean ± SE (n = 9). ## p <0.01; ### p < 0.001 vs. the HFD group; * p < 0.05; ** p < 0.01; *** p < 0.001 vs. the control.

3.6. The Ratio of Liver to Body Weight

The liver weight in mice normally falls in the 2–3 g range (3–5%/bw) [38]. Consistent with this, we showed the liver weight and the ratio of liver to body weight (in %) in the experimental control of mice to be 1.89 ± 0.25 g or 5.83 ± 0.61%, respectively (Table 2). HFD induced a higher liver weight (3.22 ± 0.48 g), hence leading to a higher ratio (8.06 ± 1.19%) (Table 2), implicating the occurrence of fatty liver. Treatment with PL-EA dose dependently alleviated these abnormal liver weight as demonstrated by the results of 2.49 ± 0.44 and 2.31 ± 0.04 g, or in the ratio of liver to body weight 7.02 ± 0.70 and 6.48 ± 0.75% by PL-EA-L and PL-EA-H, respectively (Table 2), indicating the powerful anti-obesity effect of PL-EA. Studies in the literature emphasize the importance of the percentage of total body mass to assess the metabolic or nutritional status although it has been noted that in mice livers the results are more prominent than that of rats or humans [38].

3.7. Plasma Biochemical Measurements

3.7.1. The Total Plasma Cholesterol Level

The total plasma cholesterol (TC) level in the control, the HFD, the PL-EA-L and PL-EA-H mice lay in the following range: control (100 ± 10 mg/dL); HFD (315 ± 7 mg/dL) (p < 0.001), PL-EA-L (270 ± 12 mg/dL), and PL-EA-H (260 ± 8) mg/dL, respectively (p < 0.001) (Figure 4a), implicating that PL-EA effectively reduced the TC. Statistically, adults with high TC/HDL-C or TG/HDL-C ratios, or both, have a greater risk of NAFLD, especially advanced NAFLD [39,40].

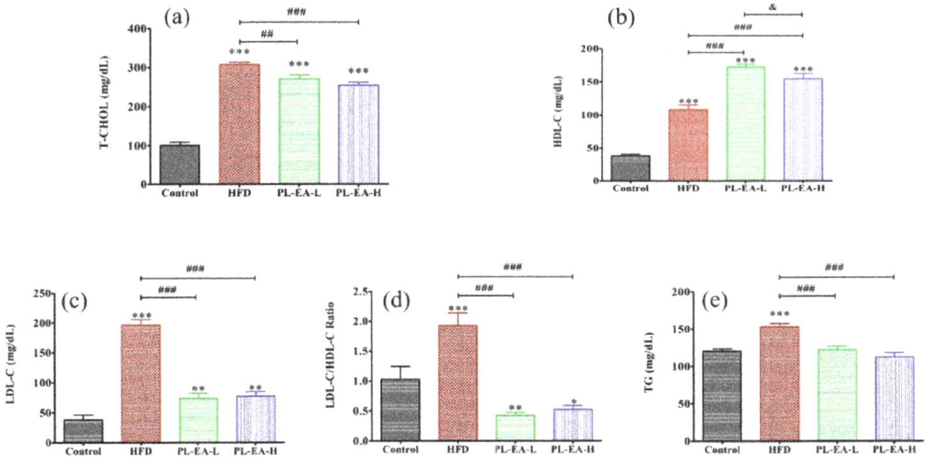

Figure 4. Effects of ethyl acetate fraction of *P. linteus* mycelia on the lipid profile. (**a**) total plasma cholesterol content, (**b**) content of plasma high density lipoprotein cholesterol, (**c**) content of plasma low density lipoprotein cholesterol, (**d**) the ratio LDL-C/HDL-C and (**e**) the plasma TG content. HFD: high-fat/high-fructose diet (HFD). PL-EA-L: HDF + low dose PL-EA (35 mg/kg). PL-EA-H: HDF + high dose PL-EA (70 mg/kg). Data are expressed as mean ± SE (n = 9). ## $p < 0.01$; ### $p < 0.001$ vs. the HFD group; * $p < 0.05$; ** $p < 0.01$; *** $p < 0.001$ vs. the control. & $p < 0.05$ significant differences between the groups.

3.7.2. The Plasma HDL-C Level

The plasma HDL-C level in the control group was 37.5 ± 4.0 mg/dL, while that of HFD was elevated at 107.0 ± 1.0 mg/dL. Interestingly, the PL-EA-L, and PL-EA-H were alleviated across the board, with higher levels of HDL-C to 175.0 ± 8.0, and 156.0 ± 14.0 mg/dL, respectively (Figure 4b). Epidemiological studies have suggested an inverse correlation between high-density lipoprotein-cholesterol (HDL-C) levels and the risk of cardiovascular diseases and atherosclerosis [41]. A 1% increase in HDL-C level is associated with a 2% decrease in CV risk [42,43].

3.7.3. The Plasma LDL-C Level

The HFD elevated the plasma LDL-C level to 193 ± 8 mg/dL compared to 40 ± 7 mg/dL in the control ($p < 0.001$). PL-EA effectively suppressed its level to 73 ± 8 and 74 ± 7 mg/dL, respectively ($p < 0.05$) (Figure 4c). Patients with higher LDL-C levels are more likely to have a higher prevalence of NAFLD than subjects with lower levels [44].

3.7.4. Ratio of HDL-C/LDL-C

A high fat/high fructose diet induced a high ratio of LDL-C/HDL-C to 1.91 ± 0.25, compared to 1.05 ± 0.22 of the control ($p < 0.001$) (Figure 4d). Interestingly, the ethyl acetate fractions, despite the low or the high level of EA, all efficiently lowered the ratio to 0.43 ± 0.07 and 0.55 ± 0.09 (Figure 4d).

More recently, Wang et al. suggested that the ratio of non-HDL-C to HDL-C would be a better predictor for new-onset NAFLD [45].

3.7.5. The Plasma Triglyceride Levels

The plasma triglycerides were significantly elevated in HFD mice, reaching 158 ± 6 mg/dL compared to 115 ± 5 mg/dL in the control, which was efficiently alleviated by the administration of PL-EA to 120 ± 7 mg/dL and 110 ± 8 mg/dL, respectively, at a high and low level of PL-EA ($p < 0.001$) (Figure 4e).

Elevated plasma TG levels are also often associated with low HDL-C levels [46]. Recently, non-HDL cholesterol (including LDL-C and remnant lipoproteins such as VLDL-c and IDL-C) has been proposed to be a better estimate of total atherogenic burden than LDL-C, especially in patients with elevated plasma TGs ranging between 200 and 500 mg/dL [47,48].

Hispidin was demonstrated to decrease the intracellular triglyceride content by 79.5 ± 1.37%, stimulate glycerol release by 276.4 ± 0.8% and inhibit lipid accumulation by 47.8 ± 0.16% [49]. Hispidin also inhibited glycerol-3-phosphate dehydrogenase (GPDH) and pancreatic lipase, representing the most potent inhibitors [49]. A similar experiment conducted by Noh et al. demonstrated davallialactone to be the active compound responsible for reducing hepatic lipid concentrations, and fat accumulation in epididymal adipocytes [37]. The mechanism of which was proposed to be partly mediated by the inhibition of enzymes associated with hepatic and intestinal lipid absorption and synthesis [37]. Suggestively, the presence of davallialactone in our ethyl acetate fraction (not shown), as previously reported elsewhere [10,26], might also synergistically contribute to the lowering of TG.

3.7.6. Plasma Level of GOT and GPT

HFD apparently elevated the level of plasma alanine aminotransferase GOT to 185 ± 35 U/L compared to 56 ± 4 U/L of the control ($p < 0.001$), which markedly alleviated to 104 ± 15 U/L and 104 ± 10 U/L, respectively by PL-EA-L and PL-EA-H ($p < 0.001$) (Figure 5a). Similarly, the GPT level in the HFD mice rose to 225 ± 10 U/L, while PL-EA-L and PL-EA-H ameliorated the level to 155 ± 13 and 85 ± 8 U/L, respectively, compared to 40 ± 4 U/L of the control (Figure 5b).

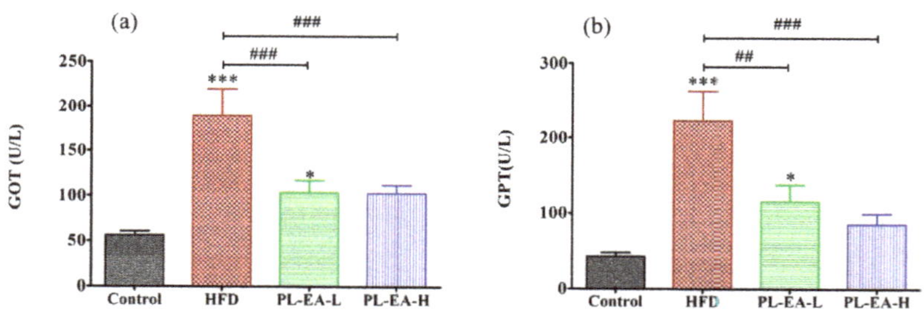

Figure 5. Effects of ethyl acetate fraction from *P. linteus* mycelia on the activities of (**a**) plasma aspartate aminotransferase and (**b**) plasma alanine aminotransferase. HFD: high-fat/high-fructose diet (HFD). PL-EA-L: HDF + low dose PL-EA (35 mg/kg). PL-EA-H: HDF + high dose PL-EA (70 mg/kg). Data are expressed as mean ± SE (n = 9). ## p <0.01; ### p < 0.001 vs. the HFD group; * p < 0.05; *** p < 0.001 vs. the control.

An increasing number of studies have demonstrated the promising hepatoprotective and antihepatotoxic effects of *P. linteus* [28,49,50]. Previously, Huang et al. [51], and recently Dong et al. [52], respectively, demonstrated the hepatoprotective bioactivity of hispidin [51] and hypholomine B [52]. Several investigations have proved the *Phellinus* species as being hepatoprotective and antihepatotoxic agents [36]. The compounds, phellinulin A [51], phellinulins D, E, F, G, H, I, K, M, and N, phenillin C, and γ-ionylideneacetic acid, when isolated from *P. linteus*, were all demonstrated to exhibit an hepatoprotective effect [51].

3.8. *Oral Glucose Tolerance Test*

The plasma glucose level reached its peak value at 30 min in all groups after the tube feeding of glucose solution at week 9 (Figure 6a). Furthermore, HFD, PL-EA-H and PL-EA-L all comparably reached a plasma glucose level of 375 ± 20 mg/dL and a slight but insignificant deviation occurred between the three groups at 120 min compared to the control (135 ± 2 mg/dL) ($p < 0.05$) (Figure 6a). After treatment with PL-EA for 4 weeks, the plasma glucose profile changed at week 14 as follows: starting from 105–120 mg/dL at zero time, increasing to 385 ± 35 mg/dL (HFD), 325 ± 20 mg/dL (PL-EA-L), and 280 ± 16 mg/dL (PL-EA-H) at 30 min, declining steadily to 295 ± 14 mg/dL, 240 ± 17 mg/dL and 225 ± 15 mg/dL at 120 min, respectively, compared to 140 mg/dL for the control (Figure 6b). The results implicated the promising antihyperlipidemic and antihyperglycemic effects of PL-EA. However, it is advisable to slightly raise the dose of PL-EA for treatment instead.

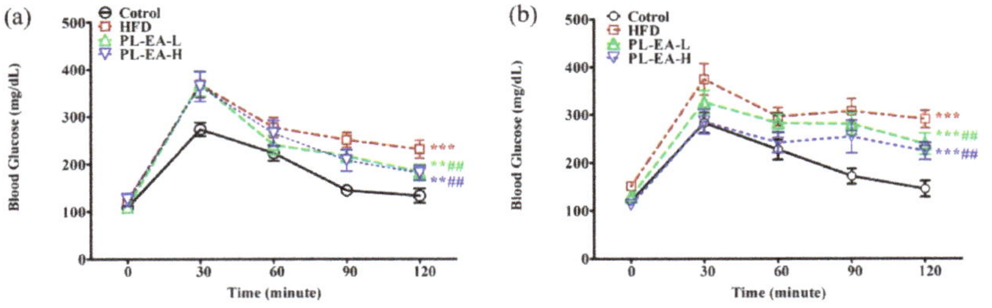

Figure 6. Effects of ethyl acetate fraction from *P. linteus* mycelia on the OGTT at week 9 (**a**) and at week 14 (**b**). HFD: high-fat/high-fructose diet (HFD). PL-EA-L: HDF + low dose PL-EA (35 mg/kg). PL-EA-H: HDF + high dose PL-EA (70 mg/kg). Data are expressed as mean ± SE ($n = 9$). ## $p < 0.01$; vs. the HFD group; ** $p < 0.01$; *** $p < 0.001$ vs. the control.

Besides antioxidant activity, hispidin displays potentially hypoglycemic effects [53].

In chronic hyperglycemia, an excessive amount of glucose is shunted to the polyol pathway, where aldose reductase reduces glucose into sorbitol at the expense of NADPH. Since NADPH is essential for the generation of reduced glutathione (GSH, intracellular antioxidant) from oxidized glutathione (GSSG), the depletion of NADPH by the aldose reductase pathway may impair intracellular antioxidant defense [30]. Sorbitol can be converted to fructose via sorbitol dehydrogenase (SDH) with the production of NADH potentially leading to increased ROS via NADH oxidase [30].

Glucotoxicity may impair the regulation of glucokinase (GK) and its inhibitory protein, the GK regulatory protein (GKRP) [54], which plays a prognostic role in acute pancreatitis [55,56].

The fruiting body of *P. linteus* showed inhibitory activity against both the aldose reductase-related polyol pathway and protein glycation, effectively preventing artherosclerosis, cardiac dysfunction, retinopathy, neuropathy and nephropathy (Lee et al., 2008a; 2008b). The active principles of davallialactone, hypholomine B, and ellagic acid present in *P. linteus* exhibited potent human recombinant aldose reductase inhibitory activity [53].

3.9. HE Staining and Oil-Red Staining of Mice Liver Tissues

To identify the effects of PL-EA on the expression of lipid accumulation in NAFLD mice, H&E staining and Oil-Red O staining of liver tissues were performed, respectively. HFD enhanced oil drop accumulation in the liver tissues (hepatic steatosis) (Figure 7(b-1,b-2)), compared to the control (Figure 7(a-1,a-2)). PL-EA dose dependently but incompletely alleviated such pathological changes (Figure 7(c-1,c-2,d-1,d-2)), suggesting a longer treatment time may be required. The HFD mice had a number of oil drops that accumulated in the tissues, most of which were reduced by feeding them PL-EA-L and PL-EA-H in a semi dose-dependent manner (Figure 7(c-1,c-2,d-1,d-2)).

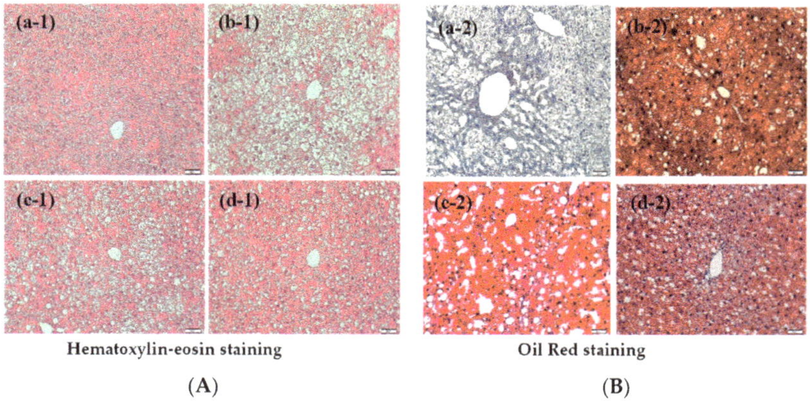

Figure 7. Effects of PL-EA on the hepatic lipogenesis in NAFLD mouse model. Representative photographs of hematoxylin-eosin (H&E) (**A**) and Oil Red O staining (**B**) of mice liver tissues for histological examination. (**a-1**,**a-2**) control. (**B**) (**b-1**,**b-2**) HFD, high-fat/high-fructose diet. (**c-1**,**c-2**) PL-EA-L: HDF + low dose PL-EA (35 mg/kg). (**d-1**,**d-2**) PL-EA-H: HDF + high dose PL-EA (70 mg/kg). Magnification, 200×.

3.10. Effects of Purified Compounds from PL-EA on Free Fatty Acids-Induced Steatosis in HepG2 Cells

The main active ingredients in the antioxidation of PL, such as hispidin and hypholomine B, have previously been reported [12]. However, it remains uncertain whether hispidin or hypholomine B are capable of antagonizing free fatty acids-induced hepatic steatosis. To investigate the effects of these two compounds on the lipid accumulation of hepatocytes, both hispidin and hypholomine B were purified from PL-EA using Hep G2 cell model. A total of 400 μM of the O/P treatment on HepG2 cells significantly induced lipid accumulation (24% increase at 24 h) compared with the control group (Figure 8A,B). Treatment with hispidin or hypholomine B (10 or 50 μM) and 400 μM of O/P significantly decreased lipid accumulation after 24 h of treatment (38% and 35% reduction with hispidin or 40% and 47% reduction with hypholomine B, at 10 and 50 μM, respectively). Furthermore, the contribution of the inhibitory activity of hypholomine B was significantly superior to that of hispidin at a concentration of 50 μM (Figure 8B). These findings suggest that the inhibition capability of hypholomine B is higher than that of hispidin, which is related to the ameliorative effect of oxidative stress according to a previous report [12].

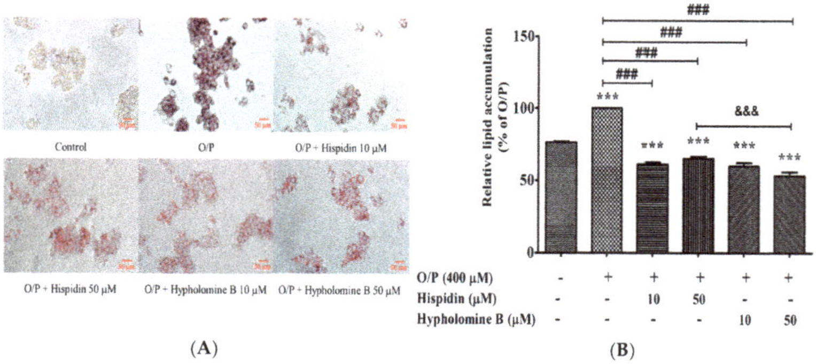

Figure 8. Effects of purified compounds from PL-EA on Oil Red O staining and lipid. accumulation in HepG2 cells. Lipid droplets in HepG2 cells were photographed by phase contrast microscopy

(original magnification ×200) (**A**). The lipid content from Oil Red O stained cells was quantified by spectrophotometric analysis at 500 nm (**B**). Bars represent mean ± SE (n = 3) *** $p < 0.001$ vs. the control; ### $p < 0.001$ vs. the O/P group; &&& $p < 0.001$. Significant differences between groups were determined using one-way ANOVA followed by Tukey's procedure.

3.11. Relative Gene Expression in Mice Liver Tissues

Some of the lipid-metabolism-related genes in the mice liver tissues were examined (Figure 9). As found, the genes PGC-1α, Sirt1, and adiponectin were all downregulated ($p < 0.001$), while SREBP-1c ($p < 0.001$) was upregulated with HFD. Feeding the mice with PL-EA apparently reversed such trends (Figure 9). Compared to that of HFD, the increments were as follows: for PGC-1α (+4.20 fold), Sirt1 (+6.53 fold), and adiponectin (+5.61 fold), following the administration of LP-EA-H ($p < 0.001$). In contrast, SREBP-1c was downregulated by 77.5% ($p < 0.05$) (Figure 9).

PGC-1α overexpression increased in markers of mitochondrial content and function; as a result, fatty acid oxidation was enhanced which was accompanied by reduced triacylglycerol accumulation and secretion [57].

SIRT1 is involved in both NAFLD and alcoholic fatty liver diseases (AFLD) [58]. An increased number of studies have provided evidence that SIRT1 acts as a key metabolic/energy sensor (via intracellular NAD^+/NADH ratio) by transferring signals to initiate transcriptional activity and gene expressions that are involved in metabolic homeostasis [58–60].

Adiponectin has been revealed to protect the liver against hepatic steatosis by decreasing serum lipid and glucose production [61]. De novo lipogenesis has an important contribution to the pathophysiology of NAFLD because it provides almost one third of the accumulated hepatic triglycerides in patients with hepatosteatosis [52,62,63]. Thus, our findings suggest that PL-EA may improve hepatic steatosis and ameliorate HFD-induced fatty liver disease through the regulation of the hepatic fatty acids metabolism.

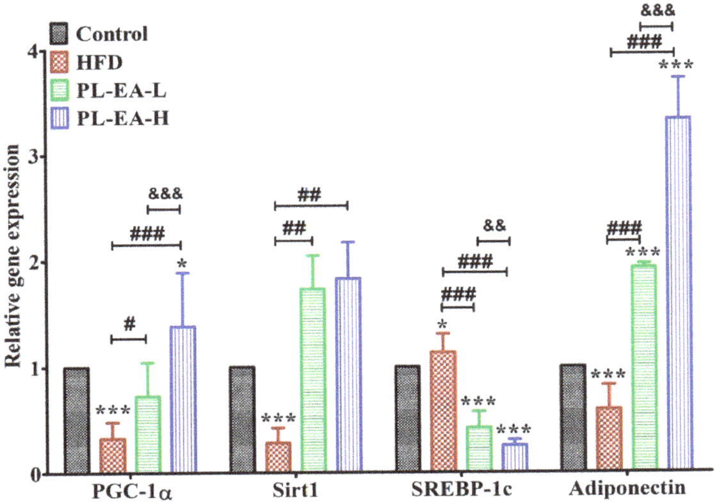

Figure 9. Effects of ethyl acetate fraction from *P. linteus* mycelia on the relative gene expression of mice livers. HFD: high-fat/high-fructose diet (HFD). PL-EA-L: HDF + low dose PL-EA (35 mg/kg). PL-EA-H: HDF + high dose PL-EA (70 mg/kg). Data are expressed as mean ± SE (n = 9). * and ***, $p < 0.05$ and 0.001, respectively vs. the control; #, ## and ###, $p < 0.05$, 0.01 and 0.001, respectively vs. the O/P group. && and &&&, $p < 0.01$ and 0.001, respectively, significant differences between the groups. PGC1-α: peroxisome proliferator-activated receptor gamma coactivator 1-alpha; Sirt-1: NAD-dependent deacetylase sirtuin-1; SREBP-1c: sterol regulatory element-binding protein 1c.

3.12. Relative Gene Expression in HepG2 Cells

To further verify the possible candidates of active compounds in the PL-EA extract on NAFLD, the purified compounds of hispidin and hypholomine B were prepared and used in the genes expression analyses with an HepG2 cell model. After the exposure of HepG2 cells to O/P induction of a mimic steatosis, the cells were treated with hispidin or hypholomine for 24 h. Hypholomine B was found to show greater up-regulated activity in PGC1-α, SIRT1 and adiponectin genes expression than hispidin at the same dose of 50 μM (Figure 10). Nevertheless, the expression of PGC1-α, SIRT1 and adiponectin genes significantly increased in hispidin and hypholomine B at 10 and 50 μM, respectively, in comparison to the O/P group (Figure 10). Furthermore, expression of the lipogenesis-related gene SREBP-1c was elevated in HepG2 cells treated with 400 μM O/P. Hispidin or hypholomine B treatment significantly ($p < 0.001$) decreased the expression of the biogenesis marker (Figure 10). These results demonstrate that the main active ingredients of PL-EA may have pharmacological effects on NAFLD in both in vitro and in vivo studies.

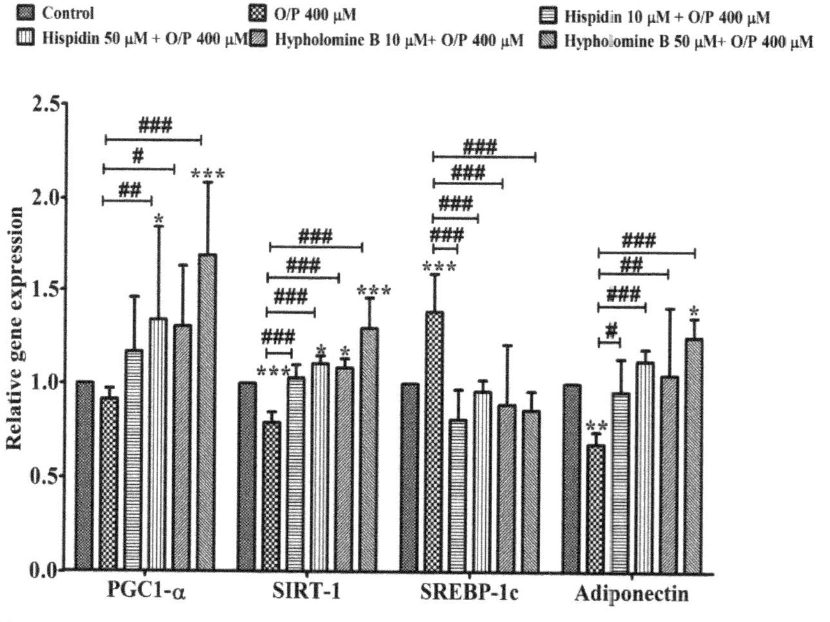

Figure 10. Effects of hispidin and hypholomine B isolated from the ethyl acetate fraction of *P. linteus* mycelia on the relative gene expression of HepG2 cell. Data are expressed as mean ± SE ($n = 3$). *, **, and ***, $p < 0.05$, 0.01 and 0.001, respectively vs. the control; #, ## and ###, $p < 0.05$, 0.01 and 0.001, respectively vs. the O/P group. PGC1-α: peroxisome proliferator-activated receptor gamma coactivator 1-alpha; Sirt-1: NAD-dependent deacetylase sirtuin-1; SREBP-1c: sterol regulatory element-binding protein 1c.

3.13. Western Blotting

HFD downregulated the expression of PPARγ, pAMPK, PGC1α, but upregulated SREBP-1 and NFκB in mice (Figure 11). PL-EA dose dependently alleviated the level of PPARγ and PGC1α. As for pAMPK, SREBP-1 and NFκB, both PL-EA-L and PL-EA-H showed very comparable effects. Compared to HDF, the PL-EA-L diet increased PPARγ 1.43 fold, p-AMPK 1.44 fold, PGC1α 1.19 fold. Conversely, it downregulated NFκB 52.5% and SREBP-1 35.8% (Figure 11). In contrast to that of PL-EA-L, PL-EA-H was found to be upregulated PPARγ 1.89 fold, p-AMPK 1.53 fold, and PGC1α 1.57 fold; it downregulated NFκB 57.5% and SREBP-1 45.2% (Figure 11).

PPARγ stimulates the expression of adiponectin and initiates signalling cascades in the liver, leading to increased β-oxidation, decreased gluconeogenesis and less insulin-resistant hepatic tissue [64]. PPARγ also induces phosphoenolpyruvate carboxykinase to facilitate the triglyceride synthesis [64], which obviously can be inhibited by PL-EA (Figure 4e).

AMPK is a master regulator of the cellular metabolism and is responsible for the overall energy balance and the activation of AMPK (pAMPK) is recognized as an important regulator in the amelioration of NAFLD [65]. In addition, the upregulation of NFκB may reflect that the NAFLD is slightly associated with inflammation, which was apparently alleviated by PL-EA (Figure 11).

To summarize, the underlying mechanisms of PL-EA for alleviating HFD-induced NAFLD are summarized in Figure 12a. Furthermore, the hypolipidemic effect with regard to the purified hispidin and hypholomine from the PL-EA is shown in Figure 12b.

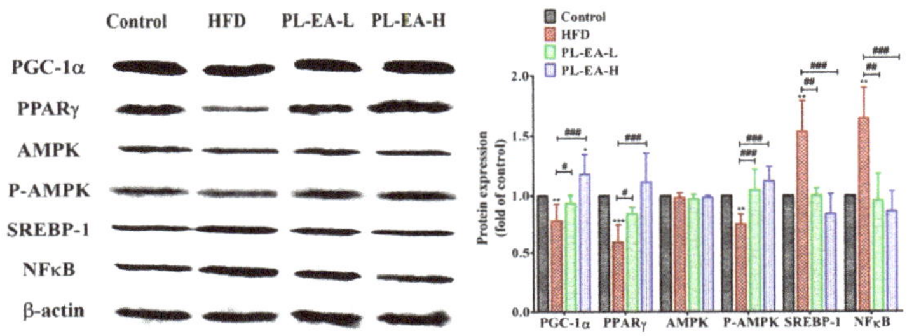

Figure 11. Effects of ethyl acetate fraction from *P. linteus* mycelia on the relative proteins expression of mice livers. Representative Western blots (left panel) and quantified bar graphs relative to each control (right panel) showing alterations among groups. HFD: high-fat/high-fructose diet (HFD). PL-EA-L: HDF + low dose PL-EA (35 mg/kg). PL-EA-H: HDF + high dose PL-EA (70 mg/kg). Data are expressed as mean ± SE (n = 9). # $p < 0.05$; ## $p < 0.01$; ### $p < 0.001$ vs. the HFD group; * $p < 0.05$; ** $p < 0.01$; *** $p < 0.001$ vs. control. PPARγ: peroxisome proliferator-activated receptor gamma. AMPK: AMP activated protein kinase. *p*-AMPK: phosphorylated AMPK. PGC1-α: peroxisome proliferator-activated receptor gamma coactivator 1-alpha. SREBP1: sterol regulatory element-binding protein 1. NF-κB: nuclear factor kappa-light-chain-enhancer of activated B cells.

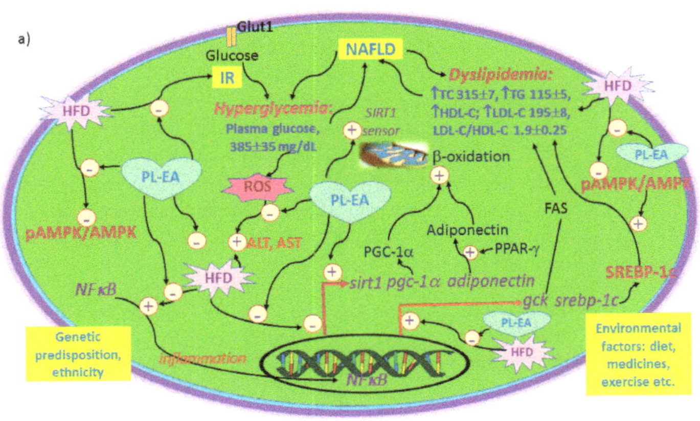

Figure 12. *Cont.*

b)

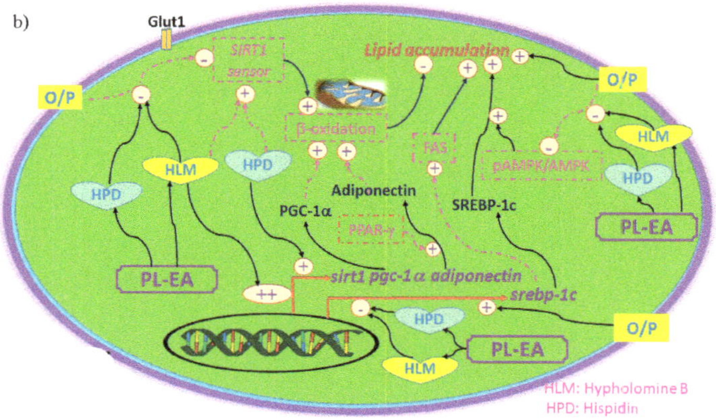

Figure 12. Mechanism of action related to the alleviative effect of NAFLD in mice with the ethyl acetate partition from *Phellinus linteus* (**a**) and that of in vitro hypolipidemic effect of hispidin and hypolomine B in HepG2 cell model (**b**). In Figure 12b, the items highlighted in pink are not shown in the HepG2 cell model but were found in the in vivo mice model. The purified active components from PL-EA used in the in vitro experiment are hispidin (HPD) and hypholomine B and isomers (HLM). ALT: alanine aminotransferase (GPT). AMPK: AMP-activated protein kinase. AST: aspartic aminotransferase (GOT). FAS: fatty acid synthase. *gck: glucokinase gene*. HFD: high fat diet. HDL-C: high density lipoprotein–cholesterol. IR: insulin resistance. LDL-C: low density lipoprotein–cholesterol. NAFLD: non-alcoholic fatty liver disease. NFκB: nuclear factor kappa-light-chain-enhancer of activated B cells. pAMPK: phosphorylated AMPK. *sirt-1*: NAD^+-dependent deacetylase sirtuin-1 gene. PGC-1α: peroxisome proliferator-activated receptor-gamma coactivator-1α. *pgc-1α*: peroxisome proliferator-activated receptor-gamma coactivator-*1α gene*. PPAR-γ: peroxisome proliferator- activated receptor gamma (PPAR-γ). ROS: reactive oxygen species. SREBP-1c: sterol regulatory element-binding protein-1c. *srebp-1c*: sterol regulatory element-binding protein-1c gene. TC: plasma total cholesterol. TG: triglycerides. PL-EA: the ethyl acetate partition from *Phellinus linteus*.

4. Conclusions

Based on the presented results of this study, it could be concluded that 75% methanol extraction and solvent partition using ethyl acetate (PL-EA) is satisfactory for the extraction of bioactive antioxidants in PL mycelia. The pronounced antioxidant activity of PL-EA was associated with a high content of hispidin and, hypholomine B and its isomer. Furthermore, the results of the mouse model study indicate that the PL-EA has anti-NAFLD effects due to its regulation of hepatic lipogenesis and the potential antihyperglycemic effect it imparts. P. linteus has been traditionally utilized in East Asian countries for more than two hundred years and most of its active components have been evidenced elsewhere without toxic symptoms and complications. The implication of the study is that the development of a practically effective PL nutraceuticals therapy may be expected and demonstrated in the future.

Supplementary Materials: The following supporting information can be downloaded at: https://www.mdpi.com/article/10.3390/antiox11050898/s1, Table S1. Specific MRM settings for the styrylpyrone compounds from PL-EA and internal standard.; Table S2. List of primers for real-time PCR analyses in mouse liver and HepG2 cell; Table S3. Partition coefficients (K) of hispidin and hypholomine B in several different solvent systems; Figure S1. The total ion chromatogram from multiple reaction monitoring (top panel) and HPLC profile using diode array detection on ethyl acetate fraction of 75% methanol extract of Phellinus linteus mycelia.

Author Contributions: J.-J.L. and C.-H.C. performed the animal study. C.-C.C. (Chin-Chu Chen) performed the fermentation work. C.-C.C. (Chun-Chao Chang) assisted in funding acquisition and conducted the project. C.-C.C. (Charng-Cherng Chyau) designed and conducted the experiments and wrote the manuscript and R.Y.P. drafted and revised the manuscript. All authors have read and agreed to the published version of the manuscript.

Funding: This research was supported by the Ministry of Science and Technology of the Republic of China (MOST 109-2320-B-241-001- and 105-2320-B-241-002-).

Institutional Review Board Statement: The study was conducted in accordance with the Declaration of Helsinki, and approved by the Ethic Committee of Animals Care and Protection of Hungkuang University (protocol code: 10419).

Informed Consent Statement: Not applicable.

Data Availability Statement: All data generated or analyzed during this study are included in the published article (and its online Supplementary Files).

Acknowledgments: We thank Shiau-Huei Huang for technical assistance.

Conflicts of Interest: The authors declare no conflict of interest.

References

1. Younossi, Z.; Anstee, Q.M.; Marietti, M.; Hardy, T.; Henry, L.; Eslam, M.; George, J.; Bugianesi, E. Global burden of NAFLD and NASH: Trends, predictions, risk factors and prevention. *Nat. Rev. Gastroenterol. Hepatol.* **2018**, *15*, 11–20. [CrossRef]
2. Younossi, Z.M.; Koenig, A.B.; Abdelatif, D.; Fazel, Y.; Henry, L.; Wymer, M. Global epidemiology of nonalcoholic fatty liver disease—Meta-analytic assessment of prevalence, incidence, and outcomes. *Hepatology* **2016**, *64*, 73–84. [CrossRef]
3. Chen, C.-H.; Huang, M.-H.; Yang, J.-C.; Nien, C.-K.; Yang, C.-C. Prevalence and risk factors of nonalcoholic fatty liver disease in an adult population of Taiwan: Metabolic significance of nonalcoholic fatty liver disease in nonobese adults. *J. Clin. Gastroenterol.* **2006**, *40*, 745–752. [CrossRef]
4. Zhang, Q.-Q.; Lu, L.-G. Nonalcoholic fatty liver disease: Dyslipidemia, risk for cardiovascular complications, and treatment strategy. *J. Clin. Transl. Hepatol.* **2015**, *3*, 78–84.
5. Lazarus, J.V.; Mark, H.E.; Anstee, Q.M.; Arab, J.P.; Batterham, R.L.; Castera, L.; Cortez-Pinto, H.; Crespo, J.; Cusi, K.; Dirac, M.A.; et al. NAFLD Consensus Consortium. Advancing the global public health agenda for NAFLD: A consensus statement. *Nat. Rev. Gastroenterol. Hepatol.* **2022**, *19*, 60–78. [CrossRef]
6. Araujo, A.R.; Rosso, N.; Bedogni, G.; Tiribelli, C.; Bellentani, S. Global epidemiology of non-alcoholic fatty liver disease/non-alcoholic steatohepatitis: What we need in the future. *Liver Int.* **2018**, *38*, 47–51. [CrossRef]
7. Kanwal, F.; Kramer, J.R.; Mapakshi, S.; Natarajan, Y.; Chayanupatkul, M.; Richardson, P.A.; Li, L.; Desiderio, R.; Thrift, A.P.; Asch, S.M.; et al. Risk of Hepatocellular cancer in patients with non-alcoholic fatty liver disease. *Gastroenterology* **2018**, *155*, 1828–1837. [CrossRef]
8. Zhu, T.; Kim, S.-H.; Chen, C.Y. A medicinal mushroom: *Phellinus linteus*. *Curr. Med. Chem.* **2008**, *15*, 1330–1335. [CrossRef]
9. Zhou, L.-W.; Ghobad-Nejhad, M.; Tian, X.-M.; Wang, Y.-F.; Wu, F. Current status of 'Sanghuang' as a group of medicinal mushrooms and their perspective in industry development. *Food Rev. Int.* **2020**. [CrossRef]
10. Lee, I.-K.; Yun, B.-S. Highly oxygenated and unsaturated metabolites providing a diversity of hispidin class antioxidants in the medicinal mushrooms *Inonotus* and *Phellinus*. *Bioorg. Med. Chem.* **2007**, *15*, 3309–3314. [CrossRef]
11. Yeom, J.H.; Lee, I.K.; Ki, D.W.; Lee, M.S.; Seok, S.J.; Yun, B.S. Neuraminidase inhibitors from the culture broth of *Phellinus linteus*. *Mycobiology* **2012**, *40*, 142–144. [CrossRef]
12. Jung, J.Y.; Lee, I.K.; Seok, S.J.; Lee, H.J.; Kim, Y.H.; Yun, B.S. Antioxidant polyphenols from the mycelial culture of the medicinal fungi *Inonotus xeranticus* and *Phellinus linteus*. *J. Appl. Microbiol.* **2008**, *104*, 1824–1832. [CrossRef]
13. Singh, S.B.; Jayasuriya, H.; Dewey, R.; Polishook, J.D.; Dombrowski, A.W.; Zink, D.L.; Guan, Z.; Collado, J.; Platas, G.; Pelaez, F.; et al. Isolation, structure, and HIV-1-integrase inhibitory activity of structurally diverse fungal metabolites. *J. Ind. Microbiol. Biotechnol.* **2003**, *30*, 721–731.
14. Park, J.M.; Lee, J.S.; Song, J.E.; Sim, Y.C.; Ha, S.; Hong, E.K. Cytoprotective effect of hispidin against pal-mitate-induced lipotoxicity in C2C12 myotubes. *Molecules* **2015**, *20*, 5456–5467. [CrossRef]
15. Park, I.H.; Jeon, S.Y.; Lee, H.J.; Kim, S.I.; Song, K.S. A beta-secretase (BACE1) inhibitor hispidin from the mycelial cultures of *Phellinus linteus*. *Planta Med.* **2004**, *70*, 143–146.
16. Liu, Y.; Wang, C.; Li, J.; Mei, Y.; Liang, Y. Hypoglycemic and hypolipidemic effects of *Phellinus Linteus* mycelial extract from solid-state culture in a rat model of type 2 Diabetes. *Nutrients* **2019**, *11*, 296. [CrossRef]
17. Pouwels, S.; Sakran, N.; Graham, Y.; Leal, A.; Pintar, T.; Yang, W.; Kassir, R.; Singhal, R.; Mahawar, K.; Ramnarain, D. Non-alcoholic fatty liver disease (NAFLD): A review of pathophysiology, clinical management and effects of weight loss. *BMC Endocr. Disord.* **2022**, *22*, 63. [CrossRef]
18. Chen, W.; Tan, H.; Liu, Q.; Zheng, X.; Zhang, H.; Liu, Y.; Xu, L. A review: The bioactivities and pharmacological applications of *Phellinus linteus*. *Molecules* **2019**, *24*, 1888. [CrossRef]
19. Li, I.C.; Chen, C.C.; Sheu, S.J.; Huang, I.H.; Chen, C.C. Optimized production and safety evaluation of hispidin-enriched Sanghuangporus sanghuang mycelia. *Food Sci. Nutr.* **2020**, *8*, 1864–1873. [CrossRef]

20. Lien, H.M.; Huang, S.H.; Chang, C.H.; Huang, C.L.; Chen, C.C.; Chyau, C.C. Innovative Purification Method of Ovatodiolide from Anisomeles indica to Induce Apoptosis in Human Gastric Cancer Cells. *Molecules* **2022**, *27*, 587. [CrossRef]
21. Chyau, C.C.; Wang, H.F.; Zhang, W.J.; Chen, C.C.; Huang, S.H.; Chang, C.C.; Peng, R.Y. Antrodan alleviates high-Fat and high-fructose diet-induced fatty liver disease in C57BL/6 mice model via AMPK/Sirt1/SREBP-1c/PPARγ pathway. *Int. J. Mol. Sci.* **2020**, *21*, 360. [CrossRef]
22. Cui, A.; Hu, Z.; Han, Y.; Yang, Y.; Li, Y. Optimized Analysis of In Vivo and In Vitro Hepatic Steatosis. *J. Vis. Exp.* **2017**, *121*, 55178.
23. Sun, Y.; Yuan, X.; Zhang, F.; Han, Y.; Chang, X.; Xu, X.; Li, Y.; Gao, X. Berberine ameliorates fatty acid-induced oxidative stress in human hepatoma cells. *Sci. Rep.* **2017**, *7*, 11340. [CrossRef]
24. Chyau, C.C.; Wu, H.L.; Peng, C.C.; Huang, S.H.; Chen, C.C.; Chen, C.H.; Peng, R.Y. Potential Protection Effect of ER Homeostasis of N^6-(2-Hydroxyethyl)adenosine Isolated from *Cordyceps cicadae* in Nonsteroidal Anti-Inflammatory Drug-Stimulated Human Proximal Tubular Cells. *Int. J. Mol. Sci.* **2021**, *22*, 1577. [CrossRef]
25. Bojczuk, M.; Zyzelewicz, D.; Hodurek, P. Centrifugal partition chromatography—A review of recent applications and some classic references. *J. Sep. Sci.* **2017**, *40*, 1597–1609. [CrossRef]
26. Min, G.-J.; Jeong, E.-U.; Yun, B.-S.; Kang, H.-W. Chemical identification and antioxidant activity of phenolic compounds extracted from the fruiting body of 'Hankyong Sanghwang', *Phellinus linteus* KACC 93057P. *J. Mushroom* **2018**, *16*, 311–317.
27. Lee, M.-S.; Hwang, B.S.; Lee, I.-K.; Seo, G.-S.; Yun, B.-S. Chemical constituents of the culture broth of *Phellinus linteus* and their antioxidant activity. *Mycobiology* **2015**, *43*, 43–48. [CrossRef]
28. Chang, H.-Y.; Ho, Y.-L.; Sheu, M.-J.; Lin, Y.-H.; Tseng, M.-C.; Wu, S.-H.; Huang, G.-J.; Chang, Y.-S. Antioxidant and free radical scavenging activities of *Phellinus merrillii* extracts. *Bot. Stud.* **2007**, *48*, 407–417.
29. Jang, J.S.; Lee, J.S.; Lee, J.H.; Kwon, D.S.; Lee, K.E.; Lee, S.Y.; Hong, E.K. Hispidin produced from *Phellinus linteus* protects pancreatic β-cells from damage by hydrogen peroxide. *Arch. Pharmacal. Res.* **2010**, *33*, 853–861. [CrossRef]
30. Lee, I.-K.; Yun, B.S. Styrylpyrone-class compounds from medicinal fungi *Phellinus* and *Inonotus* spp., and their medicinal importance. *J. Antibiot.* **2011**, *64*, 349–359. [CrossRef]
31. Park, I.-H.; Chung, S.-K.; Lee, K.-B.; Yoo, Y.-C.; Kim, S.-K.; Kim, G.-S.; Song, K.-S. An antioxidant hispidin from the mycelial cultures of *Phellinus linteus*. *Arch. Pharmacal. Res.* **2004**, *27*, 615–618. [CrossRef]
32. Jeon, Y.E.; Lee, Y.S.; Lim, S.S.; Kim, S.-J.; Jung, S.-H.; Bae, Y.-S.; Yi, J.-S.; Kang, I.-J. Evaluation of the antioxidant activity of the fruiting body of *Phellinus linteus* using the on-line HPLC-DPPH method. *J. Korean Soc. Appl. Biol. Chem* **2009**, *52*, 472–479. [CrossRef]
33. Wang, Y.; Shang, X.Y.; Wang, S.J.; Mo, S.Y.; Li, S.; Yang, Y.C.; Ye, F.; Shi, J.G.; He, L. Structures, biogenesis, and biological activities of pyrano [4,3-c] isochromen-4-one derivatives from the fungus *Phellinus igniarius*. *J. Nat. Prod.* **2007**, *70*, 296–299. [CrossRef]
34. Lahiri, S.K.; Gokania, R.H.; Shuklab, M.D.; Modic, H.A.; Santanid, D.D.; Shaha, B. Evaluation of antioxidant activity of plant-parasitic macrofungus: *Phellinus durissimus* (Lloyd) Roy. *Eurasian J. Anal. Chem.* **2010**, *5*, 32–45.
35. Reis, F.S.; Barreira, J.C.M.; Calhelha, R.C.; Griensven, L.J.I.D.V.; Ćirić, A.J.; Glamoclija, J.; Soković, M.; Ferreira, I.C.F.R. Chemical characterization of the medicinal mushroom *Phellinus linteus* (Berkeley & Curtis) Teng and contribution of different fractions to its bioactivity. *LWT-Food Sci. Technol.* **2014**, *58*, 478–485.
36. Azeem, U.; Dhingra, G.S.; Shri, R. Pharmacological potential of wood inhabiting fungi of genus *Phellinus* Quél.: An overview. *J. Pharmacogn. Phytochem.* **2018**, *7*, 1161–1171.
37. Noh, J.-R.; Lee, I.-K.; Ly, S.-Y.; Yang, K.-J.; Gang, G.-T. A *Phellinus baumii* extract reduces obesity in high-fat diet-fed mice and absorption of triglyceride in lipid-loaded mice. *J. Med. Food* **2011**, *14*, 209–218. [CrossRef]
38. Rogers, A.B.; Dintzis, R.Z. Hepatobiliary System: Gross Anatomy. In *Comparative Anatomy and Histology-A Mouse, Rat, and Human Atlas*, 2nd ed.; Treuting, P., Dintzis, S., Montine, K.S., Eds.; Academic Press: Cambridge, MA, USA, 2017.
39. Wu, K.T.; Kuo, P.L.; Su, S.B.; Chen, Y.Y.; Yeh, M.L.; Huang, C.I.; Yang, J.F.; Lin, C.I.; Hsieh, M.H.; Hsieh, M.Y.; et al. Nonalcoholic fatty liver disease severity is associated with the ratios of total cholesterol and triglycerides to high-density lipoprotein cholesterol. *J. Clin. Lipidol.* **2016**, *10*, 420–425. [CrossRef]
40. Ren, X.Y.; Shi, D.; Ding, J.; Cheng, Z.Y.; Li, H.Y.; Li, J.S.; Pu, H.Q.; Yang, A.M.; He, C.L.; Zhang, J.P.; et al. Total cholesterol to high-density lipoprotein cholesterol ratio is a significant predictor of nonalcoholic fatty liver: Jinchang cohort study. *Lipids Health Dis.* **2019**, *18*, 47. [CrossRef]
41. Estrada-Luna, D.; Ortiz-Rodriguez, M.A.; Medina-Briseño, L.; Carreón-Torres, E.; Izquierdo-Vega, J.A.; Sharma, A.; Cancino-Diaz, J.C.; Pérez-Méndez, O.; Belefant-Miller, H.; Betanzos-Cabrera, G. Current therapies focused on high-density lipoproteins associated with cardiovascular disease. *A Review. Molecules* **2018**, *23*, 2730. [CrossRef]
42. Grover, S.A.; Kaouache, M.; Joseph, L.; Barter, P.; Davignon, J. Evaluating the incremental benefits of raising high-density lipoprotein cholesterol levels during lipid therapy after adjustment for the reductions in other blood lipid levels. *Arch. Intern. Med.* **2009**, *169*, 1775–1780. [CrossRef]
43. Lee, J.M.S.; Robson, M.D.; Yu, L.-M.; Shirodaria, C.C.; Cunnington, C.; Kylintireas, I.; Digby, J.E.; Bannister, T.; Handa, A.; Wiesmann, F. Effects of high-dose modified-release nicotinic acid on atherosclerosis and vascular function: A randomized, placebo-controlled, magnetic resonance imaging study. *J. Am. Coll. Cardiol.* **2009**, *54*, 1787–1794. [CrossRef]
44. Sun, D.Q.; Liu, W.Y.; Wu, S.J.; Zhu, G.Q.; Braddock, M.; Zhang, D.C.; Shi, K.Q.; Song, D.; Zheng, M.H. Increased levels of low-density lipoprotein cholesterol within the normal range as a risk factor for nonalcoholic fatty liver disease. *Oncotarget* **2016**, *7*, 5728–5737. [CrossRef]

45. Wang, K.; Shan, S.; Zheng, H.; Zhao, X.; Chen, C.; Liu, C. Non-HDL-cholesterol to HDL-cholesterol ratio is a better predictor of new-onset non-alcoholic fatty liver disease than non-HDL-cholesterol: A cohort study. *Lipids Health Dis.* **2018**, *17*, 196. [CrossRef]
46. Yin, W.; Carballo-Jane, E.; McLaren, D.G.; Mendoza, V.H.; Gagen, K.; Geoghagen, N.S.; McNamara, L.A.; Gorski, J.N.; Eiermann, G.J.; Petrov, A.; et al. Plasma lipid profiling across species for the identification of optimal animal models of human dyslipidemia. *J. Lipid Res.* **2012**, *53*, 51–65. [CrossRef]
47. National Cholesterol Education Program (NCEP) Expert Panel on Detection, Evaluation, and Treatment of High Blood Cholesterol in Adults (Adult Treatment Panel III). Third Report of the National Cholesterol Education Program (NCEP) Expert Panel on Detection, Evaluation, and Treatment of High Blood Cholesterol in Adults (Adult Treatment Panel III) final report. *Circulation* **2002**, *106*, 3143–3421. [CrossRef]
48. Harchaoui, K.E.; Visser, M.E.; Kastelein, J.J.; Stroes, E.S.; Dallinga-Thie, G.M. Triglycerides and cardiovascular risk. *Curr. Cardiol. Rev.* **2009**, *5*, 216–222. [CrossRef]
49. Kim, S.H.; Lee, H.S.; Lee, S.; Cho, J.; Ze, K.; Sung, J.; Kim, Y.C. Mycelial culture of *Phellinus linteus* protects primary cultured rat hepatocytes against hepatotoxins. *J. Ethnopharmacol.* **2004**, *95*, 367–372. [CrossRef]
50. Jeon, T.I.; Hwang, S.G.; Lim, B.O.; Park, D.K. Extracts of *Phellinus linteus* grown on germinated brown rice suppress liver damage induced by carbon tetrachloride in rats. *Biotechnol. Lett.* **2003**, *25*, 2093–2096. [CrossRef]
51. Huang, S.C.; Wang, P.W.; Kuo, P.C.; Hung, H.Y.; Pan, T.L. Hepatoprotective principles and other chemical constituents from the mycelium of *Phellinus linteus*. *Molecules* **2018**, *23*, 1705. [CrossRef]
52. Dong, Y.; Qiu, P.; Zhao, L.; Zhang, P.; Huang, X.; Li, C.; Chai, K.; Shou, D. Metabolomics study of the hepatoprotective effect of *Phellinus igniarius* in chronic ethanol-induced liver injury mice using UPLC-Q/TOF-MS combined with ingenuity pathway analysis. *Phytomedicine* **2020**, *74*, 152697. [CrossRef]
53. Lee, Y.S.; Kang, Y.-H.; Jung, J.-Y.; Kang, I.-J.; Han, S.-N.; Chung, J.-S.; Shin, H.-K.; Lim, S.S. Inhibitory constituents of aldose reductase in the fruiting body of *Phellinus linteus*. *Biol. Pharm. Bull.* **2008**, *31*, 765–768. [CrossRef]
54. Fujimoto, Y.; Torres, T.P.; Donahue, E.P.; Masakazu Shiota, M. Glucose toxicity is responsible for the development of impaired regulation of endogenous glucose production and hepatic glucokinase in zucker diabetic fatty rats. *Diabetes* **2006**, *55*, 2479–2490. [CrossRef]
55. Campos, C. Chronic Hyperglycemia and Glucose Toxicity: Pathology and Clinical Sequelae. *Postgrad. Med.* **2012**, *124*, 90–97. [CrossRef]
56. Yoon, S.B.; Lee, I.S.; Choi, M.H.; Lee, K.; Ham, H.; Oh, H.J.; Park, S.H.; Lim, C.H.; Choi, M.G. Impact of Fatty Liver on Acute Pancreatitis Severity. *Gastroenterol. Res. Pract.* **2017**, *2017*, 4532320. [CrossRef]
57. Morris, E.M.; Meers, G.M.E.; Booth, F.W.; Fritsche, K.L.; Hardin, C.D.; Thyfault, J.P.; Ibdah, J.A. PGC-1α overexpression results in increased hepatic fatty acid oxidation with reduced triacylglycerol accumulation and secretion. *Am. J. Physiol.-Gastrointest. Liver Physiol.* **2012**, *303*, G979–G992. [CrossRef]
58. Ding, R.-B.; Bao, J.; Deng, C.-X. Emerging roles of SIRT1 in fatty liver diseases. *Int. J. Biol. Sci.* **2017**, *13*, 852–867. [CrossRef]
59. Orellana-Gavalda, J.M.; Herrero, L.; Malandrino, M.I.; Paneda, A.; Sol Rodriguez-Pena, M.; Petry, H.; Asins, G.; Van Deventer, S.; Hegardt, F.G.; Serra, D. Molecular therapy for obesity and diabetes based on a long-term increase in hepatic fatty-acid oxidation. *Hepatology* **2011**, *53*, 821–832. [CrossRef]
60. Rector, R.S.; Thyfault, J.P.; Uptergrove, G.M.; Morris, E.M.; Naples, S.P.; Borengasser, S.J.; Mikus, C.R.; Laye, M.J.; Laughlin, M.H.; Booth, F.W.; et al. Mitochondrial dysfunction precedes insulin resistance and hepatic steatosis and contributes to the natural history of non-alcoholic fatty liver disease in an obese rodent model. *J. Hepatol.* **2010**, *52*, 727–736. [CrossRef]
61. Gamberi, T.; Magherini, F.; Modesti, A.; Fiaschi, T. Adiponectin signaling pathways in liver diseases. *Biomedicines* **2018**, *6*, 52. [CrossRef]
62. Peter, A.; Stefan, N.; Cegan, A.; Walenta, M.; Wagner, S.; Königsrainer, A.; Königsrainer, I.; Machicao, F.; Schick, F.; Häring, H.-U.; et al. Hepatic glucokinase expression is associated with lipogenesis and fatty liver in humans. *J. Clin. Endocrinol. Metab.* **2011**, *96*, E1126–E1130. [CrossRef]
63. Bechmann, L.P.; Gastaldelli, A.; Vetter, D.; Patman, G.L.; Pascoe, L.; Hannivoort, R.A.; Lee, U.E.; Fiel, I.; Muñoz, U.; Ciociaro, D.; et al. Glucokinase links Krüppel-like factor 6 to the regulation of hepatic insulin sensitivity in non-alcoholic fatty liver disease. *Hepatology* **2012**, *55*, 1083–1093. [CrossRef]
64. Skat-Rørdam, J.; Ipsen, D.H.; Lykkesfeldt, J.; Tveden-Nybor, P. A role of peroxisome proliferator-activated receptor in non-alcoholic fatty liver disease. A Review. *Basic Clin. Pharmacol. Toxicol.* **2019**, *124*, 528–537. [CrossRef]
65. Esquejo, R.M.; Salatto, C.T.; Delmore, J.; Albuquerque, B.; Reyes, A.; Shi, Y.; Moccia, R.; Cokorinos, E.; Peloquin, M.; Monetti, M.; et al. Activation of liver AMPK with PF-06409577 corrects NAFLD and lowers cholesterol in rodent and primate preclinical models. *EBioMedicine* **2018**, *31*, 122–132. [CrossRef]

antioxidants

MDPI

Article

Potential for Prebiotic Stabilized *Cornus mas* L. Lyophilized Extract in the Prophylaxis of Diabetes Mellitus in Streptozotocin Diabetic Rats

Szymon Sip [1], Daria Szymanowska [2], Justyna Chanaj-Kaczmarek [1], Krystyna Skalicka-Woźniak [3], Barbara Budzyńska [4], Olga Wronikowska-Denysiuk [4], Tymoteusz Słowik [5], Piotr Szulc [6] and Judyta Cielecka-Piontek [1,*]

[1] Department of Pharmacognosy, Poznan University of Medical Sciences, Rokietnicka 3, 60-806 Poznan, Poland; szymonsip@ump.edu.pl (S.S.); justyna.chanaj-kaczmarek@ump.edu.pl (J.C.-K.)
[2] Department of Biotechnology and Food Microbiology, Poznan University of Life Sciences, 31 Wojska Polskiego Street, 60-634 Poznan, Poland; darszy@up.poznan.pl
[3] Department of Natural Products Chemistry, Medical University of Lublin, 1 Chodźki Street, 20-093 Lublin, Poland; krystyna.skalicka-wozniak@umlub.pl
[4] Independent Laboratory of Behavioral Studies, Medical University of Lublin, 20-093 Lublin, Poland; barbara.budzynska@umlub.pl (B.B.); olga.wronikowska-denysiuk@umlub.pl (O.W.-D.)
[5] Experimental Medicine Center, Medical University of Lublin, Jaczewskiego 8D, 20-090 Lublin, Poland; tymoteusz.slowik@o2.pl
[6] Department of Agronomy, Poznań University of Life Sciences, Dojazd 11, 60-632 Poznan, Poland; piotr.szulc@up.poznan.pl
* Correspondence: jpiontek@ump.edu.pl

Citation: Sip, S.; Szymanowska, D.; Chanaj-Kaczmarek, J.; Skalicka-Woźniak, K.; Budzyńska, B.; Wronikowska-Denysiuk, O.; Słowik, T.; Szulc, P.; Cielecka-Piontek, J. Potential for Prebiotic Stabilized *Cornus mas* L. Lyophilized Extract in the Prophylaxis of Diabetes Mellitus in Streptozotocin Diabetic Rats. *Antioxidants* **2022**, *11*, 380. https://doi.org/10.3390/antiox11020380

Academic Editors: Jicheng Zhan, Zhigang Liu and Hui-Min David Wang

Received: 13 January 2022
Accepted: 10 February 2022
Published: 14 February 2022

Abstract: As a systemic disease, diabetes mellitus (DM) is characterized by the disruption of many glucose metabolic pathways. Therefore, it seems critical to study new therapies to support treatment to develop therapeutic systems that can operate across a broad metabolic spectrum. The current state of knowledge indicates an essential role of the gut microbiota in the development and course of the disease. *Cornus mas* fruits have demonstrated a rich biological activity profile and potential for application in the treatment of DM. As part of a preliminary analysis, the activity of four cultivars of *Cornus mas* fruits was analyzed. The cultivar *Wydubieckij* was selected as having the highest activity in in vitro conditions for further prebiotic system preparation. The study aimed to develop a unique therapeutic system based, first of all, on the mechanism of α-glucosidase inhibition and the antioxidant effect resulting from the activity of the plant extract used, combined with the prebiotic effect of inulin. The obtained system was characterized in vitro in terms of antioxidant activity and enzyme inhibition capacity, and was then tested on diabetic rats. The study was coupled with an analysis of changes in the intestinal microflora. The system of prebiotic stabilized *Cornus mas* L. lyophilized extract with inulin offers valuable support for the prophylaxis and treatment of DM.

Keywords: *Cornus mas*; diabetes mellitus; microbiome; prebiotics; natural compounds; civilization disease

1. Introduction

Raw materials of plant origin are increasingly used to treat and support diseases of civilization, particularly diabetes mellitus (DM) [1,2]. Due to the complexity of the causes and effects of DM, most patients, in addition to polytherapy, also use dietary supplements to support glycemic control [3,4]. Plant raw materials are a rich source of active substances with high therapeutic potential. In the case of such a complex disease as DM, they will not fully replace conventional pharmacotherapy, but they create the possibility of reducing the doses used or reducing the number of drugs used in pharmacotherapy to control glycemia [5,6]. The primary method of reducing blood sugar levels is diet modification, which is always the first step in implementing a therapeutic program. Providing adequate nutrients in the diet is vital for two reasons—they can lead to a direct reduction in blood sugar and stimulate the composition of the gut microbiome, which is essential for inhibiting the development of DM.

Cornus mas L. (also known as Cornelian cherry) is one of the most valuable fruit plants with hypoglycemic potential that belongs to the Cornaceae family. The *Cornus* genus consists of about 50 species of Cornelian cherry, occurring naturally in the temperate climate zone. The Cornelian cherry can mainly be found as a shrub or a small tree, growing up to 9 m high. It grows in central and south-western Europe, to the Caucasus and western Asia [7]. The fruits of Cornelian cherry can be oval-shaped, bottle-shaped, or pear-shaped, and range in color from dark red to yellow [8].

Cornus mas fruits have a strong inhibitory effect on exocrine enzymes responsible for breaking down complex carbohydrates (α-amylase and α-glucosidase) into easily digestible simple sugars [9,10]. The inhibition of these enzymes limits the absorption of simple sugars, contributing to the hypoglycemic effect, without interfering with intracellular metabolic pathways. The early stages of DM development are characterized by an increased production of oxygen radicals which translates into more significant oxidative stress. This causes damage to pancreatic β-cells, leading to insufficient insulin production. The excessive production of oxygen radicals persists throughout the development of the disease, contributing to DM progression over time. The increased amount of oxygen radicals causes the glycation of proteins, determining the main metabolic changes resulting from DM [11]. Due to high polyphenol content, *Cornus mas* fruits show high antioxidant potential, decreasing oxidative stress, and reducing and preventing the cascade of adverse symptoms arising in the course of the disease [12,13]. *Cornus mas* fruits also contain iridoids, which may affect insulin metabolism. In particular, loganin and loganic acid are important due to their properties that reduce intraocular pressure [14], protect against autoimmune disease [15], help reduce inflammation [16], and protect blood vessels [17]. In the course of DM, patients often struggle with the occurrence of glaucoma, which is the result of damage to the optic nerve caused by high intraocular pressure. Moreover, in the course of the disease, problems are often observed with the venous system, as a secondary consequence of excess weight and little physical effort, causing swelling of the limbs and lymphatic effusions [18,19]. An essential aspect of DM is extensive inflammation resulting from improper carbohydrate metabolism [20,21]. For this reason, it seems crucial to incorporate consideration of the influence of iridoids in research. Their presence in the plant raw material directly affects the glucose metabolic pathway and can eliminate potential secondary symptoms occurring as part of a multi-level response.

Previous studies indicate the potential of *Cornus mas* fruit to inhibit lipase. In vitro studies have confirmed the beneficial effects of *Cornus mas* fruit on the lipid profile while suppressing inflammation of blood vessels [22]. Studies in rats with hypercholesterolemia have stimulated glutathione production with antioxidant properties [23]. Numerous in vivo and in vitro studies have been carried out which consider the high potential of *Cornus mas* fruit in treating type 2 DM. The fruit has also been shown to have a beneficial effect on the expression of peroxisome proliferator-activated receptor (PPAR) genes that act as transcription factors for genes involved in metabolism and inflammation [10,15,24–27].

Current knowledge indicates the significant importance of the gut microbiota in preventing DM and the effectiveness of implemented treatment [28]. Maintaining the intestinal microbiome's correct composition protects against potential pathogens and maintains the intestinal epithelium's integrity. In addition, studies have demonstrated an essential role for a healthy microbiome in glucose tolerance by enhancing tissue response to insulin [29]. This role is attributed to short-chain fatty acids formed in the fermentation of complex carbohydrates [30–32]. A critical aspect resulting from the intestinal microbiota composition is the pH of the environment in which they occur. The excessive development of specific microorganisms, and the metabolism that they lead to, may change the pH of the gastrointestinal tract. Such a change will promote the growth of selected microorganisms;

reducing the diversity of the intestinal ecosystem will also aggravate the problem. Particularly important for the proper functioning of the intestinal microflora is the presence of bacteria with pro-health properties, mainly belonging to the genera *Lactobacillus* and *Bifidobacterium*, and the appropriate supply of prebiotic substances ensuring their proper growth [32–35].

The research carried out was based on the assumption that it is necessary to combine the ability to reduce the absorption of simple sugars and the antioxidant activity produced by the plant extract, with prebiotic activity, that can alleviate the course of DM over a long period, through the use of inulin as a prebiotic carrier. The research undertaken aimed to assess the possibility of using Cornelian cherry extract in supporting DM treatment, informed by the latest literature reports indicating the intertwined development and course of the disease with the state of the intestinal microbiota. The research began with the preparation of extracts that were standardized for selected active compounds. A sequence of tests using in vitro models was then performed to assess the α-glucosidase inhibition capacity and the antioxidant potential using 2,2-diphenyl-1-picrylhydrazyl (DPPH) and ferric reducing antioxidant power assay (FRAP), leading to selection of a variety from which to prepare a prebiotic pre-formulation. As a result of the research conducted, an inulin-based pre-formulation was obtained which was further tested in vivo in rats. The effect on the change in the intestinal microflora in the tested animals was assessed.

2. Materials and Methods

2.1. Materials and Instruments

Standard compounds used in the HPLC analysis including loganic acid (≥95%), delphinidin 3-*O*-glucoside (≥95.0%), cyanidin-3-*O*-glucoside (≥95.0%), and pelargonidin 3-*O*-glucoside (≥95.0%) were supplied by Sigma-Aldrich, St. Louis, MO, USA.

Reagents used in the biological activity studies, including α-D-glucopyranoside (PNPG), α-glucosidase from *Saccharomyces cerevisiae* (Type I, lyophilized powder, ≥10 units/mg protein), acarbose, 2,2-diphenyl-1-picrylhydrazyl, TPTZ (2,4,6-tripyridyl-S-triazine) and iron (III) chloride hexahydrate ($FeCl_3{\cdot}6H_2O$) were supplied by Sigma-Aldrich, St. Louis, MO, USA. Folin–Ciocalteu reagent was supplied by Fischer Chemicals (Fisher Scientific, Waltham, MA, USA).

For the microbiological studies the following were assessed: The total number of bacteria (5% sheep blood agar, bioMerieux, Craponne, France), number of *Bifidobacterium* (agar BSM, Sigma-Merck; St. Louis, MO, USA), *Enterococcus* and *Escherichia coli* bacteria (chromogenic CPS, bioMerieux, Marcy-l'Étoile, France), including the potentially pathogenic *E. coli* form Biovare (ENDO, Heipha), anaerobic *Clostridium* (TSC, Biocorp, Issoire, France), lactobacilli (*Lactobacillus* (MRS agar, Oxoid, Basingstoke, UK)), other *Enterobacteriaceae* (*Enterobacter* spp., *Proteus* spp.) and of the genus *Pseudomonas*, and the number of yeast-like fungi of the genus *Candida* (CHROMagar Candida, CHROMagar Company, Springfield, NJ, USA).

The organic solvent evaporation process was carried out using rotavapor BÜCHI B-490. The lyophilization process was carried out in a lyophilizer (Heto PowerDry PL3000) Freeze Dryer (Thermo Scientific, Waltham, MA, USA). Qualitative and quantitative research was carried out using the ultra-high performance liquid chromatograph DionexUltiMate 3000 coupled a with DionexUltiMate 3000 RS Diode Array Detector with the computer program DionexChromeleon Version 7.12.1478. The extraction was carried out using an ultrasonic bath (Elmasonic S180H, Singen, Germany). For the biological tests the following were used: a centrifuge (Nüve NF800, Ankara, Turkey), a plate reader (Multiskan GO (Thermo Scientific), and a laboratory incubator (MaxQ 4450, Thermo Scientific). The water content of the raw material was determined using a moisture analyzer (OHAUS MB120). To measure weight, a Radwag AS 220.X2 (Radom, Poland) analytical balance was used throughout the study.

2.2. Plant Material

Ripe *Cornus mas* L. fruits were collected from the "Szynsad" fruit farm in Dąbrówka Nowa, Błędów, Mazowieckie, Poland (51°47′01″ N 20°43′04″ E) in 2018. *Cornus mas* fruit came from the third year of plantation cultivation. Fruit of *Cornus mas* of four cultivars were obtained: *Bolestraszycki, Florianka, Słowianin*, and *Wydubieckij*. A selection of fruit was made with all fruits containing blemishes and traces of possible penetration by insect larvae removed. Fully ripe, red fruits were obtained for the research.

2.3. *Corni fructus* Extract Preparation

The fruits were thoroughly washed to remove any mechanical and chemical impurities that could affect further testing. The next step was to remove the stones for easier processing in later stages of the study. The fruits were wiped dry, frozen at −22 °C, and then freeze-dried.

Freeze-dried *Cornus mas* fruit was sieved through a 0.5 mm sieve. An amount of 10 g of fruit was weighed into a 100 mL volumetric flask on an analytical balance, supplemented with distilled water until proportional, incubated for 1 h in an ultrasonic bath at 40 °C, and then centrifuged at 9000 r/min for 10 min to produce a clear supernatant. The obtained stock solution, with a concentration of 0.1 g of freeze-dried fruit/mL, was used to investigate the polyphenol, loganic acid (LA) and anthocyanin content, antioxidant activity, and α-glucosidase inhibition. HPLC analysis samples were also filtered through a 0.2 mL filter (Merck-Millipore Burlington, MA, USA) to remove mechanical impurities

2.4. Phytochemical Analysis of Aqueous Extracts

2.4.1. Determination of Anthocyanin Content

The HPLC gradient method, coupled with a UV-Vis detector, enabled the qualitative and quantitative determination of three anthocyanins in the *Corni fructus* extracts (Table S1). The analyses were carried out using a Zorbax Eclipse Plus C18 column (4.6 mm × 100 mm; 3.5 μm), the mobile phase containing 0.1% formic acid (A), and acetonitrile (B). The gradient developed to fit the requirements of the method assumed changes in the mobile phase according to the scheme: 0–45 min B = 10–20%, 45–60 min B = 20–30%, 60–70 min B = 30–40%. The method then included a 10 min re-equilibration period to establish the mobile phase equilibrium column relative to the initial mobile phase ratio. The phase flow was set at 0.3 mL/min, injection for all test samples was 10 μL, and the detection was performed at 520 nm. The anthocyanin retention times in this analytical method were, respectively, delphinidin 3-*O*-glucoside (6.16 min), cyanidin-3-*O*-glucoside (8.93 min), and pelargonidin 3-*O*-glucoside (12.15 min). Identity confirmation was based on differences in retention times, coupled with the compound's spectrum obtained relative to the standard during the analysis. The results are presented as μg/mL, where the weight refers to the dry weight of the freeze-dried fruit used to prepare the extract.

2.4.2. Determination of Loganic Acid Content

The HPLC gradient method, coupled with DAD detector, allowed the qualitative and quantitative determination of loganic acid in water extracts from *Corni fructus* (Figure S1 and Table S2). The determinations were carried out using a Zorbax SB-C18 column, Rapid Resolution 4.6 mm × 100 mm 3,5—Micron; the mobile phase contained 0.1% acetic acid (A) and methanol (B). The gradient developed for the needs of the method assumed changes in the mobile phase according to the scheme: 0–4 min B = 1%, 4–18 min B = 1–12%, 18–22 min B = 12–20%, 22–35 min B = 20–35%, 35–40 min B = 30–95%, 40–50 min B = 95%. The method considers 10 min re-equilibration time to determine the phase equilibrium column relative to the initial injection phase. The phase flow was set to level 1 mL/min, injection for all test samples was 5 μL, the detection was carried out at 240 nm. The LA retention time in this analytical method was 23.5 min. The results are presented as μg/mL, where the weight refers to the dry weight of the freeze-dried fruit used to prepare the extract.

2.4.3. Determination of Total Phenolic Content (TPC)

TPC was determined using the Folin–Ciocalteu method with minor modifications. A 50 μL plant extract solution diluted 10 times was mixed with 50 μL of Folin-Ciocalteu reagent (F.-C.) and 100 μL distilled water. The mixture was pre-incubated for 5 min at 37 °C with shaking at 100 rpm. Then 100 μL 20% Na_2CO_3 aq. solution was added and incubated for 30 min at 37 °C with shaking at 100 rpm. The absorbance was read at 750 nm against the blank sample (water instead of the extract) in sixplicate. TPC was expressed as mg of gallic acid equivalent (GAE) per g of lyophilized *Corni fructus* utilizing a standard curve of gallic acid ($y = 9.7183x - 0.2776$; $R^2 = 0.9984$) in the concentration range 0.06–0.2 mg/mL [36]. The content of TPC in the tested extract was calculated following the standard curve for gallic acid. The curve used to calculate the TPC content in the form of gallic acid as a conversion factor is presented in the Supplementary Materials (Figure S5).

2.5. In Vitro Activity of Corni fructus Extracts

2.5.1. α-Glucosidase Inhibitory Assay

A spectrophotometric method with minor modifications was used to determine the inhibition of α-glucosidase by the *Corni fructus* water extracts [37]. Briefly, 50 μL of sample solution (80–160 μg/mL *Wydubieckij*, 480–560 μg/mL *Słowianin*, 500–580 μg/mL *Florianka* and 470–550 μg/mL *Bolestraszycki*) or acarbose (positive control, 1–5 mg/mL) in different concentrations, 50 μL of 0.1 M phosphate buffer (pH 6.8) and 30 μL α-glucosidase solution (1.0 U/mL) was pre-incubated in 96 well plates at 37 °C for 15 min. Next, 20 μL of 5 mM p-nitrophenyl-α-D-glucopyranoside (pNPG) solution in a 0.1 M phosphate buffer (pH 6.8) was added and incubated at 37 °C for 20 min. The reaction was terminated by adding 100 μL of sodium carbonate (0.2 M) into the mixture. The absorbance of the liberated p-nitrophenol was measured at 405 nm. The absorbance of enzyme solution, but without plant extracts/acarbose, served as the control with total enzyme activity. The absorbance in the absence of the enzyme was used as the blind control. The enzyme inhibition rate, expressed as a percentage of inhibition, was calculated using the following formula:

$$\%\ \text{inhibition activity} = ((A_C - A_S)/A_C) \times 100$$

where A_C is the absorbance of the control (100% enzyme activity), and A_S is the absorbance of the tested sample (*Corni fructus* water extract or acarbose). For the investigated extracts, two independent experiments were carried out in triplicate. Results were expressed as means ± S.D. The results are presented as μg/mL, where the weight refers to the dry weight of the freeze-dried fruit used to prepare the extract.

2.5.2. DPPH Assay

The DPPH assay was effected according to Studzińska-Sroka et al., with modifications [38]. Briefly, 25 μL of extracts of *Corni fructus* dissolved in distilled water at different concentrations (0.25–8.0 mg/mL), was mixed with 175 μL of DPPH solution (3.9 mg in 50 mL of MeOH). The reaction mixture was shaken and incubated in the dark at room temperature for 30 min. The control contained 25 μL of distilled water and 175 μL of DPPH solution. Absorbance was measured at 517 nm. The inhibition of the DPPH radical by the sample was calculated according to the following formula:

$$\text{DPPH scavenging activity}\ (\%) = ((A_C - A_S)/A_C) \times 100\%$$

where A_C is the absorbance of the control and A_S is the absorbance of the sample. The results are presented as μg/mL, where the weight refers to the dry weight of the freeze-dried fruit used to prepare the extract.

2.5.3. FRAP Assay

Following Tiveron et al. (2012), the FRAP assay was performed with some modifications [39]. The stock solutions of FRAP reagent included 300 mM acetate buffer (pH 3.6), 10 mM TPTZ solution in 40 mM HCl, and 20 mM $FeCl_3{\cdot}6H_2O$ solution. The working FRAP solution was freshly prepared by mixing 25 mL of acetate buffer, 2.5 mL of TPTZ solution, and 2.5 mL of $FeCl_3{\cdot}6H_2O$ solution and then warmed at 37 °C before usage. Briefly, 25 μL of the tested extracts dissolved in distilled water at different concentrations (1.0–6.0 mg/mL) were mixed with 175 μL of FRAP solution, shaken, and incubated 37 °C for 30 min. in the dark condition. Then the absorbance was read at 593 nm. The results were expressed as the $IC_{0.5}$, corresponding to the extract concentration required to produce a 0.5 O.D. value. The results are presented as μg/mL, where the weight refers to the dry weight of the freeze-dried fruit used to prepare the extract.

2.6. Preparation of Corni fructus Extract System with Prebiotic Carrier

An amount of 20 g of the freeze-dried, previously sieved, fruit of *Cornus mas Wydubieckij* variety was extracted twice with 250 mL of 70% ethanol for 60 min at 45 °C on an ultrasonic bath. The obtained extracts were concentrated below 50 °C under vacuum to dryness. The extract was standardized for TPC, loganic acid, and anthocyanin content (Table S3). Following this, dry extract was mixed with 200 mL of 10% inulin aqueous solution (an amount of 20 g of inulin was added to the dry extract), frozen, and then lyophilized. The process was carried out for 72 h to remove all water from the system; the automatic procedure programmed in the freeze dryer (Heto PowerDry 3000) was used. The resulting formulation was triturated in an agate mortar to obtain a homogeneous powder.

2.7. In Vivo Antidiabetic Activity of Prebiotic Stabilized Lyophilizate of Corni fructus

The experiments were carried out on 40 naive male Wistar rats (from the Center for Experimental Medicine of the Medical University of Lublin) weighing 200–250 g. The animals were kept under standard laboratory conditions (12 h light/dark cycle, room temperature 21 ± 1 °C) with free access to tap water and a laboratory chow (Agropol, Warsaw, Poland). Each experimental group consisted of 10 animals. All experiments were being conducted according to the National Institute of Health Guidelines for the Care and Use of Laboratory Animals and in accordance with the European Community Council Directive for the Care and Use of Laboratory Animals of 22 September 2010 (2010/63/E.U.), and were approved by the local ethics committee (96/2019).

Rats were randomly divided into four groups: the control group, STZ-treated group, control group treated with *Corni fructus* extract with inulin, and STZ-treated with *Corni fructus* extract with inulin. In the DM group, animals received an intraperitoneal injection of freshly prepared streptozotocin (STZ), singe injection, 60 mg/kg, in sterile sodium citrate buffer at pH 4.5 [40]. *Corni fructus* extract with inulin at a dose of 50 mg/kg m.c (dose based on the obtained pre-formulation containing inulin and plant extract suspended in it) was administered intragastrically for three weeks once daily. Fasting blood glucose was measured using an Accu-Check blood glucose meter (Roche, Pleasanton, CA, USA) on the 14th and 21st day from day 1 of the injection of STZ.

Bodyweight and blood glucose levels were monitored. Rats with blood glucose levels higher than 270 mg/dL (>15 mmol/L) were considered diabetic [41].

2.8. Quantitative and Qualitative Analysis of Microorganisms Present in the Gastrointestinal Tract

An amount of 0.2 g of feces was collected and placed in 2.0 mL 0.9% NaCl for bacteriological studies. The procedure was repeated until a dilution of 10^{-6} was obtained. A series of appropriate dilutions were inoculated in a volume of 50 µL and spread with a sterile loop on selectively differentiating and propagating media. The cultures were grown under conditions specific for a given group of microorganisms: *Bifidobacterium*, *Clostridium*—48 h, at 37 °C, in anaerobic conditions; *Lactobacillus* 48 h, at 37 °C, under microaerophilic conditions; *Enterococcus*, *Enterobacteriaceae* rods and *Proteus* rods—24 h, at 37 °C, under aerobic conditions.

For the qualitative and quantitative determination of fungi, 0.2 g of feces was collected and placed in 2 mL of trypsin solution with the addition of 25 µg of antibiotics (penicillin and streptomycin) to inhibit bacterial growth. The samples were placed at 37 °C for 15 min to digest food residues to prevent fungal growth. The samples were washed in a buffered physiological PBS solution, then inoculated on two fungal growth media, Sabouraud, or chloramphenicol (bioMerieux). Incubation was conducted at 37 °C for 48 h and room temperature (20 °C) for 48 h, respectively. This was to distinguish fungi from the environment from potentially pathogenic fungi. Then identification was made using *Candida* chromogenic medium. Mold fungi were identified based on direct preparation and a mycological key.

2.9. Statistical Analysis

Analysis of biological activity in vitro was performed in at least six replicates. Statistical analysis was performed using Statistica 13.3 software (StatSoft Poland, Krakow, Poland). The Shapiro–Wilk test was implemented to check data distribution normality and the Levene's test assessed the equality of variances. Statistical significance was performed using a one-way analysis of variance (ANOVA) followed by the Bonferroni post hoc test. Measurements were considered significant at $p < 0.05$.

3. Results and Discussion

3.1. Quantitative and Qualitative Analysis of the Examined Extracts

The first stage of the research was the standardization of water extracts from lyophilized *Cornus mas* fruits from three Polish cultivars *Słowianin*, *Florianka*, *Bolestraszycki*, and one Ukrainian cultivar *Wydubieckij*. These were standardized for the content of anthocyanins, loganic acid, and polyphenols taking into account the water content of freeze-dried fruit (Table S4).

Ultra-high performance liquid chromatographic methods, using a diode array detector (UHPLC-DAD) with gradient elution, were developed to determine anthocyanin 3-*O*-glucosides of pelargonidin, cyanidin, delphinidin, and loganic acid (Figures 1 and 2). The UHPLC-DAD methods were validated according to the International Conference on Harmonization (ICHQ2) guidelines [42]. The validation parameters are presented in the Supplementary Materials (Tables S1 and S2). The peaks of studied compounds in the *Cornus mas* fruits water extracts were compared with the reference substances' retention times and UV spectra.

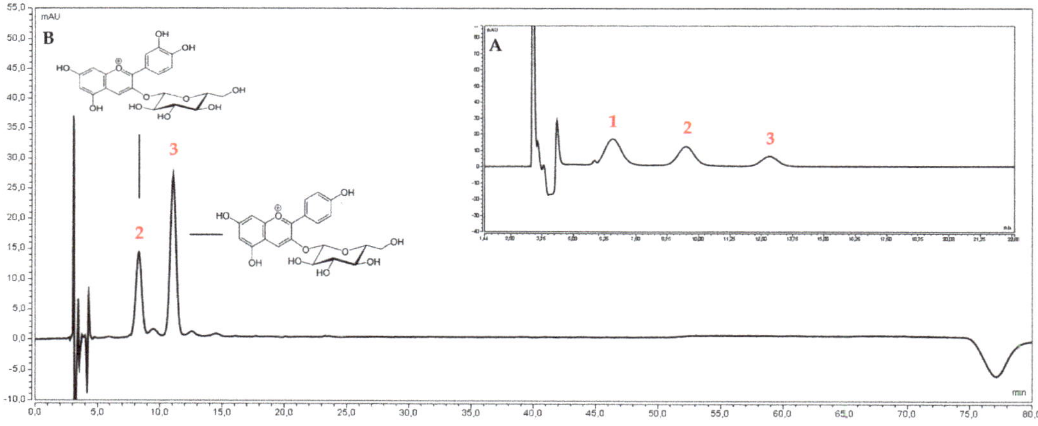

Figure 1. The chromatogram of the reference substances. (**A**) 1-delphinidin 3-*O*-glucoside, 2- cyanidin 3-*O*-glucoside, 3- pelargonidin 3-*O*-glucoside, and (**B**) water extract of *Corni fructus* cultivar *Wydubieckij*.

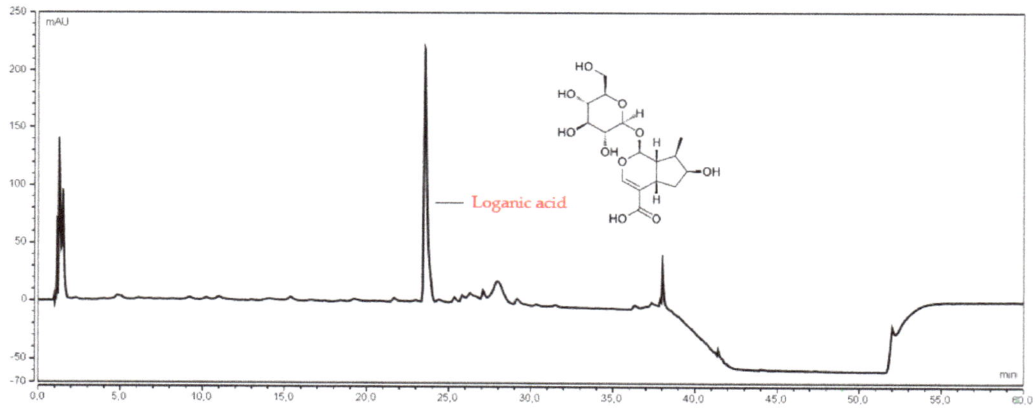

Figure 2. The chromatogram of loganic acid presented in the water extract of *Corni fructus* cultivar *Wydubieckij*.

Anthocyanin content depends on the cultivar, environmental factors and growth, and storage conditions [43]. Table 1 shows that the major anthocyanin in *Corni fructus* extracts for the cultivars *Bostraszycki*, *Florianka*, and *Słowianin* was pelargonidin 3-*O*-glucoside ranging from 2.21 ± 0.19 to 3.09 ± 0.25 mg/g. On the other hand, the *Corni fructus* cultivar *Wydubieckij* contained a similar amount of pelargonidin 3-*O*-glucoside and cyanidin 3-*O*-glucoside, which was found to be 0.10 ± 0.02 mg/g and 0.10 ± 0.00 mg/g, respectively. Delphinidin 3-*O*-glucoside was not found in any water extracts of *Cornus mas* fruit. Based on the literature, anthocyanins are usually extracted from plant material with an acidified organic solvent (mainly alcohols or alcohol-water mixtures) [44]. These solvents destroy the cell membranes of vacuoles, and dissolve and stabilize the anthocyanins [45]. However, pH reduction is not always necessary to increase these compounds' extraction efficiency [43]. The previous studies reported comparable results when determining anthocyanin content [46–48]. Similarly, the *Corni fructus* cultivars *Bostraszycki*, *Florianka*, and *Słowianin* water extracts showed a comparable content of loganic acid; values were 5.38 ± 0.52, 4.71 ± 0.45, 4.96 ± 0.12, respectively. The lowest content was found in the *Wydubieckij* variety, 2.82 ± 0.27. According to the literature data, the concentration of loganic acid in different cultivars of *Corni fructus* ranged from 2.26 to 8.20 mg per g of dry fruit [49,50].

Additionally, the method described by Blainski et al. to determine the total phenolic content, with gallic acid as the standard, was used [36]. The highest phenolic compounds (16.03 ± 1.28 mg GAE/g) were found in the water extract from the *Corni fructus* cultivar *Wydubieckij*. The phenolic compounds concentration in other cultivars ranged from 7.34 ± 0.66 to 8.94 ± 0.83 mg GAE/g of lyophilized *Corni fructus* extracts (Table 1) [46].

Table 1. The content of active compounds in the lyophilized *Cornus mas* fruit.

Cultivar	Content			
	mg GAE/g	mg/g		
	TPC	Loganic Acid	Pelargonidin 3-*O*-Glucoside	Cyanidin-3-*O*-Glucoside
Bolestraszycki	8.94 ± 0.83 *	5.38 ± 0.52 *	2.21 ± 0.19 *	1.08 ± 0.13 *
Florianka	7.34 ± 0.66 *	4.71 ± 0.45 *	3.09 ± 0.25 *	0.05 ± 0.00 *
Słowianin	8.42 ± 0.79 *	4.96 ± 0.12 *	2.28 ± 0.11 *	0.94 ± 0.09 *
Wydubieckij	16.03 ± 1.28 *	2.82 ± 0.27 *	0.10 ± 0.02 *	0.10 ± 0.00 *

Data expressed as mean ± SD; * significance with $p \leq 0.05$.

3.2. In Vitro Activity

The antidiabetic properties of *Corni fructus* water extracts were evaluated for their ability to inhibit α-glucosidase, relevant to small intestinal glucose uptake (Table 2). Several reports have indicated the α-glucosidase inhibitory properties of plant extracts and isolated compounds [51–53].

Table 2. In vitro activity of the lyophilized *Cornus mas* fruit water extracts.

Cultivar	IC_{50} (μg/mL)		$IC_{0.5}$ (μg/mL)
	Inhibition of α-Glucosidase	DPPH	FRAP
Bolestraszycki	167.94 ± 2.97 *	383.49 ± 1.54 *	0.30 ± 0.02 *
Florianka	178.26 ± 3.64 *	471.25 ± 5.20 *	0.32 ± 0.02 *
Słowianin	170.47 ± 4.92 *	343.63 ± 1.29 *	0.25 ± 0.01 *
Wydubieckij	45.23 ± 1.96 *	417.56 ± 0.81 *	0.31 ± 0.02 *

Data expressed as mean ± SD; * significance with $p \leq 0.05$.

The water extracts of *Corni fructus* showed significant α-glucosidase inhibitory activity (Figure 3). The most potent inhibitory activity was observed for the *Wydubieckij* cultivar (IC_{50} = 45.23 μg/mL), which was 30 times more potent than the acarbose standard (IC_{50} = 1.22 mg/mL). The obtained results demonstrate the significant potential for using the studied extracts in support of the treatment of DM, based on the inhibition of the digestive enzyme α-glucosidase. In particular, the obtained results show higher activity for the tested extracts than acarbose, which is successfully used in medicine and typically exhibits an inhibitory effect on the enzyme. Among compounds capable of reducing blood glucose levels, phenolic compounds, mainly anthocyanins, and iridoids, should be mentioned [54]. The conducted studies showed a strong positive correlation between glucosidase inhibitory activity and TPC (R^2 = 0.9899) (Figure 4), which indicates the effect of the synergistic action of several classes of phenolic compounds, such as phenolic acids (e.g., chlorogenic acid, gallic acid) and flavonoids (especially flavonols and flavan-3-ols, which are well-known as good glucose-lowering agents [55–59].

Moreover, the statistical analysis performed showed a significant ($p < 0.05$) negative correlation between the obtained concentration of loganic acid and pelargonidin 3-*O*-glucoside (R^2 = 0.9066 and R^2 = 0.9387, respectively) with the ability of the extract to inhibit α-glucosidase. The obtained results showed a decrease in the activity of the extract with an increasing amount of the indicated compounds. However, in the study of enzyme inhibition capacity, it was shown that higher extract concentrations showed a higher enzyme inhibition capacity. It can be concluded that the observed increasing inhibitory activity of the extract with increase in its concentration results from the presence of compounds that show much more potent enzymatic inhibition, which translates into effective increased activity. No statistically significant correlation was found between cyanidin 3-*O*-glucoside and α-glucosidase inhibition ($R^2 = 0.2344$; $p > 0.05$).

Oxidative stress results from an imbalance between systems producing radicals and neutralizing radicals, i.e., increased free radicals, reduced antioxidant defense activity, or both. Several studies have shown that diabetes is accompanied by increased free radical formation and reduced antioxidant capacity, which leads to loss of pancreatic β-cell function, aggravation of insulin resistance, and vascular complications [60,61]. The total antioxidant capacity of *Cornus mas* fruits water extracts was assessed by two spectrophotometric methods (DPPH and FRAP) that utilize the SET (single electron transfer) mechanism. Most natural antioxidants are multifunctional, so it is essential to carry out more than one type of antioxidant capacity measurement to cover different antioxidant activity mechanisms [62]. As shown in Figure 5 and Table 2, the most potent antioxidant activity was produced by water extract from the *Corni fructus* cultivar *Słowianin* with IC_{50} = 343.63 μg/mL and $IC_{0.5}$ = 0.25 μg/mL, respectively, in the DPPH and FRAP assays. The other *Cornus mas* fruit cultivars showed comparable antioxidant activity, however slightly weaker than the *Słowianin* cultivar. The correlation coefficients between TPC and antioxidant activity were not significant ($p > 0.05$), suggesting that the free-radical scavenging activity of *Corni fructus* water extracts may be attributed to differences in the activity of phenols and other active compounds, such as ascorbic acid [63], monoterpenes (especially limonene) and iridoids [64,65]. On the other hand, the antioxidant potential for the entire extract, which was tested, is an exponential result of single reactions taking place in the single electron transfer process, which does not fully reflect the reactions taking place in vivo, providing only an estimate of the antioxidant capacity [66].

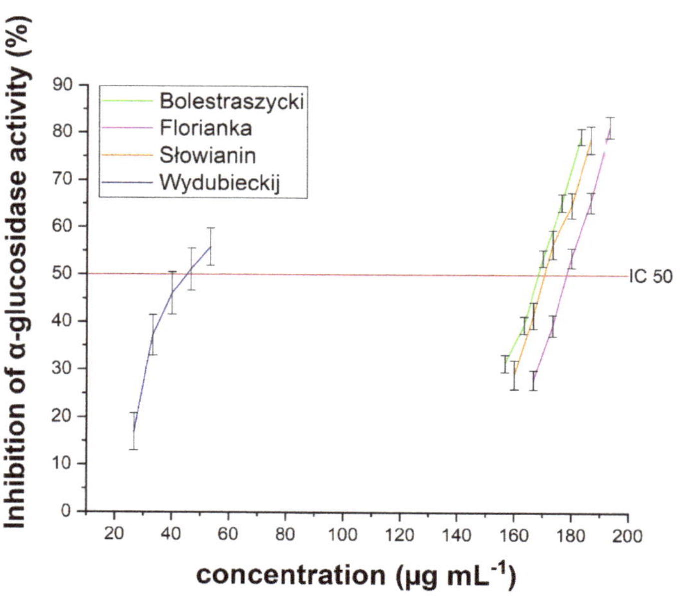

Figure 3. The α-glucosidase inhibitory activity of *Corni fructus* water extracts. Significance with $p \leq 0.05$.

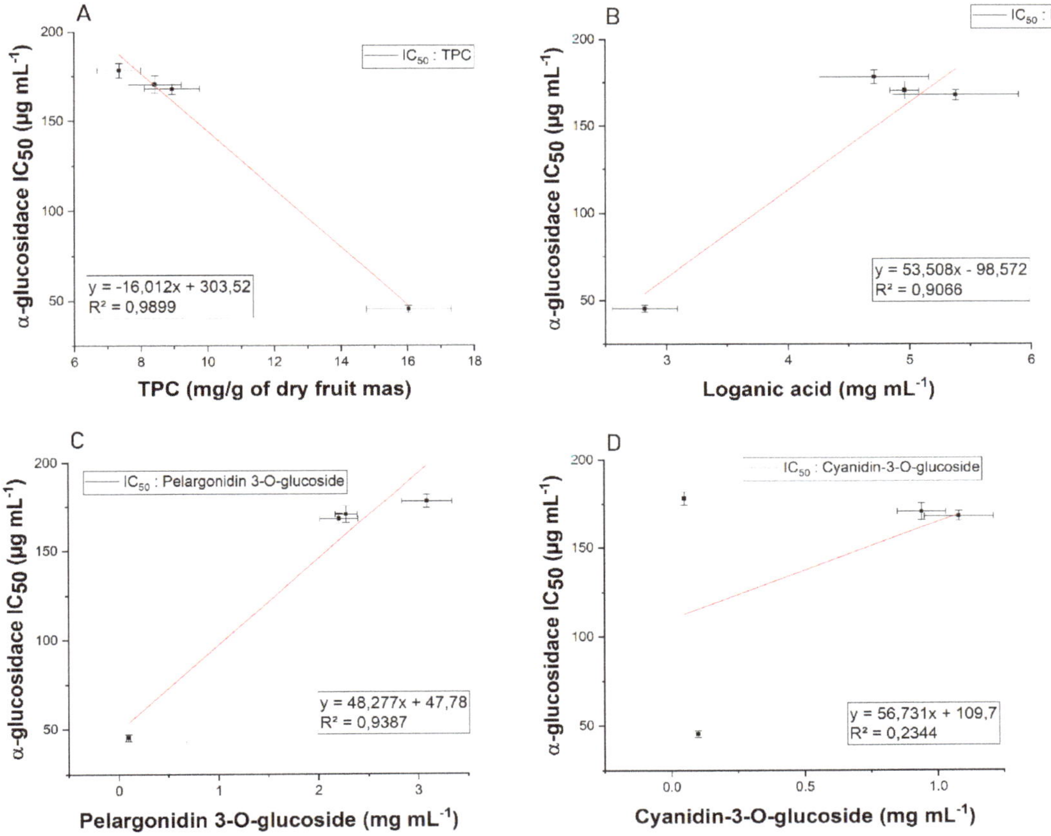

Figure 4. Correlation between α-glucosidase inhibitory activity and TPC (**A**) *, loganic acid (**B**) *, pelargonidin 3-*O*-glucoside (**C**) *, and cyanidin 3-*O*-glucoside (**D**) ** content (*—significance with $p \leq 0.05$; **- significance with $p > 0.05$).

After confirming the potential of using *Cornus mas* fruit, following the assumed research objective, the next stage of the research was to prepare a system based on a prebiotic substance capable of modifying the intestinal microbiome, the effect of which is the long-term control of blood sugar levels. Inulin was chosen as the carrier substance (Figure 6). It is a polysaccharide mainly obtained from chicory, with proven prebiotic potential against *Bifidobacterium* and *Lactobacillus* [67–69]. An essential aspect of selecting an appropriate carrier is its safety in DM. Attention should be paid to the glycemic index, which in the case of inulin is relatively low and is in the range of 8–14 [70]. Inulin is a polysaccharide, consisting of monomers linked by β-2,1-glycosidic bonds in an unbranched chain; these bonds cannot be broken down by human digestive enzymes, while the naturally occurring bacterial microflora can decompose it and utilize the resulting simple sugars for their growth [71].

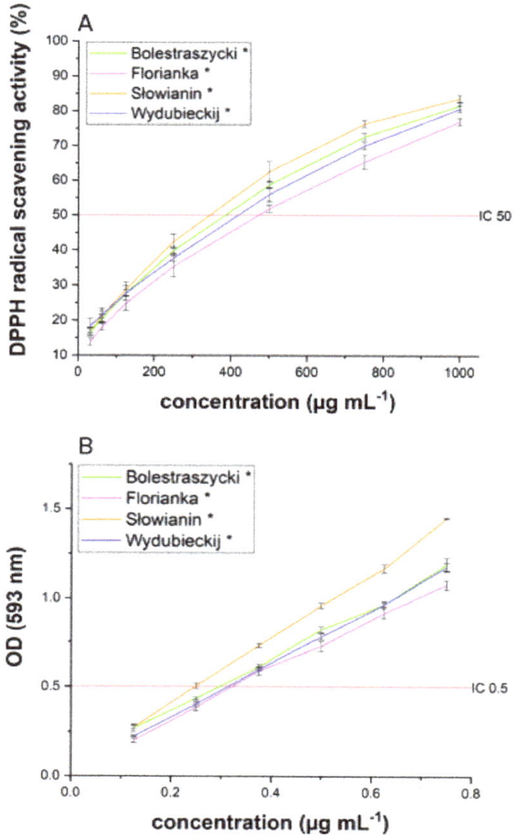

Figure 5. Antioxidant activity of *Corni fructus* water extracts by DPPH (**A**) and FRAP (**B**) assays. Data expressed as mean ± SD; * significance with $p \leq 0.05$.

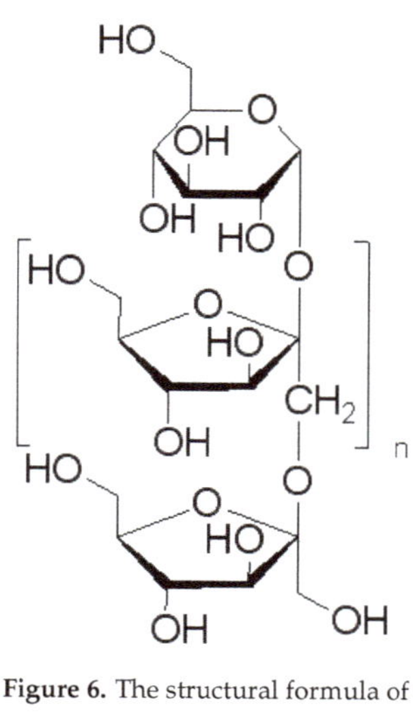

Figure 6. The structural formula of the monomer of the inulin molecule (n = 30–35).

The pre-formulation obtained by suspending fruit extract in the prebiotic carrier was standardized for the content of active substances previously tested in the plant extracts (Table 3), converted into g of the obtained system.

Table 3. The content of active compounds in the lyophilized *Cornus mas* system.

	Content			
	mg GAE/g	mg/g		
	TPC	Loganic Acid	Pelargonidin 3-*O*-Glucoside	Cyanidin-3-*O*-Glucoside
Prebiotic system	119.0 ± 2.34	5.33 ± 0.48	1.30 ± 0.11	0.32 ± 0.05

Data expressed as mean ± SD.

In addition, an activity analysis was performed to control the quality and confirm the activity of the obtained system before proceeding to the next stage of the research (Table 4). The figures showing the activity curves are provided in the Supplementary Materials (Figures S2–S4).

Table 4. In vitro activity of the lyophilized *Cornus mas* fruit system.

Cultivar	IC_{50} (mg/mL)		$IC_{0.5}$ (mg/mL)
	Inhibition of α-Glucosidase	DPPH	FRAP
Prebiotic system	0.173 ± 0.025	1.176 ± 0.066	0.905 ± 0.07

Data expressed as mean ± SD.

The obtained results indicate high activity of the obtained system; it showed almost ten times higher activity against α-glucosidase than acarbose which is traditionally used in medicine as an inhibitor of carbohydrate digestive enzymes and is also characterized by high antioxidant activity. As part of the preparation, it was possible to preserve the original properties of the plant extracts, enriching them with prebiotic activity.

3.3. *In Vivo Activity*

In the present studies, STZ induced diabetes, reflected in an increase of blood glucose (>270 mg/mL measured 14 and 21 days after STZ injection) as well as a decrease in body-weight (20% vs. control group). *Corni fructus* extract did not influence the measured parameters. The activity of *Corni fructus* in diabetes type I was also evaluated by Gao et al. They revealed that *Corni fructus*-treated diabetic rats showed significant decreases in blood glucose, urinary protein levels, water consumption, and improved lipid profile [72]. The difference between our studies and those mentioned was the time of administration (21 days vs. 40) and dose (50 mg/kg vs. 100, 200 mg/kg). Further studies, using a more extended administration scheme or higher doses of the extract, are required.

3.4. *Quantitative and Qualitative Analysis of Changes in the Intestinal Microflora*

Four feces samples of 10 rats in each group were analyzed at four measuring points (t0, t7 day, t14 day, t21 day). The samples were designated as STZ + extract with inulin, Healthy + extract with inulin, STZ + saline, and Healthy + saline. In all cases studied, the microbiological profile of the intestinal microorganism was similar at a given measuring point; the results are shown in Table 5.

Table 5. Effect of *Cornus mas* fruit with inulin on the microbiological profile of the intestinal microorganisms in rats.

STZ + Extract with Inulin	**T0**	**T7**	**T14**	**T21**
		CFU/g		
Total number of microorganisms	5.00×10^{11}	1.90×10^{10}	6.70×10^{11}	7.60×10^{12}
Bifidobacterium	6.70×10^{6}	2.30×10^{6}	7.70×10^{7}	2.30×10^{7}
Lactobacillus	5.00×10^{7}	4.50×10^{7}	5.40×10^{8}	6.40×10^{8}
Enterococcus	1.20×10^{6}	7.60×10^{7}	6.50×10^{8}	4.30×10^{8}
E. coli	1.00×10^{3}	5.40×10^{5}	3.20×10^{5}	1.30×10^{6}
E. coli Biovare	3.10×10^{5}	2.30×10^{5}	3.20×10^{4}	3.20×10^{4}
Clostridium	4.00×10^{8}	6.70×10^{5}	3.20×10^{5}	1.20×10^{4}
Proteus	<10	<10	<10	<10
Pseudomonas	<10	<10	<10	<10
Candida	<10	<10	<10	<10
Healthy + Extract with Inulin	**T0**	**T7**	**T14**	**T21**
		CFU/g		
Total number of microorganisms	5.00×10^{11}	5.60×10^{10}	5.60×10^{12}	9.40×10^{11}
Bifidobacterium	6.70×10^{6}	6.50×10^{7}	5.40×10^{7}	3.40×10^{8}
Lactobacillus	5.00×10^{7}	1.90×10^{7}	7.60×10^{7}	2.30×10^{7}
Enterococcus	1.20×10^{6}	6.60×10^{6}	2.30×10^{8}	5.40×10^{8}
E. coli	1.00×10^{3}	5.40×10^{4}	5.40×10^{4}	3.40×10^{4}
E. coli Biovare	3.10×10^{5}	3.20×10^{4}	2.10×10^{4}	3.20×10^{4}
Clostridium	4.00×10^{8}	4.50×10^{7}	3.50×10^{5}	8.70×10^{5}
Proteus	<10	<10	<10	<10
Pseudomonas	<10	<10	<10	<10
Candida	<10	<10	<10	<10
STZ + Saline	**T0**	**T7**	**T14**	**T21**
		CFU/g		
Total number of microorganisms	5.00×10^{11}	4.30×10^{10}	5.40×10^{11}	2.30×10^{11}
Bifidobacterium	6.70×10^{6}	4.50×10^{6}	4.50×10^{6}	7.60×10^{6}
Lactobacillus	5.00×10^{7}	5.40×10^{6}	3.30×10^{7}	6.70×10^{6}
Enterococcus	1.20×10^{6}	5.20×10^{6}	3.20×10^{7}	4.50×10^{7}
E. coli	1.00×10^{3}	2.20×10^{6}	1.20×10^{7}	8.90×10^{7}
E. coli Biovare	3.10×10^{5}	2.30×10^{3}	2.50×10^{4}	2.30×10^{5}
Clostridium	4.00×10^{8}	1.50×10^{7}	2.30×10^{7}	2.30×10^{5}
Proteus	<10	<10	<10	<10
Pseudomonas	<10	<10	<10	<10
Candida	<10	2.20×10^{3}	2.30×10^{4}	2.10×10^{4}
Healthy + Saline	**T0**	**T7**	**T14**	**T21**
		CFU/g		
Total number of microorganisms	5.00×10^{11}	1.40×10^{10}	2.30×10^{10}	4.30×10^{10}
Bifidobacterium	6.70×10^{6}	4.50×10^{6}	1.20×10^{6}	2.40×10^{5}
Lactobacillus	5.00×10^{7}	4.50×10^{6}	4.50×10^{6}	3.80×10^{6}
Enterococcus	1.20×10^{6}	2.30×10^{5}	2.30×10^{5}	4.50×10^{5}
E. coli	1.00×10^{3}	1.20×10^{4}	4.30×10^{4}	1.30×10^{5}
E. coli Biovare	3.10×10^{5}	1.80×10^{4}	4.50×10^{4}	8.40×10^{4}
Clostridium	4.00×10^{8}	3.40×10^{8}	4.50×10^{7}	9.80×10^{8}
Proteus	<10	<10	<10	<10
Pseudomonas	<10	<10	<10	<10
Candida	<10	4.50×10^{2}	5.20×10^{3}	7.60×10^{3}

The analyzed groups of microorganisms were divided into three main groups; microorganisms with protective functions (*Bifidobacterium*, *Lactobacillus* sp.), immunostimulatory species (*E. coli*, *Enterococcus* sp.), and proteolytic species (*Proteus* sp., *Pseudomonas* sp., *E. coli* Biovare, *Clostridium* pp.).

The total number of microorganisms in all tested samples ranged from 10^{10}–10^{12} CFU/g of feces, with the lowest on the seventh day of feeding. The number of *Bifidobacterium* and *Lactobacillus* bacteria, particularly their quantitative ratio, plays a crucial role in assessing intestinal microbial microbes' effect. The highest dynamics of *Bifidobacterium* increase were observed in the Healthy + extract with inulin variant. During the 21 days of the experiment, this group of microorganisms increased from 6.7×10^6 CFU/g to 3.4×10^8 CFU/g. In contrast, in the Healthy + saline variant, there was no reduction in *Bifidobacterium* abundance or increase. The value was constant at 10^6 CFU/g. In the same variant, the number of *Lactobacillus* bacteria did not change throughout the entire feeding period, and its value was close to that of *Bifidobacterium*. The increase in *Lactobacillus* sp. counts was demonstrated in the STZ + extract with inulin variant. In this case, the number of bacteria increased from 5.0×10^7 CFU/g before feeding to 6.4×10^8 CFU/g after 21 days of feeding. In this case, a very considerable reduction in anaerobic bacteria of the genus *Clostridium* sp. from 4.0×10^8 CFU/g to 1.2×10^4 CFU/g was also demonstrated. However, in the Healthy + saline variant, this group of microorganisms remained stable and was initially high. The increase in the number of probiotic bacteria of the genus *Bifidobacterium* sp. and bacteria of the genus *Lactobacillus* sp. was observed in two variants ("STZ + extract with inulin" and "Healthy + extract with inulin"). The probable cause of the growth of these probiotic microorganisms was the addition of inulin. Inulin is a widely used fiber that is considered a prebiotic because of its ability to be selectively used by the gut microbiota for health benefits.

The presence was not demonstrated of yeast-like fungi of the genus Candida in variants I and II. Simultaneously, in III and IV, the number increased from 7 to 21 days of feeding. This indicates the protective properties of the obtained mixture of *Cornus mas* fruit extract with inulin and confirms the lack of relationship between dietary insulin and *Candida* sp. development. Many microorganisms of the genera *Pseudomonas* and *Proteus* were found in none of the examined variants.

4. Conclusions

Considering reports in the literature indicating the potential for using *Cornus mas* fruit, studies were carried out to determine the activity profile of different cultivars. The obtained water extracts were standardized for anthocyanins, such as pelargonidin 3-*O*-glucoside and cyanidin-3-*O*-glucoside, loganic acid, and phenolic compounds. As part of the evaluation of the activity profile, inhibition of α-glucosidase was performed. The *Wydubieckij* cultivar (IC_{50} = 45.23 μg/mL) showed 30 times higher activity than the standard acarbose (IC_{50} = 1.22 mg/mL), which was correlated with the content of TPC. In addition, the most potent antioxidant activity was found using water extract from the *Corni fructus* cultivar *Słowianin* with IC_{50} = 343.63 μg/mL and $IC_{0.5}$ = 0.25 μg/mL, respectively, in the DPPH and FRAP methods.

Due to high biological activity and good antioxidant activity, the *Wydubieckij* cultivar was selected for further research. Considering the latest reports on the gut microbiome's influence on the development and course of DM2, further studies involved development of a pre-formulation containing inulin as a carrier with prebiotic potential. In order to assess the activity of the obtained pre-formulation, it was tested in an in vivo model on rats.

Although we did not observe a decrease in glucose blood level in STZ-treated diabetic rats, we are aware of the limitations of the present studies. Further in vivo research is required to consider different diabetes induction patterns and to adjust the therapeutic dose.

In addition, as part of the control of the activity of the obtained system, fecal analysis was performed to estimate quantitative and qualitative changes in the intestinal microbiome. The obtained results indicated a positive development of prebiotic microorganisms and inhibition of the growth of potentially pathogenic microorganisms.

The obtained results suggest that *Corni fructus* is a potentially valuable raw material for the prophylaxis of symptoms for type I diabetes. Combining it with inulin increases this potential by influencing the intestinal microbiome. Therefore, the tested plant material should be considered a valuable functional food component.

Supplementary Materials: The following are available online at https://www.mdpi.com/article/10.3390/antiox11020380/s1, Figure S1. Chromatogram showing the loganic acid pattern in the developed method; Table S1. Statistical assay of linear plots of the anthocyanin determined by the UHPLC-DAD method; Table S2. Statistical assay of linear plots of the loganic acid determined by the UHPLC-DAD method; Table S3. The content of active compounds in the lyophilized *Cornus mas* fruit in 70% EtOH extract; Table S4. Water content in freeze-dried fruit expressed as % of weight loss during drying in a moisture analyzer; Figure S2. The α-glucosidase inhibitory activity of *Corni fructus* prebiotic system; Figure S3. Antioxidant activity of *Corni fructus* system by DPPH assay; Figure S4. Antioxidant activity of *Corni fructus* system by FRAP assay; Figure S5. Curve for gallic acid used to calculate TPC content.

Author Contributions: Conceptualization, S.S. and J.C.-P.; Data curation, S.S.; software, S.S.; methodology, S.S., J.C.-K., D.S., K.S.-W., B.B., O.W.-D., T.S., P.S. and J.C.-P., writing—original draft preparation, S.S., J.C.-K., D.S., K.S.-W., B.B., P.S.; J.C.-P.; writing—review and editing, S.S. and J.C.-P.; supervision, J.C.-P.; funding acquisition, J.C.-P. All authors have read and agreed to the published version of the manuscript.

Funding: This study was supported by the grant OPUS from the National Science Centre Poland UMO-2020/37/B/NZ7/03975.

Institutional Review Board Statement: The study was conducted according to the guidelines of the Declaration of Helsinki, and approved by the Ethics Committee of the Medical University of Lublin (25.11.19; number: 96/2019).

Informed Consent Statement: Not applicable.

Data Availability Statement: The data is contained within the article or Supplementary Material.

Conflicts of Interest: The authors declare no conflict of interest.

References

1. Putthapiban, P.; Sukhumthammarat, W.; Sriphrapradang, C. Concealed Use of Herbal and Dietary Supplements among Thai Patients with Type 2 Diabetes Mellitus. *J. Diabetes Metab. Disord.* **2017**, *16*, 36. [CrossRef] [PubMed]
2. Yeh, G.Y.; Eisenberg, D.M.; Kaptchuk, T.J.; Phillips, R.S. Systematic Review of Herbs and Dietary Supplements for Glycemic Control in Diabetes. *Diabetes Care* **2003**, *26*, 1277–1294. [CrossRef] [PubMed]
3. Durazzo, A.; Lucarini, M.; Santini, A. Plants and Diabetes: Description, Role, Comprehension and Exploitation. *Int. J. Mol. Sci.* **2021**, *22*, 3938. [CrossRef] [PubMed]
4. Thi Bui, D.H.; Nguyen, B.X.; Truong, D.C.; Meyrowitsch, D.W.; Søndergaard, J.; Gammeltoft, T.; Bygbjerg, I.C.; Jannie, N. Polypharmacy among People Living with Type 2 Diabetes Mellitus in Rural Communes in Vietnam. *PLoS ONE* **2021**, *16*, e0249849. [CrossRef]
5. Mahdavi, A.; Bagherniya, M.; Mirenayat, M.S.; Atkin, S.L.; Sahebkar, A. Medicinal Plants and Phytochemicals Regulating Insulin Resistance and Glucose Homeostasis in Type 2 Diabetic Patients: A Clinical Review. *Adv. Exp. Med. Biol.* **2021**, *1308*, 161–183. [CrossRef]
6. Liu, X.; Zeng, X.; Liu, W.; Lu, Y.; Cheng, J.; Chen, Y. An Overview of Dietary Supplements on Obesity and Type 2 Diabetes: Efficacy and Mechanisms. *Curr. Drug Metab.* **2021**, *22*, 415–440. [CrossRef]
7. Kazimierski, M.; Regula, J.; Molska, M. Cornelian Cherry (*Cornus mas* L.)—Characteristics, Nutritional and pro-Health Properties. *Acta Sci. Pol. Technol. Aliment.* **2019**, *18*, 5–12. [CrossRef]
8. Klimenko, S. The Cornelian cherry (*Cornus mas* L.): Collection, preservation, and utilization of genetic resources. *J. Fruit Ornam. Plant Res.* **2004**, *12*, 93–98.
9. Blagojević, B.; Agić, D.; Serra, A.T.; Matić, S.; Matovina, M.; Bijelić, S.; Popović, B.M. An in Vitro and in Silico Evaluation of Bioactive Potential of Cornelian Cherry (*Cornus mas* L.) Extracts Rich in Polyphenols and Iridoids. *Food Chem.* **2021**, *335*, 127619. [CrossRef]
10. Dzydzan, O.; Bila, I.; Kucharska, A.Z.; Brodyak, I.; Sybirna, N. Antidiabetic Effects of Extracts of Red and Yellow Fruits of Cornelian Cherries (*Cornus mas* L.) on Rats with Streptozotocin-Induced Diabetes Mellitus. *Food Funct.* **2019**, *10*, 6459–6472. [CrossRef]
11. Bajaj, S.; Khan, A. Antioxidants and Diabetes. *Indian J. Endocrinol. Metab.* **2012**, *16*, S267–S271. [CrossRef] [PubMed]
12. Klymenko, S.; Kucharska, A.Z.; Sokół-Łętowska, A.; Piórecki, N.; Przybylska, D.; Grygorieva, O. Iridoids, Flavonoids, and Antioxidant Capacity of *Cornus mas*, *C. officinalis*, and *C. mas* × *C. officinalis* Fruits. *Biomolecules* **2021**, *11*, 776. [CrossRef] [PubMed]
13. Tiptiri-Kourpeti, A.; Fitsiou, E.; Spyridopoulou, K.; Vasileiadis, S.; Iliopoulos, C.; Galanis, A.; Vekiari, S.; Pappa, A.; Chlichlia, K. Evaluation of Antioxidant and Antiproliferative Properties of *Cornus mas* L. Fruit Juice. *Antioxidants* **2019**, *8*, 377. [CrossRef] [PubMed]
14. Szumny, D.; Sozański, T.; Kucharska, A.Z.; Dziewiszek, W.; Piórecki, N.; Magdalan, J.; Chlebda-Sieragowska, E.; Kupczynski, R.; Szeląg, A.; Szumny, A. Application of Cornelian Cherry Iridoid-Polyphenolic Fraction and Loganic Acid to Reduce Intraocular Pressure. *Evid. Based Complement. Alternat. Med.* **2015**, *2015*, 939402. [CrossRef]

15. Sozański, T.; Kucharska, A.Z.; Rapak, A.; Szumny, D.; Trocha, M.; Merwid-Ląd, A.; Dzimira, S.; Piasecki, T.; Piórecki, N.; Magdalan, J.; et al. Iridoid-Loganic Acid versus Anthocyanins from the *Cornus mas* Fruits (Cornelian Cherry): Common and Different Effects on Diet-Induced Atherosclerosis, PPARs Expression and Inflammation. *Atherosclerosis* **2016**, *254*, 151–160. [CrossRef]
16. Carrillo-Ocampo, D.; Bazaldúa-Gómez, S.; Bonilla-Barbosa, J.R.; Aburto-Amar, R.; Rodríguez-López, V. Anti-Inflammatory Activity of Iridoids and Verbascoside Isolated from *Castilleja tenuiflora*. *Molecules* **2013**, *18*, 12109–12118. [CrossRef]
17. Sozański, T.; Kucharska, A.Z.; Dzimira, S.; Magdalan, J.; Szumny, D.; Matuszewska, A.; Nowak, B.; Piórecki, N.; Szeląg, A.; Trocha, M. Loganic Acid and Anthocyanins from Cornelian Cherry (*Cornus mas* L.) Fruits Modulate Diet-Induced Atherosclerosis and Redox Status in Rabbits. *Adv. Clin. Exp. Med.* **2018**, *27*, 1505–1513. [CrossRef]
18. Kanapathy, M.; Portou, M.J.; Tsui, J.; Richards, T. Diabetic Foot Ulcers in Conjunction with Lower Limb Lymphedema: Pathophysiology and Treatment Procedures. *CWCMR* **2015**, *2*, 129–136. [CrossRef]
19. Zhao, Y.-X.; Chen, X.-W. Diabetes and Risk of Glaucoma: Systematic Review and a Meta-Analysis of Prospective Cohort Studies. *Int. J. Ophthalmol.* **2017**, *10*, 1430–1435. [CrossRef]
20. Tsalamandris, S.; Antonopoulos, A.S.; Oikonomou, E.; Papamikroulis, G.-A.; Vogiatzi, G.; Papaioannou, S.; Deftereos, S.; Tousoulis, D. The Role of Inflammation in Diabetes: Current Concepts and Future Perspectives. *Eur. Cardiol.* **2019**, *14*, 50–59. [CrossRef]
21. Wellen, K.E.; Hotamisligil, G.S. Inflammation, Stress, and Diabetes. *J. Clin. Investig.* **2005**, *115*, 1111–1119. [CrossRef] [PubMed]
22. Świerczewska, A.; Buchholz, T.; Melzig, M.F.; Czerwińska, M.E. In Vitro α-Amylase and Pancreatic Lipase Inhibitory Activity of *Cornus mas* L. and *Cornus alba* L. Fruit Extracts. *J. Food Drug Anal.* **2019**, *27*, 249–258. [CrossRef] [PubMed]
23. Sozański, T.; Kucharska, A.Z.; Szumny, D.; Magdalan, J.; Merwid-Ląd, A.; Nowak, B.; Piórecki, N.; Dzimira, S.; Jodkowska, A.; Szeląg, A.; et al. Cornelian Cherry Consumption Increases the L-Arginine/ADMA Ratio, Lowers ADMA and SDMA Levels in the Plasma, and Enhances the Aorta Glutathione Level in Rabbits Fed a High-Cholesterol Diet. *J. Funct. Foods* **2017**, *34*, 189–196. [CrossRef]
24. Dinda, B.; Kyriakopoulos, A.M.; Dinda, S.; Zoumpourlis, V.; Thomaidis, N.S.; Velegraki, A.; Markopoulos, C.; Dinda, M. *Cornus mas* L. (Cornelian Cherry), an Important European and Asian Traditional Food and Medicine: Ethnomedicine, Phytochemistry and Pharmacology for Its Commercial Utilization in Drug Industry. *J. Ethnopharmacol.* **2016**, *193*, 670–690. [CrossRef]
25. Jayaprakasam, B.; Olson, L.K.; Schutzki, R.E.; Tai, M.-H.; Nair, M.G. Amelioration of Obesity and Glucose Intolerance in High-Fat-Fed C57BL/6 Mice by Anthocyanins and Ursolic Acid in Cornelian Cherry (*Cornus mas*). *J. Agric. Food Chem.* **2006**, *54*, 243–248. [CrossRef]
26. Rosenson, R.S.; Wright, R.S.; Farkouh, M.; Plutzky, J. Modulating Peroxisome Proliferator–Activated Receptors for Therapeutic Benefit? Biology, Clinical Experience, and Future Prospects. *Am. Heart J.* **2012**, *164*, 672–680. [CrossRef]
27. White Fat Progenitor Cells Reside in the Adipose Vasculature | Science. Available online: https://science.sciencemag.org/content/322/5901/583 (accessed on 27 March 2020).
28. Pascale, A.; Marchesi, N.; Marelli, C.; Coppola, A.; Luzi, L.; Govoni, S.; Giustina, A.; Gazzaruso, C. Microbiota and Metabolic Diseases. *Endocrine* **2018**, *61*, 357–371. [CrossRef]
29. Utzschneider, K.M.; Kratz, M.; Damman, C.J.; Hullarg, M. Mechanisms Linking the Gut Microbiome and Glucose Metabolism. *J. Clin. Endocrinol. Metab.* **2016**, *101*, 1445–1454. [CrossRef]
30. Rowland, I.; Gibson, G.; Heinken, A.; Scott, K.; Swann, J.; Thiele, I.; Tuohy, K. Gut Microbiota Functions: Metabolism of Nutrients and Other Food Components. *Eur. J. Nutr.* **2018**, *57*, 1–24. [CrossRef]
31. Jandhyala, S.M.; Talukdar, R.; Subramanyam, C.; Vuyyuru, H.; Sasikala, M.; Reddy, D.N. Role of the Normal Gut Microbiota. *World J. Gastroenterol.* **2015**, *21*, 8787–8803. [CrossRef] [PubMed]
32. D'Argenio, V.; Salvatore, F. The Role of the Gut Microbiome in the Healthy Adult Status. *Clin. Chim. Acta* **2015**, *451*, 97–102. [CrossRef] [PubMed]
33. Dominguez-Bello, M.G.; Godoy-Vitorino, F.; Knight, R.; Blaser, M.J. Role of the Microbiome in Human Development. *Gut* **2019**, *68*, 1108–1114. [CrossRef] [PubMed]
34. Lloyd-Price, J.; Abu-Ali, G.; Huttenhower, C. The Healthy Human Microbiome. *Genome Med.* **2016**, *8*, 51. [CrossRef]
35. Ruan, W.; Engevik, M.A.; Spinler, J.K.; Versalovic, J. Healthy Human Gastrointestinal Microbiome: Composition and Function After a Decade of Exploration. *Dig. Dis. Sci.* **2020**, *65*, 695–705. [CrossRef] [PubMed]
36. Blainski, A.; Lopes, G.C.; De Mello, J.C.P. Application and Analysis of the Folin Ciocalteu Method for the Determination of the Total Phenolic Content from *Limonium brasiliense* L. *Molecules* **2013**, *18*, 6852–6865. [CrossRef]
37. Telagari, M.; Hullatti, K. In-Vitro α-Amylase and α-Glucosidase Inhibitory Activity of *Adiantum caudatum* Linn. and *Celosia argentea* Linn. Extracts and Fractions. *Indian J. Pharmacol.* **2015**, *47*, 425–429. [CrossRef]
38. Studzińska-Sroka, E.; Piotrowska, H.; Kucińska, M.; Murias, M.; Bylka, W. Cytotoxic Activity of Physodic Acid and Acetone Extract from *Hypogymnia physodes* against Breast Cancer Cell Lines. *Pharm. Biol.* **2016**, *54*, 2480–2485. [CrossRef]
39. Tiveron, A.P.; Melo, P.S.; Bergamaschi, K.B.; Vieira, T.M.F.S.; Regitano-d'Arce, M.A.B.; Alencar, S.M. Antioxidant Activity of Brazilian Vegetables and Its Relation with Phenolic Composition. *Int. J. Mol. Sci.* **2012**, *13*, 8943–8957. [CrossRef]
40. Zha, D.; Yao, T.; Bao, L.; Gao, P.; Wu, X. Telmisartan Attenuates Diabetic Nephropathy Progression by Inhibiting the Dimerization of Angiotensin Type-1 Receptor and Adiponectin Receptor-1. *Life Sci.* **2019**, *221*, 109–120. [CrossRef]
41. Essa, R.; Sadek, A.M.E.; Baset, M.E.; Rawash, M.A.; Sami, D.G.; Badawy, M.T.; Mansour, M.E.; Attia, H.; Saadeldin, M.K.; Abdellatif, A. Effects of Turmeric (*Curcuma longa*) Extract in Streptozocin-Induced Diabetic Model. *J. Food Biochem.* **2019**, *43*, e12988. [CrossRef]

42. Anonymous ICH Q2 (R1) Validation of Analytical Procedures: Text and Methodology. Available online: https://www.ema.europa.eu/en/ich-q2-r1-validation-analytical-procedures-text-methodology (accessed on 26 November 2021).
43. Vieira, L.M.; Marinho, L.M.G.; Rocha, J.d.C.G.; Barros, F.A.R.; Stringheta, P.C. Chromatic Analysis for Predicting Anthocyanin Content in Fruits and Vegetables. *Food Sci. Technol.* **2019**, *39*, 415–422. [CrossRef]
44. Martín, J.; Navas, M.J.; Jiménez-Moreno, A.-M.; Asuero, A.G. Anthocyanin Pigments: Importance, Sample Preparation and Extraction. In *Phenolic Compounds—Natural Sources, Importance and Applications*; InTech: London, UK, 2017. [CrossRef]
45. Naczk, M.; Shahidi, F. Phenolics in Cereals, Fruits and Vegetables: Occurrence, Extraction and Analysis. *J. Pharm. Biomed. Anal.* **2006**, *41*, 1523–1542. [CrossRef] [PubMed]
46. Szczepaniak, O.M.; Kobus-Cisowska, J.; Kusek, W.; Przeor, M. Functional Properties of Cornelian Cherry (*Cornus mas* L.): A Comprehensive Review. *Eur. Food Res. Technol.* **2019**, *245*, 2071–2087. [CrossRef]
47. Enache, I.M.; Vasile, A.M.; Enachi, E.; Barbu, V.; Stănciuc, N.; Vizireanu, C. Co-Microencapsulation of Anthocyanins from Cornelian Cherry Fruits and Lactic Acid Bacteria in Biopolymeric Matrices by Freeze-Drying: Evidences on Functional Properties and Applications in Food. *Polymers* **2020**, *12*, 906. [CrossRef] [PubMed]
48. Milenkovic Andjelkovic, A.; Radovanović, B.; Andjelkovic, M.; Radovanovic, A.; Nikolic, V.; Randjelovic, V. The Anthocyanin Content and Bioactivity of Cornelian Cherry (*Cornus mas*) and Wild Blackberry (*Rubus fruticosus*): Fruit Extracts from the Vlasina Region. *Adv. Technol.* **2015**, *4*, 26–31. [CrossRef]
49. Salaritabar, A.; Darvish, B.; Hadjiakhoondi, F.; Manayi, A.; Sureda, A.; Nabavi, S.; Fitzpatrick, L.; Nabavi, S.; Bishayee, A. Therapeutic Potential of Flavonoids in Inflammatory Bowel Disease: A Comprehensive Review. *World J. Gastroenterol.* **2017**, *23*, 5097. [CrossRef]
50. Szczepaniak, O.; Cielecka-Piontek, J.; Kobus-Cisowska, J. Hypoglycaemic, Antioxidative and Phytochemical Evaluation of *Cornus mas* Varieties. *Eur. Food Res. Technol.* **2021**, *247*, 183–191. [CrossRef]
51. Assefa, S.T.; Yang, E.-Y.; Chae, S.-Y.; Song, M.; Lee, J.; Cho, M.-C.; Jang, S. Alpha Glucosidase Inhibitory Activities of Plants with Focus on Common Vegetables. *Plants* **2020**, *9*, 2. [CrossRef]
52. Bhatia, A.; Singh, B.; Arora, R.; Arora, S. In Vitro Evaluation of the α-Glucosidase Inhibitory Potential of Methanolic Extracts of Traditionally Used Antidiabetic Plants. *BMC Complement. Altern. Med.* **2019**, *19*, 74. [CrossRef]
53. Kumar, S.; Narwal, S.; Kumar, V.; Prakash, O. α-Glucosidase Inhibitors from Plants: A Natural Approach to Treat Diabetes. *Pharmacogn. Rev.* **2011**, *5*, 19–29. [CrossRef]
54. Asgary, S.; Sahebkar, A.; Afshani, M.R.; Keshvari, M.; Haghjooyjavanmard, S.; Rafieian-Kopaei, M. Clinical Evaluation of Blood Pressure Lowering, Endothelial Function Improving, Hypolipidemic and Anti-Inflammatory Effects of Pomegranate Juice in Hypertensive Subjects. *Phytother. Res.* **2014**, *28*, 193–199. [CrossRef]
55. De Biaggi, M.; Donno, D.; Mellano, M.G.; Riondato, I.; Rakotoniaina, E.N.; Beccaro, G.L. *Cornus mas* (L.) Fruit as a Potential Source of Natural Health-Promoting Compounds: Physico-Chemical Characterisation of Bioactive Components. *Plant Foods Hum. Nutr.* **2018**, *73*, 89–94. [CrossRef] [PubMed]
56. Deng, S.; West, B.J.; Jensen, C.J. UPLC-TOF-MS Characterization and Identification of Bioactive Iridoids in *Cornus mas* Fruit. *J. Anal. Methods Chem.* **2013**, *2013*, 710972. [CrossRef] [PubMed]
57. Wu, Q.; Tang, S.; Zhang, L.; Xiao, J.; Luo, Q.; Chen, Y.; Zhou, M.; Feng, N.; Wang, C. The Inhibitory Effect of the Catechin Structure on Advanced Glycation End Product Formation in Alcoholic Media. *Food Funct.* **2020**, *11*, 5396–5408. [CrossRef]
58. Oboh, G.; Ogunsuyi, O.B.; Ogunbadejo, M.D.; Adefegha, S.A. Influence of Gallic Acid on α-Amylase and α-Glucosidase Inhibitory Properties of Acarbose. *J. Food Drug Anal.* **2016**, *24*, 627–634. [CrossRef]
59. Tan, C.; Wang, Q.; Luo, C.; Chen, S.; Li, Q.; Li, P. Yeast α-Glucosidase Inhibitory Phenolic Compounds Isolated from Gynura Medica Leaf. *Int. J. Mol. Sci.* **2013**, *14*, 2551–2558. [CrossRef]
60. Bashan, N.; Kovsan, J.; Kachko, I.; Ovadia, H.; Rudich, A. Positive and Negative Regulation of Insulin Signaling by Reactive Oxygen and Nitrogen Species. *Physiol. Rev.* **2009**, *89*, 27–71. [CrossRef]
61. Evans, J.L.; Goldfine, I.D.; Maddux, B.A.; Grodsky, G.M. Oxidative Stress and Stress-Activated Signaling Pathways: A Unifying Hypothesis of Type 2 Diabetes. *Endocr. Rev.* **2002**, *23*, 599–622. [CrossRef]
62. Gan, B.; Hu, J.; Jiang, S.; Liu, Y.; Sahin, E.; Zhuang, L.; Fletcher-Sananikone, E.; Colla, S.; Wang, Y.A.; Chin, L.; et al. Lkb1 Regulates Quiescence and Metabolic Homeostasis of Haematopoietic Stem Cells. *Nature* **2010**, *468*, 701–704. [CrossRef]
63. Chakraborthy, A.; Ramani, P.; Sherlin, H.J.; Premkumar, P.; Natesan, A. Antioxidant and Pro-Oxidant Activity of Vitamin C in Oral Environment. *Indian J. Dent. Res.* **2014**, *25*, 499. [CrossRef] [PubMed]
64. Nizioł-Łukaszewska, Z.; Wasilewski, T.; Bujak, T.; Osika, P. Iridoids from *Cornus mas* L. and Their Potential as Innovative Ingredients in Cosmetics. *Pol. J. Chem. Technol.* **2017**, *19*, 122–127. [CrossRef]
65. Roberto, D.; Micucci, P.; Sebastian, T.; Graciela, F.; Anesini, C. Antioxidant Activity of Limonene on Normal Murine Lymphocytes: Relation to H_2O_2 Modulation and Cell Proliferation. *Basic Clin. Pharmacol. Toxicol.* **2010**, *106*, 38–44. [CrossRef]
66. Huang, D.; Ou, B.; Prior, R.L. The Chemistry behind Antioxidant Capacity Assays. *J. Agric. Food Chem.* **2005**, *53*, 1841–1856. [CrossRef]
67. Falony, G.; Lazidou, K.; Verschaeren, A.; Weckx, S.; Maes, D.; Vuyst, L.D. In Vitro Kinetic Analysis of Fermentation of Prebiotic Inulin-Type Fructans by *Bifidobacterium* Species Reveals Four Different Phenotypes. *Appl. Environ. Microbiol.* **2009**. [CrossRef]
68. Ramirez-Farias, C.; Slezak, K.; Fuller, Z.; Duncan, A.; Holtrop, G.; Louis, P. Effect of Inulin on the Human Gut Microbiota: Stimulation of *Bifidobacterium adolescentis* and *Faecalibacterium prausnitzii*. *Br. J. Nutr.* **2009**, *101*, 541–550. [CrossRef]

69. Harmsen, H.J.M.; Raangs, G.C.; Franks, A.H.; Wildeboer-Veloo, A.C.M.; Welling, G.W. The Effect of the Prebiotic Inulin and the Probiotic *Bifidobacterium longum* on the Fecal Microflora of Healthy Volunteers Measured by FISH and DGGE. *Microb. Ecol. Health Dis.* **2002**, *14*, 212–220. [CrossRef]
70. Staffolo, M.D.; Bevilacqua, A.E.; Rodríguez, M.S.; Albertengo, L. *Dietary Fiber and Availability of Nutrients: A Case Study on Yoghurt as a Food Model*; IntechOpen: London, UK, 2012; ISBN 978-953-51-0819-1.
71. Navarro, D.M.D.L.; Abelilla, J.J.; Stein, H.H. Structures and Characteristics of Carbohydrates in Diets Fed to Pigs: A Review. *J. Anim. Sci. Biotechnol.* **2019**, *10*, 39. [CrossRef]
72. Gao, D.; Li, Q.; Gao, Z.; Wang, L. Antidiabetic Effects of Corni Fructus Extract in Streptozotocin-Induced Diabetic Rats. *Yonsei Med. J.* **2012**, *53*, 691–700. [CrossRef]

antioxidants

MDPI

Article

Antioxidant Activity of Resveratrol Diastereomeric Forms Assayed in Fluorescent-Engineered Human Keratinocytes

Ilaria Bononi [1], Paola Tedeschi [2], Vanessa Mantovani [3], Annalisa Maietti [2], Elisa Mazzoni [2], Cecilia Pancaldi [3], Vincenzo Brandolini [2] and Mauro Tognon [3,*]

1 Department of Translational Medicine and for Romagna, University of Ferrara, 44121 Ferrara, Italy; bnnlri@unife.it
2 Department of Chemical, Pharmaceutical and Agricultural Sciences-DOCPAS, University of Ferrara, 44121 Ferrara, Italy; paola.tedeschi@unife.it (P.T.); annalisa.maietti@unife.it (A.M.); elisa.mazzoni@unife.it (E.M.); vincenzo.brandolini@unife.it (V.B.)
3 Laboratories of Cell Biology and Molecular Genetics, Section of Experimental Medicine, Department of Medical Sciences, School of Medicine, University of Ferrara, 44121 Ferrara, Italy; vanessa.mantovani@unife.it (V.M.); cecilia.pancaldi@unife.it (C.P.)
* Correspondence: tgm@unife.it; Tel.: +39-0532-455538

Citation: Bononi, I.; Tedeschi, P.; Mantovani, V.; Maietti, A.; Mazzoni, E.; Pancaldi, C.; Brandolini, V.; Tognon, M. Antioxidant Activity of Resveratrol Diastereomeric Forms Assayed in Fluorescent-Engineered Human Keratinocytes. *Antioxidants* **2022**, *11*, 196. https://doi.org/10.3390/antiox11020196

Academic Editors: Jicheng Zhan, Zhigang Liu and Hui-Min David Wang

Received: 23 December 2021
Accepted: 16 January 2022
Published: 20 January 2022

Abstract: Resveratrol is a powerful antioxidant molecule. In the human diet, its most important source is in *Vitis vinifera* grape peel and leaves. Resveratrol exists in two isoforms, cis- and trans. The diastereomeric forms of many drugs have been reported as affecting their activity. The aim of this study was to set up a cellular model to investigate how far resveratrol could counteract cytotoxicity in an oxidant agent. For this purpose, a keratinocyte cell line, which was genetically engineered with jelly fish green fluorescent protein, was treated with the free radical promoter Cumene hydroperoxide. The antioxidant activity of the trans-resveratrol and its diastereomeric mixture was evaluated indirectly in these treated fluorescent-engineered keratinocytes by analyzing the cell number and cell proliferation index. Our results demonstrate that cells, which were pre-incubated with resveratrol, reverted the oxidative damage progression induced by this free radical agent. In conclusion, fluorescent-engineered human keratinocytes represent a rapid and low-cost cellular model to determine cell numbers by studying emitted fluorescence. Comparative studies carried out with fluorescent keratinocytes indicate that trans-resveratrol is more efficient than diastereomeric mixtures in protecting cells from the oxidative stress.

Keywords: oxidative stress; resveratrol; diastereomeric mixture; keratinocyte

1. Introduction

Oxidative stress is defined as a set of alterations at molecular, cellular, and tissue levels in a living organism. In the event of excessive exposure to oxidizing agents, modification in the redox balance inside cells arises between the reactive oxygen species (ROS) and the antioxidant system, which favors ROS [1].

ROS is constantly produced in cells through a variety of biochemical processes [2]. At moderate concentrations, ROS are actively involved in complex biological processes, such as control of gene expression, apoptosis, and signal transduction [3]. However, at a high concentration, ROS cause cell damage, such as aging and alterations in cellular constituents [4]. Eukaryotic cells have developed an antioxidant system including endogenous compounds, which work synergistically with exogenous molecules to neutralize free radicals [5] such as ROS. Antioxidants are molecules that are capable of protecting cells against the oxidation induced by free radicals, by blocking the initiation phase of radical production or neutralizing radicals formed during the propagation phase [5].

Resveratrol is one of the most highly investigated antioxidant molecules (Figure 1a) [6]. Since resveratrol is also a phytoalexin, its biosynthesis starts from the plant's secondary metabolism, which is stimulated by microorganism infection or abiotic elicitors [7]. While flavonoids are found throughout the plant kingdom, different resveratrol forms have been isolated from only a few plant families. The most important sources of resveratrol in the human diet are grapes, peanuts, berries, and their byproducts. In plants and their derived foods, resveratrol exists in various forms, such as aglycone, glucoside (piceid), and several oligomers. A mixture of cis and trans isomers was identified for each form [8].

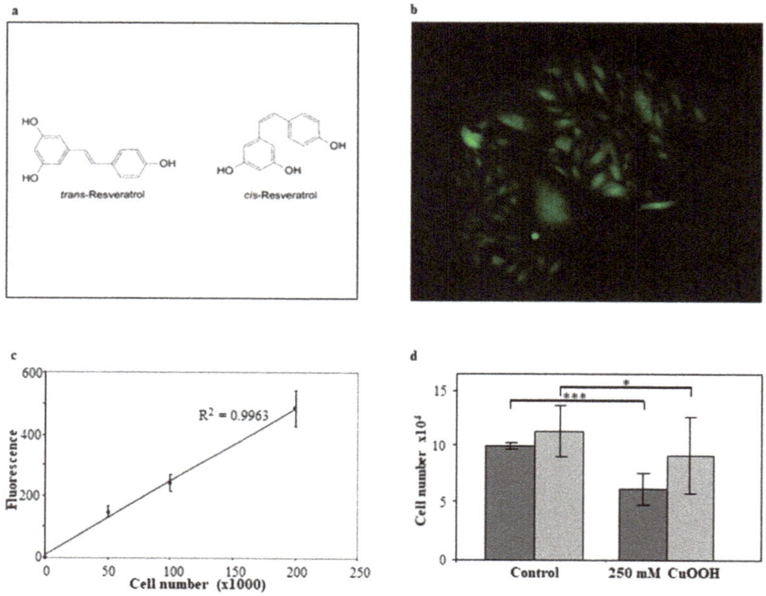

Figure 1. Resveratrol structures, florescent human keratinocyte cells, relation between the number of NCTC-GFP cells and fluorescence intensity, and the effect of CuOOH treatment: Panel (**a**): Resveratrol cis- and trans-isomers chemical structures. Panel (**b**): Microphotograph showing fluorescent NCTC-GFP (magnitude 20X). Keratinocytes exhibit small differences in morphology/shape due to the micro-environment in which they are expanding, i.e., cells inside the monolayer are more rounded, whereas cells growing at the edge of the monolayer are more elongated. Panel (**c**): Relation between the number of NCTC-GFP cells and fluorescence intensity. Samples containing an increased number of cells, ranging from 3.125×10^3 to 10^5 cells/cm^2, were analyzed by spectrofluorimetric reading. Five different sets of samples were tested at different times. Each value was reported in the graph as a function of the corresponding cell number. Bars represent the standard deviation. The linear relation between cell number and fluorescence intensity was represented by the R^2 value (0.9963). Panel (**d**): Effect of CuOOH treatment on cell number (dark gray) and proliferation (light gray). Horizontal bars indicate a statistically significant difference (*: $p < 0.05$; ***: $p < 0.0005$).

The most important source of resveratrol in the human diet is *Vitis vinifera* grape peel and leaves [9,10], which are used for the wine production. In *Vitis vinifera*, 16 resveratrol forms have been detected in grape stems, seeds, skins, and juice. Stems contain the richest source of stilbenes, followed by grape skins. These byproducts of the winemaking processes could provide useful information for determining how to enrich potential health-promoting compounds [11]. Resveratrol contains chemicals that determine its ability to inhibit the progress of certain infections. In general, stressful factors induce an accumulation of resveratrol in plants. This mechanism allows plants to resist parasites and other adverse conditions. Indeed, resveratrol is produced by more than 70 different species of plants in response to such types of stressful situations [12,13]. Given its wide variety of pharmacological properties, resveratrol was indicated as a probable explanation for the so called "French paradox". Indeed, some epidemiological studies revealed an inverse correlation between wine consumption and cardiovascular diseases in France, a country known for its high intake of saturated fatty acids [13,14] and wine. Resveratrol demonstrates several biological activities such as anti-inflammatory, vasorelaxing, anticancer, antiaging, anti-frailty, as well as antiallergenic characteristics [13,15]. However, the key relevance of resveratrol is due to its antioxidant activity and its behavior as an important radical scavenger [10].

Published studies indicate that resveratrol has a powerful antioxidant activity associated with the presence of three hydroxyl groups in its structure. Resveratrol has an inhibitory effect on excessive ROS production, aberrant mitochondrial distribution, and lipid peroxidation [16–18]. Resveratrol leads to an increase in endogenously generated glutathione, and a quantitative reduction in the cellular redox environment and endogenous ROS production [19]. In vitro studies have demonstrated that resveratrol shows a role in maintaining cellular redox homeostasis by preventing an increase in ROS production and a decrease in mitochondrial membrane potential [20]. Resveratrol decreases mitochondria fragmentation and maintains the potential of the mitochondrial membrane while preventing oxidative phosphorylation attenuation, thus exerting a protective effect against the harmful impact of ROS [16]. Resveratrol treatment protects cells from the oxidative damage induced by hydrogen peroxide. This protection is caused by reduced malondialdehyde (MDA) and intracellular ROS concentrations and increased expression levels of antioxidant enzymes [21]. In vivo studies on patients affected by diabetes demonstrate that the protective effects of chronic resveratrol administration is beneficial in: reducing advanced glycation end product (AGE)-induced oxidative stress and apoptosis [22]; normalizing antioxidant status, exacerbated by oxidative stress induced by hyperglycemia [23]; reducing ROS production, elevating membrane potential, and inhibiting cytochrome c release from the inner mitochondrial membrane [24]; and contrasting the 3-nitrotyrosine accumulation and 4-hydroxynonenal generation, increasing total antioxidant capacity [25]. Resveratrol protects the spinal cord from ischemic damage in rats by reducing plasma levels in nitrite, advanced oxidation protein products (AOPP) and MDA, and increasing the enzymatic activity of superoxide dismutase (SOD) and catalase (CAT) [26]. In vivo studies demonstrate that resveratrol also attenuates (i) oxidative stress in rats with experimental periodontitis, (ii) early Alzheimer's disease, and (iii) chronic obstructive pulmonary disease [27–29].

The above mentioned in vitro and in vivo studies indicate that resveratrol has a therapeutic effect in cells and animals suffering from increased oxidative stress, which is associated with a limitation in ROS generation and stimulation of compounds that act as an antioxidant barrier [16].

Resveratrol is characterized by two phenolic rings joined by a styrene double bond, with hydroxyls in position 3 and 5 of the first ring and 4 of the second ring [30]. Resveratrol is a low molecular weight molecule of 228 Dalton [31]. It exists in two diastereomeric forms, cis- and trans-, and its isomerization from the trans- to cis-form is favored by exposure to UV radiation [32]. A diastereomeric mixture is also known as a diastereomers blend, where the proportions of diastereomers may affect its biological activity, and its properties can significantly differ from those of its two individual diastereomers [33,34].

In order to study the effect of the isomer forms of resveratrol, a new in vitro cellular model was set up, represented by fluorescent-engineered human keratinocytes. These cells were employed for measuring in vitro viability. Indeed, cell viability may be used as an indirect parameter of oxidative damage. The fluorescent-engineered human keratinocytes developed in this investigation may represent an alternative study model, which is able to complement animal experimentations.

In our experimental model, Cumene hydroperoxide (CuOOH) was used as a free radical promoter [35]. As reported above, resveratrol as a radical scavenger is likely to contrast the effect of the CuOOH neutralizing radicals that arise from its action.

The aim of this study is to test different forms of resveratrol for their potential antioxidant activity in fluorescent-engineered human keratinocyte cells for the first time. This innovative approach allowed us to investigate whether a diastereomeric mixture shows quantitative differences in antioxidant activity compared to trans-resveratrol.

2. Materials and Methods

2.1. Cells and Medium

NCTC-2544 human keratinocyte cells (ICN Flow, Irvine, UK) were cultured in a 1:1 mixture of DMEM medium and Ham's F-12 with Hepes, L-glutamine, 10% FBS, and penicillin/streptomycin (Lonza, Basel, Switzerland).

2.2. Fluorescent-Engineered Human Keratinocytes

In order to determine adequate concentrations of the antibiotic Geneticin (G418) in human keratinocytes, NCTC-2544, cells were plated at a density of 5×10^4 cells/cm^2 with different concentrations of G418, ranging from 0 to 1200 μg/mL. The proportion of viable cells was assessed by trypan-blue staining. Subsequently, NCTC-2544 human keratinocytes parental cells were transfected with the recombinant vector named pIRES2-AcGFP1, using the FuGENE HD Transfection Reagent (Roche, Milan, Italy). The pIRES2-AcGFP1 vector contains coding sequences for the green fluorescent protein from Aequorea coerulescens (AcGFP). After 15–20 days of antibiotic treatment, resistant clones were isolated with sterile glass cylinders and expanded as independent cell lines.

2.3. Immunofluorescence Technique

In order to undertake immunofluorescence assays, cells were cultured on coverslips for 18 h, and fixed with 4% paraformaldehyde (PFA). Subsequently, cells were permeabilized and incubated for 1 h at 37 °C with TRITC-conjugated Phalloidin. (Sigma-Aldrich, Milan, Italy), diluted 0.2 μg/mL in PBS 1×. After applying the DAPI solution, the glass coverslips were mounted with glycerol/PBS 9:1. Images were obtained using a TE 2000-E fluorescent microscope. Digital images were captured using ACT-1 and ACT-2 software for DXM1200F digital cameras (Nikon Instruments, Sesto Fiorentino, Italy).

2.4. Cell Quantification

NCTC_2544_AcGFP cells were seeded in 24-well plates at different concentrations, ranging from 2.5×10^4 to 1×10^5, which were then used to evaluate the cell number by measuring the cell fluorescence, using the spectrofluorometer Wallac Victor2 multiwell instrument (Perkin-Elmer, Monza, Italy). The calibration curve was obtained by reporting the number of cells in each well on a graph with the relative spectrofluorometer reading.

2.5. Resveratrol

Standard trans-resveratrol was purchased from Sigma, Milan (Figure 1a). The trans-resveratrol diastereomeric mixture was obtained from the isomerization of trans-resveratrol in our laboratories as follows: 10 mg/mL of a standard solution of trans-resveratrol in acetonitrile/water 50:50 was isomerized by UV irradiation. UV irradiation was performed with a Helios Italquartz UV lamp at 366 nm for 3 h. This time lapse was needed to obtain a 50:50 trans/cis-resveratrol mixture. The isomerization process was controlled by capillary electrophoresis, as reported previously [36].

2.6. Determining Antioxidant Capacity

Antioxidant capacity was tested with the OxiSelect Hydrogen Peroxide Assay Kit Colorimetric (Cat. No STA-343, Cell Biolabs, Inc., San Diego, CA, USA) following the manufacturer's instructions. The assay was carried out in triplicate in 96-well plates.

2.7. Cytotoxicity Assay

Cells were seeded in triplicate in 24-well plates and incubated for 24 h. Culture cells were exposed to 1.25 and 2.5 μM trans-resveratrol or diastereomeric mixture for 18 h. The cell monolayer was exposed to 250 μM CuOOH (Sigma-Aldrich) for 10 min. Two controls were included in these experiments. One control is represented by untreated cells, i.e., without resveratrol and CuOOH, whereas the other control is represented by cells treated with CuOOH, without resveratrol pre-treatment.

2.8. Cell Counting and Proliferation

Cell number was evaluated using the Wallac Victor2 spectrofluorometer by reading the fluorescence emitted from the attached cells. The number of cells was calculated using the calibration curve obtained with wells containing a known number of cells. Cell proliferation was quantitatively measured using the Alamar Blue Cell Viability Reagent (Invitrogen). Cell proliferation within samples was evaluated using the interpolation method, by comparing the percentage of Alamar Blue reduction in the sample to the standard curve [37,38].

2.9. Statistical Analysis

Statistical analyses were performed by Prism software. Results were obtained from four independent experiments and expressed as a mean value and SD. Student's *t*-test was used to evaluate the statistical significance of the differences in mean values; p values < 0.05 were considered significant.

3. Results

3.1. Cell Clones

Parental human keratinocytes (NCTC-2544), transfected with the recombinant plasmid expressing the green fluorescent protein (pIRES2-AcGFP1), were cultured in the presence of 300 μg/mL of Geneticin. A total of 28 G418 resistant and fluorescent cell clones, i.e., NCTC/GFP, were isolated and expanded as independent cells. One out of these twenty-eight cell clones was expanded as a cell line and employed herein to perform all the experiments. This clone showed homogeneous cell characteristics. AcGFP expression was assessed by fluorescence microscopy observation, while its stability was confirmed over a period of 12 weeks as fluorescence intensity by spectrofluorimetric analysis (Figure 1b).

3.2. Cytological Characterization

Comparative molecular characterization was carried out with the parental NCTC-2544 cell line and NCTC_2544_AcGFP engineered cells. Actin fibers were visualized using Phalloidin-TRIC conjugated specifically in order to assess whether the exogenous protein could alter NCTC_2544_AcGFP cytoskeletal organization. The cytoskeletal architecture of the parental and engineered cells was indistinguishable (Figure 2). Therefore, the presence of the exogenous protein AcGFP does not affect the cytoskeletal organization of the engineered cells.

3.3. Determining Antioxidant Capacity

The antioxidant capacity of resveratrol was tested using the OxiSelect Hydrogen Peroxide Assay Kit Colorimetric. Results were shown in Figure 3.

3.4. Cell Number Counting

Since NCTC/GFP cells express green fluorescent protein constitutively, fluorescence intensity may depend exclusively on the number of cells within a sample. To confirm the linearity of fluorescence intensity with the cell number, samples sets containing 3.125×10^3, 6.25×10^3, 1.25×10^4, 2.5×10^4, 5×10^4, and 1×10^5 cells/cm^2 were analyzed by spectrofluorometer. The graphical expression of these experiments confirmed that fluorescence intensity values (absolute measure) are proportional to the number of cells within a sample. On this basis, having gained the fluorescence value, it was possible to extrapolate the number of cells contained within a sample by interpolation with the standard curve (Figure 1c).

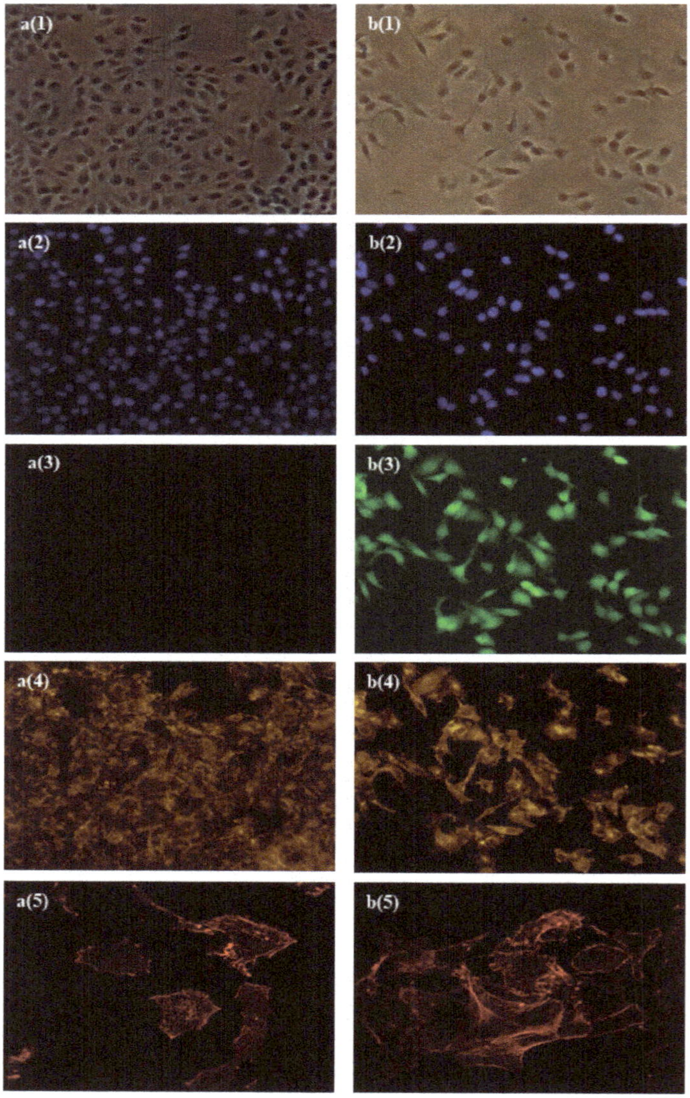

Figure 2. Microphotographs showing parental cells NCTC_2544 (**a**) and engineered cells NCTC_2544_AcGFP (**b**). Panels (**a**(**1**),**b**(**1**)) are microphotographs with white light to show cell morphology and shape; panels (**a**(**2**),**b**(**2**)) are microphotographs with nuclei counterstained with DAPI; panels (**a**(**3**),**b**(**3**)) are microphotographs with fluorescence light to visualize constitutive green fluorescence; panels (**a**(**4**),**b**(**4**)) are microphotographs with fluorescence light to visualize actin fiber architecture (magnitude 20×; panels (**a**(**5**),**b**(**5**)) are microphotographs with fluorescence light to visualize actin fiber architecture (magnitude 60×).

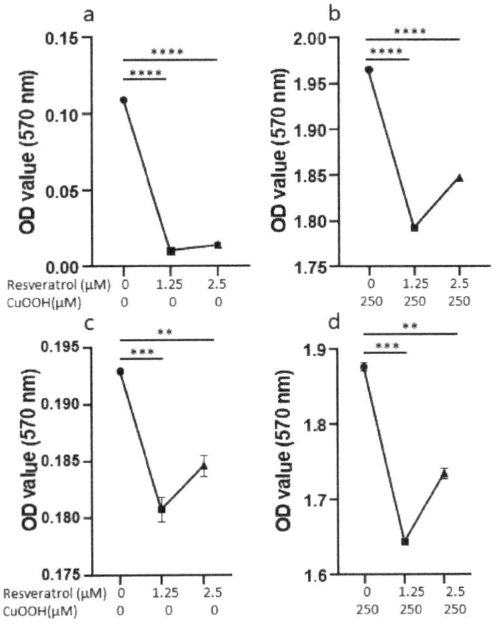

Figure 3. Effect of trans-resveratrol on oxidative stress on non-treated cells (panel (**a**)) and effect of trans-resveratrol on oxidative stress on cells treated with CuOOH (panel (**b**)). Horizontal bars indicate a statistically significant difference. Effect of resveratrol diastereomeric mixture on oxidative stress on non-treated cells (panel (**c**)) and effect of trans-resveratrol on oxidative stress on cells treated with CuOOH (panel (**d**)). Horizontal bars indicate statistically significant differences (**: $p < 0.005$; ***: $p < 0.0005$; ****: $p < 0.00005$).

3.5. Effect of CuOOH on the Human Keratinocyte NCTC/GFP Cell Line

Cells were treated with 250 μM CuOOH for 10 min. Spectrofluorimetric analysis showed that the oxidant agent caused a reduction in NCTC-GFP cells from 9.6×10^4 to 6.1×10^4 ($p < 0.0005$). Moreover, the Alamar Blue assay showed a proliferation reduction from 1.1×10^5 to 9×10^4 ($p < 0.05$). The two cell number reductions are statistically significant, as shown by the p values (Figure 1d).

3.6. Resveratrols

In general, cis-resveratrol is less pervasive in natural products than trans-resveratrol, thereby justifying a lack of analytical standards and difficulties in its identification and quantification. Notwithstanding this, the possibility of converting trans-resveratrol to cis-resveratrol by UV irradiation has previously been described in literature [39]. In a previous study we reported on the effect of irradiation time on the isomerization of trans-resveratrol to equilibrium conditions with an 80% final conversion [40].

Every 30 min, a portion of isomerized trans-resveratrol standard solution was diluted 1:100 and injected in order for the analysis to verify the isomerization percentage. The electropherogram of isomerized trans-resveratrol standard solution, after 90 min, was reported in Figure 4a. At this time the mixture was made up of 30% cis-resveratrol and 70% of trans-resveratrol. Trans-resveratrol (1) and cis-resveratrol (2) peaks were identified by comparing retention times and their absorption spectrum. The absorption spectrum maximum reading for trans-resveratrol (1) is at 315 nm, while for cis-resveratrol (2) it is at 290 (Figure 4b). After 3 h, trans-resveratrol isomerization was at 50%. The resulting solution was lyophilized overnight in a Christ α 1–2 lyophilizer, and the obtained product was used for the biological assay.

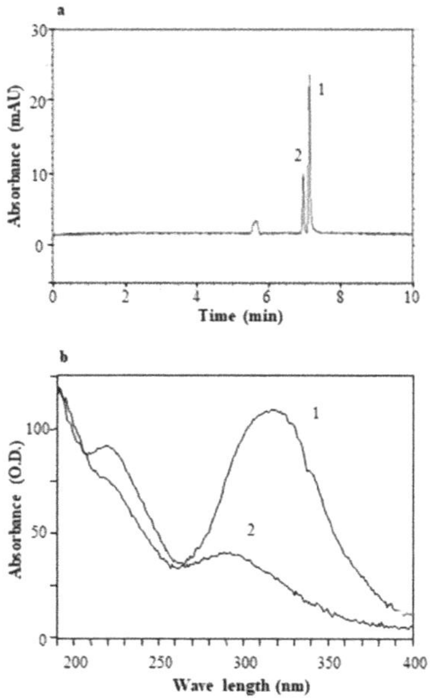

Figure 4. Electropherogram of trans-resveratrol standard solution and absorption spectrum. Panel (**a**): Electropherogram of trans-resveratrol standard solution diluted 1:100 after 90 min of isomerization (peak 1 = trans-resveratrol; peak 2 = and cis-resveratrol). Panel (**b**): Comparison between absorption spectrum of standard (trans-resveratrol (1) absorption spectrum and cis-resveratrol (2) absorption spectrum).

3.7. Effect of Resveratrol on the Cell Numbers

Cells ($n = 5 \times 10^4/\text{cm}^2$) were pre-incubated with resveratrol in our experimental conditions. Cell numbers increased both on CuOOH-treated and non-treated cells when evaluated using the spectrofluorometer.

Specifically, cell numbers increased by a statistically significant difference when 1.25 μM of trans-resveratrol was used on CuOOH-treated cells, from 6×10^4 to 1×10^5 ($p < 0.05$). On the other hand, cell numbers still increased upon using trans-resveratrol at a concentration of 2.5 μM, but the difference was not statistically significant. The same result, in terms of non-statistically significant differences in the cell number, was obtained by pre-incubating CuOOH non-treated cells with trans-resveratrol at 1.25 μM and 2.5 μM. Similarly, both CuOOH-treated and non-treated cells, when pre-incubated with 1.25 μM and 2.5 μM diastereomeric mixture, showed no statistically significant increase in cell numbers (Figure 5a,b). Higher concentrations of the diastereomeric mixture and of trans-resveratrol caused a mild increase in the cell numbers or reduced human keratinocyte number, but the difference was not statistically significant (data not shown).

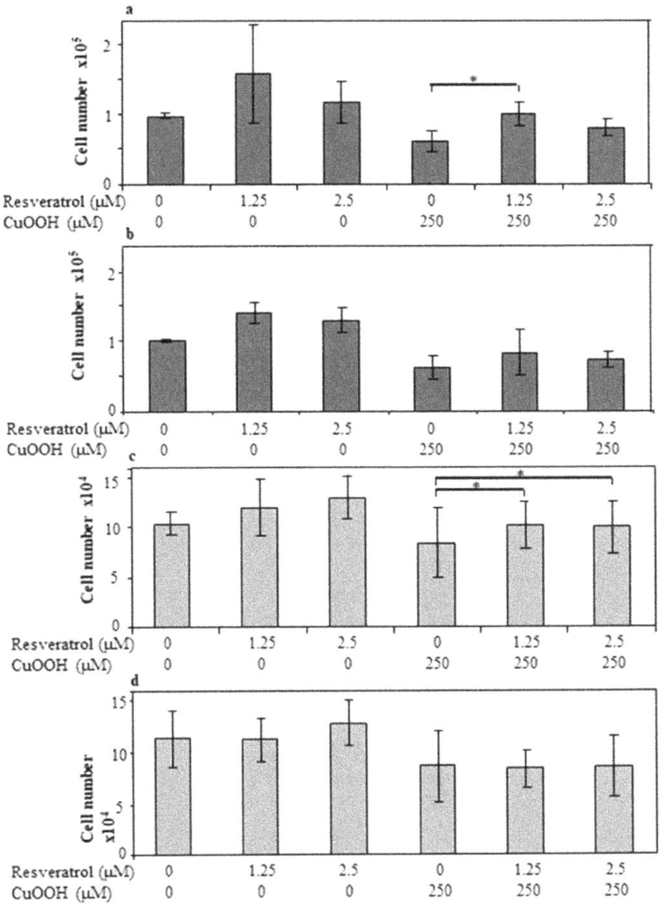

Figure 5. Effect of trans-resveratrol (panel (**a**)) and diastereomeric mixture (panel (**b**)) treatment on cell numbers evaluated by the measuring fluorescence. Horizontal bars indicate a statistically significant difference (*: $p < 0.05$). Effect of trans-resveratrol (panel (**c**)) and diastereomeric mixture (panel (**d**)) treatment on cell proliferation evaluated by Alamar Blue assay. Horizontal bars indicate a statistically significant difference (*: $p < 0.05$).

3.8. *Effect of Resveratrol on Cell Proliferation*

Proliferation increased in cells that had been pre-incubated with resveratrol, both in cells treated with CuOOH and in the control. Data were obtained using the Alamar Blue assay. The increase in cell proliferation in CuOOH-treated cells was statistically significant when two different concentrations of trans-resveratrol were employed, i.e., 1.25 μM and 2.5 μM. The increases were from 8.4×10^4 to 1×10^5 ($p < 0.05$) for 1.25 μM and from 8.4×10^4 to 9.9×10^4 ($p < 0.05$) for 2.5 μM, respectively. In the same experimental conditions, the resveratrol diastereomeric mixture caused a non-statistically significant increase in cell proliferation. The same result was obtained in CuOOH non-treated cells, which have been pre-incubated with both the trans-resveratrol and diastereomeric mixtures, at a concentration of 1.25 μM and 2.5 μM, respectively (Figure 5c,d). Cell proliferation data, obtained on cells pre-incubated with higher concentrations (5 μM) of the diastereomeric mixture and trans-resveratrol, showed a very low increase in cell proliferation or even a reduction (data not shown).

4. Discussion

In this comparative study the antioxidant effects of trans-resveratrol and its diastereomeric mixture were determined in fluorescent-engineered human keratinocytes. As reported previously, we herein show that the presence of AcGFP does not affect cytoskeletal organization [41], and that fluorescence intensity is proportional to the number of cells [41,42]. The high expression level of trans-gene in the genetically modified NCTC/GFP cell line ensured a direct proportion between fluorescence intensity and cell number [41,43].

In a recent study, human keratinocytes HaCaT, which had been irradiated with UV A to induce oxidative stress, were treated with increasing concentrations of trans-resveratrol. It was shown that both pre- and post-treatment with trans-resveratrol provided significant protection, as demonstrated by an increase in cellular activity and a decrease in ROS level resveratrol [44]. Another study conducted on HaCaT [45] showed that the toxicity of an elicitor was strongly attenuated by post-treatment with trans-resveratrol. Further studies focused their attention on the differences between the effects of the two resveratrol isomers in different experimental models. Indeed, it has been reported that the two isomers have similar levels of efficiency [46,47], while other investigations have found trans-resveratrol and cis-resveratrol to have different efficiency levels [48–51].

Our results indicate that keratinocytes incubated with CuOOH showed a reduction in cell number and proliferation index. Cells that had been pre-incubated with resveratrol were protected from damage. These data confirm and extend previous results on the effect of resveratrol on human epidermal keratinocytes [44,52,53]. Our comparative study with resveratrol isomers allowed for some differences in the behavior of trans-resveratrol and its diastereomeric mixture to be shown. The addition of trans-resveratrol to keratinocytes, protected the cells by inactivating the cytotoxic effects of CuOOH. Trans-resveratrol at a concentration of 1.25 μM caused a statistically significant decrease in oxidative stress and an increase in both cell number and proliferation. However, it should be noted that only the concentration of 2.5 μM showed a statistical significance in cell proliferation. The addition of trans-resveratrol to the control cells demonstrated an increase in cell number and proliferation at all concentrations tested, although without statistically significant differences. The diastereomeric mixture illustrated a protective effect against oxidative damage similar to that of trans-resveratrol, but with less efficacy. When the diastereomeric mixture was used on cells that had previously been treated with CuOOH, a non-statistically significant increase in both cell number and proliferation was obtained. Similar results were detected with the pre-incubation of CuOOH non-treated cells using the diastereomeric mixture, at all concentrations tested. Herein, the trans-resveratrol and diastereomeric mixtures, at concentrations of 1.25 μM and 2.5 μM, were used as reported in previous studies [44,54–56]. When higher concentration (5 μM) of both the trans-resveratrol and diastereomeric mixture was employed, a less protective effect was obtained, whereas in some cases, a negligible depletion in cell number and inhibition of proliferation was observed. These data are in agreement with the results reported by Holian et al. [43]. In our experimental model, CuOOH and resveratrol were used as a free-radical promoter and antioxidant molecule, respectively. CuOOH, a lipophilic compound, may act in the phospholipidic bilayer structure of the cell membrane where it induces lipid peroxidation and generates ROS, which may diffuse from the membrane lipid bilayer into the intracellular compartment [35]. Resveratrol has been reported as contrasting oxidative stress by acting as a radical scavenger [45,57].

Resveratrol pre-incubation of cells may eliminate, at least in part, the ROS generated by CuOOH. As a consequence, adding resveratrol led to a protective effect on the keratinocytes. In addition, previous studies demonstrated (i) the presence of resveratrol binding sites in human keratinocytes [56] and (ii) that trans-resveratrol protects these cells from oxidative stress [53]. The protective effects of resveratrol against the oxidative agent are mediated by its ability to enhance the cellular production of antioxidant enzymes, such as glutathione S-transferase (GST), glutathione peroxidase (Gpx), NAD(P)H, quinone oxidoreductase 1 (NQO1), catalase (cat), and superoxide dismutase (SOD). Resveratrol activity is mediated along the ERK pathway, which allows the transcription of genes encoding antioxidant enzymes. Resveratrol also modulates multiple signaling mechanisms, while displaying an effect on the NF-κB and sirtuin/FoxO3a pathways [58].

5. Conclusions

Herein, the innovative cellular model represented by fluorescent-engineered keratinocyte cells allowed us to confirm that resveratrol has the ability to protect human cells from oxidative stress. Specifically, trans-resveratrol proved to be more effective than the diastereomeric mixture. This characteristic could be ascribed to the spatial position of the hydroxyl groups that are of prime importance in terms of its chelating capacity [10]. Similarly, how far resveratrol activity protects cells from oxidative stress, depends on the hydroxyl group position. Our data indicate that fluorescent-engineered keratinocytes may represent a rapid, low-cost cellular model for measuring the antioxidant effects of trans-resveratrol and diastereomeric mixture indirectly.

Author Contributions: Conceptualization, M.T. and V.B.; methodology, V.M., P.T. and A.M.; validation, C.P. and E.M.; formal analysis, E.M.; investigation, I.B.; resources, P.T.; data curation, I.B.; writing—original draft preparation, I.B.; writing—review and editing, I.B. and C.P.; supervision, M.T. and V.B.; project administration, M.T.; funding acquisition, M.T. All authors have read and agreed to the published version of the manuscript.

Funding: This study was supported, in part, by grants from the Region Emilia-Romagna (POR-FESR 2020) and University of Ferrara grants FAR and FIR 2020 and 2021, Ferrara, Italy.

Institutional Review Board Statement: Not applicable.

Informed Consent Statement: Not applicable.

Data Availability Statement: All of the data are contained within the article.

Acknowledgments: We thank Georgia Emma Gili for revising the English text of the manuscript.

Conflicts of Interest: The authors declare no conflict of interest. The funders had no role in the design of the study; in the collection, analyses, or interpretation of data; in the writing of the manuscript, or in the decision to publish the results.

References

1. Comino-Sanz, I.M.; López-Franco, M.D.; Castro, B.; Pancorbo-Hidalgo, P.L. The Role of Antioxidants on Wound Healing: A Review of the Current Evidence. *J. Clin. Med.* **2021**, *10*, 3558. [CrossRef] [PubMed]
2. Khan, T.A.; Hassan, I.; Ahmad, A.; Perveen, A.; Aman, S.; Quddusi, S.; Alhazza, I.M.; Ashraf, G.M.; Aliev, G. Recent updates on the dynamic association between oxidative stress and neurodegenerative disorders. *CNS Neurol. Disord.-Drug Targets* **2016**, *15*, 1. [CrossRef] [PubMed]
3. Finkel, T. Signal transduction by reactive oxygen species. *J. Cell Biol.* **2011**, *194*, 7–15. [CrossRef] [PubMed]
4. Hemerková, P.; Vališ, M. Role of Oxidative Stress in the Pathogenesis of Amyotrophic Lateral Sclerosis: Antioxidant Metalloenzymes and Therapeutic Strategies. *Biomolecules* **2021**, *11*, 437. [CrossRef]
5. Georgieva, E.; Ivanova, D.; Zhelev, Z.; Bakalova, R.; Gulubova, M.; Aoki, I. Mitochondrial Dysfunction and Redox Imbalance as a Diagnostic Marker of "Free Radical Diseases". *Anticancer. Res.* **2017**, *37*, 5373–5381. [CrossRef]
6. Carrizzo, A.; Forte, M.; Damato, A.; Trimarco, V.; Salzano, F.; Bartolo, M.; Maciag, A.; Puca, A.A.; Vecchione, C. Antioxidant effects of resveratrol in cardiovascular, cerebral and metabolic diseases. *Food Chem. Toxicol.* **2013**, *61*, 215–226. [CrossRef]
7. Belchí-Navarro, S.; Almagro, L.; Sabater-Jara, A.B.; Fernández-Pérez, F.; Bru, R.; Pedreno, M.A. Induction of trans-resveratrol and extracellular pathogenesis-related proteins in elicited suspension cultured cells of Vitis vinifera cv Monastrell. *J. Plant Physiol.* **2013**, *170*, 258–264. [CrossRef]
8. Pugajeva, I.; Pērkons, I.; Górnaś, P. Identification and determination of stilbenes by Q-TOF in grape skins, seeds, juice and stems. *J. Food Compos. Anal.* **2018**, *74*, 44–52. [CrossRef]
9. Orallo, F. Trans-resveratrol: A magical elixir of eternal youth? *Curr. Med. Chem.* **2008**, *15*, 1887–1898. [CrossRef]
10. Frémont, L. Biological effects of resveratrol. *Life Sci.* **2000**, *66*, 663–673. [CrossRef]
11. Keylor, M.H.; Matsuura, B.S.; Stephenson, C.R.J. Chemistry and Biology of Resveratrol-Derived Natural Products. *Chem. Rev.* **2015**, *115*, 8976–9027. [CrossRef] [PubMed]
12. Maier-Salamon, A.; Böhmdorfer, M.; Thalhammer, T.; Szekeres, T.; Jaeger, W. Hepatic Glucuronidation of Resveratrol: Interspecies Comparison of Enzyme Kinetic Profiles in Human, Mouse, Rat, and Dog. *Drug Metab. Pharmacokinet.* **2011**, *26*, 364–373. [CrossRef]

13. Gambini, J.; Ingles, M.; Olaso, G.; Lopez-Grueso, R.; Bonet-Costa, V.; Gimeno-Mallench, L.; Mas-Bargues, C.; Abdelaziz, K.M.; Gomez-Cabrera, M.C.; Vina, J.; et al. Properties of Resveratrol:In VitroandIn VivoStudies about Metabolism, Bioavailability, and Biological Effects in Animal Models and Humans. *Oxid. Med. Cell. Longev.* **2015**, *2015*, 837042. [CrossRef] [PubMed]
14. Frombaum, M.; Le Clanche, S.; Bonnefont-Rousselot, D.; Borderie, D. Antioxidant effects of resveratrol and other stilbene derivatives on oxidative stress and NO bioavailability: Potential benefits to cardiovascular diseases. *Biochimie* **2012**, *94*, 269–276. [CrossRef] [PubMed]
15. Cristòfol, R.; Porquet, D.; Corpas, R.; Coto-Montes, A.; Serret, J.; Camins, A.; Pallàs, M.; Sanfeliu, C. Neurons from senescence-accelerated SAMP8 mice are protected against frailty by the sirtuin 1 promoting agents melatonin and resveratrol. *J. Pineal Res.* **2012**, *52*, 271–281. [CrossRef]
16. Galiniak, S.; Aebisher, D.; Bartusik-Aebisher, D. Health benefits of resveratrol administration. *Acta Biochim. Pol.* **2019**, *66*, 13–21. [CrossRef] [PubMed]
17. Liu, Y.; He, X.-Q.; Huang, X.; Ding, L.; Xu, L.; Shen, Y.-T.; Zhang, F.; Zhu, M.-B.; Xu, B.-H.; Qi, Z.; et al. Resveratrol Protects Mouse Oocytes from Methylglyoxal-Induced Oxidative Damage. *PLoS ONE* **2013**, *8*, e77960. [CrossRef]
18. Liguori, I.; Russo, G.; Curcio, F.; Bulli, G.; Aran, L.; Della-Morte, D.; Gargiulo, G.; Testa, G.; Cacciatore, F.; Bonaduce, D.; et al. Oxidative stress, aging, and diseases. *Clin. Interv. Aging* **2018**, *13*, 757–772. [CrossRef] [PubMed]
19. Plauth, A.; Geikowski, A.; Cichon, S.; Wowro, S.; Liedgens, L.; Rousseau, M.; Weidner, C.; Fuhr, L.; Kliem, M.; Jenkins, G.; et al. Hormetic shifting of redox environment by pro-oxidative resveratrol protects cells against stress. *Free Radic. Biol. Med.* **2016**, *99*, 608–622. [CrossRef]
20. Bobermin, L.D.; Souza, D.; Gonçalves, C.-A.; Quincozes-Santos, A. Resveratrol prevents ammonia-induced mitochondrial dysfunction and cellular redox imbalance in C6 astroglial cells. *Nutr. Neurosci.* **2017**, *21*, 276–285. [CrossRef]
21. Wang, N.; Han, Q.; Wang, G.; Ma, W.-P.; Wang, J.; Wu, W.-X.; Guo, Y.; Liu, L.; Jiang, X.-Y.; Xie, X.-L.; et al. Resveratrol Protects Oxidative Stress-Induced Intestinal Epithelial Barrier Dysfunction by Upregulating Heme Oxygenase-1 Expression. *Dig. Dis. Sci.* **2016**, *61*, 2522–2534. [CrossRef]
22. Ginés, C.; Cuesta, S.; Kireev, R.; García, C.; Rancan, L.; Paredes, S.D.; Vara, E.; Tresguerres, J.A. Protective effect of resveratrol against inflammation, oxidative stress and apoptosis in pancreas of aged SAMP8 mice. *Exp. Gerontol.* **2017**, *90*, 61–70. [CrossRef] [PubMed]
23. Hussein, M.M.; Mahfouz, M.K. Effect of resveratrol and rosuvastatin on experimental diabetic nephropathy in rats. *Biomed. Pharmacother.* **2016**, *82*, 685–692. [CrossRef]
24. Zhang, T.; Chi, Y.; Kang, Y.; Lu, H.; Niu, H.; Liu, W.; Li, Y. Resveratrol ameliorates podocyte damage in diabetic mice via SIRT1/PGC-1α mediated attenuation of mitochondrial oxidative stress. *J. Cell. Physiol.* **2019**, *234*, 5033–5043. [CrossRef] [PubMed]
25. Khazaei, M.; Karimi, J.; Sheikh, N.; Goodarzi, M.T.; Saidijam, M.; Khodadadi, I.; Moridi, H. Effects of Resveratrol on Receptor for Advanced Glycation End Products (RAGE) Expression and Oxidative Stress in the Liver of Rats with Type 2 Diabetes. *Phytotherapy Res.* **2016**, *30*, 66–71. [CrossRef]
26. Fu, S.; Lv, R.; Wang, L.; Hou, H.; Liu, H.; Shao, S. Resveratrol, an antioxidant, protects spinal cord injury in rats by suppressing MAPK pathway. *Saudi J. Biol. Sci.* **2018**, *25*, 259–266. [CrossRef] [PubMed]
27. Corrêa, M.G.; Absy, S.; Tenenbaum, H.; Ribeiro, F.V.; Cirano, F.R.; Casati, M.Z.; Pimentel, S.P. Resveratrol attenuates oxidative stress during experimental periodontitis in rats exposed to cigarette smoke inhalation. *J. Periodontal Res.* **2018**, *54*, 225–232. [CrossRef] [PubMed]
28. Lin, Y.-T.; Wu, Y.-C.; Sun, G.-C.; Ho, C.-Y.; Wong, T.-Y.; Lin, C.-H.; Chen, H.-H.; Yeh, T.-C.; Li, C.-J.; Tseng, C.-J.; et al. Effect of Resveratrol on Reactive Oxygen Species-Induced Cognitive Impairment in Rats with Angiotensin II-Induced Early Alzheimer's Disease. *J. Clin. Med.* **2018**, *7*, 329. [CrossRef]
29. Wang, X.-L.; Li, T.; Li, J.-H.; Miao, S.-Y.; Xiao, X.-Z. The Effects of Resveratrol on Inflammation and Oxidative Stress in a Rat Model of Chronic Obstructive Pulmonary Disease. *Molecules* **2017**, *22*, 1529. [CrossRef]
30. Hashim, S.N.; Schwarz, L.J.; Boysen, R.I.; Yang, Y.; Danylec, B.; Hearn, M.T. Rapid solid-phase extraction and analysis of resveratrol and other polyphenols in red wine. *J. Chromatogr. A* **2013**, *1313*, 284–290. [CrossRef] [PubMed]
31. Riccio, B.V.F.; Spósito, L.; Carvalho, G.C.; Ferrari, P.C.; Chorilli, M. Resveratrol isoforms and conjugates: A review from biosynthesis in plants to elimination from the human body. *Arch. Der Pharm.* **2020**, *353*, 353. [CrossRef] [PubMed]
32. Orallo, F. Comparative Studies of the Antioxidant Effects of Cis- and Trans- Resveratrol. *Curr. Med. Chem.* **2006**, *13*, 87–98. [CrossRef] [PubMed]
33. Chen, F.; Bai, Q.; Wang, Q.; Chen, S.; Ma, X.; Cai, C.; Wang, D.; Waqas, A.; Gong, P. Stereoselective Pharmacokinetics and Chiral Inversions of Some Chiral Hydroxy Group Drugs. *Curr. Pharm. Biotechnol.* **2020**, *21*, 1632–1644. [CrossRef] [PubMed]
34. Skalova, L.; Szotakova, B. Chiral Inversion of Drugs: Coincidence or Principle? *Curr. Drug Metab.* **2004**, *5*, 517–533. [CrossRef]
35. Luo, S.; Yin, J.; Peng, Y.; Xie, J.; Wu, H.; He, D.; Li, X.; Cheng, G. Glutathione is Involved in Detoxification of Peroxide and Root Nodule Symbiosis of Mesorhizobium huakuii. *Curr. Microbiol.* **2020**, *77*, 1–10. [CrossRef]
36. Maietti, A.; Menziani, E.; Mazzotta, D. Determinazione del Resveratrolo in Vini del Commercio. In Proceedings of the Atti del IV Congresso di Chimica degli Alimenti, Ferrara, Italy, 28–30 June 2000; pp. 276–281.
37. Goegan, P.; Johnson, G.; Vincent, R. Effects of serum protein and colloid on the alamarBlue assay in cell cultures. *Toxicol. Vitr.* **1995**, *9*, 257–266. [CrossRef]

38. Ahmed, S.A.; Gogal, R.M.; Walsh, J.E. A new rapid and simple non-radioactive assay to monitor and determine the proliferation of lymphocytes: An alternative to [^{3}H]thymidine incorporation assay. *J. Immunol. Methods* **1994**, *170*, 211–224. [CrossRef]
39. Chen, X.; He, H.; Wang, G.; Yang, B.; Ren, W.; Ma, L.; Yu, Q. Stereospecific determination ofcis- andtrans-resveratrol in rat plasma by HPLC: Application to pharmacokinetic studies. *Biomed. Chromatogr.* **2007**, *21*, 257–265. [CrossRef]
40. Brandolini, V.; Maietti, A.; Tedeschi, P.; Durini, E.; Vertuani, A.S.; Manfredini, S. Capillary Electrophoresis Determination, Synthesis, and Stability of Resveratrol and Related 3-O-β-d-Glucopyranosides. *J. Agric. Food Chem.* **2002**, *50*, 7407–7411. [CrossRef]
41. Bergoglio, V.; Larcher, F.; Chevallier-Lagente, O.; Bernheim, A.; Danos, O.; Sarasin, A.; del Rio, M.; Magnaldo, T. Safe Selection of Genetically Manipulated Human Primary Keratinocytes with Very High Growth Potential Using CD24. *Mol. Ther.* **2007**, *15*, 2186–2193. [CrossRef]
42. Cinelli, R.A.G.; Ferrari, A.; Pellegrini, V.; Tyagi, M.; Giacca, M.; Beltram, F. The Enhanced Green Fluorescent Protein as a Tool for the Analysis of Protein Dynamics and Localization: Local Fluorescence Study at the Single-Molecule Level. *Photochem. Photobiol.* **2000**, *71*, 771. [CrossRef]
43. Van Overstraeten-Schlögel, N.; Ho-Shim, Y.; Tevel, V.; Bontems, S.; Dubois, P.; Raes, M. Transfection of Immortalized Keratinocytes by Low Toxic Poly(2-(Dimethylamino)Ethyl Methacrylate)-Based Polymers. *J. Biomater. Sci. Polym. Ed.* **2012**, *23*, 739–761. [CrossRef] [PubMed]
44. Liu, Y.; Chan, F.; Sun, H.; Yan, J.; Fan, D.; Zhao, D.; An, J.; Zhou, D. Resveratrol protects human keratinocytes HaCaT cells from UVA-induced oxidative stress damage by downregulating Keap1 expression. *Eur. J. Pharmacol.* **2011**, *650*, 130–137. [CrossRef] [PubMed]
45. Moldzio, R.; Radad, K.; Krewenka, C.; Kranner, B.; Duvigneau, J.C.; Rausch, W.-D. Protective effects of resveratrol on glutamate-induced damages in murine brain cultures. *J. Neural Transm.* **2013**, *120*, 1271–1280. [CrossRef] [PubMed]
46. Belguendouz, L.; Fremont, L.; Linard, A. Resveratrol inhibits metal ion-dependent and independent peroxidation of porcine low-density lipoproteins. *Biochem. Pharmacol.* **1997**, *53*, 1347–1355. [CrossRef]
47. Yáñez, M.; Fraiz, N.; Cano, E.; Orallo, F. Inhibitory effects of cis- and trans-resveratrol on noradrenaline and 5-hydroxytryptamine uptake and on monoamine oxidase activity. *Biochem. Biophys. Res. Commun.* **2006**, *344*, 688–695. [CrossRef] [PubMed]
48. Leiro, J.; Álvarez, E.; Arranz, J.A.; Laguna, R.; Uriarte, E.; Orallo, F. Effects of cis -resveratrol on inflammatory murine macrophages: Antioxidant activity and down-regulation of inflammatory genes. *J. Leukoc. Biol.* **2004**, *75*, 1156–1165. [CrossRef]
49. Rius, C.; Abu-Taha, M.; Hermenegildo, C.; Piqueras, L.; Cerda-Nicolas, J.-M.; Issekutz, A.C.; Estañ, L.; Cortijo, J.; Morcillo, E.J.; Orallo, F.; et al. Trans- but Not Cis-Resveratrol Impairs Angiotensin-II–Mediated Vascular Inflammation through Inhibition of NF-κB Activation and Peroxisome Proliferator-Activated Receptor-γ Upregulation. *J. Immunol.* **2010**, *185*, 3718–3727. [CrossRef]
50. Campos-Toimil, M.; Elíes, J.; Álvarez, E.; Verde, I.; Orallo, F. Effects of trans- and cis-resveratrol on Ca2+ handling in A7r5 vascular myocytes. *Eur. J. Pharmacol.* **2007**, *577*, 91–99. [CrossRef]
51. VanAntwerp, I.R.; Phelps, L.E.; Peuler, J.D.; Kopf, P.G. Effects of trans- versus cis-resveratrol on adrenergic contractions of the rat tail artery and role of endothelium. *Physiol. Rep.* **2020**, *8*, e14666. [CrossRef]
52. Fabbrocini, G.; Kisslinger, A.; Iannelli, P.; Vitale, N.; Procaccini, C.; Sparaneo, G.; Chieffi, P.; Ayala, F.; Mancini, F.P.; Tramontano, D. Resveratrol regulates p66Shc activation in HaCaT cells. *Exp. Dermatol.* **2010**, *19*, 895–903. [CrossRef]
53. Wen, S.; Zhang, J.; Yang, B.; Elias, P.M.; Man, M.-Q. Role of Resveratrol in Regulating Cutaneous Functions. *Evidence-Based Complement. Altern. Med.* **2020**, *2020*, 2416837. [CrossRef]
54. Olas, B.; Nowak, P.; Ponczek, M.; Wachowicz, B. Resveratrol, a natural phenolic compound may reduce carbonylation proteins induced by peroxynitrite in blood platelets. *Gen. Physiol. Biophys.* **2006**, *25*, 215–222. [PubMed]
55. Liu, Y.; Xiong, W.; Wang, C.-W.; Shi, J.-P.; Shi, Z.-Q.; Zhou, J.-D. Resveratrol promotes skin wound healing by regulating the miR-212/CASP8 axis. *Lab. Investig.* **2021**, *101*, 1363–1370. [CrossRef]
56. Bastianetto, S.; Dumont, Y.; Duranton, A.; Vercauteren, F.; Breton, L.; Quirion, R. Protective Action of Resveratrol in Human Skin: Possible Involvement of Specific Receptor Binding Sites. *PLoS ONE* **2010**, *5*, e12935. [CrossRef] [PubMed]
57. Iuga, C.; Alvarez-Idaboy, J.R.; Russo, N. Antioxidant Activity of trans-Resveratrol toward Hydroxyl and Hydroperoxyl Radicals: A Quantum Chemical and Computational Kinetics Study. *J. Org. Chem.* **2012**, *77*, 3868–3877. [CrossRef] [PubMed]
58. Mehta, J.; Rayalam, S.; Wang, X. Cytoprotective Effects of Natural Compounds against Oxidative Stress. *Antioxidants* **2018**, *7*, 147. [CrossRef]

antioxidants

MDPI

Article

Genistein Promotes Anti-Heat Stress and Antioxidant Effects via the Coordinated Regulation of IIS, HSP, MAPK, DR, and Mitochondrial Pathways in *Caenorhabditis elegans*

Sai-Ya Zhang †, Zi-Chen Qin †, Yi-Yang Sun, Yu-Si Chen, Wen-Bo Chen, Hong-Gang Wang, Di An, Dan Sun and Yan-Qiang Liu *

College of Life Sciences, Nankai University, Tianjin 300071, China

* Correspondence: liuyanq@nankai.edu.cn or liuyanq2@126.com; Tel.: +86-22-23508378

† These authors contributed equally to this work.

Abstract: To determine the anti-heat stress and antioxidant effects of genistein and the underlying mechanisms, lipofuscin, reactive oxygen species (ROS), and survival under stress were first detected in *Caenorhabditis elegans (C. elegans)*; then the localization and quantification of the fluorescent protein was determined by detecting the fluorescently labeled protein mutant strain; in addition, the aging-related mRNAs were detected by applying real-time fluorescent quantitative PCR in *C. elegans*. The results indicate that genistein substantially extended the lifespan of *C. elegans* under oxidative stress and heat conditions; and remarkably reduced the accumulation of lipofuscin in *C. elegans* under hydrogen peroxide (H_2O_2) and 35 °C stress conditions; in addition, it reduced the generation of ROS caused by H_2O_2 and upregulated the expression of *daf-16*, *ctl-1*, *hsf-1*, *hsp-16.2*, *sip-1*, *sek-1*, *pmk-1*, and *eat-2*, whereas it downregulated the expression of *age-1* and *daf-2* in *C. elegans*; similarly, it upregulated the expression of *daf-16*, *sod-3*, *ctl-1*, *hsf-1*, *hsp-16.2*, *sip-1*, *sek-1*, *pmk-1*, *jnk-1 skn-1*, and *eat-2*, whereas it downregulated the expression of *age-1*, *daf-2*, *gst-4*, and *hsp-12.6* in *C. elegans* at 35 °C; moreover, it increased the accumulation of HSP-16.2 and SKN-1 proteins in nematodes under 35 °C and H_2O_2 conditions; however, it failed to prolong the survival time in the deleted mutant MQ130 nematodes under 35 °C and H_2O_2 conditions. These results suggest that genistein promote anti-heat stress and antioxidant effects in *C. elegans* via insulin/-insulin-like growth factor signaling (IIS), heat shock protein (HSP), mitogen-activated protein kinase (MAPK), dietary restriction (DR), and mitochondrial pathways.

Keywords: genistein; anti-heat stress; antioxidant; *Caenorhabditis elegans*

Citation: Zhang, S.-Y.; Qin, Z.-C.; Sun, Y.-Y.; Chen, Y.-S.; Chen, W.-B.; Wang, H.-G.; An, D.; Sun, D.; Liu, Y.-Q. Genistein Promotes Anti-Heat Stress and Antioxidant Effects via the Coordinated Regulation of IIS, HSP, MAPK, DR, and Mitochondrial Pathways in *Caenorhabditis elegans*. *Antioxidants* **2023**, *12*, 125. https://doi.org/10.3390/antiox12010125

Academic Editor: Stanley Omaye

Received: 27 October 2022
Revised: 26 December 2022
Accepted: 27 December 2022
Published: 4 January 2023

1. Introduction

With the increase in aging population and weak environmental resistance, the incidence of age-related health disorders, including degenerative diseases, exponentially increased. The aging process is involved in many variables, including oxidative stress and thermotolerance [1–4]. The rise of anti-aging medicine puts forward a new type of health concept known as "healthy aging" [5]. Anti-oxidative and anti-heat stress may be effective methods for "healthy aging" [6,7]. Flavonoids [8], tannins, and other polyphenols [9] showed strong antioxidant activity. Genistein is a plant secondary metabolite flavonoid, widely found in legumes such as soybeans and has a variety of physiological effects, such as anti-inflammatory and antioxidant, anti-tumor [10], anti-cardiovascular disease, and anti-osteoporosis effects [11]. Because its structure is similar to that of 17β-estradiol, it can bind to estrogen receptors, it is often called a phytoestrogen, and it has a marked estradiol-related effect [12–14]. Genistein is produced by intestinal microbial enzymatic hydrolysis of soy products and can bind and trans-activate all three subtypes of peroxisome proliferator-activated receptors (α, δ, and γ) [15]; thus, it produces wide physiological and biochemical effects, including delaying aging and increasing life expectancy.

C. elegans has a short life cycle and is easy to culture; therefore, it is helpful for studying the relative issues of aging and anti-aging. Because longevity-associated signaling pathways are highly conserved, some similarity might be detected between these pathways in *C. elegans* and higher mammals, including humans [16]. It is estimated that more than 83% of the proteins in the proteome of *C. elegans* have homologs in humans [17]. Existing research shows that multiple signaling pathways, such as the mitochondrial signaling pathway, insulin/insulin-like growth factor signaling (IIS), heat shock transcription factor (HSF-1), mitogen-activated protein kinase (MAPK), dietary restriction (DR) pathways, and mitochondrial pathways, can regulate aging and anti-stress ability in *C. elegans* [17]. The IIS pathway regulates the lifespan of nematodes and mammals by regulating the conserved transcription factor DAF-16 [18]. The HSF-1 pathway prolongs lifespan by activating the expression of heat shock genes, while MAPK upregulates the longevity transcription factor (SKN-1) to combat stress-induced senescence [19,20]. Diet restriction (DR) prolongs the lifespan of *C. elegans*, depending on *eat-2* [21]. Meanwhile, CLK-1 acts in the nucleus to control the mitochondrial stress response [22]. These transcription factors are critical for stress resistance and longevity [23]. Previous studies reported that flavonoids from *Scutellariae Barbatae Herba* delay the aging of *C. elegans* through IIS [24]. A dihydroflavonoid, naringin extends the lifespan of *C. elegans* via the forkhead box (FOXO) transcription factor-16 (DAF-16) pathway [25], and proanthocyanidins alleviate β-amyloid peptide-induced toxicity by improving proteostasis through the homeostasis network regulator HSF-1 pathway in *C. elegans* [26]. In addition, genistein increasing the stress tolerance of a nematode is involved in superoxide dismutase (SOD-3) and heat shock protein (HSP-16.2) based on the experiment of transgenic strains [27]. Thus, it can be concluded that flavonoids have anti-aging effects.

Genistein is thought to have anti-stress effects and increase life expectancy; however, the underlying mechanism is still unclear. In this study, we first evaluated the effects of genistein on heat stress and oxidative stress in *C. elegans*, based on the change in ROS, lipofuscin, and survival rate. Then, we observed the nuclear localization of the aging-related transcription factor, DAF-16, and assessed the aging-related proteins, *daf-2* and HSP-16.2, and the transient expression changes of aging-related genes under heat stress and oxidative stress. It will provide mechanistic insights into the potential ability of genistein reducing stress and increasing life expectancy.

2. Materials and Methods

2.1. Materials

Peptone, agar, and hypochlorite solutions were obtained from Beijing Soleibao Technology Co., Ltd. (Beijing, China) Genistein (99%, Catalog No. 253493) was purchased from Beijing Bailingwei Technology Co., Ltd. (Beijing, China) Yeast extract, sodium hydroxide, glycerol, isopropanol, anhydrous calcium chloride, anhydrous magnesium sulfate, sodium chloride, sodium dihydrogen phosphate, potassium dihydrogen phosphate, dimethyl sulfoxide, absolute ethanol, tryptone, and chloroform were obtained from Tianjin Fengchuan Chemical Reagent Technology Co., Ltd. (Tianjin, China) Cholesterol was purchased from the Shanghai Aladdin Biochemical Technology Co., Ltd. (Shanghai, China) Levamisole hydrochloride was purchased from Nanjing Dulai Biotechnology Co., Ltd. (Nanjing, China) DNase/RNase-free water was obtained from the DEPC Department Beijing Solebold Technology Co., Ltd. (Beijing, China) RNAiso Plus and PrimeScript™ 1st Strand cDNA synthesis kits were purchased from Takara, Japan. SuperReal PreMix Plus was purchased from Beijing Tiangen Biochemical Technology Co., Ltd. (Beijing, China) Phenylmethanesulfonylfluoride (PMSF), ROS kit, and superoxide dismutase (SOD) kit were purchased from Tianjin Dingguo Technology Co., Ltd. (Tianjin, China).

2.2. *C. elegans* Strains and Maintenance

Bristol N2 (WT), EU1 [*skn-1*(zu67)], TJ375 (gpIs1 [*hsp-16.2*p::GFP]), TJ356 (zIs356 [P*daf-16*::*daf-16*a/b-gfp; rol-6]), MQ130[*clk-1*(qm30)], and LG333 [(zu135)*skn-1*::GFP] strains were provided by the Caenorhabditis Genetics Center, University of Minnesota, USA. They were maintained at 20 °C on a solid nematode growth medium (NGM) seeded with *Escherichia coli* OP50 (uracil auxotroph). OP50 (200 μL) was dropped on the center of 70 mm NGM plates, which were allowed to dry overnight and grow into a suitable colony.

2.3. Synchronization of C. elegans

Nematodes in the oviposition period, which contains a large number of hermaphrodite nematodes, were chosen and washed with M9 buffer. The eluent was collected and centrifuged at 400× *g* for 2 min, and the process was repeated three times. Then, lysis buffer (0.3 mL 4% sodium hypochlorite and 0.1 mL 5 M NaOH) was added to 0.6 mL sample solution, and the lysed liquid was centrifuged at 400× *g* for 2 min. The pellet was washed three times with M9 buffer and transferred to solid NGM medium for 48 h at 20 °C, and synchronous L4 (the fourth-stage larvae of nematodes) stage larvae were obtained.

2.4. Oxidative Stress and Heat Stress Test

The synchronized nematodes were collected and randomly divided into a blank group and an experimental group containing 200 μM genistein [the concentration was based on the result of previous studies in our laboratory [28]; and genistein was first dissolved in a little dimethyl sulfoxide, then was diluted to the required solvent with distilled water], with three parallel plates in each group, 30 per plate. After culturing for 72 h to the stage L4 at 20 °C, each group of nematodes was transferred to a new solid NGM medium for the assay of oxidative stress and heat stress. For the oxidative stress assay, 10 μL of 30% H_2O_2 was added to every 10 mL of NGM medium, and the number of surviving nematodes was recorded every hour until all the nematodes died [29]. For the heat stress assay, the nematodes of each group were transferred to an incubator at 35 °C for culture, and the number of surviving nematodes was recorded every hour until all the nematodes died [30]. The nematodes were considered dead when they did not respond to a gentle touch with a platinum wire on their bodies. Survival curves were drawn based on the time of nematode survival and death. These determinations were performed in three independent experiments, with each experimental group including at least 90 nematodes.

2.5. Determination of Lipofuscin Accumulation in C. elegans

The synchronized nematodes were collected and randomly divided into a blank group and an experimental group containing 200 μM genistein, with three parallel plates in each group, 30 per plate. Then, the fluorescence intensity of lipofuscin in *C. elegans* was observed on days number 5, 11, and 17 (from the time the eggs hatched). Then, the fluorescence intensity of lipofuscin in nematodes treated with H_2O_2 and 35 °C for 5 days was observed and recorded using a Leica DFC420 fluorescence microscope (excitation wavelength 340–380, emission wavelength 425 nm). The nematodes were anesthetized using levamisole hydrochloride (0.25 M) to facilitate microscopic photography. On the data processing, ImageJ software was used to open the image and extract a single channel (image–color–split channels); in the blue channel, the threshold value was adjusted, the appropriate region was selected, and the whole worm was quantified. The threshold value of the control group and genistein group of the same batch should be consistent. This experiment was performed three times, and the number of images in each experimental group was more than six.

2.6. Determination of ROS in C. elegans

Synchronized nematode eggs were placed on NGM plates with or without genistein. Nematodes were cultivated for 72 h to the stage L4 at 20 °C and then washed three times with M9 buffer. The nematodes were then transferred to 180 μL of M9 buffer containing 20 μL of 10 mM 2′,7′-dichlorofluorescein diacetate. The mixed liquor was incubated in the dark at 25 °C for 30 min. Then, the mixture was centrifuged at 400× *g* for 2 min. The nematode pellet was washed three times with M9 buffer and detected with a fluorescence intensity microplate reader at excitation and emission wavelengths of 485 nm and 530 nm, respectively. This experiment was performed three times with at least 500 nematodes per group.

2.7. Survival Determination of Mutant Strains of EU1 [skn-1(zu67)] and MQ130[clk-1(qm30)] under Oxidative and Heat Stress Conditions

Synchronized EU1 and MQ130 nematodes were collected and randomly divided into a blank group, and an experimental group with 200 μM genistein, with three parallel plates in each group, 30 per plate. After culturing for 72 h to the stage L4 at 20 °C, each group of mutant nematodes was transferred to a new solid NGM medium for the assay of oxidative stress and heat stress. For the oxidative stress assay, 30% H_2O_2 (10 μL) was added to every 10 mL of NGM medium (the final concentration 0.03% of H_2O_2), and the number of surviving EU1 and MQ130 nematodes was recorded every hour until all the nematodes died. For the heat stress assay, the EU1 nematodes and MQ130 nematodes of each group were transferred to an incubator at 35 °C for culture, and the number of surviving EU1 and MQ130 nematodes was recorded every hour until all the nematodes died. These determinations were performed in three independent experiments, with each experimental group including at least 90 nematodes.

2.8. Observation of the Nuclear Translocation of daf-16::GFP

The nuclear entry of DAF-16 was analyzed based on the accumulation of green fluorescence in the strain TJ356 nematodes. Synchronized nematode eggs were placed on NGM plates with or without genistein. Nematodes were cultivated to the L4 stage at 20 °C, then they were transferred to 30% H_2O_2 (the final concentration 0.03% of H_2O_2) and 35 °C for incubation for one hour. Next, the nematodes were anesthetized with levamisole hydrochloride (0.25 M), observed, and photographed under a fluorescence microscope (excitation wavelength 450–490 nm, emission wavelength 510 nm). This experiment was performed three times, and the number of shots in each experimental group was not less than six.

2.9. Quantification of hsp-16.2::GFP and skn-1::GFP Expression

Transgenic strains TJ375 (gpIs1 [*hsp-16.2*p::GFP]) and LG333 [(zu135)*skn-1*::GFP] were used for the analysis of *hsp-16.2*::GFP and *skn-1*::GFP expression. Synchronized nematode eggs were placed on NGM plates with or without genistein. Nematodes were cultivated to the L4 stage at 20 °C, then they were transferred to 30% H_2O_2 and 35 °C for incubation for one hour. Then they were anesthetized with levamisole hydrochloride (0.25 M). Finally, nematodes were observed and photographed under a fluorescence microscope (excitation wavelength 450–490 nm, emission wavelength 510 nm). On the data processing: ImageJ software was used to open the image and extract a single channel (image–color–split channels); in the green channel, the threshold value was adjusted, the appropriate region was selected, and the whole worm was quantified. The threshold value of the control group and genistein group of the same batch should be consistent. This experiment was performed three times, and the number of shots in each experimental group was not less than six.

2.10. Measurement of SOD Enzyme Activity

Synchronized nematodes were collected and randomly divided into blank and experimental groups with 200 μM genistein. After culturing for 72 h to the stage L4 at 20 °C, each group of mutant nematodes was ground with the lysis solution, and the activity was determined applying the SOD ELISA kit. The 560 nm absorption wavelength was used to detect the light absorption intensity using a microplate reader. This experiment was performed three times with at least 500 nematodes per group.

2.11. Extraction of C. elegans RNA

Synchronized nematode eggs were transferred to plates containing genistein or control plates for hatching. The hatched nematodes were grown to L4 at 20 °C and then collected by washing with deionized water. The collected nematode sample was added to 1 mL of RNAiso Plus (Takara, Kusatsu, Japan) and placed in a –80 °C refrigerator overnight. Then, the sample was slowly thawed in a 37 °C water bath, repeatedly frozen, and melted 20 times in liquid nitrogen and a 37 °C water bath, respectively. This was followed by the addition of 200 μL of chloroform, vigorous shaking for 30 s, resting at 20 °C for 10 min, and centrifugation at 3200 g for 10 min at 4 °C. The RNA containing supernatant was carefully transferred to a new vial, and an equal volume of isopropanol was added and mixed well. The mixture was incubated at 25 °C for 10 min and then centrifuged at 12,000× *g* for 10 min at 4 °C. The supernatant was carefully discarded, and 1 mL of 75% ethanol was added to the sediment, gently mixed, and centrifuged at 12,000× *g* for 5 min at 4 °C. The supernatant was discarded, and the sediment sample was air-dried for 2–5 min, followed by the addition of approximately 30 μL of DEPC water to dissolve the precipitate, and the RNA solution was stored at −80 °C. A NanoDrop 2000 spectrophotometer (Thermo, Waltham, MA, USA) was used to determine the concentration of total RNA, and RNA purity was assessed using OD260/OD280 ratios. This experiment was performed three times with at least 500 nematodes per group.

2.12. mRNA Analysis of Aging-Related Genes Using Quantitative Polymerase Chain Reaction (qPCR)

Using the PrimeScript™ 1st Strand cDNA Synthesis Kit, cDNA was synthesized. The reverse transcription reaction conditions were as follows: 25 °C for 10 min, 42 °C for 60 min, and 70 °C for 15 min. Quantitative PCR (qPCR) was performed using the SuperReal PreMix Plus kit. The qPCR primers used are shown in Table 1. The qPCR conditions were as follows: 95 °C for 10 min, followed by 40 cycles of 10 s at 95 °C and 30 s at 60 °C. Gene expression data were analyzed using the comparative $2^{-\triangle\triangle Ct}$ method, with *act-1* as the reference gene [23,31]. This experiment was performed in triplicates.

Table 1. The primer sequence of the genes.

Genes	Forward Primer	Reverse Primer
daf-16(NM_001381205.1)	AGGAGTCGAAGCCGATTAAGAC	GGTAGTGGCATTGGCTTGAAG
daf-2(NM_065249.7)	TACTTGAATCGGGCGTCGTT	GACGACTTCAACAACCGCTG
age-1(NM_064061.6)	CTCCTGAACCGACTGCCAAT	AAATGCGAGTTCGGAGAGCA
gst-4(NM_069447.8)	CCCATTTTACAAGTCGATGG	CTTCCTCTGCAGTTTTTCCA
sod-3(NM_078363.9)	TGGCTAAGGATGGTGGAGAA	GCCTTGAACCGCAATAGTGAT
ctl-1 (NM_064578.6)	CGATACCGTACTCGTGATGAT	CCAAACAGCCACCCAAATCA
skn-1(NM_171345.6)	CGTCCAACCAACCACATCATCTC	ATCTTCCAATTCGGCTTTT
sir-2.1(NM_001268555.5)	AAATCTTCCCAGGACAGTTCGTA	ATGGGCAACACGCATAGCA
eat-2(NM_064558.6)	ACCATGGGGAATTTGCAACG	GGATTTGCGTGAGGGGTATGA
hsp-16.2(NM_001392482.1)	CTGCAGAATCTCTCCATCTGAGTC	AGATTCGAAGCAACTGCACC
hsf-1(NM_060630.7)	ATGTACGGCTTCCGAAAGATGA	TCTTGCCGATTGCTTTCTCTTAA
hsp-12.6(NM_069267.4)	TGGAGTTGTCAATGTCCTCG	GACTTCAATCTCTTTTGGGAGG
sip-1(NM_066915.5)	CGAGCACGGGTTCAGCAAGAG	CAGCGTGTCCAGCAGAAGTGTG
jnk-1(NM_001026099.6)	TGGAACCAGCCAATTCCCAA	TCACAACACTCTGCTCGCAT
nsy-1(NM_001383826.1)	AGCGGCTCGATCAACAAGAA	CCCATTCCACCGATATGCGA
sek-1(NM_076921.8)	CACTGTTTGGCGACGATGAG	ATTCCGTCCACGTTGCTGAT
pmk-1(NM_068964.7)	CCAAAAAATGACTCGCCGTGA	CTTTTGCAGTTGGACGACGA

2.13. Data Treatment and Statistical Analyses

Microsoft Excel (Microsoft Corporation, Redmond, WA, USA), JMP (SAS Institute, Cary, NC, USA), GraphPad Prism 8.0 (GraphPad, La Jolla, CA, USA), and ImageJ (NIH, Bethesda, MA, USA) were used for data treatment and statistical analyses. The experimental results are presented as mean ± standard error of the mean (SEM). Significant differences between the two groups were assessed using the Student's t-test. A log-rank test was performed for stress assays. Statistical significance was set at $p < 0.05$.

3. Results

3.1. Genistein Improves Stress Resistance of Wild-Type Nematodes Exposed to Oxidative and Heat Stress

The effect of genistein on the survival rate of the H_2O_2- and 35 °C-treated nematodes is shown in Figure 1 and Table 2. Treatment with 200 μM genistein increased the mean survival rate of nematodes treated with H_2O_2 and 35 °C by 56.7% and 76.7%, respectively ($p < 0.01$).

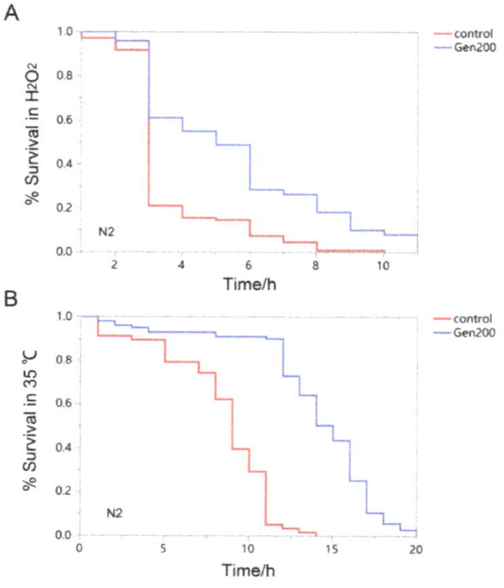

Figure 1. The effect of 200 μM genistein on the survival rate of *C. elegans* under stress. (**A**) The change in survival rate of *C. elegans* exposed to H_2O_2-induced oxidation in the genistein and control groups. (**B**) The change in survival rate of *C. elegans* exposed to heat shock at 35 °C in the genistein and control groups.

Table 2. The effect of 200 μM genistein on the lifespan of *C. elegans* under H_2O_2 and 35 °C stress.

Group	Number	Mean Lifespan (h)	Median Survival (Days)	Maximum Lifespan (Days)	*p* Value
Control(H_2O_2)	109	3.541 ± 0.15	3	10	–
Gen (H_2O_2)	102	5.531 ± 0.39	5	11	<0.01
Control (35 °C)	103	8.362 ± 0.41	9	14	–
Gen (35 °C)	107	14.02 ± 0.39	15	20	<0.01

$p < 0.01$, compared with Control(H_2O_2) and Control (35 °C), respectively.

3.2. Genistein Lowered Lipofuscin Accumulation in C. elegans under Control, Heat and Oxidative Conditions

The results of lipofuscin accumulation in *C. elegans* under control, H_2O_2, and 35 °C conditions are shown in Figure 2. Under control conditions, the accumulation of lipofuscin increased with an increase in the number of days, and the accumulation of lipofuscin in the nematodes on the 11th and 17th day was 1.39 and 2.12 times that on the 5th day (Figure 2A,B). Genistein significantly reduced the accumulation of lipofuscin in nematodes by 32.6% ($p < 0.05$) and 79.0% ($p < 0.01$) on days 11 and 17, respectively (Figure 2A,B). In addition, on day 5, genistein reduced the accumulation of lipofuscin by 52.5% and 44.4% in *C. elegans* under heat and oxidative conditions, respectively (Figure 2C,D) ($p < 0.01$).

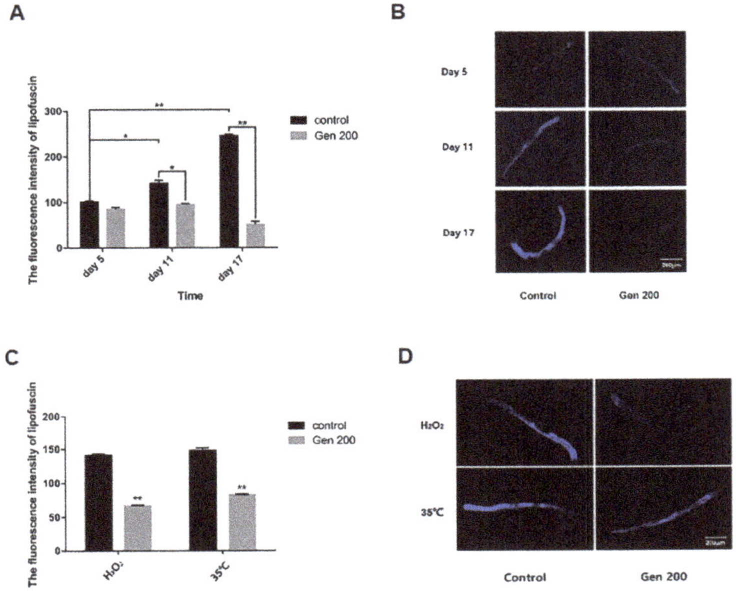

Figure 2. The effect of 200 μM genistein on lipofuscin accumulation in *C. elegans* under control, H_2O_2 and 35 °C conditions. (**A**) Relative fluorescence intensity of lipofuscin in *C. elegans* (% of control) on days 5, 7, and 11 under control conditions. (**B**) Photomicrographs of lipofuscin fluorescence in *C. elegans* on days 5, 7, and 11 under control conditions. (**C**) Relative fluorescence intensity of lipofuscin in *C. elegans* on day 5 in H_2O_2 and 35 °C conditions. (**D**) Photomicrograph of lipofuscin fluorescence in *C. elegans* on day 5 in H_2O_2 and 35 °C conditions (* $p < 0.05$; ** $p < 0.01$).

3.3. Genistein Lowered ROS Accumulation in H_2O_2-Treated Nematodes, and Increased the SOD Activity of Nematodes under Control, 35 °C and H_2O_2 Conditions

The effect of genistein on ROS accumulation and SOD activity of the differently treated nematodes are shown in Figure 3. Genistein reduced the accumulation of ROS by 47.9% in H_2O_2-treated nematodes ($p < 0.01$). However, genistein produced no obvious effect in nematodes treated at 35 °C (Figure 3A). Genistein increased SOD activity by 34.1% ($p < 0.05$), 67.5% ($p < 0.05$), and 117.4% ($p < 0.01$) under control, H_2O_2, and 35 °C conditions, respectively (Figure 3B).

3.4. Genistein Partially Improved the Survival Rate of EU1 [skn-1(zu67)] Mutants, but It Did Not Significantly Change the Survival Rate of MQ130[clk-1(qm30)] Mutant under Heat and Oxidative Conditions

From Figure 4, genistein increased the mean survival rate of *skn-1* (EU1) mutant strains by 93.4% under oxidative conditions (Figure 4A) ($p < 0.01$). However, genistein did not influence the survival rate curves of EU1 mutants under heat stress conditions (Figure 4B). Additionally, genistein did not significantly change the survival rate curves of the MQ130 mutants under heat and oxidative conditions (Figure 4C,D).

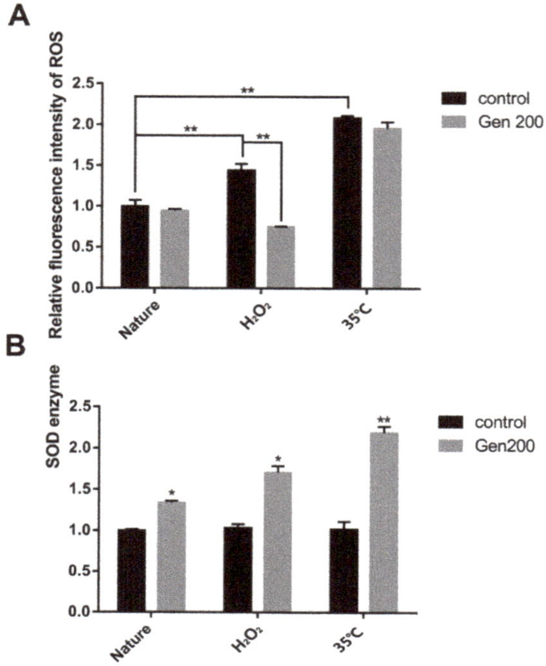

Figure 3. Under control, H_2O_2, and 35 °C conditions, 200 µM genistein influences ROS accumulation and the activity of superoxide dismutase (SOD) in the stage L4 *C. elegans*. (**A**) The effect of genistein on the accumulation of ROS in *C. elegans* under control, H_2O_2, and 35 °C conditions. (**B**) The effect of genistein on the activity of SOD in *C. elegans* under control, H_2O_2, and 35 °C conditions (* $p < 0.05$; ** $p < 0.01$).

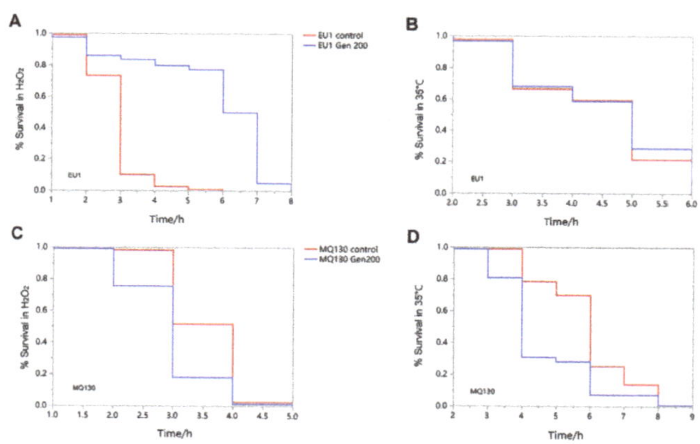

Figure 4. The effect of 200 µM genistein on the survival rate of EU1 and MQ130 mutant nematodes at L4 under H_2O_2 and 35 °C conditions. (**A**) The effect of genistein on the survival rate of EU1 mutant exposed to H_2O_2-induced oxidation. (**B**) The effect of genistein on the survival rate of EU1 mutant exposed to heat shock at 35 °C. (**C**) The effect of genistein on the survival rate of MQ130 mutant exposed to H_2O_2-induced oxidation. (**D**) The effect of genistein on the survival rate of MQ130 mutant exposed to heat shock at 35 °C.

3.5. Genistein Increased the Nuclear Translocation of DAF-16 Transcription Factor under 35 °C and H_2O_2 Conditions

The effects of genistein on the nuclear translocation of the DAF-16 transcription factor are shown in Figure 5. From Figure 5A, the DAF-16::GFP fluorescence of the genistein-treated group under H_2O_2 conditions was concentrated in the nucleus, while the DAF-16::GFP fluorescence of the control group was concentrated in the cytoplasm and diffusely distributed. Moreover, under 35 °C conditions, the DAF-16::GFP fluorescence of the genistein-treated group was concentrated in the nucleus, and the DAF-16::GFP fluorescence of the control group was concentrated and dispersed in the cytoplasm (Figure 5B).

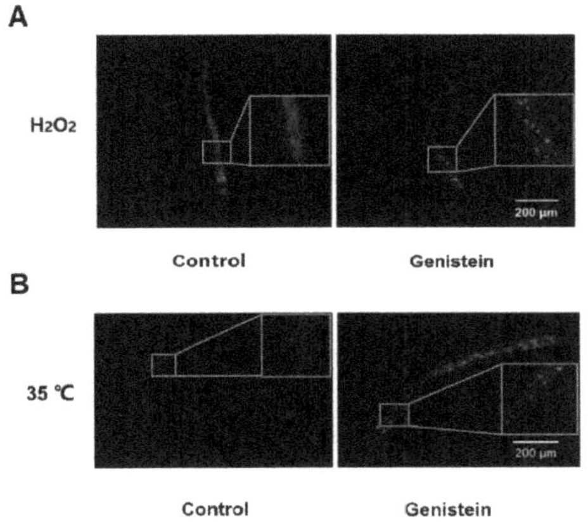

Figure 5. The effect of 200 µM genistein on the nuclear localization of DAF-16::GFP in the stage L4 *C. elegans* under H_2O_2 and 35 °C conditions for one hour. (**A**) The effect of genistein on the nuclear localization of DAF-16::GFP in *C. elegans* exposed to H_2O_2-induced oxidation. (**B**) The effect of genistein on the nuclear localization of DAF-16::GFP in *C. elegans* exposed to heat shock at 35 °C.

3.6. Genistein Enhanced the Accumulation of HSP-16.2 Protein in Nematodes under Control, 35 °C, and H_2O_2 Conditions

We assessed the accumulation of HSP-16.2 using the *C. elegans* strain TJ375 (HSP-16.2::GFP), which expresses the GFP-fused HSP-16.2 (Figure 6). From Figure 6B, under control, 35 °C, and H_2O_2 conditions, the green fluorescence of HSP-16.2 protein in the genistein-treated group was significantly higher than that in the control group. Based on Figure 6A, compared with the control group, the fluorescence intensity of HSP-16.2::GFP in the genistein-treated group was significantly enhanced by 41.8%, 42.5%, and 52.3% under natural, oxidative stress, and heat stress conditions, respectively ($p < 0.01$).

3.7. Genistein Enhanced the Accumulation of SKN-1 Protein in Nematodes under 35 °C and H_2O_2 Conditions

Compared with the control group, the intensity of the green fluorescence of SKN-1 protein in the genistein-treated group did not significantly change under control conditions (Figure 7A,B). From Figure 7B, under oxidative and heat stress conditions, the green fluorescence of SKN-1 protein in the genistein-treated group was significantly higher than that of the control group, and compared with the control group, the fluorescence intensity of the genistein-treated was increased by 100.2% and 122.7%, respectively (Figure 7A) ($p < 0.01$).

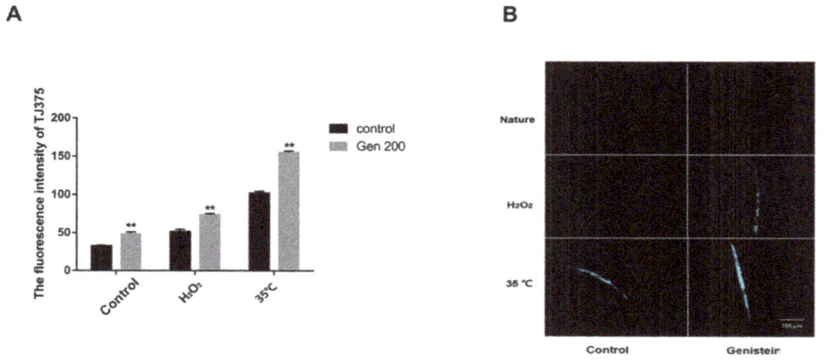

Figure 6. Under control, H_2O_2, and 35 °C conditions, 200 µM genistein influences the accumulation of HSP-16.2 protein in the stage L4 *C. elegans*. (**A**) The effect of genistein on the relative fluorescence intensity of HSP-16.2::GFP in *C. elegans*. (**B**) Micrograph of HSP-16.2::GFP fluorescence in *C. elegans* (** $p < 0.01$).

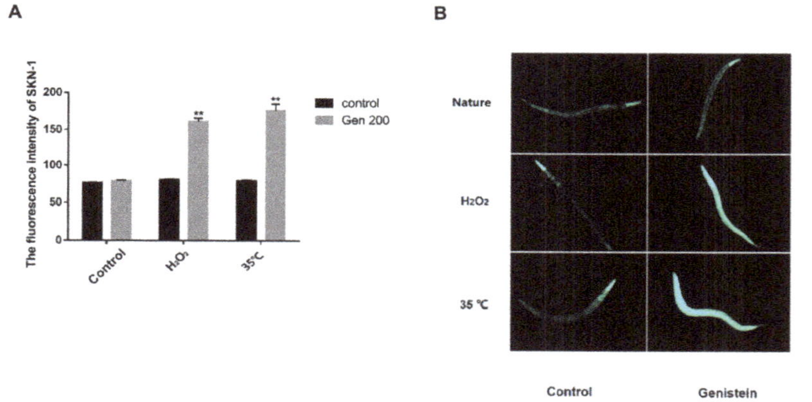

Figure 7. The accumulation of SKN-1 protein in the stage L4 *C. elegans* under control, H_2O_2, and 35 °C conditions is influenced by 200 µM genistein. (**A**) The effect of genistein on the relative fluorescence intensity of SKN-1::GFP in *C. elegans*. (**B**) Micrograph of SKN-1::GFP fluorescence in *C. elegans* (** $p < 0.01$).

3.8. Genistein Regulated the Expression of Messenger RNA (mRNA) in C. elegans under 35 °C and H_2O_2 Conditions

The results of the effect of genistein on the mRNA expression of daf-2, age-1, daf-16, sod-3, gst-4, ctl-1, hsf-1, hsp-16.2, hsp-12.6, sir-2.1, sek-1, skn-1, pmk-1, nsy-1, jnk-1, sip-1, and eat-2 in C. elegans at 35 °C and H_2O_2 conditions are shown in Figure 8. Under H_2O_2 conditions, genistein treatment upregulated the relative expression levels of daf-16, ctl-1, hsf-1, hsp-16.2, sip-1, sek-1, pmk-1, and eat-2 by 84.4%($p < 0.01$), 81.6%($p < 0.01$), 60.4% ($p < 0.05$), 86.6% ($p < 0.01$), 66.7% ($p < 0.01$), 81.8% ($p < 0.05$), 44.6% ($p < 0.01$), and 40.7% ($p < 0.05$), respectively, but downregulated the relative expression levels of daf-2 and age-1 by 37.8% ($p < 0.01$), and 44.9% ($p < 0.01$), respectively, and no significant effect was observed on sod-3, gst-4, hsp-12.6, nsy-1, jnk-1, skn-1, and sir-2.1 compared with the control. Similarly, genistein treatment upregulated the relative expression levels of daf-16, sod-3, ctl-1, hsf-1, hsp-16.2, sip-1, sek-1, pmk-1, jnk-1, skn-1, and eat-2 by 35.3% ($p < 0.05$), 84.1% ($p < 0.01$), 64.4% ($p < 0.01$), 33.7% ($p < 0.05$), 29.0% ($p < 0.05$), 77.7% ($p < 0.01$), 58.7% ($p < 0.01$), 97.7% ($p < 0.01$), 18.0% ($p < 0.05$), 182.3% ($p < 0.05$), and 177.2%($p < 0.05$), but downregulated the relative expression levels of daf-2, age-1, gst-4, and hsp-12.6 by 86.4%

($p < 0.01$), 39.9% ($p < 0.01$), 60.9% ($p < 0.01$), and 49.4% ($p < 0.01$), respectively, and no significant effect was observed on nsy-1 and sir-2.1 at 35 °C compared with the control.

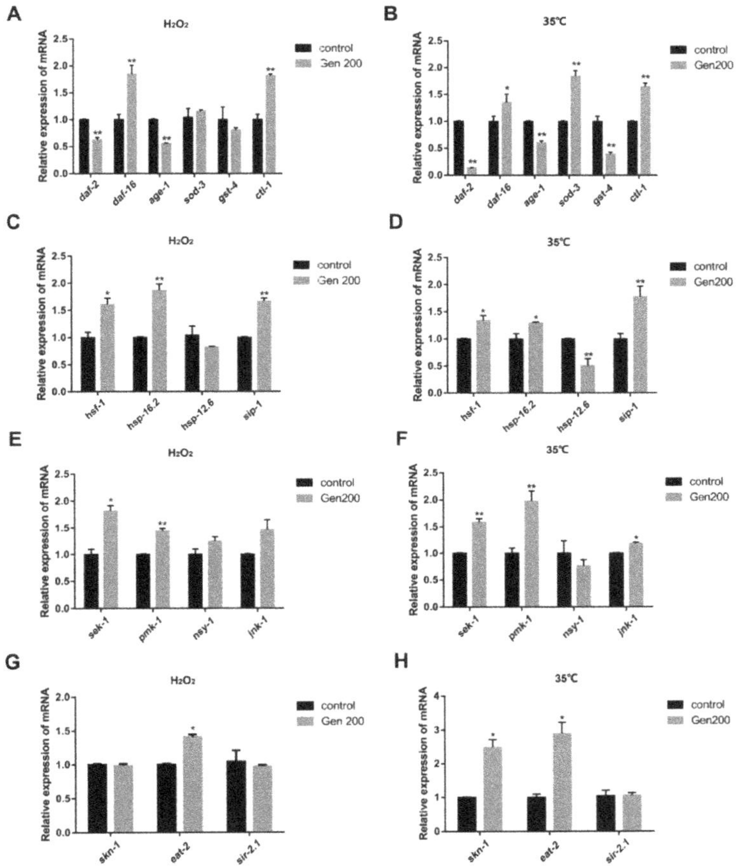

Figure 8. Under H_2O_2 and 35 °C conditions, 200 μM genistein influences the mRNA level of stress resistance genes in the stage L4 C. elegans. (**A**) The effect of genistein on the mRNA of *daf-16*, *sod-3*, *ctl-1*, *daf-2*, *age-1*, and *gst-4* in *C. elegans* under H_2O_2 conditions. (**B**) The effect of genistein on the mRNA of *daf-16*, *sod-3*, *ctl-1*, *daf-2*, *age-1*, and *gst-4* in *C. elegans* under 35 °C conditions. (**C**) The effect of genistein on the mRNA of *hsf-1*, *hsp-16.2*, *hsp-12.6*, and *sip-1* in *C. elegans* under H_2O_2 conditions. (**D**) The effect of genistein on the mRNA of *hsf-1*, *hsp-16.2*, *hsp-12.6*, and *sip-1* in *C. elegans* under 35 °C conditions. (**E**) The effect of genistein on the mRNA of *sek-1*, *pmk-1*, *nsy-1*, and *jnk-1* in *C. elegans* under H_2O_2 conditions. (**F**) The effect of genistein on the mRNA of *sek-1*, *pmk-1*, *nsy-1*, and *jnk-1* in *C. elegans* under 35 °C conditions. (**G**) The effect of genistein on the mRNA of *skn-1*, *eat-2*, and *sir-2.1* in *C. elegans* under H_2O_2 conditions. (**H**) The effect of genistein on the mRNA of *skn-1*, *eat-2*, and *sir-2.1* in *C. elegans* under 35 °C conditions (* $p < 0.05$; ** $p < 0.01$).

4. Discussion

Previous studies showed that genistein inhibits the aging of human umbilical vein endothelial cells [32] and attenuates the senescence of vascular smooth muscles [33]. The increase in lifespan was accompanied by anti-stress effects in *C. elegans* [34–36]. In this study, we evaluated whether genistein treatment improved heat stress and oxidative stress in *C. elegans*. The results show that genistein could protect the nematodes and prolong the average life and maximum life of nematodes treated with H_2O_2 at 35 °C. Genistein significantly decreased the accumulation of lipofuscin in *C. elegans* under H_2O_2 and 35 °C stress conditions. As for the quantitative determination of lipofuscin, it was pointed out in the literature that only detection of blue fluorescence changes at low levels before and after death, which may reflect the proportion of dead or nearly dead individuals in the sample; at the same time, blue fluorescence should be accepted as reliable in the current paper, and we will make more attempts including determining red autofluorescence and adding the number of worms [37–40].

The IIS pathway is an evolutionarily conserved mechanism that is involved in the longevity and metabolism of different species [41]. Several genes are involved in the IIS pathway; *daf-2* encodes a receptor tyrosine kinase homolog, and *age-1* encodes a phosphatidylinositol 3-kinase (PI3K) homolog, *daf-16* is an ortholog of human FOXO1, which encodes DAF-16, *sod-3* encodes a superoxide dismutase gene, *gst-4* encodes a glutathione transferase, and *ctl-1* encodes a catalase [18,42]. The initiation of the kinase cascade depends on the phosphorylation of the insulin/IGF-1 transmembrane receptor (IGFR) ortholog, the forkhead box (FOXO) transcription factor-2 (DAF-2), and then regulates its downstream signaling molecule, AGE-1 (phosphatidylinositol kinase) [43]. This signaling regulated the nuclear translocation of DAF-16 [44]. This further activated the expression of a large number of anti-stress and longevity genes, such as *sod-3*, *gst-4*, and *ctl-1*, which eventually led to the enhancement of the anti-aging ability of *C. elegans* [18,45]. Our results show that genistein treatment significantly increased the mRNA level of *sod-3* at 35 °C and increased the mRNA levels of *ctl-1* and SOD enzyme activity under H_2O_2 and 35 °C stress conditions in *C. elegans*. These results suggest that the effects of genistein, such as prolonging the lifespan and improving the anti-heat and antioxidant capacity, might result from the activation of SOD-3 and CTL-1 through the IIS pathway in *C. elegans*.

Some heat-related proteins can protect nematodes from stress [46,47]. The HSP protein family was used as a predictor of longevity in *C. elegans*, because it is closely related to heat tolerance and longevity [48,49]. The representatives of the HSP protein family were HSP-12.6, HSP-16.2,SIP-1 (the PDZ-domain containing protein), and HSF-1 [50]. HSP-12.6 is a heat shock protein located in the cytoplasm and is an ortholog of human HSPB2 (heat shock protein family B (small) member 2) [51]. HSP-16.2 is another heat shock protein located in the cytoplasm, which is an ortholog of human CRYAB (crystallin alpha B) [52]. SIP-1 is a member of the alpha-crystallin/HSP20 family and is involved in thermal defense and aging processes [53]. To be precise, these protein expressions interact with each other and are mainly regulated by HSF-1 [54], an ortholog of human HSF-2 (heat shock transcription factor 2), which protects the proteins from the damage caused by extrinsic environmental stress or intrinsic age-related deterioration in *C. elegans* [46,55–57]. Our results show that the resistance of nematodes and the mRNA levels of *hsp-16.2*, *sip-1*, and *hsf-1* remarkably increased after genistein treatment under H_2O_2 condition; and the mRNA levels of *hsp-16.2*, *sip-1*, and *hsf-1* remarkably increased after genistein treatment, while *hsp-12.6* was significantly downregulated at 35 °C. Moreover, genistein treatment significantly enhanced the expression of HSP-16.2 under control, H_2O_2, and 35 °C conditions; indicating that genistein, which improves resistance under H_2O_2 and 35 °C stress conditions, might regulate in the HSP pathway.

The JNK and p38/MAPK signaling pathways were also implicated in a variety of biological functions in *C. elegans*, including response to stress [58]. ROS reflects the level of oxidative free radicals in the cells and are involved in cell senescence and epigenetic regulation of senescence [59,60]. The above results show that genistein significantly reduced ROS production in the cells under oxidative stress, and thus improved the survival rate of nematodes under oxidative stress. Several genes are involved in the MAPK pathway. For example, *nsy-1* encodes a homolog of the human apoptosis signal-regulating kinase (ASK1, MAPKKK) and *jnk-1*, a JUN kinase, which is an ortholog of human MAPK10 and MAPK8 [61–63]; *sek-1*, an ortholog of human MAP2K6, has several functions, including MAPK binding activity, and *pmk-1* encodes a p38 MAPK homolog [64]. SKN-1 is the *C. elegans* ortholog of mammalian Nrf/CNC proteins and plays a role in a wide range of detoxification processes, as well as in immunity, oxidative stress defense, and metabolism [65]. Their regulatory mechanism might be mediated through *nsy-1*; it activates *jnk-1* and *sek-1*, and the latter further activates *pmk-1* [66]. Activation of the MAPK pathway leads to the activation of *skn-1*, which further regulates the expression of downstream antioxidant protein detoxification enzymes, such as the SOD enzyme [67,68]. This experiment showed that genistein elevated the expression of genes such as *sek-1*, *pmk-1*, *skn-1*, and *sod* in H_2O_2-treated nematodes. Similarly, genistein elevated the expression of genes such as *sek-1*, *pmk-1*, *jnk-1*, *skn-1*, *skn-1*, and *sod* at 35 °C. The interesting thing is, genistein could also increase SOD under control conditions. However, there was a mismatch between gene expression levels and enzyme activity. Studies show that the mRNA expression might not be synchronized with protein expression of the enzyme, and microRNA regulation in transcription, ubiquitination, and phosphorylation in protein activation might be an uncertain factor [69]. However, we will confirm this idea with more data or experiments. These results indicate that the effect of genistein on the survival rate of *C. elegans* under oxidative and heat stress conditions might be attributed to the MAPK pathway. On the other hand, genistein increased the survival time of the *skn-1* (EU1) mutant under oxidative conditions but did not increase survival time under heat stress conditions, which indicated that *skn-1* was essential for genistein to improve the survival of *C. elegans* under heat stress, but was not essential under oxidative stress conditions. These results indicate that the mechanism by which genistein adjusted the resistance of *C. elegans* to stress was not exactly the same under different stress conditions.

In addition, some studies showed that the DR pathway can induce longevity in *C. elegans*, and that the longevity effect depends on *eat-2* [52]; *eat-2* enabled acetylcholine-gated cation-selective channel activity and was involved in several processes in *C. elegans*, including feeding behavior. Additionally, *eat-2* is an ortholog of human CHRFAM7A and FAM7A. The results show that genistein increased the mRNA expression of *eat-2* under oxidative and heat stress conditions. These results indicate that the effects of genistein in improving the resistance of *C. elegans* might benefit from the DR pathway.

Furthermore, *sir-2.1*, enabled deacetylase activity, is an ortholog of human SIRT1 (sirtuin 1), which is involved in several processes, including determination of the adult lifespan, histone modification, and the intrinsic apoptotic signaling pathway in response to DNA damage in *C. elegans* [70]. Studies show that *sir-2.1* plays an important role in regulating the genes related to longevity in *C. elegans* [71]; *sir-2.1* and 14-3-3 act in parallel to the insulin-like pathway to activate DAF-16 and extend the lifespan [72]. However, no significant difference was observed in the mRNA expression of *sir-2.1* in *C. elegans* under H_2O_2 and 35 °C conditions after genistein treatment. One possible interpretation is that *sir-2.1* did not play a role in genistein-enhanced resistance in *C. elegans* under H_2O_2 and 35 °C conditions.

Lastly, the monooxygenase CLK-1, which is homologous to human COQ7, was previously reported to act in the nucleus to control the mitochondrial stress response and lifespan [73]. This pathway is conserved from *C. elegans* to humans, and regulates mitochondrial reactive oxygen metabolism and mitochondrial unfolded protein response [22]. We found that genistein failed to prolong the survival time of nematodes in the deletion mutant MQ130, which lacks *clk-1* under oxidative and heat stress conditions. This indicated that the mitochondrial pathway was essential for genistein-enhanced resistance in *C. elegans* under H_2O_2 and 35 °C conditions.

Previous data suggest that these pathways interact with each other [74]. For example, the expression of *hsp-16.2* is mainly regulated by HSF-1. However, it is also affected by the IIS pathway in *C. elegans* [75]. In addition, SKN-1, which regulates the MAPK pathway, is also an important direct target of IIS, and DAF-16 regulates its activity by reducing the nuclear localization of SKN-1 [76]. Our results suggest that genistein resistance to H_2O_2 and 35 °C is related to IIS, HSP, MAPK, DR, and mitochondrial signaling pathways, and is also involved in the interactions of these signaling pathways.

5. Conclusions

In summary, genistein improved resistance to oxidant and heat stress and increased the lifespan in *C. elegans* through the coordinated regulation of IIS, HSP, MAPK, DR, and mitochondrial signaling pathways. This points out the possible applications of genistein as an anti-heat, antioxidant, and anti-aging drug. However, more convincing data from in-depth experiments using mammalian models are required to extrapolate the biotransformation pathways of genistein in mammals, including humans.

Author Contributions: Conceptualization, Y.-Q.L. and S.-Y.Z.; methodology, S.-Y.Z.; software, Z.-C.Q.; validation, Z.-C.Q., Y.-Y.S., Y.-S.C. and W.-B.C.; formal analysis, S.-Y.Z.; investigation, S.-Y.Z.; resources, H.-G.W., D.A. and D.S.; data curation, S.-Y.Z.; writing—original draft preparation, S.-Y.Z.; writing—review and editing, Y.-Q.L.; S.-Y.Z. and Z.-C.Q.; visualization, S.-Y.Z. and Z.-C.Q.; supervision, Y.-Q.L.; project administration, Y.-Q.L.; funding acquisition, Y.-Q.L. All authors have read and agreed to the published version of the manuscript.

Funding: This work was supported by grants from the National Natural Science Foundation of China (31272317), the Natural Science Foundation of Tianjin City (20JCYBJC01370), the Fundamental Research Funds for the Central Universities of Nankai University (BE123081).

Institutional Review Board Statement: Not applicable.

Informed Consent Statement: Not applicable.

Data Availability Statement: No other data be available except the published.

Conflicts of Interest: The authors declare no conflict of interest.

Abbreviations

CAT, catalase; C. elegans, Caenorhabditis elegans; DNA, deoxyribonucleic acid; DR, dietary restriction pathway; FOXO1, Recombinant Forkhead Box Protein O1: HSF-1, heat shock factor-1; HSPB2, Recombinant Human Heat shock protein beta-2; HSPs, heat shock proteins; IGF-1, insulin-like growth factors-1; IIS, insulin-like growth factor1 signaling pathway; JNK, Jun kinase; MAPK, the generic mitogen-activated protein kinases signaling pathway; mRNA, messenger RNA; Nrf2, NF-E2-related factor 2; ROS, reactive oxygen species; and SOD, superoxide dismutase.

References

1. da Costa, J.P.; Vitorino, R.; Silva, G.M.; Vogel, C.; Duarte, A.C.; Rocha-Santos, T. A synopsis on aging-Theories, mechanisms and future prospects. *Ageing Res. Rev.* **2016**, *29*, 90–112. [CrossRef] [PubMed]
2. Wickens, A.P. Ageing and the free radical theory. *Respir. Physiol.* **2001**, *128*, 379–391. [CrossRef] [PubMed]
3. Zhang, H.; Davies, K.J.A.; Forman, H.J. Oxidative stress response and Nrf2 signaling in aging. *Free Radic. Biol. Med.* **2015**, *88*, 314–336. [CrossRef] [PubMed]
4. Muñoz, M.J. Longevity and heat stress regulation in *Caenorhabditis elegans*. *Mech. Ageing Dev.* **2003**, *124*, 43–48. [CrossRef] [PubMed]
5. Jin, K. New perspectives on healthy aging. *Prog. Neurobiol.* **2017**, *157*, 1. [CrossRef] [PubMed]
6. Kim, B.; Lee, J.; Kim, Y.; Lee, S.V. Regulatory systems that mediate the effects of temperature on the lifespan of *Caenorhabditis elegans*. *J. Neurogenet.* **2020**, *34*, 518–526. [CrossRef]
7. Senoner, T.; Dichtl, W. Oxidative Stress in Cardiovascular Diseases: Still a Therapeutic Target? *Nutrients* **2019**, *11*, 2090. [CrossRef]
8. Schmitt-Schillig, S.; Schaffer, S.; Weber, C.C.; Eckert, G.P.; Müller, W.E. Flavonoids and the aging brain. *J. Physiol. Pharmacol.* **2005**, *56* (Suppl. 1), 23–36.
9. Mukherjee, P.K.; Banerjee, S.; Biswas, S.; Das, B.; Kar, A.; Katiyar, C.K. *Withania somnifera* (L.) Dunal—Modern perspectives of an ancient Rasayana from Ayurveda. *J. Ethnopharmacol.* **2021**, *264*, 113157. [CrossRef]
10. Spagnuolo, C.; Russo, G.L.; Orhan, I.E.; Habtemariam, S.; Daglia, M.; Sureda, A.; Nabavi, S.F.; Devi, K.P.; Loizzo, M.R.; Tundis, R.; et al. Genistein and cancer: Current status, challenges, and future directions. *Adv. Nutr.* **2015**, *6*, 408–419. [CrossRef]
11. Martiniakova, M.; Babikova, M.; Omelka, R. Pharmacological agents and natural compounds: Available treatments for osteoporosis. *J. Physiol. Pharmacol.* **2020**, *71*, 307–320.
12. Křížová, L.; Dadáková, K.; Kašparovská, J.; Kašparovský, T. Isoflavones. *Molecules* **2019**, *24*, 1076. [CrossRef]

13. Thangavel, P.; Puga-Olguín, A.; Rodríguez-Landa, J.F.; Zepeda, R.C. Genistein as Potential Therapeutic Candidate for Menopausal Symptoms and Other Related Diseases. *Molecules* **2019**, *24*, 3892. [CrossRef]
14. Vona, R.; Gambardella, L.; Cittadini, C.; Straface, E.; Pietraforte, D. Biomarkers of Oxidative Stress in Metabolic Syndrome and Associated Diseases. *Oxidative Med. Cell. Longev.* **2019**, *2019*, 8267234. [CrossRef]
15. Wang, L.; Waltenberger, B.; Pferschy-Wenzig, E.M.; Blunder, M.; Liu, X.; Malainer, C.; Blazevic, T.; Schwaiger, S.; Rollinger, J.M.; Heiss, E.H.; et al. Natural product agonists of peroxisome proliferator-activated receptor gamma (PPARγ): A review. *Biochem. Pharmacol.* **2014**, *92*, 73–89. [CrossRef]
16. Murphy, C.T.; Hu, P.J. Insulin/insulin-like growth factor signaling in *C. elegans*. *WormBook* **2018**, 1–43. [CrossRef]
17. Zhang, S.; Li, F.; Zhou, T.; Wang, G.; Li, Z. *Caenorhabditis elegans* as a Useful Model for Studying Aging Mutations. *Front. Endocrinol.* **2020**, *11*, 554994. [CrossRef]
18. Zečić, A.; Braeckman, B.P. DAF-16/FoxO in *Caenorhabditis elegans* and Its Role in Metabolic Remodeling. *Cells* **2020**, *9*, 109. [CrossRef]
19. Seo, K.; Choi, E.; Lee, D.; Jeong, D.E.; Jang, S.K.; Lee, S.J. Heat shock factor 1 mediates the longevity conferred by inhibition of TOR and insulin/IGF-1 signaling pathways in C. elegans. *Aging Cell* **2013**, *12*, 1073–1081. [CrossRef]
20. Yan, F.; Chen, Y.; Azat, R.; Zheng, X. Mulberry Anthocyanin Extract Ameliorates Oxidative Damage in HepG2 Cells and Prolongs the Lifespan of *Caenorhabditis elegans* through MAPK and Nrf2 Pathways. *Oxidative Med. Cell. Longev.* **2017**, *2017*, 7956158. [CrossRef]
21. Lakowski, B.; Hekimi, S. The genetics of caloric restriction in *Caenorhabditis elegans*. *Proc. Natl. Acad. Sci. USA* **1998**, *95*, 13091–13096. [CrossRef] [PubMed]
22. Monaghan, R.M.; Barnes, R.G.; Fisher, K.; Andreou, T.; Rooney, N.; Poulin, G.B.; Whitmarsh, A.J. A nuclear role for the respiratory enzyme CLK-1 in regulating mitochondrial stress responses and longevity. *Nat. Cell Biol.* **2015**, *17*, 782–792. [CrossRef] [PubMed]
23. Qi, J.H.; Dong, F.X. The relevant targets of anti-oxidative stress: A review. *J. Drug Target.* **2021**, *29*, 677–686. [CrossRef] [PubMed]
24. Zhou, W.; Lin, J.M.; Wang, S.L.; Lin, H.Z. Effect of Herba Scutellariae Barbatae flavonoids in delaying aging of *Caenorhabditis elegans* and human umbilical vein endothelial cells in vitro. *Nan Fang Yi Ke Da Xue Xue Bao* **2017**, *37*, 821–826. [CrossRef] [PubMed]
25. Zhu, Q.; Qu, Y.; Zhou, X.G.; Chen, J.N.; Luo, H.R.; Wu, G.S. A Dihydroflavonoid Naringin Extends the Lifespan of *C. elegans* and Delays the Progression of Aging-Related Diseases in PD/AD Models via DAF-16. *Oxidative Med. Cell. Longev.* **2020**, *2020*, 6069354. [CrossRef]
26. Guo, H.; Cao, M.; Zou, S.; Ye, B.; Dong, Y. Cranberry Extract Standardized for Proanthocyanidins Alleviates β-Amyloid Peptide Toxicity by Improving Proteostasis Through HSF-1 in *Caenorhabditis elegans* Model of Alzheimer's Disease. *J. Gerontol. A Biol. Sci. Med. Sci.* **2016**, *71*, 1564–1573. [CrossRef]
27. Lee, E.B.; Ahn, D.; Kim, B.J.; Lee, S.Y.; Seo, H.W.; Cha, Y.S.; Jeon, H.; Eun, J.S.; Cha, D.S.; Kim, D.K. Genistein from Vigna angularis Extends Lifespan in *Caenorhabditis elegans*. *Biomol. Ther.* **2015**, *23*, 77–83. [CrossRef]
28. Li, K.; Zhang, S.; Sun, Y.; Chen, Y.; Chen, W.; Ruan, W.; Liu, Y. Antiamyloid β toxicity effect of genistein via activation of DAF-16 and HSP-16.2 signal pathways in *Caenorhabditis elegans*. *J. Biochem. Mol. Toxicol.* **2022**, *36*, e23055. [CrossRef]
29. Wu, Z.; Tan, B.; Liu, Y.; Dunn, J.; Martorell Guerola, P.; Tortajada, M.; Cao, Z.; Ji, P. Chemical Composition and Antioxidant Properties of Essential Oils from Peppermint, Native Spearmint and Scotch Spearmint. *Molecules* **2019**, *24*, 2825. [CrossRef]
30. Doshi, S.; Braganza, V. Ameliorative effect of Argyreia boseana Sant. & Pat. on stress in *C. elegans*. *J. Ayurveda Integr. Med.* **2020**, *11*, 147–152.
31. Phulara, S.C.; Pandey, S.; Jha, A.; Chauhan, P.S.; Gupta, P.; Shukla, V. Hemiterpene Compound, 3,3-Dimethylallyl Alcohol Promotes Longevity and Neuroprotection in Caenorhabditis Elegans. *Geroscience* **2021**, *43*, 791–807. [CrossRef]
32. Zhang, H.; Yang, X.; Pang, X.; Zhao, Z.; Yu, H.; Zhou, H. Genistein protects against ox-LDL-induced senescence through enhancing SIRT1/LKB1/AMPK-mediated autophagy flux in HUVECs. *Mol. Cell. Biochem.* **2019**, *455*, 127–134. [CrossRef]
33. Lee, K.Y.; Kim, J.R.; Choi, H.C. Genistein-induced LKB1-AMPK activation inhibits senescence of VSMC through autophagy induction. *Vasc. Pharmacol.* **2016**, *81*, 75–82. [CrossRef]
34. Wang, H.; Liu, J.; Li, T.; Liu, R.H. Blueberry extract promotes longevity and stress tolerance via DAF-16 in *Caenorhabditis elegans*. *Food Funct.* **2018**, *9*, 5273–5282. [CrossRef]
35. Song, B.; Zheng, B.; Li, T.; Liu, R.H. Raspberry extract promoted longevity and stress tolerance via the insulin/IGF signaling pathway and DAF-16 in *Caenorhabditis elegans*. *Food Funct.* **2020**, *11*, 3598–3609. [CrossRef]
36. Vatner, S.F.; Zhang, J.; Oydanich, M.; Berkman, T.; Naftalovich, R.; Vatner, D.E. Healthful aging mediated by inhibition of oxidative stress. *Ageing Res. Rev.* **2020**, *64*, 101194. [CrossRef]
37. Kittimongkolsuk, P.; Roxo, M.; Li, H.; Chuchawankul, S.; Wink, M.; Tencomnao, T. Extracts of the Tiger Milk Mushroom (*Lignosus Rhinocerus*) Enhance Stress Resistance and Extend Lifespan in Caenorhabditis Elegans Via the Daf-16/Foxo Signaling Pathway. *Pharmaceuticals.* **2021**, *14*, 93. [CrossRef]
38. Lee, E.B.; Kim, J.H.; An, C.W.; Kim, Y.J.; Noh, Y.J.; Kim, S.J.; Kim, J.E.; Shrestha, A.C.; Ham, H.N.; Leem, J.Y.; et al. Longevity and Stress Resistant Property of 6-Gingerol from Zingiber officinale Roscoe in *Caenorhabditis elegans*. *Biomol. Ther.* **2018**, *26*, 568–575. [CrossRef]
39. Pincus, Z.; Mazer, T.C.; Slack, F.J. Autofluorescence as a measure of senescence in *C. elegans*: Look to red, not blue or green. *Aging* **2016**, *8*, 889–898. [CrossRef]

40. Yang, X.; Wang, H.; Li, T.; Chen, L.; Zheng, B.; Liu, R.H. Nobiletin Delays Aging and Enhances Stress Resistance of *Caenorhabditis elegans*. *Int. J. Mol. Sci.* **2020**, *21*, 341. [CrossRef]
41. Grönke, S.; Clarke, D.F.; Broughton, S.; Andrews, T.D.; Partridge, L. Molecular evolution and functional characterization of Drosophila insulin-like peptides. *PLoS Genet.* **2010**, *6*, e1000857. [CrossRef] [PubMed]
42. Zhou, L.; Liu, J.; Bu, L.L.; Liao, D.F.; Cheng, S.W.; Zheng, X.L. Curcumin Acetylsalicylate Extends the Lifespan of *Caenorhabditis elegans*. *Molecules* **2021**, *26*, 6609. [CrossRef] [PubMed]
43. Vanfleteren, J.R.; Braeckman, B.P. Mechanisms of life span determination in *Caenorhabditis elegans*. *Neurobiol. Aging* **1999**, *20*, 487–502. [CrossRef] [PubMed]
44. Lin, K.; Hsin, H.; Libina, N.; Kenyon, C. Regulation of the *Caenorhabditis elegans* longevity protein DAF-16 by insulin/IGF-1 and germline signaling. *Nat. Genet.* **2001**, *28*, 139–145. [CrossRef] [PubMed]
45. Ke, T.; Antunes Soares, F.A.; Santamaría, A.; Bowman, A.B.; Skalny, A.V.; Aschner, M. N,N′ bis-(2-mercaptoethyl) isophthalamide induces developmental delay in *Caenorhabditis elegans* by promoting DAF-16 nuclear localization. *Toxicol. Rep.* **2020**, *7*, 930–937. [CrossRef] [PubMed]
46. Baird, N.A.; Douglas, P.M.; Simic, M.S.; Grant, A.R.; Moresco, J.J.; Wolff, S.C.; Yates, J.R.; Manning, G.; Dillin, A. HSF-1–mediated cytoskeletal integrity determines thermotolerance and life span. *Science* **2014**, *346*, 360–363. [CrossRef]
47. Jagdale, G.B.; Grewal, P.S.; Salminen, S.O. Both heat-shock and cold-shock influence trehalose metabolism in an entomopathogenic nematode. *J. Parasitol.* **2005**, *91*, 988–994. [CrossRef]
48. Lin, C.; Chen, Y.; Lin, Y.; Wang, X.; Hu, L.; Cao, Y.; Chen, Y. Antistress and anti-aging activities of *Caenorhabditis elegans* were enhanced by Momordica saponin extract. *Eur. J. Nutr.* **2021**, *60*, 1819–1832. [CrossRef]
49. Prithika, U.; Deepa, V.; Balamurugan, K. External induction of heat shock stimulates the immune response and longevity of *Caenorhabditis elegans* towards pathogen exposure. *Innate Immun.* **2016**, *22*, 466–478. [CrossRef]
50. Kumsta, C.; Chang, J.T.; Schmalz, J.; Hansen, M. Hormetic heat stress and HSF-1 induce autophagy to improve survival and proteostasis in C. elegans. *Nat. Commun.* **2017**, *8*, 14337. [CrossRef]
51. Sugawara, T.; Sakamoto, K. Killed Bifidobacterium longum enhanced stress tolerance and prolonged life span of *Caenorhabditis elegans* via DAF-16. *Br. J. Nutr.* **2018**, *120*, 872–880. [CrossRef]
52. Smita, S.S.; Trivedi, S.; Pandey, T.; Trivedi, M.; Pandey, R. A Bioactive compound Shatavarin IV-mediated longevity as revealed by dietary restriction-induced autophagy in *Caenorhabditis elegans*. *Biogerontology* **2020**, *21*, 827–844. [CrossRef]
53. Snutch, T.P.; Heschl, M.F.; Baillie, D.L. The *Caenorhabditis elegans* hsp70 gene family: A molecular genetic characterization. *Gene* **1988**, *64*, 241–255. [CrossRef]
54. Hsu, A.L.; Murphy, C.T.; Kenyon, C. Regulation of aging and age-related disease by DAF-16 and heat-shock factor. *Science* **2003**, *300*, 1142–1145. [CrossRef]
55. Linder, B.; Jin, Z.; Freedman, J.H.; Rubin, C.S. Molecular characterization of a novel, developmentally regulated small embryonic chaperone from *Caenorhabditis elegans*. *J. Biol. Chem.* **1996**, *271*, 30158–30166. [CrossRef]
56. Walther, D.M.; Kasturi, P.; Zheng, M.; Pinkert, S.; Vecchi, G.; Ciryam, P.; Morimoto, R.I.; Dobson, C.M.; Vendruscolo, M.; Mann, M.; et al. Widespread Proteome Remodeling and Aggregation in Aging C. elegans. *Cell* **2017**, *168*, 944. [CrossRef]
57. Deonarine, A.; Walker, M.W.G.; Westerheide, S.D. HSF-1 displays nuclear stress body formation in multiple tissues in *Caenorhabditis elegans* upon stress and following the transition to adulthood. *Cell Stress Chaperones* **2021**, *26*, 417–431. [CrossRef]
58. Prasanth, M.I.; Gayathri, S.; Bhaskar, J.P.; Krishnan, V.; Balamurugan, K. Understanding the role of p38 and JNK mediated MAPK pathway in response to UV-A induced photoaging in *Caenorhabditis elegans*. *J. Photochem. Photobiol. B* **2020**, *205*, 111844. [CrossRef]
59. Radi, E.; Formichi, P.; Battisti, C.; Federico, A. Apoptosis and oxidative stress in neurodegenerative diseases. *J. Alzheimer's Dis.* **2014**, *42* (Suppl. 3), S125–S152. [CrossRef]
60. Davalli, P.; Mitic, T.; Caporali, A.; Lauriola, A.; D'Arca, D. ROS, Cell Senescence, and Novel Molecular Mechanisms in Aging and Age-Related Diseases. *Oxidative Med. Cell. Longev.* **2016**, *2016*, 3565127. [CrossRef]
61. Kawasaki, M.; Hisamoto, N.; Iino, Y.; Yamamoto, M.; Ninomiya-Tsuji, J.; Matsumoto, K. A *Caenorhabditis elegans* JNK signal transduction pathway regulates coordinated movement via type-D GABAergic motor neurons. *EMBO J.* **1999**, *18*, 3604–3615. [CrossRef] [PubMed]
62. Sagasti, A.; Hisamoto, N.; Hyodo, J.; Tanaka-Hino, M.; Matsumoto, K.; Bargmann, C.I. The CaMKII UNC-43 activates the MAPKKK NSY-1 to execute a lateral signaling decision required for asymmetric olfactory neuron fates. *Cell* **2001**, *105*, 221–232. [CrossRef] [PubMed]
63. Wang, J.; Du, H.; Nie, Y.; Wang, Y.; Dai, H.; Wang, M.; Wang, D.; Xu, A. Mitochondria and MAPK cascades modulate endosulfan-induced germline apoptosis in *Caenorhabditis elegans*. *Toxicol. Res.* **2017**, *6*, 412–419. [CrossRef] [PubMed]
64. Berman, K.; Mckay, J.; Avery, L.; Cobb, M. Isolation and characterization of pmk-(1-3): Three p38 homologs in *Caenorhabditis elegans*. *Mol. Cell Biol. Res. Commun.* **2001**, *4*, 337–344. [CrossRef] [PubMed]
65. Blackwell, T.K.; Steinbaugh, M.J.; Hourihan, J.M.; Ewald, C.Y.; Isik, M. SKN-1/Nrf, stress responses, and aging in *Caenorhabditis elegans*. *Free Radic Biol. Med.* **2015**, *88*, 290–301. [CrossRef]
66. Ichijo, H.; Nishida, E.; Irie, K.; ten Dijke, P.; Saitoh, M.; Moriguchi, T.; Takagi, M.; Matsumoto, K.; Miyazono, K.; Gotoh, Y. Induction of apoptosis by ASK1, a mammalian MAPKKK that activates SAPK/JNK and p38 signaling pathways. *Science* **1997**, *275*, 90–94. [CrossRef]

67. Back, P.; Matthijssens, F.; Vlaeminck, C.; Braeckman, B.P.; Vanfleteren, J.R. Effects of sod gene overexpression and deletion mutation on the expression profiles of reporter genes of major detoxification pathways in *Caenorhabditis elegans*. *Exp. Gerontol.* **2010**, *45*, 603–610. [CrossRef]
68. Wu, C.W.; Deonarine, A.; Przybysz, A.; Strange, K.; Choe, K.P. The Skp1 Homologs SKR-1/2 Are Required for the *Caenorhabditis elegans* SKN-1 Antioxidant/Detoxification Response Independently of p38 MAPK. *PLoS Genet.* **2016**, *12*, e1006361. [CrossRef]
69. Buccitelli, C.; Selbach, M. mRNAs, proteins and the emerging principles of gene expression control. *Nat. Rev. Genet.* **2020**, *21*, 630–644. [CrossRef]
70. Kobayashi, Y.; Furukawa-Hibi, Y.; Chen, C.; Horio, Y.; Isobe, K.; Ikeda, K.; Motoyama, N. SIRT1 is critical regulator of FOXO-mediated transcription in response to oxidative stress. *Int. J. Mol. Med.* **2005**, *16*, 237–243. [CrossRef]
71. Viswanathan, M.; Kim, S.K.; Berdichevsky, A.; Guarente, L. A role for SIR-2.1 regulation of ER stress response genes in determining *C. elegans* life span. *Dev. Cell* **2005**, *9*, 605–615. [CrossRef]
72. Berdichevsky, A.; Viswanathan, M.; Horvitz, H.R.; Guarente, L. *C. elegans* SIR-2.1 interacts with 14-3-3 proteins to activate DAF-16 and extend life span. *Cell* **2006**, *125*, 1165–1177. [CrossRef]
73. Vajo, Z.; King, L.M.; Jonassen, T.; Wilkin, D.J.; Ho, N.; Munnich, A.; Clarke, C.F.; Francomano, C.A. Conservation of the *Caenorhabditis elegans* timing gene clk-1 from yeast to human: A gene required for ubiquinone biosynthesis with potential implications for aging. *Mamm. Genome* **1999**, *10*, 1000–1004. [CrossRef]
74. Yanase, S.; Yasuda, K.; Ishii, N. Interaction between the ins/IGF-1 and p38 MAPK signaling pathways in molecular compensation of sod genes and modulation related to intracellular ROS levels in C. elegans. *Biochem. Biophys. Rep.* **2020**, *23*, 100796. [CrossRef]
75. Lu, L.; Zhao, X.; Zhang, J.; Li, M.; Qi, Y.; Zhou, L. Calycosin promotes lifespan in *Caenorhabditis elegans* through insulin signaling pathway via daf-16, age-1 and daf-2. *J. Biosci. Bioeng.* **2017**, *124*, 1–7. [CrossRef]
76. Tullet, J.M.; Hertweck, M.; An, J.H.; Baker, J.; Hwang, J.Y.; Liu, S.; Oliveira, R.P.; Baumeister, R.; Blackwell, T.K. Direct inhibition of the longevity-promoting factor SKN-1 by insulin-like signaling in C. elegans. *Cell* **2008**, *132*, 1025–1038. [CrossRef]

antioxidants

MDPI

Article

Safety Assessment of 3S, 3′S Astaxanthin Derived from Metabolically Engineered *K. marxianus*

Sabrina Yeo Samuel [1,2,3,†], Hui-Min David Wang [4,5,6,†], Meng-Yuan Huang [7], Yu-Shen Cheng [8,9], Juine-Ruey Chen [10], Wen-Hsiung Li [11,12] and Jui-Jen Chang [13,14,*]

1 Trade Wind Biotech Co., Ltd., Taipei 11574, Taiwan
2 Institute of Molecular Medicine, National Taiwan University, Taipei 10051, Taiwan
3 Institute of Biomedical Sciences, Academia Sinica, Taipei 11529, Taiwan
4 Graduate Institute of Biomedical Engineering, National Chung Hsing University, Taichung 40227, Taiwan
5 Graduate Institute of Medicine, College of Medicine, Kaohsiung Medical University, Kaohsiung 80708, Taiwan
6 Department of Medical Laboratory Science and Biotechnology, China Medical University, Taichung 40447, Taiwan
7 Department of Life Science, National Chung Hsing University, Taichung 40227, Taiwan
8 Department of Chemical and Materials Engineering, National Yunlin University of Science and Technology, Yunlin 64002, Taiwan
9 College of Future, National Yunlin University of Science and Technology, Yunlin 64002, Taiwan
10 RuenHuei Biopharmaceuticals Inc., Taipei 10050, Taiwan
11 Biodiversity Research Center, Academia Sinica, Taipei 11529, Taiwan
12 Department of Ecology and Evolution. University of Chicago, Chicago, IL 60637, USA
13 Graduate Institute of Integrated Medicine, China Medical University, Taichung 40447, Taiwan
14 Department of Medical Research, China Medical University Hospital, Taichung 40447, Taiwan
* Correspondence: t25046@mail.cmuh.org.tw
† These authors contributed equally to this work.

Citation: Samuel, S.Y.; Wang, H.-M.D.; Huang, M.-Y.; Cheng, Y.-S.; Chen, J.-R.; Li, W.-H.; Chang, J.-J. Safety Assessment of 3S, 3′S Astaxanthin Derived from Metabolically Engineered *K. marxianus*. *Antioxidants* **2022**, *11*, 2288. https://doi.org/10.3390/antiox11112288

Academic Editor: Ren-You Gan

Received: 21 October 2022
Accepted: 15 November 2022
Published: 18 November 2022

Abstract: Previous reviews have already explored the safety and bioavailability of astaxanthin, as well as its beneficial effects on human body. The great commercial potential in a variety of industries, such as the pharmaceutical and health supplement industries, has led to a skyrocketing demand for natural astaxanthin. In this study, we have successfully optimized the astaxanthin yield up to 12.8 mg/g DCW in a probiotic yeast and purity to 97%. We also verified that it is the desired free-form 3S, 3′S configurational stereoisomer by NMR and FITR that can significantly increase the bioavailability of astaxanthin. In addition, we have proven that our extracted astaxanthin crystals have higher antioxidant capabilities compared with natural esterified astaxanthin from *H. pluvialis*. We also screened for potential adverse effects of the pure astaxanthin crystals extracted from the engineered probiotic yeast by dosing SD rats with 6, 12, and 24 mg/kg/day of astaxanthin crystals via oral gavages for a 13-week period and have found no significant biological differences between the control and treatment groups in rats of both genders, further confirming the safety of astaxanthin crystals. This study demonstrates that developing metabolically engineered microorganisms provides a safe and feasible approach for the bio-based production of many beneficial compounds, including astaxanthin.

Keywords: antioxidant; probiotics; 3S; 3S′ astaxanthin; *Kluyveromyces marxianus*; food safety

1. Introduction

Astaxanthin, or 3,3′-dihydroxy-β,β-carotene-4,4′-dione, belongs to the family of xanthophylls, which are the oxygenated derivatives of naturally occurring organic pigments called carotenoids [1]. Humans are unable to synthesize carotenoids on their own; hence, they depend on vegetables, fruits, and animal products to attain these valuable compounds. For astaxanthin, in particular, humans rely primarily on seafood, especially salmon. Other natural sources of astaxanthin include microalgae, yeasts, and different types of seafood, including krill and shrimps.

Recent research has indicated many beneficial effects of astaxanthin on human health, with astaxanthin being found to possess strong antioxidant [2], cardioprotective [3], neuroprotective [4], immunoprotective [5], and anti-cancer [6] effects. The great commercial potential in a variety of industries, such as the pharmaceutical and health supplement industries, has led to skyrocketing demands for natural astaxanthin. With the amount of research backing the beneficial functions of astaxanthin for the human body, there is now a developing market for nutraceutical astaxanthin that is expected to reach a market of several hundred million dollars in the near future.

Microalgae have long been utilized by humans as a vital source of nutrient-rich food, feed, and health-promoting substances. *Haematococcus pluvialis*, a commercial microalgae, is considered the richest source of natural astaxanthin, with 2–4% astaxanthin content in *Haematococcus* under stress conditions [7], which was why it was first used for the industrial-scale production of natural astaxanthin [8]. In addition, *H. pluvialis* biosynthesis predominantly produces the most valuable 3*S*, 3′*S* stereoisomers [9]. However, the industrial production of astaxanthin via microalgae is limited by the need for long cultivation periods, large surface areas, and significant capital needed for maintenance and infrastructure.

Natural astaxanthin can also be extracted from various micro-organisms, including the red yeast *Phaffia rhodozyma* (*Xanthophyllomyces dendrorhous*) [10]. However, despite their ability to reach high densities after brief propagation periods, these microbes possess low astaxanthin/carotenoid ratios, which decrease their economic feasibility, especially when compared with microalgal production.

Astaxanthin can also be synthesized chemically, with synthetic astaxanthin chemically synthesized from petrochemicals in a highly involved, multistep process, where the molecules assume different forms before attaining the same chemical formula as natural astaxanthin [11]. However, despite sharing the same chemical formula, synthetic and natural astaxanthin differ substantially in three aspects: esterification, stereochemistry, and the presence of other naturally occurring carotenoids in natural astaxanthin. In addition, the bioavailability of synthetic astaxanthin is found to be much lower than that of natural astaxanthin [12]. Due to these differences, synthetic astaxanthin is currently not allowed for human use due to safety issues [13]. Instead, it is mostly utilized in fish feed to pigment the flesh of certain commercially farmed fish, primarily salmonids such as Atlantic salmon and trout.

To keep up with the increasing demand for natural astaxanthin in a sustainable way, efforts have been made to increase astaxanthin production in microorganisms through metabolic engineering due to the simplicity of genome engineering and short cell cycles. The construction of the metabolic pathway for astaxanthin production has been previously established in *Escherichia coli* K-12, a strain that has long been considered GRAS (Generally Recognized as Safe) and has a long history of safe commercial use in food applications. Genes encoding the enzymes needed for astaxanthin biosynthesis were cloned from the bacteria *Erwinia uredovora* and *Brevundimonas aurantiaca* into *E. coli* K-12 and had resulted in the generation of 5.8 mg/g DCW astaxanthin [14,15]. The astaxanthin extracted from *E. coli* K-12 has also been assessed via a 13-week study on subchronic toxicity in rats and accepted by both US Food and Drug Administration (USFDA) and Taiwan Food and Drug Administration (TFDA), further proving its safety for consumption [16], demonstrating that the biosynthesis of astaxanthin in metabolically engineered microorganisms can be a safe and feasible approach for the mass production of astaxanthin.

Recent trends have used yeast cells such as *Pichia pastoris* or *Saccharomyces cerevisiae*, for the production of recombinant proteins in view of their increased yield and ability to synthesize functional eukaryotic proteins. Attempts at introducing the astaxanthin biosynthesis pathway into yeast have yielded considerable results, with 8.10 mg/g DCW being the highest astaxanthin yield reported to date in *S. cerevisiae* containing overexpressed genes [17].

With enhanced biomass production via the supplementation of excessive carbon sources [18], thermo-tolerant characteristics [19], appropriate glycosylation [20], and strong signal peptides [20], *K. marxianus* can potentially yield higher astaxanthin levels compared with *S. cerevisiae*. The construction of a more efficient astaxanthin biosynthesis pathway was attempted in *K. marxianus*, this time by expressing enzymes from algae, which yield the highest astaxanthin levels in nature. In addition, β-carotene ketolase (*bkt*) and hydroxylase (*hpchyb*) were also included to promote the conversion of beta-carotene into astaxanthin more efficiently. This resulted in the construction of an astaxanthin yield-improved yeast strain, S3−2, which can yield up to 3.125 mg/g DCW in a YPL medium and 5.701 mg/g DCW in a YPG medium, which is the highest astaxanthin content in an engineered host to date [21]. Biofunctional assessment of the astaxanthin extracted from the S3−2 *K. marxianus* strain, termed YEAST-astaxanthin, was performed to investigate the safety of the product, which was achieved by in vitro assays and two animal models. DPPH (2,2′-diphenyl-1-picrylhydrazyl radical) scavenging analysis, ferrous ion chelating ability, reducing power assessment, and mushroom tyrosinase inhibition evaluation indicated that YEAST-astaxanthin showed both antioxidative and tyrosine suppressive properties. The immersion of zebrafish larvae in YEAST-astaxanthin solution showed no significant toxic effects, while rats fed with YEAST-astaxanthin showed neither visible abnormalities nor substantial changes in body weight or blood biochemistry tests. In addition, rats fed with YEAST-astaxanthin also reported the inhibition of metastasis in lung melanoma cells and an increased survival rate [22].

In this study, we have successfully optimized the production of astaxanthin in *K. marxianus*, with a current yield of 12.9 mg/g DCW, and validated its purity and configuration. We also performed experimental tests to assess the biosafety of astaxanthin extracted from the metabolically engineered *K. marxianus*. A safety assessment was achieved by confirming the genome stability of astaxanthin-producing genes in the yeast and by performing a 13-week repeated dose oral toxicity study in rats. Our results indicate that the astaxanthin extracted from genetically engineered *K. marxianus* cells is indeed safe for consumption.

2. Materials and Methods

2.1. HPLC

The following mobile phases were employed with the Nomura Chemical Develosil C30-UG Column (Interlink Scientific Services, London, UK): buffer A, composed of methanol/MtBE/water (81:15:4 *v*/*v*/*v*), and buffer B, composed of methanol/MtBE/water (7:90:3 *v*/*v*/*v*). The flow rate of the mobile phase was 1 mL/ min., and the solvent gradient was as follows: from 0 to 45 min, 100% buffer A from 0 to 20 min, 100% buffer B from 20 to 21 min, and then 100% buffer A from 21 to 45 min. Samples were observed using a Jasco870-UV intelligent UV-VIS detector (JASCO International Co., Ltd., Tokyo, Japan). Using chromatography, commercially available reference chemicals, and a comparison of their spectra, all carotenoids were identified. Additionally, standards were used in combination with the extinction coefficients for quantification.

2.2. UV Spectrometry

The general pattern, maximum absorbance peaks, and wavelength range of the UV spectra of the isolated astaxanthin crystals from this study were compared. The extract was measured in the UV region (477.6 nm) with a PREMA PRO-739 spectrophotometer (Chuan Hua Precision, Taipei, Taiwan).

2.3. Nuclear Magnetic Resonance Spectroscopy (NMR)

A Jasco870-UV intelligent UV-VIS detector (JASCO International Co., Ltd., Tokyo, Japan) was previously used to identify extracted and purified astaxanthin. Ten milligrams of the pure astaxanthin was loaded into the Bruker Avance III HD 400 MHz NMR Spectrometer (Bruker Corporation, Massachusetts, USA) after being dissolved in 600 uL of DMSO-d6. The experiment's parameters were (1) pulse program zg30; (2) number of scans, 64; and

(3) relaxation time, 2 s. Finally, TopSpin software was used to examine the results. All NMR spectra were recorded on a Bruker Advance III 400 MHz. According to the residual proton resonances of the appropriate deuterated solvent, 1H NMR chemical shifts were reported relative to TMS.

2.4. ABTS Assay

The ABTS assay was conducted according to Arnao et al. (2001) [23], with a few adjustments. The stock solutions included both 2.6 mM potassium persulfate solution and 7.4 mM ABTSradical dot+ solution, which were combined in equal parts to create the working solution. It was then left to react for 12 h at room temperature in the dark before being diluted by combining 1mL of ABTSradical dot+ solution with 60 mL of methanol. This yielded an absorbance of 1.1 $\pm$ 0.02 units at 734 nm when checked by the spectrophotometer. For each assay, fresh ABTS radical dot+ solution was prepared. Fruit extracts (150 μL) were combined with 2850 L of the ABTS radical dot+ solution in the dark for 2 h. The spectrophotometer was then used to measure the absorbance at 734 nm Between 25 and 600 M Trolox, the standard curve was linear. The units of measurement for the results were μM Trolox equivalents (TE)/g fresh mass. If the measured ABTS value was over the linear range of the standard curve, more dilution was required.

2.5. DPPH Assay

The DPPH assay was performed according to the method of Brand-Williams et al. (1995) [24], with a few adjustments. An amount of 24 mg DPPH was combined with 100 mL of methanol to create the stock solution, which was then stored at −20 °C until needed. A stock solution of 10 mL was then mixed with 45 mL of methanol to obtain the working solution. This yielded an absorbance of 1.1 $\pm$ 0.02 units at 515 nm when checked by the spectrophotometer. Fruit extracts (150 μL) were combined with 2850 L of the ABTS radical dot+ solution in the dark for 24 h. The spectrophotometer was then used to measure the absorbance at 515 nm. Between 25 and 600 M Trolox, the standard curve was linear. The units of measurement for the results were μM Trolox equivalents (TE)/g fresh mass. If the measured ABTS value was over the linear range of the standard curve, more dilution was required.

2.6. Reducing Power Assay

The testing samples were assessed for reducing property in accordance with a prior study [25]. By using a ferric 2,4,6-tripyridyl-S-triazine Fe(III)-TPTZ complex to produce a ferrous Fe(II)-TPTZ complex with a dark blue color using an adopted reductant, this technique measured the antioxidant's decreasing ability. The positive control, 3-tert-butyl-4-hydroxyanisole (BHA) at 100 M, was utilized to detect the colorization at 700 nm. About 85 mL of phosphate-buffered saline (PBS) (67 mM, pH 6.8) and 2.5 L of 20% potassium ferricyanide ($K_3Fe(CN)_6$) were added to the test samples to slightly increase their volume. After 20 min of reaction time at 50 °C, 160 mL of 10% trichloroacetic acid was added to the reactants, and the mixture was centrifuged at 3000× *g* for 10 min. An amount of 25 uL $FeCl_3$ (2%) was mixed with 75 mL of upper layer solution, and the optical density was measured at 700 nm. A higher optical absorption indicates a higher reductive property.

2.7. Test Facility

This study is designed according to the Safety Evaluation Methods for Health Food (2020), Ministry of Health and Welfare, Taiwan. In addition, the thirteen-week repeated dose oral toxicity study was conducted in compliance with the principles of Good Laboratory Practice for Non-clinical Laboratory Studies (FDA, 21 CFR, Part 58), the Good Laboratory Practice for Non-clinical Laboratory Studies (Ministry of Health and Welfare, ROC, 3rd ed., 2006), the OECD Principles on Good Laboratory Practice (TAF OECD GLP Compliance, No. 1, 1997), and the Regulations for Application of Health Food Permit (TFDA, 2016) outlined by the Ministry of Health and Welfare, Taiwan at the Preclinical Testing Center of Level Biotechnology Inc, New Taipei City, Taiwan.

2.8. Test Substance

The test substance, astaxanthin crystal (Asta-S Crystal, provided by Trade Wind Biotech Co. Ltd., Taipei, Taiwan), was extracted and purified from engineered *K. marxianus* biomass [22]. After fermentation, yeast cells were killed by heating at 90 °C and centrifuged to remove the spent medium. The concentrated astaxanthin-containing *K. marxianus* cells were then oven-dried before using ethyl acetate to extract the astaxanthin. Subsequently, residual biomass was then filtered out and crystallized by the addition of ethanol. Pure astaxanthin crystals as a food product were then collected using filter paper and analyzed to ensure quality control by performing assays for the astaxanthin content, solvent residue, heavy metals, DNA contamination, and microbiological safety. The astaxanthin crystal utilized in the toxicity study had a red crystalline appearance; was made of more than 98% total carotenoids according to a UV test; and included 97% (w/w) astaxanthin, as determined by HPLC.

2.9. Animals and Treatment Regimen

All experimental protocols were approved by the Institutional Animal Care and Use Committee (IACUC number: 210310) of Level Biotechnology Inc., New Taipei City, Taiwan. CD® (Sprague Dawley) IGS rats were purchased from BioLASCO Ltd. (Yilan, Taiwan). The animals were housed in the AAALAC International accredited facility of Level Biotech. Inc. A total of 80 rats were randomly assigned to four study groups (vehicle control, low, mid, and high-dose groups; 10 animals/sex/group). The body weight variation of all animals used fell into an interval within ±20 percent of the mean weight. The basic design is presented in Table S1.

According to the information provided by the sponsor, the human dose is 0.012 g per day (0.2 mg/kg for a 60 kg human). According to body weight conversion, the three dose levels selected in this study were approximately 30, 60, or 120-fold the human dose. The proposed human administration route is oral. Therefore, oral gavage was utilized in this study. The accuracies of the 1 or 3 mL syringes used in this study were the closest to 0.01 and 0.1 mL, respectively. The animals in all groups were dosed once daily for 91 consecutive days with the designated dose formulations (suspension) listed in the table. The first dosing day was denoted as Study Day 1 (Day 1). No animal replacement was performed.

Throughout the experimental period, all animals were subjected to daily clinical monitoring following administration. The animals' physical health, any relevant behavioral changes, and any overt poisoning symptoms were all noted. Observations were made on all animals' overall health, skin, fur, eyes, and mucous membranes; the presence of fluids and excretions; and their autonomic activity (e.g., lacrimation, piloerection, and unusual respiratory pattern), among other things. The occurrence of chronic or tonic movements, stereotypies (such as excessive grooming or repetitive circling), and unusual behavior (such as self-mutilation or walking backward) were also noted, along with alterations in stride, posture, and responsiveness to handling.

On the day of grouping and Day 91, an ophthalmologic examination was carried out on every animal. Using an ophthalmoscope, the cornea, conjunctiva, anterior chamber, iris, and lens were all inspected. Before necropsy, females had a single vaginal smear, and the estrus cycle was noted. After the predetermined dose time, hematology (including coagulation), serum chemistry, and urinalysis were carried out on all surviving animals. Blood was drawn through the abdominal aorta on the day of necropsy and divided into three tubes: one containing K2 EDTA for the analysis of the complete blood count, one containing sodium citrate for the analysis of the coagulation factors, and one containing no anticoagulant for the analysis of the serum chemistry. Prior to terminal sacrifice, urine samples were obtained by employing metabolism cages for 12–16 h, and the urine volume was recorded. The items analyzed can be found in Table S2.

The surviving animals on the day of sacrifice were anesthetized with the ketamine and xylazine mixture solution (80 mg/mL and 8 mg/mL, respectively), followed by blood collection, exsanguination, and necropsy. The gross necropsy includes examinations of the external surface of the body, all thoracic and abdominal cavities, intestines, and visceral organs.

The sampled tissues and organs were kept after gross necropsy in 10% neutral buffered formalin or other suitable fixatives for a further histopathological investigation. Trimmed, embedded, sectioned, and H&E stained tissues from groups 1 (the vehicle control group) and 4 (the high-dose group) were chosen for microscopy analysis. These examinations were not extended to animals in group 2 (low-dose group) and group 3 (mid-dose group) because no treatment group-related findings were noted in the high-dose group.

3. Results

3.1. Astaxanthin Compound Analysis

We first cultured the astaxanthin-producing strain *K. marxianus* S3−2 and then streaked it on a YPG medium plate to screen for a high-performing colony, named 6−13−a6. To ensure the stability of *K. marxianus* 6−13−a6, we first cultured it in YPG medium before sub-culturing it in 10 different medium mixtures for 10 generations and spreading them on YPG medium plates to obtain single colonies. These colonies were then screened to find the highest-performing colony, which was then used to conduct further experiments. This optimized strain was capable of producing up to 12.8 mg/g DCW of natural astaxanthin. In addition, we had also successfully achieved mass cultivation of the *K. marxianus* 6-13-a6 in commercial 10,000-L tanks, from which we were able to extract and purify astaxanthin crystals for further testing. In order to ensure the stability and purity of our extracted astaxanthin crystals, we conducted a series of experiments. Previously, we had already compared both freshly extracted astaxanthin crystals and extracted astaxanthin crystals stored for a year and had observed no noticeable differences in the color and luster of the crystals (data not shown).

Astaxanthin exists in many different forms, ranging from stereoisomers, geometric isomers, and free to be esterified forms [2], and can be found in natural sources. The stereoisomers (3S, 3′S) and (3R 3′R) are the most abundant in nature. The (3S, 3′S) isomer is primarily found in *Haematococcus* algae, while the (3R, 3′R) isomer is mainly found in *Xanthophyllomyces dendrorhous*. However, a combination of all three isomers—(3S, 3′S), (3R, 3′S), and (3R, 3′R)—can be found in synthetic astaxanthin [8]. One of the most powerful known antioxidant properties is possessed by natural astaxanthin, which is regarded as a super antioxidant. Natural astaxanthin is 55× more potent than synthetic astaxanthin in trapping free radicals in our system [10,11]. Astaxanthin does, however, exist in *H. pluvialis* in three distinct forms that can be categorized as free (5%), monoesters (70%), and diesters (25%). Since esterified products are less ready to be digested and absorbed by the human body, this would affect the bioavailability of astaxanthin. In contrast, astaxanthin extracted from engineered probiotic yeast is mainly composed of free-form 3S, 3′S astaxanthin, which can significantly increase the bioavailability of astaxanthin [22]. Previous studies have also verified that the extracted astaxanthin is of the desired 3S, 3′S configurational stereoisomers by HPLC separation [22]. Therefore, to further ascertain the purity of the extracted astaxanthin crystals, we performed UV spectroscopy (477.6 nm) (Figure 1A) and High-Performance Liquid Chromatography (HPLC) (Figure 1B), where we were able to confirm its purity at 98.26% and 97.14%, respectively.

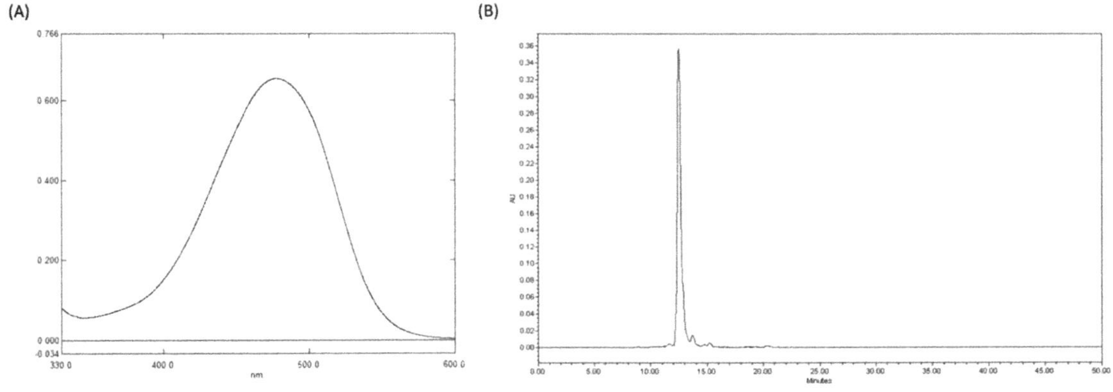

Figure 1. Validation of astaxanthin purity extracted and purified from the strain *K. marxianus* 6−13−a6. (**A**) Evaluation of astaxanthin purity by UV spectrometer. (**B**) Evaluation of astaxanthin purity by HPLC.

The ^{1}H NMR (nuclear magnetic resonance) spectra and FTIR (Fourier transform infrared) spectra of the extracted astaxanthin are shown in Figure 2. As shown in Figure 2A,B, the ^{1}H NMR spectrum analysis of astaxanthin indicates that there are multiple peaks in the chemical shift δ between 1.0 and 2.0, which are the characteristic absorption peaks of hydrogen in the -CH3, -CH2, and -CH configurations, while the chemical shifts around 2.0–3.0 and 5.05 are the characteristic absorption peaks of hydrogen in alkene's -CH=CH2 configuration, and the chemical shifts at 6.0–7.0 are the characteristic absorption peaks of hydrogen on the benzene ring. Therefore, the test substance contains multiple structures such as -CH3, -CH2, -CH, -CH=CH2, and benzene rings. After analyzing the ^{1}H NMR spectra of the astaxanthin standard (Sigma-Aldrich) and the extracted astaxanthin crystals, we are able to confirm that both materials possess the same characteristic absorption peak and, thus, have the same structural material composition.

In Figure 2C, a Fourier transform infrared spectroscopy (FTIR) analysis of the astaxanthin material structure shows peaks at 3031, 2918, and 2862 cm^{-1}, which correspond to the symmetric and antisymmetric stretching vibration peaks of the -CH3, -CH2, and -CH configurations, while peaks at 1466 and 1386 cm^{-1} correspond to the stretching vibration peaks of the saturated C-H configuration, which indicates the presence of -CH3, -CH2, and -CH configurations in the sample. Peaks at 1733, 1648, and 1602 cm^{-1} are the stretching vibration peaks of C=O and alkene -CH=CH2 configurations, and the strong peaks generated at 974 and 953 cm^{-1} indicate that the samples contain multiple polyene structures. These analyses suggest that the extracted astaxanthin crystal has the same characteristic structures present in the astaxanthin standard, such as -CH3, -CH2, -CH, C=O, and polyene.

3.2. Antioxidant Activity Determinations

Many previous studies have already shown astaxanthin's powerful antioxidant ability, which is essential for the absorption and neutralization of free radicals. In this study, we also verified the antioxidative capabilities of our extracted astaxanthin crystals using the ABTS assay, reducing power assay, and DPPH assay, which can detect the eradication of free radicals by observing color changes. The aforementioned assays are capable of distinguishing the different mechanisms used to eliminate reactive oxygen species (ROS): quenching of singlet and triplet oxygen or decomposing peroxides. Table 1 confirmed that every type of astaxanthin had dose-dependent competencies in the three experiments above, especially an ABTS scavenging ability, but barely had any differences in terms of reducing power.

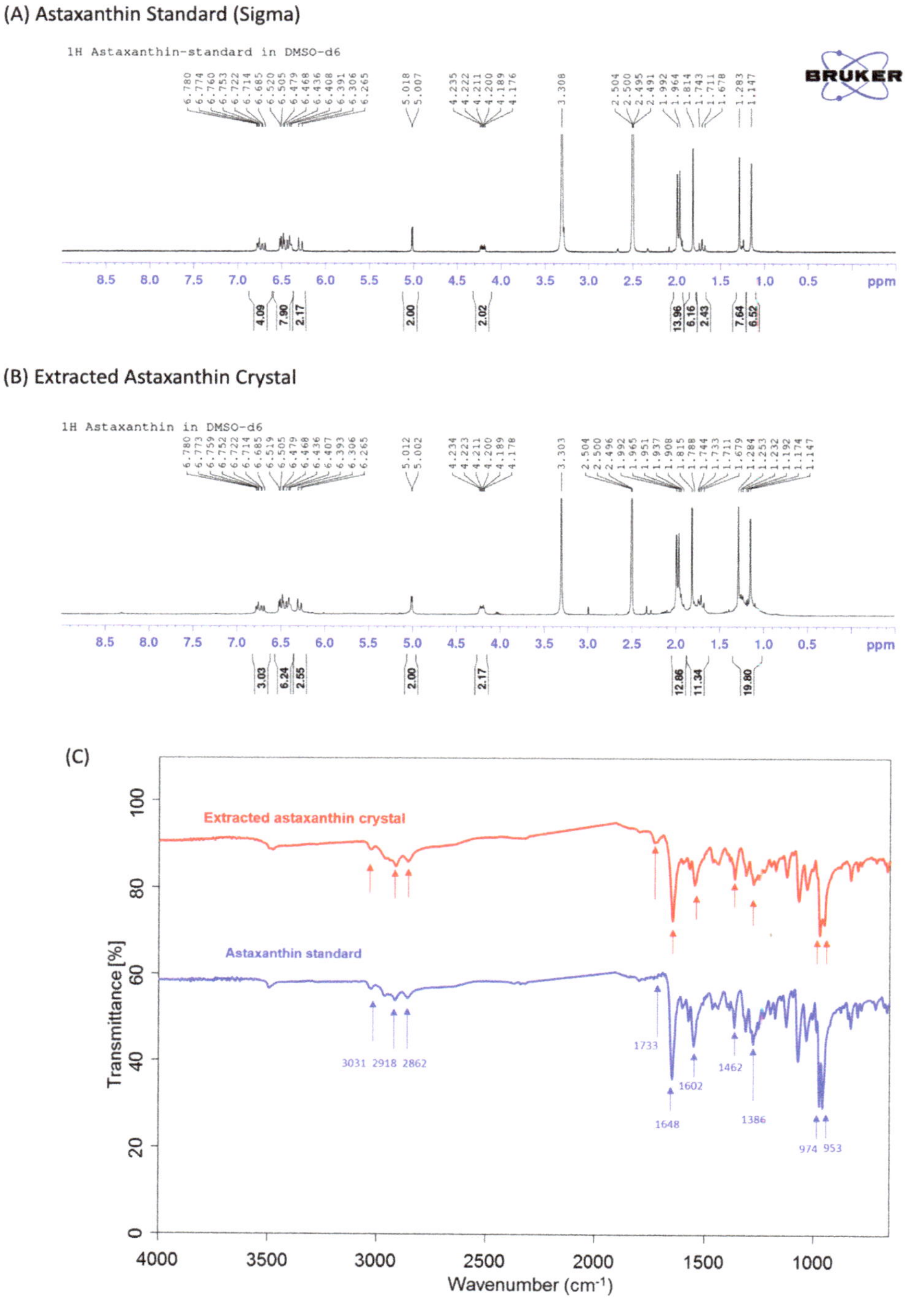

Figure 2. Verification of astaxanthin chemical composition extracted from *K. marxianus* 6−13−a6 by NMR (**A**,**B**) and FTIR (**C**).

Table 1. ABTS, reducing power, and DPPH were used to test the ROS reduce abilities in various astaxanthin. Three types of astaxanthin: Sigma-Aldrich Co., Burlington, MA, USA; extracted astaxanthin; and algae.

Concentration (μg/mL)		Antioxidant Capacity of Astaxanthin		
		ABTS (%)	DPPH (%)	Reducing Power (Absorbance at 700 nm)
Positive control [a,b,c]		93.4 ± 0.003	88.14 ± 0.008	0.66 ± 0.012
Standard astaxanthin (Sigma-Aldrich)	0.05	25.10 ± 020	24.20 ± 0.001	0.107 ± 0.001
	0.1	39.54 ± 0.006	32.21 ± 0.005	0.109 ± 0.002
	1	55.10 ± 0.0033	40.23 ± 0.003	0.122 ± 0.005
Extracted astaxanthin crystals	0.05	32.68 ± 0.034	30.22 ± 0.035	0.109 ± 0.002
	0.1	47.42 ± 0.028	42.30 ± 0.006	0.112 ± 0.006
	1	60.31 ± 0.06	60.23 ± 0.012	0.223 ± 0.005
Algae astaxanthin	0.05	26.86 ± 0.006	25.04 ± 0.002	0.109 ± 0.003
	0.1	40.21 ± 0.017	38.11 ± 0.014	0.108 ± 0.002
	1	51.08 ± 0.023	43.68 ± 0.002	0.133 ± 0.001

[a]. Vitamin C was used as a positive control on ABTS at 2 mM. [b]. Vitamin C was used as a positive control on the DPPH assay at 10 mM. [c]. BHA was used as a positive control on reducing power at 10 mM.

The principle of ABTS assays in determining the antioxidant capacity is that ABTS (2,2′-azino-bis(3-ethylbenzothiazoline-6-sulfonic acid)) could be oxidized into green ABTS free radicals under the action of appropriate oxidants. Consequently, the generation of ABTS free radicals will be inhibited in the presence of antioxidants. Therefore, the change in the absorbance of ABTS can be measured and calculated as the total antioxidant capacity of the sample. Three types of astaxanthin showed the capabilities of eliminating ROS in ABTS assay. In general, in various concentrations, the extracted astaxanthin crystals were more potent than standard astaxanthin and algae astaxanthin. For example, the extracted astaxanthin crystals displayed a 60% scavenging rate at 1 mg/mL. Moreover, in low concentrations, the extracted astaxanthin crystals showed better capabilities, with a 32.68% oxidant removing speed while that for the standard astaxanthin was 25.1% and that for algae astaxanthin was 26.86%.

An established test for determining a substance's ability to act as a hydrogen donor to eliminate the stable radical DPPH and to transform it into diphenyl-picrylhydrazine is the DPPH free radical scavenging system. When the element can absorb the free radicals, the assay color will change from deep blue to light yellow. Astaxanthin doses were directly correlated with the moderate antioxidant capacity displayed by all astaxanthin forms. For instance, the free radical emission capability was 50% in 1 mg/mL extracted astaxanthin crystals, which was higher than the exact dosage of standard astaxanthin (40.23%) and algae astaxanthin (43.68%). At 0.05 mg/mL of astaxanthin, the antioxidative ability was around 25% in three types of astaxanthin.

The reducing power assay is a common technique for determining the ability of materials in losing electrons by changing the color from yellow to green. The ability of the tested antioxidants is determined by the degree of color diversity. The presence of a suitable material allows for the reduction of the Fe^{3+}/ferricyanide complex into a ferrous state. Table 1, which indicated no significant difference in the reducing power of the control group and various concentrations of astaxanthin, showed that no astaxanthin type exhibited any increased antioxidant reduction power.

ROS promotes the secretion of a variety of inflammatory chemicals from cancer cells, which may increase the capacity of cancer cells to metastasize, boost the growth of vascular endothelium, and suppress the immune system. Due to ongoing oxidant production, ROS may also foster an immunosuppressive microenvironment for malignancies. Cancer cells can consequently continue to grow and degrade. Therefore, the biological role of antioxidants in reducing chronic inflammation and preventing cancer was implicated.

The aforementioned tests demonstrated astaxanthin's capacity to neutralize ROS, and it was superior in both the ABTS and DPPH assays. The reducing power likewise had a milder ability to reduce oxidation. Compared to the oxidative-eliminating abilities of all three types of astaxanthin, the extracted astaxanthin crystals performed much better than the standard and algae in both ABTS and DPPH in every concentration.

3.3. *Thirteen Week Toxicity Study*

In another study using astaxanthin produced by engineered *Escherichia coli* K-12, no significant biological differences between test and control groups was observed [15]. These results were consistent with prior toxicity studies using natural astaxanthin derived from bacteria or microalgae and even synthetic astaxanthin produced chemically. To further demonstrate the safety of astaxanthin derived from the engineered *K. marxianus* 6-13-a6, we conducted a 13-week repeated dose toxicity study to evaluate the potential toxicity likely to arise from repeated daily exposure of astaxanthin crystal in rats via oral gavage. The study's findings provide information on human exposure safety.

3.3.1. Mortality, Body Weight, Food Consumption, and Ophthalmological Examination

All of the animals survived the testing period and showed no toxicity-related symptoms (Table 2). When compared to the vehicle control group, neither male nor female rats in the treatment group showed statistically significant changes in body weight (Table S3). The average food consumption for each group is shown in Table S4, and for female rats, there was no statistically significant difference between the treatment groups and the vehicle control group. The mean food intake in male rats was considerably lower than the vehicle control over Days 50–57 at a dosage level of 12 mg/kg/day. Nonetheless, all values were within the historical control data range. Therefore, it was not considered a treatment group-related finding. No significant treatment group-related clinical signs were noted in male and female rats during the study period (Table 3). Several observed clinical signs such as chromodacryorrhea (1/10, high-dose males), hair loss (1/10 in the vehicle control females, 4/10 in the mid-dose females, and 1/10 in the high-dose females), and wound on the skin (1/10 in the low-dose females, 2/10 in the mid-dose females, and 1/10 in the high-dose females) were noted during the study period. Nonetheless, these clinical signs might be caused by social activities (i.e., fighting and grooming), and the incidence was not dose-dependent. Therefore, these findings were not considered a treatment group-related findings. In addition, ophthalmological examinations revealed no changes in all animals before grouping and terminal sacrifice (Table S5).

Table 2. Mortality of rats receiving astaxanthin over a 13-week period.

Gender	Dose (mg/kg/day)	Mortality (N/N) [1]
Male	0	0/10
	6	0/10
	12	0/10
	24	0/10
Female	0	0/10
	6	0/10
	12	0/10
	24	0/10

[1] N/N: Number of animals found dead/total number of animals.

3.3.2. Clinical Pathology

No treatment-related toxicologically significant changes in any of the hematological parameters were seen compared with the control group (Table 4). In contrast to female rats, the value of PT (Prothrombin Time) was statistically higher at dose levels of 12 and 24 mg/kg/day in male rats, at 14.55 ± 1.57 s and 13.75 ± 2.06 s, respectively. In female rats, no statistical difference was observed when comparing test article-treated groups to the vehicle control group. In addition, APTT (activated Partial Thromboplastin Time) values were significantly higher at dose levels 6, 12, and 24 mg/kg/day compared with the vehicle control group, at 19.11 ± 1.24, 19.95 ± 1.19, and 19.30 ± 1.18, respectively. However, none of the values fell outside the range of the historical control data and were not regarded as treatment group-related observations.

Table 3. Clinical observations of rats receiving astaxanthin over a 13-week period.

Gender	Clinical Signs	Incidence (n′/n′) [1]			
		Vehicle Control 0 mg/kg/day	Low-Dose 6 mg/kg/day	Mid-Dose 12 mg/kg/day	High-Dose 24 mg/kg/day
		$n = 10$	$n = 10$	$n = 10$	$n = 10$
Male	Chromodacryorrhea	0/10	0/10	0/10	1/10
Female	Hair loss	1/10	0/10	4/10	1/10
	Wound, skin	0/10	1/10	2/10	1/10

[1] Total number of animals with observable sign/total number of animals examined.

Table 4. (A) Coagulation analysis of rats receiving astaxanthin over a 13-week period. (B) Hematology values of rats receiving astaxanthin over a 13-week period. (C) Serum chemistry of rats receiving astaxanthin over a 13-week period. (D) Urinalysis of rats receiving astaxanthin over a 13-week period.

A

Parameters	Coagulation (Mean ± SD)							
Group	Vehicle Control 0 mg/kg/day	Low-Dose 6 mg/kg/day	Mid-Dose 12 mg/kg/day	High-Dose 24 mg/kg/day	Vehicle Control 0 mg/kg/day	Low-Dose 6 mg/kg/day	Mid-Dose 12 mg/kg/day	High-Dose 24 mg/kg/day
	Male				Female			
Number of Animals	10	10	10	10	10	10	10	10
PT (sec)	11.95 ± 1.30	13.27 ± 1.26	14.55 ± 1.57 *	13.75 ± 2.06 *	9.40 ± 0.51	9.11 ± 0.25	9.28 ± 0.27	9.25 ± 0.15
APTT (sec)	17.33 ± 2.12	19.11 ± 1.24 *	19.95 ± 1.19 *	19.30 ± 1.18 *	15.58 ± 1.00	15.58 ± 1.06	16.23 ± 0.63	16.14 ± 1.12
FIB (mg/dL)	250.33 ± 20.57	238.66 ± 24.42	237.01 ± 17.83	230.77 ± 14.C0	169.19 ± 26.81	179.75 ± 9.73	179.30 ± 11.35	184.23 ± 6.15

*: $p \leq 0.05$ (compared to vehicle control group)

B

Parameters	Hematology (Mean ± SD)							
Group	Vehicle Control 0 mg/kg/day	Low-Dose 6 mg/kg/day	Mid-Dose 12 mg/kg/day	High-Dose 24 mg/kg/day	Vehicle Control 0 mg/kg/day	Low-Dose 6 mg/kg/day	Mid-Dose 12 mg/kg/day	High-Dose 24 mg/kg/day
	Male				Female			
Number of Animals	10	10	10	10	10	10	10	10
WBC (10^3/mL)	6.795 ± 1.627	6.146 ± 1.513	6.666 ± 1.470	5.904 ± 1.592	4.692 ± 1.857	4.364 ± 1.767	5.243 ± 1.342	4.668 ± 0.543
RBC (10^6/mL)	9.002 ± 0.389	9.360 ± 0.440	9.279 ± 0.326	9.039 ± 0.364	7.985 ± 0.340	7.845 ± 0.135	8.189 ± 0.315	7.899 ± 0.256
HGB (g/dL)	15.77 ± 0.70	16.36 ± 0.61	16.40 ± 0.28	16.16 ± 0.70	14.64 ± 0.53	14.42 ± 0.34	14.89 ± 0.42	14.63 ± 0.50
HCT (%)	45.59 ± 1.91	46.94 ± 1.56	47.31 ± 1.10	46.12 ± 1.92	41.59 ± 1.61	41.31 ± 0.70	42.92 ± 1.27	42.10 ± 1.25
MCV (fL)	50.68 ± 1.43	50.20 ± 1.40	51.04 ± 1.66	51.04 ± 1.46	52.10 ± 1.27	52.67 ± 1.21	52.45 ± 1.56	53.33 ± 0.97
MCH (pg)	17.53 ± 0.53	17.48 ± 0.61	17.68 ± 0.53	17.88 ± 0.52	18.33 ± 0.46	18.38 ± 0.58	18.18 ± 0.50	18.53 ± 0.35
MCHC (g/dL)	34.59 ± 0.24	34.85 ± 0.39	34.66 ± 0.30	35.04 ± 0.37 *	35.20 ± 0.32	34.90 ± 0.44	34.70 ± 0.35*	34.75 ± 0.25 *
PLT (10^3/mL)	1065.3 ± 81.9	1075.9 ± 166.1	1057.7 ± 106.9	1084.3 ± 80 6	1015.7 ± 138.0	940.4 ± 228.5	1027.8 ± 79.9	962.5 ± 110.6
NEU (%)	26.76 ± 11.05	27.61 ± 5.12	24.08 ± 9.26	28.85 ± 9.82	15.85 ± 5.84	12.41 ± 3.26	15.49 ± 5.00	13.04 ± 5.28
LYM (%)	65.57 ± 10.33	65.52 ± 5.25	68.80 ± 9.51	63.74 ± 9.73	78.13 ± 6.61	81.97 ± 4.62	79.32 ± 5.72	81.51 ± 6.03
MON (%)	6.81 ± 1.10	6.24 ± 0.65	6.64 ± 1.11	6.94 ± 1.16	5.42 ± 1.91	5.21 ± 1.37	4.92 ± 1.68	5.03 ± 1.09
EOS (%)	0.61 ± 0.47	0.40 ± 0.24	0.26 ± 0.12 *	0.28 ± 0.14 *	0.40 ± 0.33	0.34 ± 0.26	0.16 ± 0.14	0.28 ± 0.21
BAS (%)	0.25 ± 0.08	0.23 ± 0.11	0.22 ± 0.12	0.19 ± 0.09	0.20 ± 0.13	0.07 ± 0.19	0.11 ± 0.11	0.14 ± 0.10
RET (%)	2.214 ± 0.28	2.143 ± 0.382	2.206 ± 0.173	2.162 ± 0.309	2.088 ± 0.472	2.261 ± 0.413	1.982 ± 0.338	2.321 ± 0.362

*: $p \leq 0.05$ (compared to vehicle control group)

Table 4. *Cont.*

C								
Parameters	**Serum Chemistry (Mean ± SD)**							
Group	**Vehicle Control 0 mg/kg/day**	**Low-Dose 6 mg/kg/day**	**Mid-Dose 12 mg/kg/day**	**High-Dose 24 mg/kg/day**	**Vehicle Control 0 mg/kg/day**	**Low-Dose 6 mg/kg/day**	**Mid-Dose 12 mg/kg/day**	**High-Dose 24 mg/kg/day**
	Male				**Female**			
Number of Animals	10	10	10	10	10	10	10	10
AST (U/L)	99.25 ± 18.36	103.85 ± 22.32	93.14 ± 25.58	88.91 ± 12.18	77.66 ± 11.13	81.93 ± 16.99	79.52 ± 13.35	73.12 ± 12.26
ALT (U/L)	28.82 ± 3.59	31.71 ± 4.92	27.02 ± 6.61	28.38 ± 3.02	25.57 ± 3.51	25.57 ± 8.28	22.64 ± 3.33	22.01 ± 5.33
GLU (mg/dL)	222.80 ± 51.18	201.05 ± 47.60	200.08 ± 20.95	193.29 ± 40.03	171.20 ± 29.06	188.49 ± 43.42	171.36 ± 16.88	180.13 ± 32.94
TP (g/dL)	5.92 ± 0.43	5.93 ± 0.40	5.84 ± 0.24	5.85 ± 0.33	6.22 ± 0.34	6.10 ± 0.36	6.38 ± 0.41	6.15 ± 0.32
ALB (g/dL)	3.94 ± 0.27	3.97 ± 0.23	3.92 ± 0.20	3.82 ± 0.21	4.55 ± 0.28	4.48 ± 0.34	4.52 ± 0.37	4.42 ± 0.24
TBIL (mg/dL)	0.063 ± 0.015 (n = 9 [a])	0.063 ± 0.019 (n = 9 [a])	0.063 ± 0.013 (n = 9 [a])	0.062 ± 0.017 (n = 9 [a])	0.071 ± 0.019	0.082 ± 0.020 (n = 9 [a])	0.063 ± 0.013 (n = 8 [a])	0.071 ± 0.014
BUN (mg/dL)	13.90 ± 1.97	14.28 ± 1.52	13.43 ± 1.86	13.60 ± 0.76	14.31 ± 2.42	14.77 ± 2.26	13.48 ± 1.20	13.38 ± 1.58
CREA (mg/dL)	0.34 ± 0.07	0.35 ± 0.07	0.33 ± 0.05	0.33 ± 0.05	0.41 ± 0.06	0.44 ± 0.07	0.40 ± 0.07	0.40 ± 0.05
GGT (U/L)	<2	<2	<2	<2	<2	<2	<2	<2
ALP (U/L)	221.78 ± 29.19	226.35 ± 32.10	246.24 ± 38.71	247.44 ± 56.76	122.43 ± 28.33	114.42 ± 33.98	131.72 ± 52.40	125.49 ± 46.07
CHOL (mg/dL)	68.44 ± 18.01	53.91 ± 10.23 *	55.24 ± 7.85	53.31 ± 11.14 *	52.22 ± 10.91	62.32 ± 18.28	61.47 ± 13.94	53.50 ± 14.25
TG (mg/dL)	37.60 ± 19.36	22.68 ± 6.82 *	25.64 ± 9.37	21.51 ± 6.26 *	18.59 ± 6.85	20.89 ± 6.94	15.05 ± 4.51	18.59 ± 8.67
Ca (mg/dL)	9.93 ± 0.32	9.73 ± 0.32	9.84 ± 0.32	9.71 ± 0.29	10.07 ± 0.26	9.99 ± 0.35	10.02 ± 0.21	10.03 ± 0.24
P (mg/dL)	6.50 ± 0.58	6.63 ± 0.64	6.59 ± 0.28	6.32 ± 0.45	5.41 ± 0.79	5.56 ± 0.57	5.74 ± 0.84	5.92 ± 0.69
CK (U/L)	447.55 ± 326.04	433.81 ± 305.08	394.79 ± 340.79	367.31 ± 185.55	270.22 ± 150.71	285.67 ± 133.62	276.82 ± 182.04	249.60 ± 97.66
AMY (U/L)	1557.9 ± 240.9	1395.3 ± 269.0	1368.0 ± 148.0	1338.0 ± 182.3	872.4 ± 159.7	943.9 ± 210.7	991.5 ± 149.2	948.4 ± 309.0
Na (mmol/L)	143.41 ± 2.04	143.59 ± 1.28	143.76 ± 1.47	143.50 ± 0.96	142.17 ± 0.56	142.25 ± 1.16	142.61 ± 1.44	142.55 ± 1.49
K (mmol/L)	4.480 ± 0.432	4.390 ± 0.396	4.380 ± 0.204	4.350 ± 0.310	4.330 ± 0.340	4.410 ± 0.307	4.490 ± 0.446	4.280 ± 0.193
Cl (mmol/L)	105.50 ± 1.34	105.75 ± 1.49	105.62 ± 1.31	106.90 ± 1.93	105.79 ± 1.36	106.58 ± 1.26	106.31 ± 0.92	107.04 ± 1.72
LDH (U/L)	737.2 ± 459.1	653.7 ± 375.6	737.4 ± 683.5	673.0 ± 389.8	528.5 ± 315.9	513.6 ± 241.8	611.6 ± 440.1	502.8 ± 228.2
TBA (umol/L)	6.9 ± 2.6	10.7 ± 7.7	9.1 ± 3.6	12.2 ± 10.5	9.4 ± 3.2	18.0 ± 22.4	11.0 ± 3.3	8.9 ± 1.7

[a] The values of TBIL in other animals were lower than the detection limit.
*: $p \leq 0.05$ (compared to vehicle control group)

D					
Gender	**Parameters**	**Urine Quantitative Analysis (Mean ± SD)**			
	Group	**Vehicle Control 0 mg/kg/day**	**Low-Dose 6 mg/kg/day**	**Mid-Dose 12 mg/kg/day**	**High-Dose 24 mg/kg/day**
	Number of Animals	10	10	10	10
Male	Volume (mL)	17.70 ± 7.65	17.50 ± 5.85	16.60 ± 7.06	16.20 ± 9.58
	SG	1.0178 ± 0.0036 (n = 9 [a])	1.0180 ± 0.0035	1.0172 ± 0.0036 (n = 9 [a])	1.0190 ± 0.0046
	pH	7.35 ± 0.63	7.00 ± 0.24	7.30 ± 0.35	7.20 ± 0.26
	UURO (EU/dL)	0.28 ± 0.25	0.20 ± 0.00	0.20 ± 0.00	0.20 ± 0.00
Female	Volume (mL)	12.00 ± 5.29	8.30 ± 4.11	9.20 ± 3.46	12.00 ± 6.99
	SG	1.0170 ± 0.0042	1.0189 ± 0.0042 (n = 9 [a])	1.0183 ± 0.0035 (n = 9 [a])	1.0185 ± 0.0047
	pH	6.65 ± 0.47	6.95 ± 0.55	6.75 ± 0.59	7.00 ± 0.24
	UURO (EU/dL)	0.20 ± 0.00	0.20 ± 0.00	0.20 ± 0.00	0.20 ± 0.00

[a] The values of specific gravity from other animals were higher than the detection limit.

The MCHC (mean corpuscular hemoglobin concentration) value was found to be statistically higher in male rats at the dose level of 24 mg/kg/day, with a value of 35.04 ± 0.37 g/dL, while EOS (eosinophil count) percentages were statistically lower in 12 and 24 mg/kg/day, with 0.26 ± 0.12% and 0.28 ± 0.14%, respectively. In contrast, in female rats, the MCHC value was statistically lower at dose levels of 12 and 24 mg/kg/day at 34.70 ± 0.35 g/dL and 34.75 ± 0.25 g/dL, respectively. Nonetheless, all the values were within the historical control data range. Since the changes were of a small magnitude, with no dose-dependency or correlated findings noted in HCT and HGB, it was not considered treatment group-related findings.

Treatment with astaxanthin over a period of time showed no adverse effects on the clinical chemistry parameters (Table 4). Nevertheless, statistically significant differences were observed between vehicle control groups and treatment groups in male rats: the values of CHOL (cholesterol) and TG (triglycerides) at dose levels of 6 and 24 mg/kg/day were statistically lower than those in the vehicle control group at 53.31 ± 11.14 mg/dL and 21.51 ± 6.26 mg/dL, respectively. However, all values were within the historical control data range and were not considered treatment group-related findings. No statistical difference was noted when comparing the treatment group to the vehicle control group in female rats.

Both male and female rats' urinalysis results showed no treatment-related toxicologically significant changes (Table 4).

3.3.3. Relative Organ Weights

In both male and female rats, no treatment group-related finding was noted in relative organ weight (organ-to-brain weight) data (Table S6).

3.3.4. Gross Necropsy Findings

The gross necropsy findings are summarized in Table S7. In males, gross findings such as discoloration in the thymus (1/10) and adrenal glands (1/10) and abnormal shape in the brain (1/10) were noted in the vehicle control on Day 92. In females, discoloration in the mandibular lymph node (1/10) and nodule in the liver (1/10) were indicated in the vehicle control on Day 92. Nonetheless, these findings were not treatment-related due to the lack of dose dependency.

3.3.5. Histopathology Evaluation

No signs of necropsy or other histopathologic findings (Table S8) related to astaxanthin exposure were found in the control or treatment groups. While several findings, including congestion of the thymus and mandibular lymph node, mononuclear cell infiltration and fatty change in the liver, tubular epithelial cell degeneration, interstitial mononuclear cell infiltration, and mineralization of the kidneys were noted, these findings were incidental and were not considered treatment-related findings.

4. Discussion

With the rapid development of revolutionary technology, synthetic biology will soon play an essential role in future industries; whether it is energy, medicine, chemistry, food, biodegradable consumable products, or even medical testing, it will soon be deeply involved in the industry supply chain. However, many challenges remain before genetically modified organisms and its products can become mainstream on the market. One of the biggest challenges is the acceptance of anything genetically modified by the general public. Despite numerous studies proving that GM products are similar if not even better than the conventional non-GM products, many people are still wary.

Thus, we aim to alleviate these worries by the public. We used *K. marxianus*, an emerging non-conventional food-grade yeast, as our chassis. Additionally, since this particular yeast has aerobic-respiring characteristics [26], which will not produce any ethanol byproducts, our product can also be marketed as halal. To alleviate any concerns about genetic modification of the yeast, we also extracted and purified the desired astaxanthin. In our study, we have validated the purity and configuration of our extracted astaxanthin crystals and found that it still maintains the highly sought after high antioxidative function. In addition, we also assessed the safety of consuming our extracted astaxanthin crystals. Minor changes can be seen in mean food consumption, clinical observations, PT values, APTT values, MCHC values, EOS percentages, cholesterol values, triglycerides values, and gross necropsy observations. Nonetheless, these findings were not treatment-related in view of the lack of dose dependency. Our results were also comparable with several other coordinated studies on the subchronic toxicity of astaxanthin conducted by Katsumata et al. (2014) [27] and Lin et al. (2017) [15], which employed astaxanthin extracted from native bacteria *P. carotinifaciens* and engineered *E. coli*, respectively. In addition, subchronic toxicity of chemically synthesized astaxanthin and *H. pluvialis* biomass, which is rich in astaxanthin esters, were assessed by Vega et al. (2015) [28] and Stewart et al. (2008) [29], respectively, which revealed no significant biological changes.

Previous reviews have already explored the safety and bioavailability of astaxanthin, as well as the beneficial effects it has on human body [30,31]. Our findings have further confirmed the safety of astaxanthin produced by probiotic yeast.

5. Conclusions

This study provided an example that developing metabolically engineered microorganisms provides a safe and feasible approach for the bio-based mass production of beneficial compounds. Optimizing and engineering the *K. marxianus* strain for astaxanthin production has indeed allowed for record breaking astaxanthin production, which trumps even the natural astaxanthin producer *H. pluvialis* while maintaining the desired configuration and function. Screening of the potential adverse effects of astaxanthin crystals extracted from engineered *K. marxianus* 6-13-a6 by dosing SD rats with 6, 12, and 24 mg/kg/day of astaxanthin crystals by oral gavages for a 13-week period has indicated no significant biological changes between control and treatment groups in rats of both genders, further confirming the safety of the astaxanthin crystals.

Supplementary Materials: The following supporting information can be downloaded at: https://www.mdpi.com/article/10.3390/antiox11112288/s1.

Author Contributions: Conceptualization, J.-J.C. and J.-R.C.; methodology, J.-J.C.; software, H.-M.D.W.; validation, M.-Y.H., Y.-S.C. and S.Y.S.; formal analysis, M.-Y.H.; investigation, Y.-S.C.; resources, J.-J.C.; data curation, J.-R.C.; writing—original draft preparation, S.Y.S. and H.-M.D.W.; writing—review and editing, S.Y.S., J.-J.C. and W.-H.L.; visualization, H.-M.D.W.; supervision, J.-J.C.; project administration, S.Y.S.; funding acquisition, J.-J.C. All authors have read and agreed to the published version of the manuscript.

Funding: This research was funded by China Medical University (CMU-108-MF-35), China Medical University Hospital (DMR-110-142), Ministry of Science and Technology (109-2621-M-039-001 and 108-2621-M-039-002) and Taipei city government (1Z1100894), and the Ministry of Economic Affairs (TBA210129003) in Taiwan.

Institutional Review Board Statement: The study was conducted according to the guidelines of the Declaration of Helsinki and approved by the Institutional Animal Care and Use Committee (IACUC number: 210310) of Level Biotechnology Inc., New Taipei City, Taiwan.

Informed Consent Statement: Not applicable.

Data Availability Statement: All data generated or analyzed during this study are included in the published article (and its online Supplementary Files).

Acknowledgments: The authors thank Yu-Ju Lin for her technical support while she was the CTO of Trade Wind Biotech, Taiwan. The authors also thank the CRO-Preclinical Testing Center of Level Biotechnology Inc. for helping in this 13-week repeated dose oral toxicity study in SD rats. The authors would also like to thank Trade Wind Biotech Co. Ltd. for providing Asta-S Crystal.

Conflicts of Interest: The authors declare that they have no known competing financial interests or personal relationships that could have appeared to influence the work reported in this paper.

References

1. Higuera-Ciapara, I.; Félix-Valenzuela, L.; Goycoolea, F.M. Astaxanthin: A review of its chemistry and applications. *Crit. Rev. Food Sci. Nutr.* **2006**, *46*, 185–196. [CrossRef]
2. Miki, W. Biological functions and activities of animal carotenoids. *Pure Appl. Chem.* **1991**, *63*, 141–146. [CrossRef]
3. Pashkow, F.J.; Watumull, D.G.; Campbell, C.L. Astaxanthin: A Novel Potential Treatment for Oxidative Stress and Inflammation in Cardiovascular Disease. *Am. J. Cardiol.* **2008**, *101*, S58–S68. [CrossRef]
4. Wang, X.-J.; Chen, W.; Fu, X.-T.; Ma, J.-K.; Wang, M.-H.; Hou, Y.-J.; Tian, D.-C.; Fu, X.-Y.; Fan, C.-D Reversal of homocysteine-induced neurotoxicity in rat hippocampal neurons by astaxanthin: Evidences for mitochondrial dysfunction and signaling crosstalk. *Cell Death Discov.* **2018**, *4*, 50. [CrossRef]

5. Satoh, A.; Tsuji, S.; Okada, Y.; Murakami, N.; Urami, M.; Nakagawa, K.; Ishikura, M.; Katagiri, M.; Koga, Y.; Shirasawa, T. Preliminary Clinical Evaluation of Toxicity and Efficacy of a New Astaxanthin-rich *Haematococcus pluvialis* Extract. *J. Clin. Biochem. Nutr.* **2009**, *44*, 280–284. [CrossRef]
6. Tanaka, T.; Shnimizu, M.; Moriwaki, H. Cancer Chemoprevention by Carotenoids. *Molecules* **2012**, *17*, 3202–3242. [CrossRef] [PubMed]
7. Wayama, M.; Ota, S.; Matsuura, H.; Nango, N.; Hirata, A.; Kawano, S. Three-Dimensional Ultrastructural Study of Oil and Astaxanthin Accumulation during Encystment in the Green Alga *Haematococcus pluvialis*. *PLoS ONE* **2013**, *8*, e53618. [CrossRef] [PubMed]
8. Johnson, E.A.; An, G.-H. Astaxanthin from Microbial Sources. *Crit. Rev. Biotechnol.* **1991**, *11*, 297–326. [CrossRef]
9. Hussein, G.; Sankawa, U.; Goto, H.; Matsumoto, K.; Watanabe, H. Astaxanthin, a Carotenoid with Potential in Human Health and Nutrition. *J. Nat. Prod.* **2006**, *69*, 443–449. [CrossRef]
10. Authority, E.F.S. Opinion of the Scientific Panel on additives and products or substances used in animal feed (FEEDAP) on the safety and efficacy of the product AQUASTA, an Astaxanthin-rich Phaffia rhodozyma ATCC SD-5340 for salmon and trout. *EFSA J.* **2006**, *4*, 320. [CrossRef]
11. Capelli, B.; Bagchi, D.; Cysewski, G.R. Synthetic astaxanthin is significantly inferior to algal-based astaxanthin as an antioxidant and may not be suitable as a human nutraceutical supplement. *Nutrafoods* **2013**, *12*, 145–152. [CrossRef]
12. Rüfer, C.E.; Moeseneder, J.; Briviba, K.; Rechkemmer, G.; Bub, A. Bioavailability of astaxanthin stereoisomers from wild (*Oncorhynchus* spp.) and aquacultured (*Salmo salar*) salmon in healthy men: A randomised, double-blind study. *Br. J. Nutr.* **2008**, *99*, 1048–1054. [CrossRef]
13. EUR-Lex-32006R1925-EN-EUR-Lex. Europa.eu. 2021. Available online: https://eur-lex.europa.eu/legal-content/EN/ALL/?uri=celex%3A32006R1925 (accessed on 17 November 2022).
14. Chiou, M.-H.; Wu, M.-C.; Yang, M.-H.; Lin, C.-N.; Lu, Y.-C. A Strain of Genetically Reengineered *Escherichia coli* for Biosynthesis of High Yield Carotenoids after Mutation Screening. (European Patent Office Patent). 2009. Available online: https://patents.google.com/patent/EP2088199A1/en (accessed on 17 November 2022).
15. Wang, C.W.; Oh, M.K.; Liao, J.C. Engineered isoprenoid pathway enhances astaxanthin production in Escherichia coli. *Biotechnol. Bioeng.* **1999**, *62*, 235–241. [CrossRef]
16. Lin, Y.-J.; Lin, J.-Y.; Wang, D.-S.; Chen, C.-H.; Chiou, M.-H. Safety assessment of astaxanthin derived from engineered Escherichia coli K-12 using a 13-week repeated dose oral toxicity study and a prenatal developmental toxicity study in rats. *Regul. Toxicol. Pharmacol.* **2017**, *87*, 95–105. [CrossRef] [PubMed]
17. Zhou, P.; Xie, W.; Li, A.; Wang, F.; Yao, Z.; Bian, Q.; Zhu, Y.; Yu, H.; Ye, L. Alleviation of metabolic bottleneck by combinatorial engineering enhanced astaxanthin synthesis in Saccharomyces cerevisiae. *Enzyme Microb. Technol.* **2017**, *100*, 28–36. [CrossRef] [PubMed]
18. Chang, J.-J.; Ho, C.-Y.; Mao, C.-T.; Barham, N.; Huang, Y.-R.; Ho, F.-J.; Wu, Y.-C.; Hou, Y.-H.; Shih, M.-C.; Li, W.-H.; et al. A thermo- and toxin-tolerant kefir yeast for biorefinery and biofuel production. *Appl. Energy* **2014**, *132*, 465–474. [CrossRef]
19. Banat, I.M.; Nigam, P.; Marchant, R. Isolation of thermotolerant, fermentative yeasts growing at 52 °C and producing ethanol at 45 °C and 50 °C. *World J. Microbiol. Biotechnol.* **1992**, *8*, 259–263. [CrossRef]
20. Raimondi, S.; Zanni, E.; Amaretti, A.; Palleschi, C.; Uccelletti, D.; Rossi, M. Thermal adaptability of Kluyveromyces marxianus in recombinant protein production. *Microb. Cell Factories* **2013**, *12*, 34. [CrossRef]
21. Lin, Y.-J.; Chang, J.-J.; Lin, H.-Y.; Thia, C.; Kao, Y.-Y.; Huang, C.-C.; Li, W.-H. Metabolic engineering a yeast to produce astaxanthin. *Bioresour. Technol.* **2017**, *245*, 899–905. [CrossRef]
22. Tseng, C.-C.; Lin, Y.-J.; Liu, W.; Lin, H.-Y.; Chou, H.-Y.; Thia, C.; Wu, J.H.; Chang, J.-S.; Wen, Z.-H.; Chang, J.-J.; et al. Metabolic engineering probiotic yeast produces 3S, 3′S-astaxanthin to inhibit B16F10 metastasis. *Food Chem. Toxicol.* **2020**, *135*, 110993. [CrossRef]
23. Arnao, M.B.; Cano, A.; Acosta, M. The hydrophilic and lipophilic contribution to total antioxidant activity. *Food Chem.* **2001**, *73*, 239–244. [CrossRef]
24. Brand-Williams, W.; Cuvelier, M.E.; Berset, C. Use of a free radical method to evaluate antioxidant activity. *LWT - Food Sci. Technol.* **1995**, *28*, 25–30. [CrossRef]
25. Wang, W.; Bostic, T.R.; Gu, L. Antioxidant capacities, procyanidins and pigments in avocados of different strains and cultivars. *Food Chem.* **2010**, *122*, 1193–1198. [CrossRef]
26. Wagner, J.M.; Alper, H.S. Synthetic biology and molecular genetics in non-conventional yeasts: Current tools and future advances. *Fungal Genet. Biol.* **2016**, *89*, 126–136. [CrossRef] [PubMed]
27. Katsumata, T.; Ishibashi, T.; Kyle, D. A sub-chronic toxicity evaluation of a natural astaxanthin-rich carotenoid extract of Paracoccus carotinifaciens in rats. *Toxicol. Rep.* **2014**, *1*, 582–588. [CrossRef]
28. Vega, K.; Edwards, J.; Beilstein, P. Subchronic (13-week) toxicity and prenatal developmental toxicity studies of dietary astaxanthin in rats. *Regul. Toxicol. Pharmacol.* **2015**, *73*, 819–828. [CrossRef]
29. Stewart, J.S.; Lignell, Å.; Pettersson, A.; Elfving, E.; Soni, M.G. Safety assessment of astaxanthin-rich microalgae biomass: Acute and subchronic toxicity studies in rats. *Food Chem. Toxicol.* **2008**, *46*, 3030–3036. [CrossRef]

30. Haung, H.-Y.; Wang, Y.-C.; Cheng, Y.-C.; Kang, W.; Hu, S.-H.; Liu, D.; Xiao, C.; Wang, H.-M.D. A Novel Oral Astaxanthin Nanoemulsion from *Haematococcus pluvialis* Induces Apoptosis in Lung Metastatic Melanoma. *Oxid. Med. Cell. Longev.* **2020**, *2020*, 2647670. [CrossRef]
31. Chen, Y.-T.; Kao, C.-J.; Huang, H.-Y.; Huang, S.-Y.; Chen, C.-Y.; Lin, Y.-S.; Wen, Z.-H.; Wang, H.-M.D. Astaxanthin reduces MMP expressions, suppresses cancer cell migrations, and triggers apoptotic caspases of in vitro and in vivo models in melanoma. *J. Funct. Foods* **2017**, *31*, 20–31. [CrossRef]

antioxidants

MDPI

Article

Apple Polyphenol Diet Extends Lifespan, Slows down Mitotic Rate and Reduces Morphometric Parameters in Drosophila Melanogaster: A Comparison between Three Different Apple Cultivars

Silvia Bongiorni [1], Ivan Arisi [2], Brunella Ceccantoni [3], Cristina Rossi [1], Camilla Cresta [1], Simona Castellani [4], Ivano Forgione [4], Sara Rinaldducci [1], Rosario Muleo [4,*] and Giorgio Prantera [1,*]

1. Department of Ecological and Biological Sciences (DEB), University of Tuscia, 01100 Viterbo, Italy
2. Bioinformatics Facility European Brain Research Institute (EBRI) "Rita Levi-Montalcini", and Institute of Translational Pharmacology National Research Council (CNR), Viale Regina Elena 295, 00161 Roma, Italy
3. Department for Innovation in Biological, Agro-Food and Forest Systems (DIBAF), University of Tuscia, 01100 Viterbo, Italy
4. Tree Physiology and Fruit Crop Biotechnology Laboratory, Department of Agriculture and Forest Sciences (DAFNE), University of Tuscia, 01100 Viterbo, Italy

* Correspondence: muleo@unitus.it (R.M.); prantera@unitus.it (G.P.)

Abstract: Plant-derived polyphenols exhibit beneficial effects on physiological and pathological processes, including cancer and neurodegenerative disorders, mainly because of their antioxidant activity. Apples are highly enriched in these compounds, mainly in their peel. The Tuscia Red (TR) apple variety exhibits the peculiar characteristic of depositing high quantities of polyphenols in the pulp, the edible part of the fruit. Since polyphenols, as any natural product, cannot be considered a panacea per se, in this paper, we propose to assess the biological effects of TR flesh extracts, in comparison with two commercial varieties, in a model system, the insect *Drosophila melanogaster*, largely recognized as a reliable system to test the in vivo effects of natural and synthetic compounds. We performed a comparative, qualitative and quantitative analysis of the polyphenol compositions of the three cultivars and found that TR flesh shows the highest content of polyphenols, and markedly, anthocyanins. Then, we focused on their effects on a panel of physiological, morphometrical, cellular and behavioral phenotypes in wild-type *D. melanogaster*. We found that all the apple polyphenol extracts showed dose-dependent effects on most of the phenotypes we considered. Remarkably, all the varieties induced a strong relenting of the cell division rate.

Keywords: apple polyphenols; *Drosophila melanogaster*; model organisms; red flesh apple; lifespan; mitotic index; climbing assay

Citation: Bongiorni, S.; Arisi, I.; Ceccantoni, B.; Rossi, C.; Cresta, C.; Castellani, S.; Forgione, I.; Rinalducci, S.; Muleo, R.; Prantera, G. Apple Polyphenol Diet Extends Lifespan, Slows down Mitotic Rate and Reduces Morphometric Parameters in Drosophila Melanogaster: A Comparison between Three Different Apple Cultivars. *Antioxidants* **2022**, *11*, 2086. https://doi.org/10.3390/antiox11112086

Academic Editors: Jicheng Zhan, Zhigang Liu and Hui-Min David Wang

Received: 30 September 2022
Accepted: 19 October 2022
Published: 22 October 2022

1. Introduction

Redox homeostasis represents a delicate balance between the production and removal of free radicals that plays critical roles in human health and is constantly regulated by oxidative stress and antioxidative defense systems [1] The overproduction of free radicals and reactive oxygen species (ROS), known as oxidative stress, can damage biomolecules, such as lipids, nucleic acids, proteins, causes cellular membrane peroxidation [2–4] and attracts various inflammatory mediators [5,6], leading directly or indirectly to severe diseases, such as cancer, atherosclerosis and neurological disorders [1,7,8]. Although the organisms possess endogenous systems to prevent radical-induced oxidative damage, they are sometimes insufficient to counteract extensive damage. Numerous epidemiological investigations evidenced the protection that increased consumption of fruit and vegetables exerts against cardiovascular diseases, various kinds of cancer, aging-related cognitive decline and other chronic diseases [9–12]. These properties were ascribed to the presence of phytochemicals, such as phenols [13]. Phenolic compounds are among the approximately 8000 secondary metabolites of plants that are synthesized in response to several adverse stimuli during plant development [14]. The beneficial effects of plant-derived polyphenols on a plethora of physiological and/or pathological processes, including cancer and neurodegenerative disorders, are well-known [15,16]. The main property of polyphenols is represented by the antioxidant activity they exert by scavenging free radicals normally produced by cell metabolism or in response to external factors [2,17–19]. Due to their richness in flavonoids and phenolic acids, apple fruits are considered one of the healthier fruits [20]. However, generally, apple polyphenols are enriched in the peel, and scarce and unstable in the pulp, which represents most of the edible part of the fruit. The red coloration of the apple fruit was ascribed to the accumulation of anthocyanins under the control of a MYB transcription factor [21–23].

The Tuscia Red apple variety (TR) was isolated in the experimental farm of the University of Tuscia by selection among a pool of seedlings generated from an open-pollinated old variety of Central Italy, having a red coloration of flesh and peel [24,25]. The origin of the old, red-fleshed variety could be ascribed to a lineage from the partially domesticated form of *M. sieversii* f. niedzwetzkyana, which contributed to give rise to the *M. domestica* var. niedzwetzkyana [23]. Apple species moved West along the ancient Silk Road, and the Romans had a preeminent role in the diffusion to Western Europe, from where they diversified and flourished worldwide [26]. TR combines the good taste of the domestic cultivar and the high polyphenol content of the pulp of *M. niedzwetzkyana*.

In the present paper, we preliminary assessed the effects onto an in vivo model system of the polyphenol-rich TR apple pulp in comparison with the pulps of two commercial apple varieties, Annurca and Fuji. The fruit fly, *Drosophila melanogaster*, is a widely used model system in biomedical research [27–29] and an excellent model organism for the evaluation of drug toxicity and for the identification of pharmacological properties of plant-derived compounds [30–32].

The effects of the polyphenol pool extracted from the flesh of the three apple varieties were assessed on a group of biological parameters in *D. melanogaster*, including development timing, lifespan, starvation, morphometrical measures, cell division and locomotor abilities.

2. Materials and Methods

2.1. Experimental Site and Plant Materials

The apple fruits were collected from an apple orchard located at the experimental farm of the Tuscia University in Viterbo, Italy. The apple orchard had Tuscia Red, Annurca and Fuji cultivars, grafted onto MM111 rootstock and grown as a spindle training system. The trees were planted in 2009, with intra- and inter-row distances of 3.0 m and 4.5 m, respectively. The harvesting time window was determined by visually inspecting the fruit exocarp color and maturation. The harvest was performed for each cultivar at apple commercial maturity, taking into consideration the assessment of fruit color, fruit firmness and starch index.

2.2. Apple Polyphenols Extraction

Polyphenols were extracted from three apple varieties, consisting of two commercial varieties, Annurca and Fuji, and one variety, the TR, obtained from open-pollination at the experimental agricultural farm at the Tuscia University. Three biological replicates were used for each extraction procedure. Three apples for each cultivar for each replicate were used.

Fresh fruit, deprived of peels, kept at 4 °C, were ground in a mortar under liquid nitrogen until a fine powder was obtained. The apple powder was immediately freeze-dried (EDWARDS modulo, Akribis Scientific Limited, Chester, UK), weighted and extracted in 5 mL of slightly acidified 70% methanol (pH = 4.00) for each gram of freeze-dried apple powder. The solution was sonicated for 10 s, vortexed and centrifuged at 15.000 rpm for

15 min. The pellet was separated from the supernatant, which was re-extracted in 5 mL per gram of slightly acidified 70% methanol (pH = 4.00), vortexed and centrifugated as above. The resulting supernatant was added to the supernatant previously obtained, filtered with 0.45 µm Whatman filters and evaporated by Rota-vapor at 35 °C under low pressure conditions. The remaining aqueous part of the solution was freeze-dried. Freeze-dried polyphenols were re-suspended in 3 mL of pure methanol for each gram of initial freeze-dried apple powder. The total phenolic content (TPC) of extracts was determined using the Folin–Ciocalteu (F–C) method. TPC was calculated from a calibration curve, using Gallic acids as a standard. Results are expressed as milligrams of Gallic acid equivalents (GAE) per 100 g of fresh weight of apple.

2.3. Total Anthocyanin Assay Method

Anthocyanins assay (AA) of the apple pulp was conducted according to Mancinelli et al. (1975) [33]. The grinded fresh sample (5 g), under liquid nitrogen, was placed in a test tube, and the anthocyanin fraction was extracted with 50 mL of 1% (*w*/*v*) HCl in methanol. The mixture was placed in the dark at 4 °C for 24 h, with continuous shaking. The extracts were cleared by filtration, and their absorbance at 530 nm was detected with a spectrophotometer and compared to a cyanidin-glucoside standard purchased by Extrasynthese (Genay, France).

2.4. HPLC Analysis

The total polyphenolic extract components, including phenolic acids (chlorogenic acid), dihydro-chalcones (phloretin-glucosides), anthocyanins (cyanidin-glucosides), flavonols (quercetin 3-O-ramnoside, quercetin 3-O-glucoside), flavanols (catechin and epicatechin) and procyanidins, were identified and quantified with reversed-phase liquid chromatography using a Dionex chromatograph (Dionex Corporation, Sunnyvale, CA, USA), equipped with a P680 quaternary pump, manual injector (Rheodyne) with 20 µL loop, TCC-100 thermostat oven and PDA-100 photodiode array detector, and controlled using Chromeleon software (version 6.50). The separation was carried out at 30 °C with a Dionex Acclaim® 120 C18 column, 5 µm, 120 Å (4.6 × 250 mm). Optimized conditions were determined by flow injection analysis (FIA) of standard solutions at three different concentrations, ranging from 0 to 100 mg/L. Separations were performed by the mobile phase consisting of a ternary gradient: solvent A = 50 mM ammonium dihydrogen phosphate (99% purity) acidified to pH 2.6 with phosphoric acid (≥85% purity); solvent B = 20% solvent A and 80% acetonitrile (99.9% purity); solvent C = 0.2 M orthophosphoric acid (85% purity) acidified to pH 1.5 with NaOH (99.99% purity). All the HPLC reagents were supplied by the Sigma-Aldrich Chemical Co. (St. Louis, MO, USA). Samples were introduced into the column using a single injection with a 20 µL sample loop. The phenolic compounds were identified based on their elution order and the retention times of pure compounds and the characteristics of their UV–visible spectra. Catechin, epicatechin, procyanidin B1, procyanidin B2, quercetin-glucoside, cyanidin-glucoside, cyanidin 3-galactoside, phlorizin, chlorogenic acid, p-Coumaryc acid and caffeic acid standards were purchased at Extrasynthese (Genay, France).

2.5. Fruit Fly Strains and Treatments

All assays were performed using OrR wild-type stock obtained from the Bloomington Drosophila Stock Center (BDSC) at Indiana University (Bloomington, IN, USA). Flies were raised on a standard corn meal agar food (pH 5.5) at 25 °C. For the assays, which involved the exposure of embryos and larvae to apple polyphenols (APs), the fly diet was prepared by mixing the above standard food with 1.5, 3.0 or 5.0 mg/mL APs. For all the treatments, vials containing 10 mL of standard or AP-enriched diet were used. Experiments were conducted with only 1.5 and 3.0 mg/mL AP concentrations since it was preliminary assessed that the 5.0 AP concentration was harmful to the normal development and survival of flies.

One- to five-day-old flies were used in the experiments involving adult animals.

2.6. Developmental Assay

Larval development in all sets of treatments was checked every 24 h, and the number of pupae and emerging offspring, with dates of occurrence, were recorded.

2.7. Lifespan Assay

Flies were grown in either standard food or APs-enriched food from egg deposition to eclosure [34]. A total of 100 emerged offspring per treatment were collected and used for lifespan assays. Newly eclosed flies from each treatment were divided and housed in 10 vials (10 flies per vial). The flies were no longer exposed to APs-enriched food throughout their life. Every 2–3 days, flies were transferred to fresh food and deaths were recorded until all flies died.

2.8. Stress Assay

A total of 100 flies per treatment were used for the starvation assay. Adult flies were housed in vials containing 2% agarose, to provide moisture, without any energy source. Deaths were recorded every 4 h.

2.9. Size and Weight of Emerged Offspring

For size measurements, 100 flies per treatment were randomly selected from the population of emerged offspring. Fly length was measured from the top of the head to the end of the abdomen on graph paper. To determine weight, female and male adult individuals were separately weighed using a Sartorius SE2 Ultra Micro Balance (Bradford, MA, USA) in groups of 10 (n = 100 per treatment, per sex).

2.10. Mitotic Index

Brains from third instar larvae were dissected in hypotonic solution (0.8% sodium citrate) and fixed at room temperature in a freshly prepared mixture of acetic acid/methanol/H_2O (11:12:2) for 30 s. Single fixed brains were then transferred into small drops of 45% acetic acid on a clean, dust-free non-siliconized coverslip for 2 min. A clean slide was lowered onto the coverslip and gently squashed. The slide was frozen in liquid nitrogen, the coverslip was removed with a razor blade, and the slide was immediately immersed in absolute ethanol at −20 °C for 15 min. Slides were air-dried and stained with 0.2 µg/mL DAPI fluorochrome (Sigma-Aldrich Chemical Co, St. Louis, MO, USA) in 2x saline sodium citrate (20x = 0.15 M NaCl, 0.015 M sodium citrate) to detect chromatin and mitotic figures [35]. The mitotic index was defined as the number of mitotic cells per optical field [36]. The optical field was the circular area defined by a 100× Zeiss objective/1.30 Plan-NEOFLUAR, using 10× oculars and the Optovar set at 1.25 (Zeiss Axiophot microscope). Every optical field occupied by brain tissue was scored (three brains each slide, from at least 50–100 optical fields per slide per treatment). DAPI fluorescence was observed using 375/28× nm excitation filter (Chroma Technology Corp., Bellows Falls, VT, USA).

2.11. Climbing Assay

The climbing ability test was conducted on young adult fruit flies. For each experiment, a group of 10 flies was placed in an empty vial and given 30 s to climb up. A horizontal line was drawn 17 cm above the bottom of the vial. The number of flies per treatment that could climb from the bottom to above the 17 cm mark in 10, 20 and 30 s was recorded as percentage of total flies. The test was performed three times in young adults between 1 and 5 days after eclosure.

2.12. Statistical Analysis

Different statistical tests were applied to different experimental assays.

Total polyphenol content: Anova + post hoc LSD test; Anthocyanin content: Two-tailed Mann–Whitney U test; HPLC analysis: t test and two-tailed t test; Lifespan: Pairwise log-rank test, fdr correction; Starvation: Two-tailed Mann–Whitney U test, fdr correction; Larval length: Mann–Whitney U test, fdr correction; Larval weight: Two-tailed t test; Mitotic index: Fisher's exact test, fdr correction; Climbing assay: Fisher's exact test.

3. Results

The experiments reported in this study aimed to evaluate the effects of polyphenols extracted from three different apple varieties, TR, Annurca and Fuji, on several biological parameters. The lyophilized polyphenol extracts were diluted in the *Drosophila* standard food. The reported polyphenol concentrations refer to the final dilution of the extracts in 10 mL of standard corn meal agar food (see Material and Methods). In preliminary experiments, we assessed that the extract concentration of 5 mg/mL was harmful to the normal development and survival of flies, showing toxic effects that caused massive lethality. For this reason, we reported in detail only the results obtained with the concentrations of 1.5 and 3 mg/mL.

3.1. Apple Polyphenol Extracts Characterization

The scenario depicted by the total polyphenol content detected in the fruit flesh of the three apple cultivars (Table 1), highlighted that the TR flesh accumulates higher amounts of these compounds, more than 1.5 times that of the Annurca flesh content and three times that of Fuji flesh. In particular, the differences are even greater if we take into consideration the anthocyanin content detected in the fruit flesh of the three cultivars; therefore, the TR anthocyanin content (4 mg per 100 g of fresh weight) was 16 times greater than that of Annurca, whereas in Fuji, no anthocyanin content was detectable. The anthocyanin enrichment of TR fruit flesh is likely responsible of the red variegation of the pulp that can be clearly visualized in the longitudinal fruit sections of TR that is absent in the Fuji and Annurca pulp (Supplementary Figure S1).

Table 1. Total polyphenol and anthocyanin detected in the flesh of apple fruit of the cultivars studied. The content of polyphenol is expressed as mg of Gallic Acid equivalent for 100 g of fresh weight of flesh. The content of anthocyanin is expressed as mg equivalent of cyanidin 3-glucoside for 100 g of fresh weight of flesh. Values represent the average ± S.D. For total polyphenol content, Anova + post hoc LSD test was used to assess the significance of the observed differences in the three cultivars. Anova: F = 89.621, $p < 0.001$; a vs. b, a vs. c, and b vs. c: LSD $p \leq 0.001$; For the difference in Anthocyanin content: d vs. e $p < 0.001$, Mann–Whitney U test.

Compound	Fuji	Tuscia Red	Annurca
Total Polyphenol (mg GAE/100 g fw)	101.7 ± 6.8 a	316.8 ± 28.4 b	183.6 ± 39.7 c
Anthocyanin content (mg Cy3Glu/100 g fw)	N.D.	4.234 ± 0.643 d	0.256 ± 0.078 e

Given the low total polyphenol content of Fuji flesh, further analytical analyses were limited to TR and Annurca apples. HPLC-DAD analysis results showed relevant differences in phenolic compound assortments of apple fruit flesh between TR and Annurca (Table 2). Hydroxy-cinnamic acids predominated in both varieties, principally due to the high levels of chlorogenic acid, whose concentration in TR fruit attains about double that of Annurca. Epicatechin and procyanidin B2 showed the highest concentrations in comparison to the other flavanols, while the catechin content was very low in both varieties. As a whole, the TR red fleshed fruits showed higher levels of total flavanols in comparison to Annurca (more than three folds).

Table 2. Polyphenolic content and identification by HPLC analysis in Annurca and Tuscia Red fruit flesh. Numbers following the ± symbol indicate the standard deviation. Values are expressed as mg of compound per 1 kg of fresh weight. Values significantly different ($p < 0.05$) between Tuscia Red and Annurca are indicated by * (two-tailed t test) or by ** (t test), which are placed to the side of the higher value. Examples of chromatograms of three representative compounds in the two cultivar fleshes are reported in Supplemental Figure S2.

Compounds (mg/Kg fw)	Apple Variety	
	Annurca	**Tuscia Red**
Cholorogenic acid	1439.2 ± 225.7	2427.5 ± 574.2
Coumaryl-quinic acid	203.4 ± 69.6	491.2 ± 160.9 *
Other hydroxyl-cinnamics	39.4 ± 15.4	68.4 ± 28.8
Total hydroxyl-cinnamics	1682.1± 310.7	2987.1 ± 763.9
Catechin	3.7 ± 1.2	17.2 ± 7.3 *
Epicatechin	107.5 ± 24.6	317.8 ± 110.2 *
Procyanidin B1	71.4 ± 21.4	194.7± 56.8 *
Procyanidin B2	146.5 ± 30.7	498.8 ± 169.0 *
Other procyanidins	81.4 ± 2.2	465.8 ± 33.7 *
Toal flavanols and procyanidins	410.5 ± 80.2	1494.3 ± 377.0 *
Cyanidin 3′-5′ diglucoside	0.0	12.6 ± 6.4 **
Cyanidin 3′-galactoside	0.0	64.1 ± 27.6 **
Other cyanidins	0.0	6.6 ± 0.3 **
Total anthocyanin	0.0	83.3 ± 34.2 **
Quercitin 3′ glucoside	2.8 ± 2.3	11.5 ± 6.6
Quercitin 3′ ramnoside	5.5 ± 4.0	11.6 ± 6.9
Total flavonols	8.4 ± 6.4	23.1 ± 13. 0
3′ hydroxyl-phloretin-glucoside	52.5 ± 10.2	42.2 ± 11.8
3′ phloretin-xyloglucoside	119.5 ± 42.4 *	23.5 ± 12.8
Phlorizin	31.4 ± 16.9	62.1 ± 27.0
Total di-hydro chalcones	203.4 ± 69.5	127.8 ± 41.7
Total polyphenols	2326.2 ± 466.7	4617.9 ± 1166.3 *

Flavonols represent only the smallest component (<1%) of the total polyphenolic content, with quercetin-glycosides the only flavonols found in the flesh of the two cultivars.

As expected, phloretin-glycosides represented only a small part of the phenolic composition of the apple flesh in both varieties, but with interesting differences; phloretin-xyloglucoside showed much higher levels (5:1) in Annurca than in TR, while phlorizin (phloretin-glucoside) content in TR was double that of Annurca.

Exclusive to TR was the accumulation of anthocyanin in fruit flesh. The main accumulated anthocyanins were cyanidin-3-galactoside, and to a minor extent cyanidin 3′-5′ diglucoside.

3.2. *Apple Polyphenols Do Not Impact Drosophila Development*

The effect of apple polyphenol (APs) from Fuji, Annurca and TR apple on Drosophila development was evaluated by feeding embryos and larval stages with a normal diet (basal diet BD) supplemented with APs at two doses. We monitored the duration of the following stages: from egg laying until the emergence of the 3rd instar larvae, from the 3rd instar larvae to pupation, from pupation until the eclosion of adults. APs extracts from all cultivars at both stages produced no significant effects on larvae development, pupation and emergence of adults (data not shown).

3.3. *Apple Polyphenols Extend Adult Fly Lifespan*

To test the effect of apple APs on lifespan, flies were fed with APs from the three different cultivars at 1.5 and 3 mg/mL, during larval development only. The emerged offspring was transferred to normal food and the life duration was recorded as (i) day of the last surviving fly death, (ii) day of 50% survivors and (iii) mean lifespan. APs extracts from the three different cultivars at both concentrations showed a lifespan extension of the last surviving individual as compared to normally-fed flies (Figure 1), with the 3 mg/mL APs diet (Figure 1B) effect greater than that of 1.5 mg/mL diet (Figure 1A).

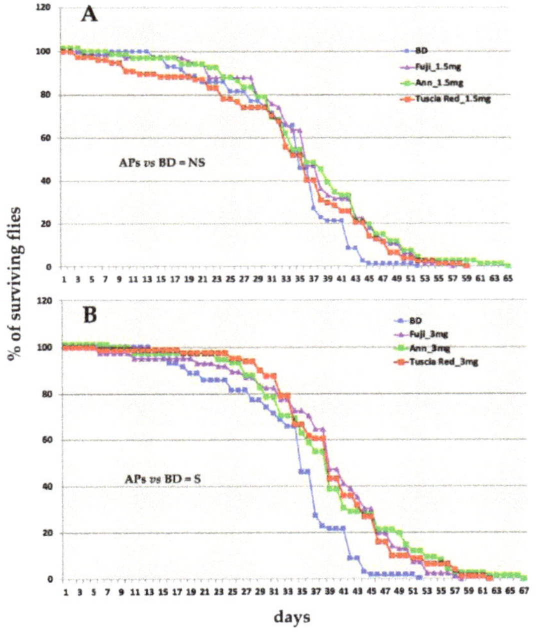

C

	max lifespan of last surviving fly (days)	max lifespan of last surviving fly (days)	50% survival (days)	50% survival (days)	mean lifespan (days)	mean lifespan (days)
BD	52		35		33.71	
	APs 1.5 mg	APs 3 mg	APs 1.5 mg	APs 3 mg	APs 1.5 mg	APs 3 mg
Fuji	57	58	36	39	36.56	38.86
Annurca	65	67	36	37	36.31	38.56
Tuscia Red	59	62	35	39	33.68	38.56

Figure 1. (**A**,**B**) The effects of 1.5 (**A**) and 3.0 (**B**) mg/mL APs from the three different apple cultivars on *D. melanogaster* lifespan. In the vertical axis, the number of surviving flies is reported; in the horizontal axis, days from the adult emergence until the fly death are reported. (**C**) In the table, the max lifespan of the last surviving individual, the day at which 50% of the original population still survived and the mean lifespan are reported. BD = basal diet with no supplement. NS = non-significant; S = significant. Statistical analysis (pairwise log-rank test, fdr correction) was applied to the differences in survival curves (3.0 mg/mL) between BD-fed and APs-fed flies. Tuscia Red vs. BD, $p < 0.01$; Fuji vs. BD, $p < 0.001$; Annurca vs. BD, $p = 0.01$.

However, statistical analysis showed that only the 3 mg/mL APs diet produced a significant increase in this parameter as compared to the basal diet (Figure 1C).

3.4. Apple Polyphenols Ameliorate Fly Resistance to Caloric Restriction

The survival to food deprivation of 100 animals for each experimental point was recorded daily.

Flies that emerged from APs food at both 1.5 (Figure 2A) and 3 mg/mL (Figure 2B were significantly more resistant to starvation than normally-fed flies, except for 1.5 mg/mL Annurca polyphenol extracts, which significantly reduced the resistance to starvation as compared to normally-fed animals. Moreover, comparing the effects on starvation of the different apple varieties with each other showed significant differences ($p < 0.001$, pairwise comparisons) and a different ranking within the two concentrations, with Fuji > TR > Annurca with the lowest concentration and Annurca > Fuji > TR with the highest.

A

STARVATION

	% survival 1.5 mg/ml APs			
Days	**BD**	**Fuji**	**Annurca**	**Tuscia Red**
1	100.00	100.00	100.00	100.00
2	90.70	100.00	66.67	100.00
3	70.93	100.00	55.56	100.00
4	33.72	73.68	22.22	48.15
5	0.00	26.32	0.00	0.00
6		0.00		
P values* APs *vs* BD		<0.001	<0.0001**	<0.001

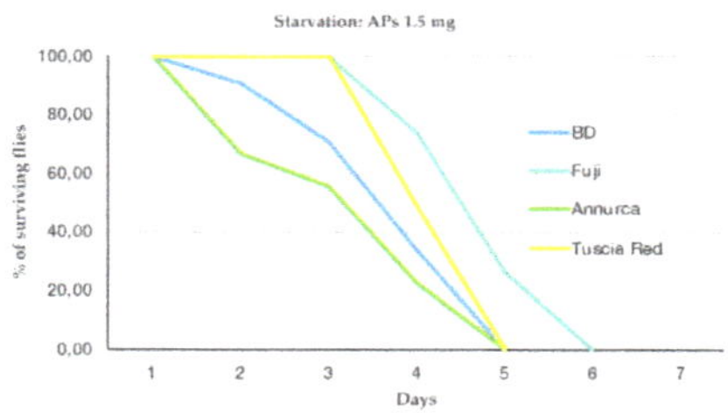

B

STARVATION

	% survival 3.0 mg/ml APs			
Days	**BD**	**Fuji**	**Annurca**	**Tuscia Red**
1	100.00	100.00	100.00	100.00
2	90.70	100.00	100.00	97.14
3	70.93	100.00	100.00	97.14
4	33.72	100.00	100.00	48.57
5	0.00	57.14	100.00	20.00
6		14.29	50.00	5.71
7		0.00	0.00	0.00
P values * APs *vs* BD		<0.0001	<0.0001	<0.0001

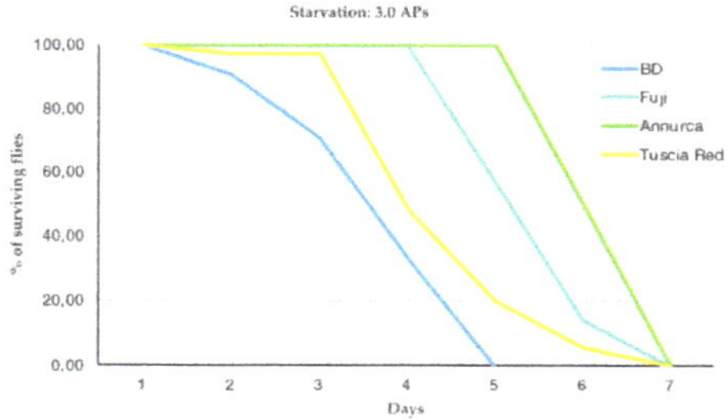

Figure 2. The effect of 1.5 (**A**) and 3.0 (**B**) mg/mL APs from the three different apple cultivars on *D. melanogaster* survival to starvation. Graphic representation of the data is on the right of each table. BD = basal diet with no supplement. All the differences between APs-supplemented diet vs. BD-fed animals were significant. * Mann–Whitney U test, fdr correction. ** Note that the Annurca extract *reduced* the resistance to starvation.

3.5. Apple Polyphenol Effects on Offspring Size and Weight

To test the effects of APs from the three apple cultivars on *Drosophila* body morphological parameters, we measured the body length of larvae and the weight of adults. With regards to length, at least 60 larvae were measured (mm) for each experimental point. Third instar larvae grown on APs-supplemented food, showed a significant reduction in body length as compared to controls at both APs concentrations of the three apple cultivars (Table 3).

Table 3. The effect of 1.5 mg/mL (A) and 3.0 mg/mL (B) APs from the three different apple cultivars on *D. melanogaster* 3rd instar larvae length. BD = basal diet with no supplement. * Mann–Whitney U test, fdr correction: APs *vs* BD.

A				
	BD	**Fuji**	**Tuscia Red**	**Annurca**
Larval length	3.9	3.7	3.8	2.9
p values * APs vs. BD		<<0.001	<0.001	<<0.001
B				
	BD	**Fuji**	**Tuscia Red**	**Annurca**
Larval length	3.9	3.0	2.7	2.6
p values * APs *vs* BD		<<0.001	<<0.001	<<0.001

Concerning weight, a significant reduction was observed in females fed with APs extracts from the three cultivars at both concentrations ($p \leq 0.01$), except for TR polyphenol extracts at the lowest concentration (Table 4). Notably, no significant effect on the weight of adult males was observed with APs at either dose. As expected, weight scores of females were significantly higher than that of males ($p < 0.001$) in both BD and APs series.

Table 4. The effect of 1.5 (A) and 3.0 (B) mg/mL APs from the three different apple cultivars on *D. melanogaster* adult weight. BD = basal diet with no supplement. * The weight of groups of ten male or female individuals is reported. At least 8 groups for each experimental point were recorded. ^{s} and ^{ns} denote whether or not the mean weight difference between the BD-fed and APs-fed flies is significant ($p \leq 0.01$, two-tailed t test). Male mean weight is significantly lower than that of females ($p < 0.001$, two-tailed t test).

A

	BD		**Fuji**		**Tuscia Red**		**Annurca**	
	Weight (g) *							
	females	males	females	males	females	males	females	males
Average	0.0085	0.0055	0.0070 ^{s}	0.0048 ^{ns}	0.0079 ^{ns}	0.0055 ^{ns}	0.0062 ^{s}	0.0052 ^{ns}
SD	0.0010	0.0014	0.0014	0.0008	0.0018	0.0009	0.0009	0.0015

B

	BD		**Fuji**		**Tuscia Red**		**Annurca**	
	Weight (g) *							
	females	males	females	males	females	males	females	males
Average	0.0085	0.0055	0.0069 ^{s}	0.0053 ^{ns}	0.0071 ^{s}	0.0052 ^{ns}	0.0069 ^{s}	0.0049 ^{ns}
SD	0.0010	0.0014	0.0014	0.0016	0.0005	0.0014	0.0010	0.0004

3.6. Apple Polyphenols Affect Drosophila Mitotic Cell Cycle

To analyze the mitotic cell cycle, brains from the 3rd instar larvae were dissected and fixed for subsequent cytological observation. The mitotic index was defined as the number of mitotic cells per optical field (see M&M). We observed a highly significantly ($p < 0.001$) reduced proliferation rate of neuroblasts in the larvae fed with both concentrations of APs, when compared to larvae fed with the basal diet (Table 5). The higher APs concentration showed a significantly more pronounced effect than the lower concentration of Annurca and TR APs ($p < 0.05$), but not of Fuji.

Table 5. The effect of 1.5 (A) and 3.0 (B) mg/mL APs from the three different apple cultivars on *D. melanogaster larval* brain mitotic index. BD—basal diet with no supplement. *P*—Fisher's exact test, fdr correction.

A

APs 1.5 mg	**Mitotic Divisions**	**Optical Fields**	**Mitotic Index**	***p* APs vs. BD**
BD	216	200	1.08	
Fuji	51	220	0.23	<<0.001
Annurca	79	308	0.26	<<0.001
Tuscia Red	88	306	0.29	<<0.001

B

APs 3 mg	**Mitotic Divisions**	**Optical Fields**	**Mitotic Index**	***p*APs vs. BD**
BD	216	200	1.08	
Fuji	56	290	0.19	<<0.001
Annurca	31	200	0.15	<<0.001
Tuscia Red	37	200	0.18	<<0.001

3.7. Apple Polyphenol Impact on Drosophila Climbing Ability

The climbing ability assay was conducted to evaluate the effect of APs extracts on locomotor function in fruit flies. The data showed that APs had only a slight effect on the motor skills of same-age adult flies, except for Annurca AP-fed flies, which showed a significantly reduced climbing ability, more pronounced with 3.0 mg/mL extracts (Table 6).

Table 6. The effect of 1.5 (A) and 3.0 (B) mg/mL APs from the three different apple cultivars on adult *D. melanogaster* climbing ability. T is indicated as the time in seconds from the start of the experiment when all the flies were at the bottom of a graduated cylinder. The % values refer to the percentages of flies achieving the height of 17.5 cm in the indicated time. The test was reproduced three times with flies of the same age. BD—basal diet with no supplement. p value APs vs. BD, Fisher's exact test (*) $p < 0.05$, (**) $p < 0.01$, (***) $p < 0.001$.

A			
APs 1.5 mg	**T10%**	**T20%**	**T30%**
BD	85	90	100
Fuji	80	95	100
Annurca	40 (*)	95	100
Tuscia Red	90	90	100
B			
APs 3.0 mg	**T10%**	**T20%**	**T30%**
BD	85	90	100
Fuji	35 (**)	90	100
Annurca	5 (***)	50 (*)	65 (*)
Tuscia Red	70	100	100

4. Discussion

The total polyphenol content in the TR apple variety was by far higher than that in Annurca and Fuji. This is particularly true for anthocyanin content (Tables 1 and 2). As our analyses were performed on fruit flesh only, deprived of the peel, our observations clearly explain why the TR pulp is reddish (Figure S1) in comparison to the white pulp of the two other varieties.

Hydroxy-cinnamic acid predominance in the TR and Annurca varieties, as revealed by HPLC, is partially in contrast with other works [37,38], where the predominant class was flavanols. This difference could be partially explained by the slightly acidified extracting phase used, which could have major affinity to Hydroxy-cinnamics rather than to flavanols and also to the absence of peel in our fresh material for extracts. The accumulation of these phenolic acids is mainly responsible for the increased total phenolic content of TR and Annurca with respect to other white fleshed varieties. The predominance of flavanols among the flavonoid class in both TR and Annurca was in accordance with that found in other cultivars [37].

The main action of polyphenols in eukaryotic cells is that of scavenging reactive oxygen species (ROS), which may cause an imbalance in the redox homeostasis that finally results in oxidative stress [39].

The present results show that when apple polyphenols were added to the diet of laboratory populations of *D. melanogaster*, they affected a series of biological parameters, such as lifespan, resistance to starvation, morphometrical measures and mitotic index.

The APs-rich diet caused a slight but not significant delay in the sequential stages of animal development for a total of one day for each stage. In other words, the one-day delay accumulated by the 3rd instar larvae (the first monitored developmental stage) is maintained in later stages without any further retardation. The cause of this development delay is likely due to the slow-down of the cell cycle (see below).

Regarding the APs effects on the *Drosophila* lifespan (Figure 1), all the parameters we tested, i.e., mean life duration, the day at which 50% of flies still survived, and the day of death of the last surviving individual, were increased by the higher concentration (3.0 mg/mL) of polyphenols from all the apple cultivars we tested. The lower APs concentration (1.5 mg/mL) caused an extension of the lifespan of the last survivors but did not significantly affect the other two parameters. These results support those of Peng's research group who demonstrated the effect of blueberry [40] and apple [41] polyphenols on the extension of the *Drosophila* lifespan.

Resistance to starvation and to oxidative stress are positively correlated [42]. Thus, it is not surprising that the AP-rich diet prolonged the fly survival to starvation (Figure 2). Intriguingly, Annurca extracts at the lowest concentration reduced the resistance to starvation with respect to normally-fed individuals, whereas at the highest ones, proved to be the most efficient AP to counteract starvation. Contrary to our results, Lopez et al. [34] reported that 10 mg/mL of green tea polyphenols sensitized flies to starvation. This discrepancy might be well-ascribed to the high dose of polyphenols used in Lopez's experiments as compared to ours. In fact, we showed that our polyphenol extracts at the concentration of 5 mg/mL severely affected fly development and viability.

Morphometrical parameters were clearly affected by the APs-rich diet, which had an impact on both larval length and weight.

Flies fed with the APs-rich diet showed a reduction in the body length of the 3rd instar larvae, with respect to the normally-fed ones (Table 3). The difference is significant at both AP concentrations for all three cultivars. The two APs concentrations exhibited a clear dose–effect correlation ($p < 0.001$). The reduced larval size is fully consistent with the reduction in the body weight of adults, although with a significant difference between female and male individuals (Table 4).

In fact, both females and males showed a weight reduction with respect to basal diet-fed flies, but the difference was significant only for females with both concentrations, except for the lowest concentration of TR extracts. It is well-known that *D. melanogaster* male individuals have a body size consistently smaller than that of females and, hence, a body weight less than that of females [43]; thus, it makes sense that the APs-induced reduction in body weight is less pronounced in males than in females, since in the former the body weight cannot likely lower beyond a given survival threshold.

These results confirm and extend the observation of Lopez et al. [34], who reported a similar reduction in body size in flies fed with green tea polyphenol extracts, even though they used a higher concentration (10 mg/mL) than ours (1.5 and 3 mg/mL). Moreover, making these comparisons, it must be taken into account that the composition and content of polyphenols varies in green tea vs. apple extracts. By indirect evidence, the authors argue that the polyphenol-induced reduction in fly body size was due to a relenting of cell division rate and, hence, to the reduction in the body cell number [34]. Our results strongly reinforce this hypothesis that the decrease in cell division rate may be the cause of fly body size reduction since we actually observed a dramatic, statistically highly significant, drop-down of mitotic index in flies fed with the APs diet at both APs concentrations (Table 5). In this context, it must be emphasized that in *D. melanogaster*, the strict dependence of the diminution of body size from the decline in cell division rate is a well-known, common characteristics shared by *Minute* mutants [44], which harbor mutations in one of the genes encoding for ribosomal proteins [45,46]. The action of polyphenols on the mitotic cycle led many laboratories to investigate the effects of these compounds on cancer cells [47]. The inhibitory effect on the cell cycle progression was demonstrated in cancer cells in culture for green and black tea and rosemary polyphenols [48–50]. Many studies demonstrated that polyphenols derived by different plants may arrest cell cycle progression in metaphase or cause mitotic catastrophe [51–53]. Our observations contrast with the above results, since polyphenols from all three apple cultivars induced a relenting not an arrest of the cell cycle, as shown by a dramatic reduction in the number of mitotic figures instead of an accumulation of metaphases. Additionally, the few mitotic figures we observed were appeared normal, with no evidence of mitotic catastrophe. Moreover, as far as we know, the demonstration of a similar effect of polyphenols on the cell cycle of normal cells in vivo was never reported in the literature.

As a behavioral test of the APs diet effect, we took into consideration the climbing ability of *D. melanogaster* adults (Table 6). Only Annurca APs negatively impacted the climbing ability. The literature data on the effects of polyphenols on climbing ability were mostly focused on the rescue of pathological, locomotory phenotypes of neurodegenerative disorders that were mimicked in the fruit flies either by chemical treatment [40,41] or by transgenic manipulation [54]. Studies on wild-type *D. melanogaster* adults demonstrated that the apple-derived polyphenol Phlorizin is able to partially rescue the age-related climbing ability decline [55]. However, the effect of phlorizin did not appear before 30 days of age of the animals; thus, it is not surprising that the locomotor skills of the young animals (1–5 days old) used in our experiments were not influenced by APs extract feeding.

A comparative analysis of the effects of the three cultivars (see synoptic Table 7), evidenced a general uniformity of effects on the considered characters, the most striking difference being the just mentioned negative effect exerted by Annurca polyphenol extracts on locomotory skills. This adverse effect could be ascribed to some cultivar-specific metabolite, which can be resolved by a high throughput, differential metabolomic analysis (manuscript in preparation).

Table 7. Synopsis of the effects of the three different apple varieties at two concentrations (1.5 and 3.0 mg/mL) on various *D. melanogaster* biological parameters. Data were compared to those of basal diet-fed animals and only significant values were considered. + indicates an incremental by APs diet effect on the considered parameter; – indicates a lowering and – – a highly lowering effect on the considered parameter; 0 indicates no effect on the considered parameter.

	Annurca		Fuji		Tuscia Red	
	1.5	3.0	1.5	3.0	1.5	3.0
Development timing	0	0	0	0	0	0
Mean lifespan	0	+	0	+	0	+
Resistance to starvation	–	+	+	+	+	+
Larval size	–	–	–	–	–	–
Weight (females)	–	–	–	–	0	–
Mitotic index	– –	– –	– –	– –	– –	– –
Climbing ability	–	–	0	0	0	0

In conclusion, looking for a possible primary effect underlying the other effects exerted by apple polyphenols, it can be considered that the drastic reduction in mitotic index may well be responsible for the diminution in both body size and adult weight. The correlation between mitotic index and lifespan is more obscure. However, evidence was reported in yeast that the cell division rate is inversely related to the lifespan of the individual [56,57]. Even though considering that yeast is a unicellular eukaryote, the above observations support our ones in that APs induced both a dramatic slowing of the cell cycle and a significant increase in the lifespan.

Supplementary Materials: The following supporting information can be downloaded at: https://www.mdpi.com/article/10.3390/antiox11112086/s1, Figure S1: Comparison of the fruits of the three apple varieties used in the study; Figure S2: Examples of chromatograms for three representative compounds extracted from the pulp of Tuscia Red and Annurca apple varieties.

Author Contributions: Conceptualization, S.B., R.M. and G.P.; Formal analysis and Software, I.A.; Investigation: S.B., B.C., C.R., C.C., S.C. and I.F.; Project administration: S.B., S.R., R.M. and G.P.; Writing—original draft: S.B., I.F., R.M. and G.P.; Writing—review and editing, S.B., I.A., S.R., R.M. and G.P. All authors have read and agreed to the published version of the manuscript.

Funding: This research was in part supported by Fondo Ordinario Enti (FOE D.M 865/2019) in the framework of a collaboration agreement between the Italian National Research Council and EBRI (2019–2021) to I.A.

Institutional Review Board Statement: Not applicable.

Informed Consent Statement: Not applicable.

Data Availability Statement: Data are contained within the article and supplementary material.

Conflicts of Interest: The authors declare no conflict of interest.

References

1. Carocho, M.; Ferreira, I.C.F.R.; Morales, P.; Soković, M. Antioxidants and Prooxidants: Effects on Health and Aging. *Oxid. Med. Cell. Longev.* **2018**, *2018*, 1472708. [CrossRef]
2. de Groot, H. Reactive oxygen species in tissue injury. *Hepatogastroenterology* **1994**, *4*, 328–332.
3. Grace, P.A. Ischaemia-reperfusion injury. *Br. J. Surg.* **1994**, *81*, 637–647. [CrossRef] [PubMed]
4. Griffiths, K.; Aggarwal, B.B.; Singh, R.B.; Buttar, H.S.; Wilson, D.; De Meester, F. Food Antioxidants and Their Anti-Inflammatory Properties: A Potential Role in Cardiovascular Diseases and Cancer Prevention. *Diseases* **2016**, *4*, 28. [CrossRef]
5. McCord, J.M. Free radicals and inflammation. *Science* **1974**, *185*, 529–531. [CrossRef]
6. Halliwell, B. Oxygen radicals, nitric oxide and human inflammatory joint disease. *Ann. Rheum. Dis.* **1995**, *54*, 505–510. [CrossRef] [PubMed]
7. Poprac, P.; Jomova, K.; Simunkova, M.; Kollar, V.; Rhodes, C.J.; Valko, M. Targeting free radicals in oxidative stress-related human diseases. *Trends Pharmacol. Sci.* **2017**, *38*, 592–607. [CrossRef]
8. Iakovou, E.; Kourti, M. A comprehensive overview of the complex role of oxidative stress in aging, the contributing environmental stressors and emerging antioxidant therapeutic interventions. *Front. Aging Neurosci.* **2022**, *14*, 827900. [CrossRef]
9. Eberhardt, M.V.; Lee, C.Y.; Liu, R.H. Antioxidant activity of fresh apples. *Nature* **2000**, *405*, 903–904. [CrossRef]
10. Williamson, G.; Manach, C. Bioavailability and bioefficacy of polyphenols in humans. II. Review of 93 intervention studies. *Am. J. Clin. Nutr.* **2005**, *81*, 243S–255S. [CrossRef]
11. Hyson, D.A. A comprehensive review of apples and apple components and their relationship to human health. *Adv. Nutr.* **2011**, *2*, 408–420. [CrossRef] [PubMed]
12. Bondonno, C.P.; Yang, X.; Croft, K.D.; Considine, M.J.; Ward, N.C.; Rich, L.; Puddey, I.B.; Swinny, E.; Mubarak, A.; Hodgson, J.M. Flavonoid-rich apples and nitrate-rich spinach augment nitric oxide status and improve endothelial function in healthy men and women: A randomized controlled trial. *Free Radic. Biol. Med.* **2012**, *52*, 95–102. [CrossRef]
13. Van Duyn, M.A.; Pivonka, E. Overview of the health benefits of fruit and vegetable consumption for the dietetics professional: Selected literature. *J. Am. Diet Assoc.* **2000**, *100*, 1511–1521. [CrossRef]
14. Stalikas, C.D. Extraction, Separation, and Detection Methods for Phenolic Acids and Flavonoids. *J. Sep. Sci.* **2007**, *30*, 3268–3295. [CrossRef]
15. Cömert, E.D.; Gökmen, V. Evolution of food antioxidants as a core topic of food science for a century. *Food Res. Int.* **2018**, *105*, 76–93. [CrossRef]
16. Neha, K.; Haider, M.R.; Pathak, A.; Yar, M.S. Medicinal prospects of antioxidants: A review. *Eur. J. Med. Chem.* **2019**, *178*, 687–704. [CrossRef] [PubMed]
17. Afanas'ev, I.B.; Dorozhko, A.I.; Brodskii, A.V.; Kostyuk, V.A.; Potapovitch, A.I. Chelating and Free Radical Scavenging Mechanisms of Inhibitory Action of Rutin and Quercetin in Lipid Peroxidation. *Biochem. Pharmacol.* **1989**, *38*, 1763–1769. [CrossRef]
18. Ohnishi, M.; Morishita, H.; Iwahashi, H.; Toda, S.; Shirataki, Y.; Kimura, M.; Kido, R. Inhibitory effects of chlorogenic acids on linoleic-acid peroxidation and hemolysis. *Phitochemistry* **1994**, *36*, 579–583. [CrossRef]
19. Castelluccio, C.; Paganga, G.; Melikian, N.; Bolwell, G.P.; Pridham, J.; Sampson, J.; Rice-Evans, C. Antioxidant potential of intermediates in phenylpropanoid metabolism in higher plants. *FEBS Lett.* **1995**, *368*, 188–192. [CrossRef]
20. Patocka, J.; Bhardwaj, K.; Klimova, B.; Nepovimova, E.; Wu, Q.; Landi, M.; Kamil Kuca, K.; Valis, M.; Wu, W. Malus domestica: A Review on Nutritional Features, Chemical Composition, Traditional and Medicinal Value. *Plants* **2020**, *9*, 1408. [CrossRef]
21. Espley, R.V.; Hellens, R.P.; Putterill, J.; Stevenson, D.E.; Kutty-Amma, S.; Allan, A.C. Red colouration in apple fruit is due to the activity of the MYB transcription factor, MdMYB10. *Plant J.* **2007**, *49*, 414–427. [CrossRef] [PubMed]
22. Allan, A.C.; Hellens, R.P.; Laing, W.A. MYB transcription factors that colour our fruit. *Trends Plant Sci.* **2008**, *13*, 99–102. [CrossRef]
23. Wang, N.; Jiang, S.; Zhang, Z.; Fang, H.; Xu, H.; Wang, Y.; Chen, X. *Malus sieversii*: The origin, flavonoid synthesis mechanism, and breeding of red-skinned and red-fleshed apples. *Hortic. Res.* **2018**, *5*, 70. [CrossRef] [PubMed]
24. Cirilli, M.; Latini, G.; Cristofori, V.; Ceccantoni, B.; Luziatelli, F.; Zecchini, M.; Muleo, R.; Ruzzi, M. Polyphenol traits, antimicrobial property and consumer preference of 'Italian Red Passion' apple genotypes and cultivar 'Annurca'. *Acta Hortic.* **2015**, *1106*, 185–190. [CrossRef]
25. Silvestri, C.; Cirilli, M.; Zecchini, M.; Muleo, R.; Ruggieri, A. Consumer acceptance of the new red-fleshed apple variety. *J. Food Prod. Mark.* **2016**, *24*, 1–21. [CrossRef]
26. Harris, S.A.; Robinson, J.P.; Juniper, B.E. Genetic clues to the origin of the apple. *Trends Genet.* **2002**, *18*, 426–430. [CrossRef]
27. Jeibmann, A.; Paulus, W. Drosophila melanogaster as a model organism of brain diseases. *Int. J. Mol. Sci.* **2009**, *10*, 407–440. [CrossRef]
28. Ugur, B.; Chen, K.; Bellen, H.J. *Drosophila* tools and assays for the study of human diseases. *Dis. Models Mech.* **2016**, *9*, 235–244. [CrossRef]
29. Mirzoyan, Z.; Sollazzo, M.; Allocca, M.; Valenza, A.M.; Grifoni, D.; Bellosta, P. *Drosophila melanogaster*: A Model Organism to Study Cancer. *Front. Genet.* **2019**, *10*, 51. [CrossRef]

30. Pandey, U.B.; Nichols, C.D. Human disease models in *Drosophila melanogaster* and the role of the fly in therapeutic drug discovery. *Pharmacol. Rev.* **2011**, *63*, 411–436. [CrossRef] [PubMed]
31. Panchal, K.; Tiwari, A.K. *Drosophila melanogaster* "a potential model organism" for identification of pharmacological properties of plants/plant-derived components. *Biomed. Pharmacother.* **2017**, *89*, 1331–1345. [CrossRef] [PubMed]
32. Yi, Y.; Xu, W.; Fan, Y.; Wang, H.-X. *Drosophila* as an emerging model organism for studies of food-derived antioxidants. *Food Res. Int.* **2021**, *143*, 110307. [CrossRef] [PubMed]
33. Mancinelli, A.L.; Yang, C.P.H.; Lindquist, P.; Anderson, O.R.; Rabino, I. Photocontrol of anthocyanin synthesis. III. The action of streptomycin on the synthesis of chlorophyll and anthocyanin. *Plant Physiol.* **1975**, *55*, 251–257. [CrossRef]
34. Lopez, T.E.; Pham, H.M.; Barbour, J.; Tran, P.; Van Nguyen, B.; Hogan, S.P.; Homo, R.L.; Coskun, V.; Schriner, S.E.; Jafari, M. The impact of green tea polyphenols on development and reproduction in *Drosophila melanogaster*. *J. Funct. Foods* **2016**, *20*, 556–566. [CrossRef]
35. Fabbretti, F.; Iannetti, I.; Guglielmi, L.; Perconti, S.; Evangelistella, C.; Proietti De Santis, L.; Bongiorni, S.; Prantera, G. Confocal analysis of nuclear lamina behavior during male meiosis and spermatogenesis in *Drosophila melanogaster*. *PLoS ONE* **2016**, *11*, e0151231. [CrossRef]
36. Volpi, S.; Bongiorni, S.; Fabbretti, F.; Wakimoto, B.T.; Prantera, G. *Drosophila rae1* is required for male meiosis and spermatogenesis *J. Cell Sci.* **2013**, *126*, 3541–3551. [CrossRef] [PubMed]
37. Tsao, R.; Yang, R.; Young, J.C.; Zhu, H. Polyphenolic Profiles in Eight Apple Cultivars Using High-Performance Liquid Chromatography (HPLC). *J. Agric. Food Chem.* **2003**, *51*, 6347–6353. [CrossRef] [PubMed]
38. Vanzani, P.; Rossetto, M.; Rigo, A.; Vrhovsek, U.; Mattivi, F.; D'amato, E.; Scarpa, A.M. Major Phytochemicals in Apple Cultivars: Contribution to Peroxyl Radical Trapping Efficiency. *J. Agric. Food Chem.* **2005**, *53*, 3377–3382. [CrossRef]
39. Leopoldini, M.; Russo, N.; Toscano, M. The molecular basis of working mechanism of natural polyphenolic antioxidants. *Food Chem.* **2011**, *125*, 288–306. [CrossRef]
40. Peng, C.; Zuo, Y.; Kwan, K.M.; Liang, Y.; Ma, K.Y.; Chan, H.Y.E.; Huang, Y.; Yu, H.; Chen, Z.-Y. Blueberry extract prolongs lifespan of Drosophila melanogaster. *Exp. Gerontol.* **2012**, *47*, 170–178. [CrossRef]
41. Peng, C.; Chan, H.Y.E.; Huang, Y.; Yu, H.; Chen, Z.-Y. Apple polyphenols extend the mean lifespan of *Drosophila melanogaster*. *J. Agric. Food Chem.* **2011**, *59*, 2097–2106. [CrossRef] [PubMed]
42. Rion, S.; Kawecki, T.J. Evolutionary biology of starvation resistance: What we have learned from Drosophila. *J. Evol. Biol.* **2007**, *20*, 1655–1664. [CrossRef] [PubMed]
43. Burggren, W.; Souder, B.M.; Ho, D.H. Metabolic rate and hypoxia tolerance are affected by group interactions and sex in the fruit fly (*Drosophila melanogaster*): New data and a literature survey. *Biol. Open* **2017**, *6*, 471–480. [CrossRef] [PubMed]
44. Morata, G.; Ripoll, P. Minutes: Mutants of *Drosophila* autonomously affecting cell division rate. *Dev. Biol.* **1975**, *42*, 211–221. [CrossRef]
45. Kongsuwan, K.; Yu, Q.; Vincent, A.; Frisardi, M.C.; Rosbash, M.; Lengyel, J.A.; Merriam, J. A *Drosophila Minute* gene encodes a ribosomal protein. *Nature* **1985**, *317*, 555–558. [CrossRef]
46. Sæbøe-Larssen, S.; Lyamouri, M.; Merriam, J.; Oksvold, M.P.; Lambertsson, A. Ribosomal Protein Insufficiency and the Minute Syndrome in Drosophila: A Dose-Response Relationship. *Genetics* **1998**, *148*, 1215–1224. [CrossRef]
47. Przystupski, D.; Niemczura, M.J.; Górska, A.; Supplitt, S.; Kotowski, K.; Wawryka, P.; Rozborska, P.; Woźniak, K.; Michel, O.; Kiełbik, A.; et al. In Search of Panacea—Review of Recent Studies Concerning Nature-Derived Anticancer Agents. *Nutrients* **2019**, *11*, 1426. [CrossRef] [PubMed]
48. Visanji, J.M.; Thompson, D.G.; Padfield, P.J. Induction of G2/M phase cell cycle arrest by carnosol and carnosic acid is associated with alteration of cyclin A and cyclin B1 levels. *Cancer Lett.* **2006**, *237*, 130–136. [CrossRef]
49. Thangapazham, R.L.; Singh, A.K.; Sharma, A.; Warren, J.; Gaddipati, J.P.; Maheshwari, R.K. Green tea polyphenols and its constituent epigallocatechin gallate inhibits proliferation of human breast cancer cells in vitro and in vivo. *Cancer Lett.* **2007**, *245*, 232–241. [CrossRef]
50. Halder, B.; Das Gupta, S.; Gomes, A. Black tea polyphenols induce human leukemic cell cycle arrest by inhibiting Akt signaling: Possible involvement of Hsp90, Wnt/beta-catenin signaling and FOXO1. *FEBS J.* **2012**, *279*, 2876–2891. [CrossRef]
51. Delmas, D.; Solary, E.; Latruffe, N. Resveratrol, a phytochemical inducer of multiple cell death pathways: Apoptosis, autophagy and mitotic catastrophe. *Curr. Med. Chem.* **2011**, *18*, 1100–1121. [CrossRef] [PubMed]
52. Tan, S.; Guan, X.; Grün, C.; Zhou, Z.; Schepers, U.; Nick, P. Gallic acid induces mitotic catastrophe and inhibits centrosomal clustering in HeLa cells. *Toxicol. In Vitro* **2015**, *30*, 506–513. [CrossRef] [PubMed]
53. Mosieniak, G.; Sliwinska, M.A.; Przybylska, D.; Grabowska, W.; Sunderland, P.; Bielak-Zmijewska, A.; Sikora, E. Curcumin-treated cancer cells show mitotic disturbances leading to growth arrest and induction of senescence phenotype. *Int. J. Biochem. Cell Biol.* **2016**, *74*, 33–43. [CrossRef] [PubMed]
54. Wu, Y.L.; Chang, J.C.; Lin, W.Y.; Li, C.C.; Hsieh, M.; Chen, H.W.; Wang, T.S.; Wu, W.T.; Liu, C.S.; Liu, K.L. Caffeic acid and resveratrol ameliorate cellular damage in cell and *Drosophila* models of spinocerebellar ataxia type 3 through upregulation of Nrf2 pathway. *Free Radic. Biol. Med.* **2018**, *115*, 309–317. [CrossRef]
55. Wang, H.; Sun, Z.; Liu, D.; Li, X.; Rehman, R.-U.; Wang, H.; Wu, Z. Apple phlorizin attenuates oxidative stress in Drosophila melanogaster. *J. Food Biochem.* **2019**, *43*, e12744.

56. Yang, J.; Dungrawala, H.; Hua, H.; Manukyan, A.; Abraham, L.; Lane, W.; Mead, H.; Wright, J.; Schneider, B.L. Cell size and growth rate are major determinants of replicative lifespan. *Cell Cycle* **2011**, *10*, 144–155. [CrossRef]
57. Jiménez, J.; Bru, S.; Ribeiro, M.; Clotet, J. Live fast, die soon: Cell cycle progression and lifespan in yeast cells. *Microb. Cell* **2015**, *2*, 62–67. [CrossRef]

antioxidants

MDPI

Article

Cordyceps militaris Reduces Oxidative Stress and Regulates Immune T Cells to Inhibit Metastatic Melanoma Invasion

Yuan-Hong Lan [1,†], Yun-Sheng Lu [2,†], Ju-Yu Wu [3,†], Hsu-Tung Lee [4,†], Penjit Srinophakun [5], Gizem Naz Canko [6], Chien-Chih Chiu [7] and Hui-Min David Wang [6,8,9,*]

1 Department of Medical Laboratory Science and Biotechnology, ASIA University, Taichung 413, Taiwan; brianblue98@gmail.com
2 Taiwan Agriculture Research Institute, Council of Agriculture, Taichung 413, Taiwan; yunsheng@tari.gov.tw
3 Program in Tissue Engineering and Regenerative Medicine, National Chung Hsing University, Taichung 402, Taiwan; xytmilk@gmail.com
4 The Department of Neurological Institute, Taichung Veterans General Hospital, Taichung 402, Taiwan; sdlee@vghtc.gov.tw
5 Chemical Engineering Department, Faculty of Engineering, Kasetsart University, 50 Ngamwongwan Rd., Ladyao, Jatujak, Bangkok 10900, Thailand; fengpjs@ku.ac.th
6 Graduate Institute of Biomedical Engineering, National Chung Hsing University, Taichung 402, Taiwan; gizemcanko@gmail.com
7 Department of Biotechnology, Kaohsiung Medical University, Kaohsiung 807, Taiwan; cchiu@kmu.edu.tw
8 Graduate Institute of Medicine, College of Medicine, Kaohsiung Medical University, Kaohsiung 807, Taiwan
9 Department of Medical Laboratory Science and Biotechnology, China Medical University, Taichung 404, Taiwan
* Correspondence: davidw@dragon.nchu.edu.tw; Tel.: +886-935-753-718 or +886-4-2284-0733 (ext. 651); Fax: +886-4-228-522-42
† These authors contributed equally to this work.

Abstract: In this study, the water extract of *Cordyceps militaris* (Linn.) Link (CM) was used as a functional material to investigate the inhibitory mechanisms on B16F10 and lung metastatic melanoma (LMM) cells. Reducing power, chelating ability, and 2,2-diphenyl-2-picrylhydrazyl (DPPH) assays were applied for antioxidative capacities, and we obtained positive results from the proper concentrations of CM. To examine the ability of CM in melanoma proliferation inhibition and to substantiate the previous outcomes, three cellular experiments were performed via (3-(4,5-dimethylthiazol-2-yl)-2,5-diphenyltetrazolium bromide, MTT, a tetrazole) assay, cell migration, and invasion evaluation. The addition of CM to the incubation medium increased the number of CD8+ T cells significantly, which improved the immunogenicity. This study showed that CM exhibits various biological capabilities, including antioxidation, anti-tumor, tumor invasion suppression, and T cytotoxic cell activity promotion.

Keywords: *Cordyceps militaris*; B16F10; lung metastatic melanoma; tumor microenvironment; T cell

Citation: Lan, Y.-H.; Lu, Y.-S.; Wu, J.-Y.; Lee, H.-T.; Srinophakun, P.; Canko, G.N.; Chiu, C.-C.; Wang, H.-M.D. *Cordyceps militaris* Reduces Oxidative Stress and Regulates Immune T Cells to Inhibit Metastatic Melanoma Invasion. *Antioxidants* **2022**, *11*, 1502. https://doi.org/10.3390/antiox11081502

Academic Editor: Stanley Omaye

Received: 19 July 2022
Accepted: 26 July 2022
Published: 30 July 2022

1. Introduction

Many studies show that the reactive oxygen species (ROS) in the tumor microenvironment is a double-edged sword for cancer curing. Different ROS levels lead to different outcomes to various aspects of cancer. When the tumor microenvironment has a higher ROS level, it could induce inflammatory responses and prolonged ROS production could lead to chronic inflammation. It can also cause a series of reactions to trigger tumor progression and metastasis [1]. Moreover, the amount of oxidative stress increase can also elevate the resistance of cancer cells to drugs [2]. Still, the excessive rise of oxidative stress could damage cells to death directly by promoting apoptosis. In contrast, the cancer cell could also activate the apoptosis process when the tumor microenvironment has a mild ROS level, which is lower than the minimum requirement for the cellular response. Some studies confirm that a sharp ROS level decrease could significantly inhibit cell proliferation [3,4].

Melanoma is an aggressive tumor that quickly metastasizes and strongly resists chemotherapy drugs. This characteristic has caused the survival rate of patients to remain low in recent years [5] Even though the current medical treatments are not effective enough to treat metastatic melanoma, stimulation of the cancer cells' death is an effective strategy for cancer chemotherapy. The process of causing cell death is named programmed cell death, which has two forms: autophagy and apoptosis. The main differences between autophagy and apoptosis are the cell morphology and the metabolic pathway. Autophagy is an evolutionary self-preservation mechanism through the dysfunctional cells removing and recycling parts of them toward cellular repair [6]. Apoptosis changes in cell morphological characteristics: cell contraction, roundness, and blistering of cell nuclei, condensation, marginalization of chromatin, and undetectable transformations of cellular organelles, and those changes cause cell death [7].

The water extract of *Cordyceps militaris* (Linn.) Link (CM, orange caterpillar fungus or Yong Chong Cao in Chinese) is a medicinal fungus that belongs to the family of Cordycipitaceae. Various insects as a host can be infected by CM in the field, including lepidopteran larvae and pupae. In addition, CM can be artificially cultivated on a rice medium to harvest the sporocarp. Because the medicinal properties of CM are like those of *Cordyceps sinensis* (CS), it is used widely as an alternative to CS in health-promoting supplements. *Cordyceps sinensis* (CS) is a fungal parasite from lepidopteran larvae. The fungus attacks the caterpillar in late autumn and devours the host leisurely. By early summer, the fungus-infected caterpillar is dead, and the fruiting body sticks out of its head. Moreover, its particular life cycle is called Cordyceps Sinensis in Chinese [8]. Furthermore, recent reports have revealed that both CM and CS have anti-inflammatory and antioxidative effects. Cordyceps has a significant impact. Cordycepin, as a biologically active compound of *Cordyceps militaris*, is very similar in structure to cellular nucleosides and adenosine and contains a purine (adenosine) nucleoside molecule linked to a ribose moiety, which acts as a nucleoside analogue [9,10]. In addition, also because cordycepin alone is widely explored for its anticancer/antioxidant activity, it has strong pharmacological and therapeutic potential to cure many dreaded diseases in the future, such as respiratory [11], renal [12], hepatic, and nervous system diseases [13]. Therefore, we assume a couple of combinations in CM could provide antioxidation functions [14]. Many valuable chemicals in natural products could interact to show anti-cancer ability, activate the immune system, and promote the progress of apoptosis [2,15]. Moreover, the interactions between the chemicals in natural products are not precise yet, and it is challenging to figure out the process.

T cells play an essential role in the immune system because they have an exact and circumspect mechanism to adjust the programmed cell death of cancer cells. Therefore, T cells can ensure no harm to the host during the process at any stage during the apoptosis [16]. Furthermore, some studies confirm that activating T cells could be an excellent strategy to cure malignant tumors, especially melanoma because a robust immune response could be mobilized by melanoma generally [12,16]. Therefore, this study will investigate the function of CM in antioxidant, anti-melanoma, and stimulating the capability of CD8+ T cells.

2. Materials and Methods

2.1. Reagents

1,1-Diphenyl-2-picrylhydrazyl, dimethyl sulfoxide (DMSO), potassium ferricyanide, trichloroacetic acid, ethylene diamine tetra-acetic acid, horseradish-peroxidase, $FeCl_3$, $FeCl_2 \cdot 4H_2O$, fetal bovine serum (FBS), and Vitamin C were obtained from GIBCO BRL (Gaithersburg, MD). 3-(4,5-dimethylthiazol-2-yl)-2,5-diphenyltetrazolium bromide (MTT), and 5(6)-carboxy-2′,7′-dichlorofluorescein diacetate (DCFDA) were purchased from Sigma Chemical.

2.2. Cell Lines and Cell Culture Materials

B16F10 melanoma cells were purchased from the Bioresource Collection and Research Center (BCRC, Hsinchu, Taiwan). The generation of LMM is based on the injection of B16F10 from the tail vein of mice. After waiting for 3–4 weeks to establish the tumor, it can extract the cancer from the lung and culture, which is LMM after subculture. B16F10 and LMM were grown in Dulbecco's Modified Eagle's Medium (DMEM, Sigma-Aldrich Corp., Burlington Township, NJ, USA) supplemented with 10% heat-inactivated fetal bovine serum (FBS, Sigma-Aldrich Corp. Burlington Township, NJ, USA), and 1% antibiotic-antimycotic (100X) (Thermo Fisher Scientific Inc., Waltham, MA, USA) and maintained at 37 °C in a 5% CO_2 humidified incubator.

2.3. CM Extraction Process and Preparation

The CM was taken from the Mushroom Laboratory of the Taiwan Agricultural Research Institute (TARI), Executive Yuan. *Cordyceps militaris* (Linn.) Link (strain number: CM03) was kindly provided by Dr. Jin-Torng Peng and Yuegi Co. (Taipei, Taiwan). CM was authenticated by Dr. Yun-Sheng Lu later cultivated under controlled temperature (23 ± 4 °C) and humidity (80 ± 5%) at the Mushroom Laboratory of TARI, Taichung, Taiwan.

Water extraction of CM was carried out by the collaborator Qiqi Industrial Co., Ltd. The resulting product was freeze-dried at −40 °C and then pulverized with super micron. After the vacuum sterilization, the vacuum decompression fusion was performed in an environment below −60 °C, vacuum filtration and ultraviolet sterilization were used, and the semi-finished products were finally stored at low temperature. The water was partially concentrated and freeze-dried to obtain a stock solution (200 mg/mL). The diluted solution was then used for the DMEM dilution (Figure 1).

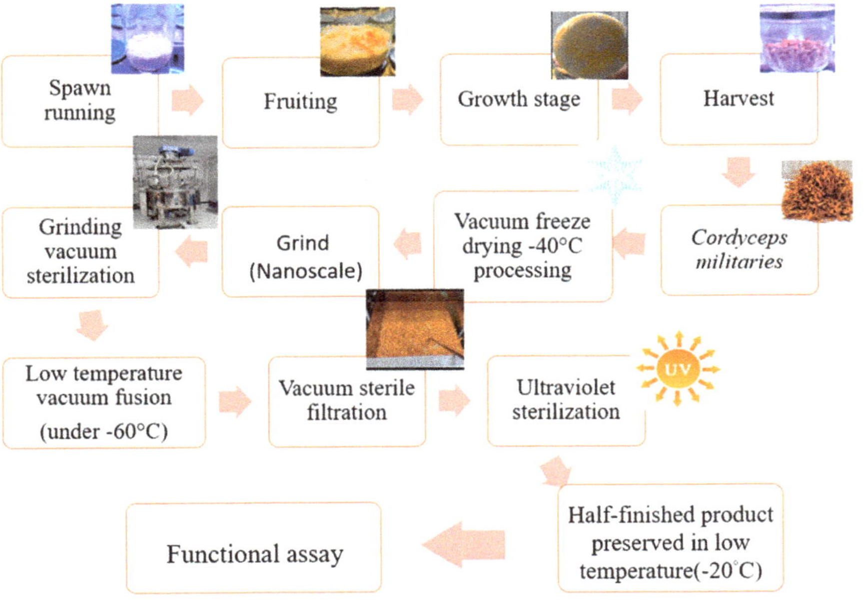

Figure 1. The extraction process using freeze-drying technology.

2.4. Reducing Power Assay

This system was used to quantify the reduction ability of antioxidants. First, 100 µL hydroxy anisole (BHA) was chosen to be a positive control. Then, 2.5 µL of the sample solutions of different concentrations (1–1000 µg/mL) mixed with 85 µL potassium and

2.5 μL potassium ferricyanide (20%, *w/v*) and phosphate buffer (0.2 M, pH 6.6), then let stand at 50 °C for 20 min. The reaction was terminated by adding 160 μL of trichloroacetic acid solution (10%, *w/v*) to the extractant. Later, 75 μL from the supernatant was removed and added into a 96-well plate, then 25 μL of ferric chloride (0.2%, *w/v*) was added, and the mixture was placed in the dark for 30 min. Finally, the absorbance value was measured at 700 nm. Results showed that when the absorbance level of the reaction mixture increases, the reducing power of the sample gets stronger, and they are directly proportional [17].

2.5. Ferrous Ion Chelating Assay

The antioxidant properties of iron were analyzed by the results of iron chelation; however, removal of iron ions is one of the things that can be achieved. In the medium of chelating agent, the structure cause complex was destroyed, which caused the complex ferrous ion to change from crimson restoration. The determination was based on the color reaction of the complex of ferrous ion and ferrozine at 562 nm. The lower the absorbance, the better the metal chelating activity. Then, 1 μL of CM of different concentrations (1–1000 μg/mL) was taken and mixed with 10 μL of $FeCl_2$ (2 mM), let stand for 5 min, then 69 μL of methanol and 20 μL (5 mM) of ferrozine were added. The sample was vigorously shaken to start the reaction, later the mixture was kept at 25 °C for 10 min. The absorbance value of the mixture was measured at 562 nm. EDTA was used as a positive standard [15]. The calculation formula was:

$$\text{Scavenging activity } (\%) = \frac{\left(A_{\text{control}} - A_{\text{sample}}\right)}{A_{\text{control}}} \times 100\%$$

where A_{control} = the absorbance of the control and A_{sample} = the absorbance of the sample.

2.6. DPPH Radical Scavenging Activity Assay

The antioxidant activity of the samples was measured by using DPPH. A DPPH solution (60 mM; 99 μL) in methanol was added to each compound's solution (1 μL). Vitamin C at 100 μg was set as a positive control. After 30 min, we measured the absorption at optical density (OD) of 517 nm [4]. Free radical scavenging capacity (percentage) was calculated according to the given formula above.

2.7. Cell Proliferation Assay by MTT Assay

The method used for the evaluation of the cell viability, proliferation, and cytotoxicity was MTT assay. This method was based on the succinate dehydrogenase of the mitochondria of living cells to reduce MTT to purple crystalline formazan (insoluble in water) and deposit it in the cells, while dead cells do not react the same way as threatening cells. First, in a 96-well plate, the cells were incubated at 1×10^4 cells/well for 24 h. After incubation, the medium was removed then replaced with fresh medium and *Cordyceps* water extract (1, 5, 25, 125, 500, and 1000 μg/well). After one more day of incubation, the supernatant was removed, and 100 μL from 0.5 μg/mL MTT/DMEM solution was added to each well and placed in the incubator for 2 h. Carefully the solution was removed, 100 μL DMSO was added into the wells to dissolve the crystals, the well plate was placed on the shaker plate, and the plate was gently shaken for 10 min in the dark. The absorbance value was measured at 570 nm (BioTek, Shoreline, WA, USA) [18].

2.8. Detect Reactive Oxygen Species by DCFDA Stain

To detect that intracellular PMA upregulates the level of intracellular oxidative stress, we used 2′,7′-dichlorofluorescein diacetate (DCFDA, Sigma-Aldrich Chemical Corp., D6883); The DCFDA stock solution was prepared by mixing it with DMSO, then diluted into a working solution (20 μMol) containing PBS. The cells were cultured in a 6-well plate at a density of 1×10^5 for 1 day, later treated with three different concentrations (5, 25, and 125 μg/mL) of water extraction of CM. The harvested cells were mixed with the working solution at 37 °C for 30 min and analyzed with a flow cytometer at 488 nm laser (Guava® easy yet 5HT, Merck KGaA, Darmstadt, Germany). The fluorescence intensity of the cells should be detectable at 530 nm [12].

2.9. Inhibiting Effects on Melanoma Cell Migration by Wound Healing Test

The cell migration experiment was completed mainly to detect the inhibitory effect of the threatening analyte through the wound healing test, with minor modifications. In short, 5×10^5 cells were seeded in a 6-well plate and incubated until grown to complete confluence. Then, a 200 µL plastic pipette tip was used to form a clear wound. The area was washed with PBS and then treated with the prepared sample. The cell crawling was checked after 24 h of wound formation. Cell migration and movement through the wound area were calculated by using the free software ImageJ [18].

2.10. Boyden Invasion Assay

The Transwell Invasion Test was used to study the influence of the analyte on the invasion of melanoma cells. It was necessary to inoculate the top of Boyden's Chamber with a serum-free medium to be able to nurture B16F10 and LMM cells. The 8-micron holes in the upper insert (ThinCert™, Greiner, Frickenhausen, Germany) were pre-coated with the base matrix which was Matrigel. The cells were stored in three different concentrations (5, 25, and 125 µg/mL) of CM and a solvent containing 10% FBS, which was regarded as a control cell. Cells on the lower surface of the filter membrane needed to be fixed by using paraformaldehyde and Giemsa staining counted to analyze the results [4].

2.11. Quantitative Real-Time Reverse Transcription Polymerase Chain Reaction (qRT-PCR) Analysis

The fluorescent signal was generated by a unique probe formed as qRT-PCR, which is a detection method using the StepOnePlus™ system (Thermo Fisher Scientific Inc., USA) fluorescence detection technology to sense each cycle. It detects and records the fluorescence intensity of each cycle and calculates real-time quantitative data. For qRT-PCR, the reaction mixture uses SYBR Green Master Mix (Qiagen, Valencia, CA, USA) using primers and templates. We used TRIzol (Invitrogen, Waltham, MA, USA) to extract RNA from B16F10 and LMM cells and then used a reverse transcription kit (Takara, Shiga, Japan) to generate DNA. For qRT-PCR using primers, single-stranded DNA was formed from the sample, then the primers were combined for dsDNA formation, and then SYBR Green dsDNA (Roche, Basel, Switzerland) was added. All steps were repeated for 40 amplification cycles, and the resultant was passed through the fluorescence detection system [18].

2.12. Western Blotting

In this step, 1×10^6 B16F10 cells and lung metastasis cells were seeded in 12 wells and treated for 24 h. After treatment with *Cordyceps* overnight, the cells were collected and lysed with lysis buffer (Thermo Scientific Pierce RIPA buffer) for cellular protein extraction. The lysate was centrifuged at 1200 rpm for 15 min, and the protein concentration was measured using the bicinchoninic acid (BCA) protein assay kit (Pierce, Rockford, IL, USA). Sodium dodecyl sulfate-polyacrylamide gel electrophoresis (SDS-PAGE) was used for the protein separation then the proteins were electro-transferred on the gel to a polyvinylidene fluoride (PVDF) membrane (Pall Life Science, Ann Arbor, MI, USA). The membrane was blocked with blocking buffer (Pierce TOOL Speed PLUS blocking reagent) and washed three times with TBST containing Tween-20 (Tris-buffered saline, pH 8.0) for 5 min each. The membrane and the corresponding primary antibody were left on a shaker overnight at 4 °C, later washed 3 times, blocked again, and the corresponding primary and secondary antibodies were shaken at room temperature for 60 min. West Femto Maximum Sensitivity Substrate Kit (SuperSignal, Rockford, IL, USA) was used to visually detect the signal through enhanced chemiluminescence (ECL) detection [12].

2.13. Statistical Analysis

All experiments were performed in triplicate in each assay and expressed as mean ± standard error. The difference between the control and *Cordyceps*-treated cells in the in vitro assay was analyzed by Student's *t*-test because p value < 0.05 was considered statistically significant [12].

3. Results

3.1. Antioxidant Activities in CM

In this study, the antioxidative capabilities of CM were evaluated by reducing power, ferrous ion chelating activity, and DPPH scavenging ability, which was used frequently as detection for the eradication of free radical quantities by observing the color-change. One of the CM bio functions was assumed as a powerful antioxidant, which played an essential role in absorbing and neutralizing free radicals. The cell composition, such as lipid and protein, would be oxidated because of the excessive construction and accumulation of free radicals. There are different mechanisms to eliminate ROS: quenching singlet, triplet oxygen, or decomposing peroxides, and those three assays could distinguish cases. Table 1 confirmed that CM had dose-dependent manner competencies in the three experiments above, especially reducing power.

Table 1. Reducing power, ferrous ion chelating, and DPPH were used to test the ROS removing ability in CM.

Concentration (μg/mL)	Antioxidant Capacity		
	Reducing Power (Absorbance at 700 nm)	Ferrous Ion Chelating (%)	DPPH (%)
Positive control [a,b,c]	0.77 ± 0.16	96.86 ± 7.90	89.43 ± 0.65
1	0.23 ± 0.14	18.95 ± 2.13	12.53 ± 1.78
5	0.25 ± 0.08	26.26 ± 3.05	20.31 ± 0.24
25	0.29 ± 0.09	36.53 ± 3.61	31.73 ± 2.09
125	0.32 ± 0.16	52.99 ± 1.85	40.72 ± 3.67
500	0.54 ± 0.06	69.12 ± 2.59	50.87 ± 9.44
1000	0.60 ± 0.07	77.78 ± 6.80	69.70 ± 2.75

[a] BHA was used as a positive control on reducing power at 100 mM. [b] EDTA was used as a positive control on ferrous ion chelating ability. [c] Vitamin C was used as a positive control on the DPPH assay.

The reducing power assay is a standard method for detecting the ability to lose electronics in the materials by altering color from yellow to green. The level of color diversity depended upon the capability of the examined antioxidants. The Fe^{3+}/ferricyanide complex could be reduced into a ferrous form by the presence of a suitable substance. The reducing power of various concentrations of CM is demonstrated in Table 1, which showed a strong potential of reduction compared to the positive group. The consequence revealed a linear bio-functional response in different suitable CM concentrations. For example, at 500 μg/mL CM, the antioxidant level exhibited higher than 50% of the antioxidant properties. The experiment demonstrated CM provided a strong potential in ROS elimination.

The ferrous ion chelating assay could examine the compounds as chelating agents to disrupt the ROS construction by lightening the red complex. In addition, different concentrations of CM would compare to the metal removal capability of EDTA, which is the positive control. The ferrous ion chelating ability of CM presented a concentration-dependent trend. At 125 μg/mL CM, the chelating power of metal exceeded 50%, which was precisely 52.9% compared to the EDTA group. The chelating level rose slightly when the consistency of CM escalated, and it was about 77.7% at 1000 μg/mL. However, the lower concentration of CM could not expose enough antioxidative capacity. At the concentration of 5 μg/mL, CM had a minor level (26.6%) compared to the positive control.

DPPH free radical scavenging system is a proven assay for evaluating the substance's capability as a hydrogen donor to remove the stable radical DPPH and turn it into diphenylpicrylhydrazine. Therefore, the assay color will shift from deep blue to light yellow when the element can absorb the free radicals. CM showed a limited antioxidant capacity, which directly correlated with the CM dosages. The free radical emission capability was under 50% when 125 μg/mL CM was conducted, only 40.7% and 1000 μg/mL CM was 69.7%; nevertheless, the antioxidative ability was proportional to the concentration of CM.

ROS promoted cancer cells to secrete various inflammatory factors, which could strengthen the metastasis capability of cancer cells, increase the proliferation of vascular endothelium, and inhibit the immune system. Furthermore, ROS could create the immunosuppression microenvironment of tumors because of the continuous oxidant generation. As a result, cancer cells could continually develop and deteriorate. Therefore, the bio function of alleviating chronic inflammation and preventing cancer was related to the existence of antioxidants. The experiments above confirmed the adequate ability to eliminate ROS in CM, which was more effective in reducing power. The ferrous ion chelating and DPPH assay likewise had a mild ability in reducing oxidant.

3.2. *Effects of CM on the Proliferative Viability of B16F10 and LMM Cells*

In this study, two melanoma-related cell lines, B16F10 and lung metastatic melanoma (LMM) cells, were incubated 24 h in 1, 5, 25, 125, 500, and 1000 μg/mL of CM independently. The cell viabilities were tested by MTT assay, which could monitor the cell proliferative viability in observing the color of chemical alternated into purple. Figure 2A,B illustrated the results of the MTT assay for B16F10 and LMM separately in different CM concentrations. In general, the increase of CM containing enhanced cell death rates in both cell lines. For instance, the IC50 of B16F10 and LMM was around 25 μg/mL and 15 μg/mL CM in Figure 2A,B. However, at the higher CM concentration, such as 500 μg/mL and 1000 μg/mL, the cell viabilities in both types of melanomas were at ground levels, roughly 30% and the cell death rates reduced insignificantly at 1 μg/mL CM. Thus, 1 μg/mL, 500 μg/mL, and 1000 μg/mL CM would be excluded, and 5 μg/mL, 25 μg/mL, and 125 μg/mL CM would be conducted in the following experiments because the proliferative viability of B16F10 and LMM decreased recognizably between 5 and 125 μg/mL CM. The results distinguished that B16F10 cells were more sensitive to CM than LMM and approved the practical bio function of CM as a dosage-related cell growth inhibitor in B16F10 and LMM.

3.3. *CM Causes Apoptosis in B16F10 and LMM Cells*

Many anti-cancer mechanisms are related to apoptosis, including the caspase pathways and Bax/BCL2 proteins. Previous experiments in this study confirmed that CM could diminish the cell viability in B16F10 and LMM, and the reason was assumed as apoptosis. Therefore, the mRNA and protein of cleaved caspase-9, Bax, and BCL2 were detected in Figure 2. The caspases were a group of intracellular proteases responsible for helping the formation of apoptotic bodies during apoptosis, and the cleaved caspase-9s were situated at critical junctions in the apoptosis pathway. When cytochrome c from mitochondria was released, it triggered the activation of cleaved caspase-9s. B-cell lymphoma protein 2 (BCL2) and B-cell lymphoma protein 2-associated X (Bax) were two proteins related to the cytochrome c release. BCL2 could prevent Bax from releasing cytochrome c and restrict apoptosis downstream progress, resulting in cell survival. Some studies confirmed that a high Bax/BCL2 ratio favors apoptosis.

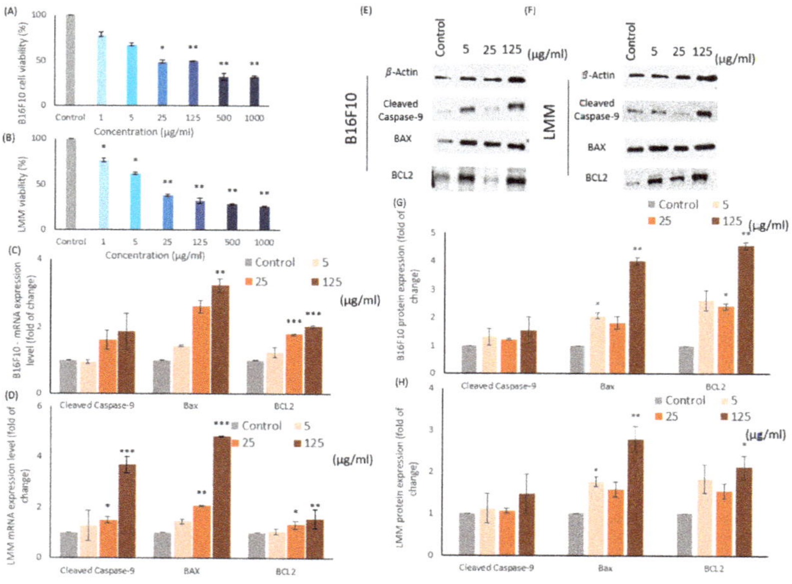

Figure 2. The cell viability of (**A**) B16F10 and (**B**) LMM was examined by MTT assay in different concentrations of CM. The mRNA levels of cleaved caspase-9, Bax, and BCL2 in (**C**) B16F10 and (**D**) LMM were tested after incubation in 5, 25, and 125 μg/mL CM. Protein levels of cleaved caspase-9, Bax, and BCL2 in (**E**) B16F10 and (**F**) LMM were detected by Western blotting after 5, 25, and 125 μg/mL CM were conducted. The protein quantity of cleaved caspase-9, Bax, and BCL2 in (**G**) B16F10 and (**H**) LMM were analyzed by ImageJ. *$p < 0.05$, ** $p < 0.01$, *** $p < 0.005$.

Figure 2C,D demonstrate the mRNA expression rate of cleaved caspase-9, Bax, and BCL2 in B16F10 and LMM individually. Figure 2E,F show the Western blots of protein quantity in cleaved caspase-9, Bax, and BCL2 in B16F10 and LMM and the protein accumulation rate illustrated in Figure 2G,H. In Figure 2C,D, the mRNA expression rate of cleaved caspase-9, Bax, and BCL2 in B16F10 and LMM enhanced because of the increasing of CM concentration. The cleaved caspase-9 and Bax were activated sharply by CM in both cell strings, but the BCL2 had only a mild increase. Figure 2G,H revealed that the protein expression in both cell strains demonstrated the same tendency, and the Bax/BCL2 protein ratio escalated in different CM concentrations. The Bax/BCL2 ratio in the control group of B16F10 and LMM were both approximately 1.3, but the percentage in both cell lines increased over 1.5 when the 125 μg/mL CM was adopted; thus, CM could induce apoptosis in tumor cells.

3.4. The Intracellular ROS Was Detected by DCFDA Staining

Intracellular oxidant stress was considered an essential motivation in inducing carcinogenesis. However, the relationship between cancer and ROS is complex. Some studies proved that the apoptosis process in cancer cells could result from extreme ROS levels, which means the superabundant or insubstantial level of ROS could lead to apoptosis. Figure 3 illustrates the ROS level in B16F10 and LMM by DCFDA staining. Figure 3A displays the cellular morphological photos of B16F10 and LMM, and the analyzed results are presented in Figure 3B. DCFDA can evaluate the intracellular ROS levels by the disclosed fluorescent brightness, and the more ROS represented, the greater fluorescence produced. Figure 3B shows that the ROS level decreased significantly because of the CM density increase. Compared to the effects of CM toward B16F10 and LMM, B16F10 is more sensitive to CM existence, and the ROS level had a significant drop in the experiments. In contrast, the LMM had more resistance toward CM.

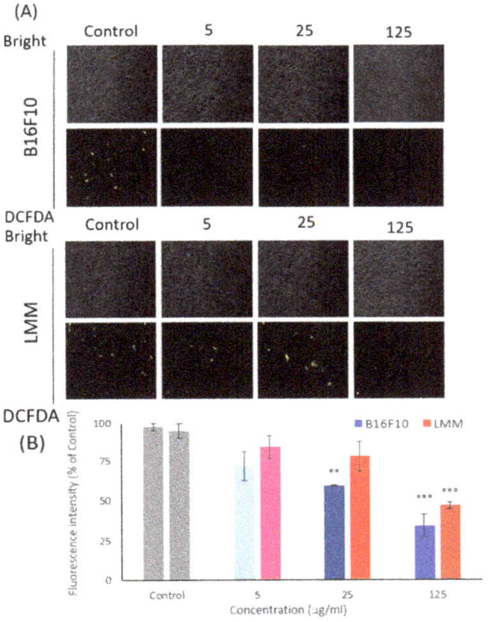

Figure 3. (**A**) The oxidative pressure inner B16F10 and LMM were tested by DCFDA fluorescent staining under a fluorescent microscope after conducted 5, 25, and 125 μg/mL CM. (**B**) The fluorescence intensity was analyzed in B16F10 and LMM after adopted 5, 25, and 125 μg/mL CM. ** $p < 0.01$, *** $p < 0.005$.

3.5. CM Inhibits LMM Migration

The main reason for cancer-associated death is cancer metastasis. Cancer cells can escape from the primary tumors and invade neighboring tissues as metastases because of the deregulation of cell migration. This research used the cell wound healing test and the Boyden invasion assay to investigate horizontal and vertical aggression rates. Figure 4A showed B16F10 and LMM in the microscopical photo of horizontal cell migration from the beginning and 24 h after being treated by different concentrations of CM and analyzed the results. In control, 5, 10, and 15 μg/mL CM, B16F10 mobility rates were 67.3%, 21.9%, and 12.56% and the cell mobility rates in LMM were 62.8%, 19.3%, and 14.5%. The experiment showed CM could reduce the horizontal migration in B16F10 and LMM, and the inhibiting rates were proportional to the CM concentrations. There were insignificant differences in the horizontal cell migration rate between B16F10 and LMM.

Figure 4B shows the results of the Boyden invasion assay, which is evidence that CM affected the vertical invaded rate in B16F10 and LMM. Different thicknesses of CM treated the cells on the above plate, and they would invade the blow plate through the semi-permeable membrane. Then, the cell quantity on the lower plate could be counted, representing the vertical invasion level of tumors. CM increased the concentration while the vertical aggressive rate in B16F10 and LMM regulated downward, around 25% of migration rate was restrained in 125 μg/mL CM. Figure 4B confirms that CM could prevent cell metastasis effectively in both B16F10 and LMM in the horizontal and vertical directions. However, CM had better inhibition ability in the horizontal migration than in vertical movement in B16F10 and LMM.

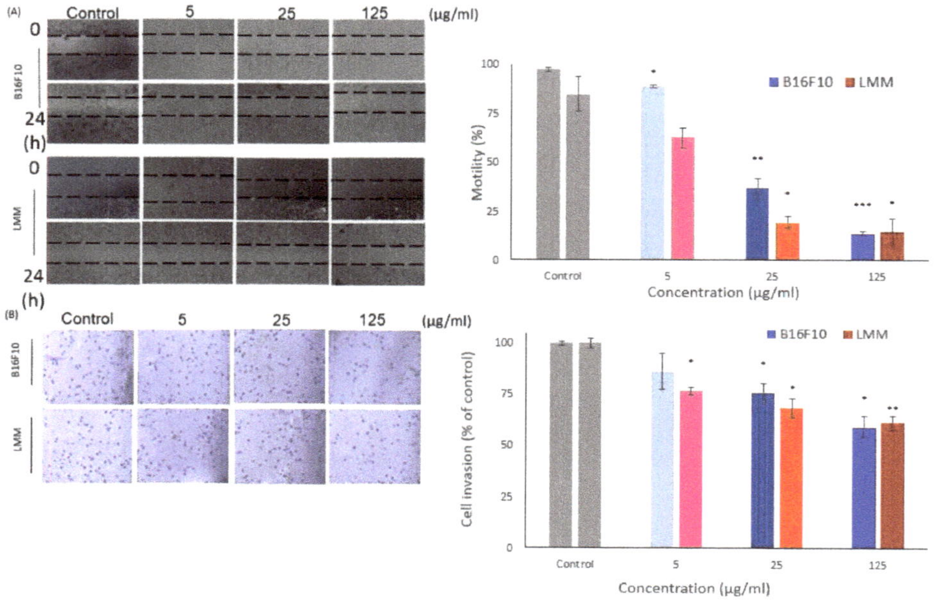

Figure 4. (**A**) The horizontal migration ability of B16F10 and LMM were tested by wound healing assay after co-incubating in different concentrations of CM. The results are shown in a microscope photo and analyzed in a histogram. (**B**) The vertical invasion ability of B16F10 and LMM was examined by Boyden invasion assay. The results are revealed in microscope photos and analyzed in a histogram. * $p < 0.05$, ** $p < 0.01$, *** $p < 0.005$.

3.6. CM Increased the Efficacy of CD8+ T Cells

It was essential to generate durable memory CD8+ T cells in the tumor microenvironment for curing cancer. However, producing and regulating a functional CD8+ T cell remained unknown [15]. This study revealed the relationship between CM, melanoma, and T cell. The CD8+ and CD4+ cell markers were double-stained and divided the experiment groups into four groups (Figure 5A). After 24 h, the ratio of CD8+/CD4+ was analyzed and measured by a flow cytometer. The CD8+ and CD4+ were the markers of T cells, which CD8+ and CD4+ cell markers could distinguish the difference between the cytotoxic T cell (CD8+) and other types of T cell (CD4+). Therefore, the CD8+/CD4+ could represent the related quantity of activated cytotoxic T cells. The T cells only cultured with CM had a slightly higher CD8+/CD4+ ratio than the control group, but there was no significant difference between those two groups (Figure 5B). In Figure 5C, T cells were incubated with LMM, the number of cytotoxic T cells increased as twice the control group, which meant LMM could activate the cytotoxic T cell. The last group included T cell, LMM, and CM into the same plate, and the CD8+ ratio was sharply boosted, which was almost four times the control group (Figure 5D). Those experiments confirmed the CM function in increasing the T cell activities.

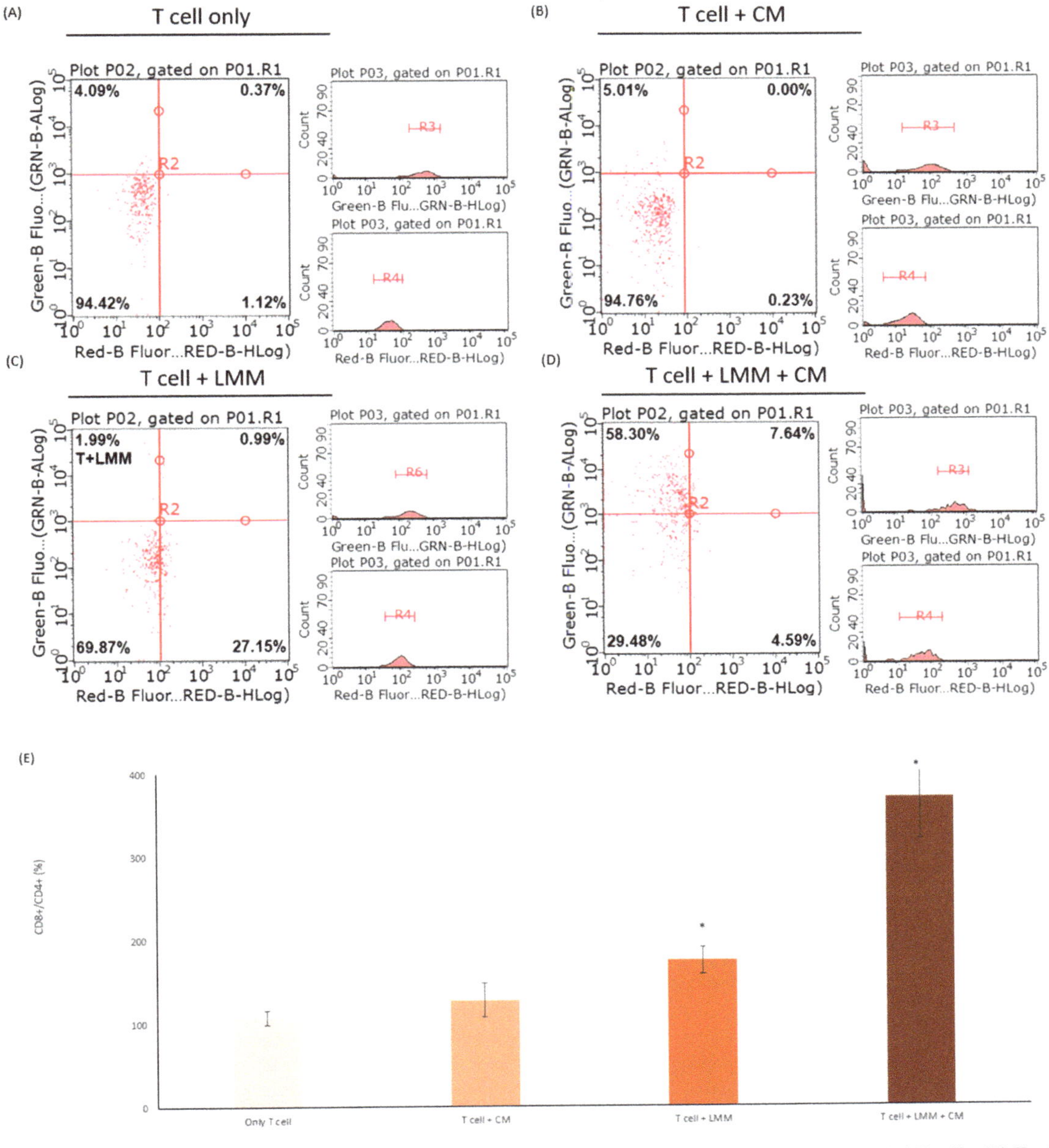

Figure 5. The cells with CD8+ and CD4+ cell markers were measured in (**A**) T cells, (**B**) T cells incubated with CM, (**C**) T cells cultured with LMM, and (**D**) T cells co-incubated with LMM and CM by flow cytometry. (**E**) The histogram of the ratio of CD8+/CD4+ in the four experiment groups above. * $p < 0.05$.

4. Discussion

Free radicals in the body are substances produced by oxygen metabolism and react exceptionally with any armamentarium. This means ROS in cells needed to stay at a dynamic level to proliferate cells and the superabundant and meagerly level of ROS could promote cell apoptosis either way [19]. When the free radical levels in the human body exceed the normal levels, the chain reactions that may facilitate the oxidation of essential substances could be triggered to eliminate the ROS into the average level. The healthy human body contains chemicals protecting from free radicals and reduce possible harm. These chemicals can reduce the rate of cell division and repair the damage caused by free radicals in many cases [20]. Furthermore, tumor progression and metastasis result from an overload of free radicals in the tumor microenvironment. Therefore, justifying the level of free radicals could be a great strategy to cure cancer.

In this study, CM showed an excellent capability for removing ROS, different capacities in reducing power, ferrous ion chelating, and DPPH free radical scavenging test and was influential in eliminating intracellular ROS. In the experiment, CM could reduce the inner cell ROS in both B16F10 and LMM even though LMM had more resistance than B16F10 in the lower level of free radical. The ROS levels sharply declined in cancer cells provoked the mRNA and protein level of cleaved caspase-9 and Bax. In contrast, the mRNA and protein level of BCL2 increased slightly. It resulted in the cleaved caspase-9 and Bax/BCL2 being higher than the regular quantity, which reduced the cell viability of B16F10 and LMM. CM could trigger cell apoptosis in melanoma, which was proven in this experiment.

Melanoma is one of the most aggressive tumors and exposure to ultraviolet (UV) radiation is one of the main causes of skin cancer. Some studies show that the spread of melanoma is rapid and systemic treatments for melanoma and metastatic melanoma are usually ineffective. Those properties of melanoma develop into the main reason for death in skin cancer. In clinical studies, the development of various tumors was linked to overloaded oxidative stress, which can directly cause DNA damage [17,21]. More-than-sufficient ROS in the tumor microenvironment could stimulate resistance to cancer-curing drugs. Melanoma also showed immune escaping and migration ability enhancement when the tumor microenvironment had a higher level of ROS. This study showed CM could reduce the speed of horizontal and vertical invasion in B16F10 and LMM effectively and had a dose-dependent trend.

Apoptosis is one of the most effective defense methods in the immune system, and the lack or excessive accumulation of lymphocytes causes profound consequences [22]. Autoimmunity is a condition in which the repose of immunocytes acts out of control. T cells are in the center of the apoptosis process in the human immune system, especially against the generation of cancer cells. However, excessive or premature apoptosis can lead to the development of cancer [23]. Thus, the regulation of apoptosis by regulating T cells could be an effective strategy for cancer treatment. The immune cells themselves will not change from CD4+ to CD8+. If there are a lot of changes after adding the drug, it should be verified whether there will be immune solid side effects after adding the compound. We know that adding the component itself will not promote a large amount of CD4+ converting to CD8+. We suspect that the addition of cancer cells will inhibit immune cells from transforming into killer T cells, and the addition of pharmaceuticals will help immune cells to identify cancer cells and transform into killer T cells to attack cancer cells [24,25]. The effects of CM in promoting T cells and apoptosis have been explored in these experiments and showed a significant result. When T cells were incubating with LMM, the CD8+ T cells were more active with CM conducted into the medium. The outcome may be giving new therapy options in treating carcinoma in situ and metastatic melanoma.

5. Conclusions

Many studies proved that CM benefits health in many ways. In this study, three important conclusions have been confirmed, in which CM had antioxidant, anti-carcinogenic, and CD8+ T cell activation properties. First, CM had a significant ability in dismissing ROS by reducing power, ferrous ion chelating, and DPPH elimination, and CM could also decrease the intracellular ROS level. Second, CM could be the cancer treatment by enhancing the apoptosis progress, depressing tumor cell viability, and suppressing the migration rate of melanoma. CM could shrink the horizontal invasion and vertical migration abilities in

B16F10 and LMM. It could decrease the cell viability of B16F10 and LMM by activating the apoptosis program, and this study confirmed the increase of the mRNA and protein level of cleaved caspase-9 and Bax/BCL2 in both cells. Finally, adding cancer cells will inhibit immune cells from transforming into killer T cells, and the addition of CM will help immune cells to identify cancer cells and transform them into killer T cells to attack cancer cells. The ratio of CD8+/CD4+ in T cells, which CM adopted, increased significantly (Figure 6). Those three pieces of evidence indicate that CM could inhibit and treat melanoma.

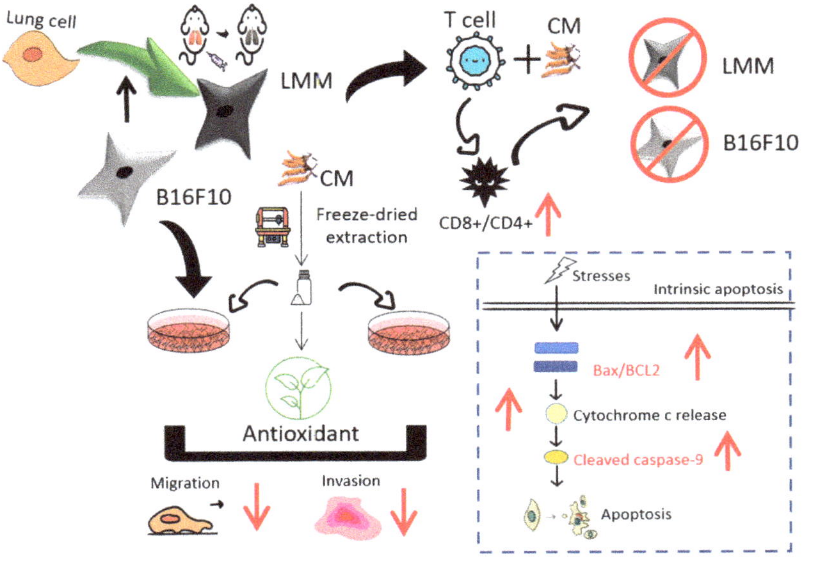

Figure 6. The process and results of this study.

Author Contributions: Experiments conduction, Y.-H.L., Y.-S.L., J.-Y.W. and H.-T.L.; Conceptualization, Y.-H.L. and H.-T.L.; Data curation, Y.-S.L., G.N.C. and H.-M.D.W.; Formal analysis, Y.-H.L., G.N.C., C.-C.C. and H.-M.D.W.; Funding acquisition, Y.-S.L., P.S. and H.-M.D.W.; Investigation, Y.-H.L. and C.-C.C.; Methodology, P.S., G.N.C. and H.-M.D.W.; Project administration, J.-Y.W.; Resources, Y.-S.L., P.S. and G.N.C.; Software, J.-Y.W.; Supervision, J.-Y.W.; Writing—original draft, Y.-S.L.; Writing re-view & editing, H.-T.L. All authors have read and agreed to the published version of the manuscript.

Funding: This research was supported by the Ministry of Science and Technology (MOST), Taiwan under grant nos. MOST 111-2221-E-005-026-MY3, 111-2221-E-005-009 and 110-2221-E-029-005. The authors also thank Kaohsiung Armed Forces General Hospital (KAFGH_D_111021 and KAFGH_D_111022) and NCHU-SCMH Research Project (110D257).

Institutional Review Board Statement: Not applicable.

Informed Consent Statement: Not applicable.

Data Availability Statement: Data are contained within the article materials.

Acknowledgments: We thank the members of the Taiwan Agriculture Research Institute, Council of Agriculture for their valuable suggestions, technical support, and materials. Ju-Yu Wu carried out this research with funding support in part by the Ph.D Program in Tissue Engineering and Regenerative Medi-cine of National Chung Hsing University and National Health Research Institutes.

Conflicts of Interest: The authors declare no conflict of interest.

References

1. Kuo, C.L.; Chou, H.Y.; Chiu, Y.C.; Cheng, A.N.; Fan, C.C.; Chang, Y.N.; Chen, C.H.; Jiang, S.S.; Chen, N.J.; Lee, A.Y. Mitochondrial oxidative stress by Lon-PYCR1 maintains an immunosuppressive tumor microenvironment that promotes cancer progression and metastasis. *Cancer Lett.* **2020**, *474*, 138–150. [CrossRef] [PubMed]
2. Chou, H.Y.; Liu, L.H.; Chen, C.Y.; Lin, I.F.; Ali, D.; Lee, A.Y.-L.; David Wang, H.M. Bifunctional mechanisms of autophagy and apoptosis regulations in melanoma from Bacillus subtilis natto fermentation extract. *Food Chem. Toxicol.* **2021**, *150*, 112020. [CrossRef] [PubMed]
3. Liu, B.; Tan, X.; Liang, J.; Wu, S.; Liu, J.; Zhang, Q.; Zhu, R. A reduction in reactive oxygen species contributes to dihydromyricetin-induced apoptosis in human hepatocellular carcinoma cells. *Sci. Rep.* **2014**, *4*, 7041. [CrossRef] [PubMed]
4. Li, J.; Huang, S.Y.; Deng, Q.; Li, G.; Su, G.; Liu, J.; Wang, H.M.D. Extraction and characterization of phenolic compounds with antioxidant and antimicrobial activities from pickled radish. *Food Chem. Toxicol.* **2020**, *136*, 111050. [CrossRef] [PubMed]
5. Wu, K.J.; Ho, S.H.; Wu, C.; Wang, H.D.; Ma, D.L.; Leung, C.H. Simultaneous blocking of the pan-RAF and S100B pathways as a synergistic therapeutic strategy against malignant melanoma. *J. Cell Mol. Med.* **2021**, *25*, 1972–1981. [CrossRef]
6. Wu, P.F.; Chiu, C.C.; Chen, C.Y.; Wang, H.M. 7-Hydroxydehydronuciferine induces human melanoma death via triggering autophagy and apoptosis. *Exp. Dermatol.* **2015**, *24*, 930–935. [CrossRef]
7. Haung, H.Y.; Wang, Y.C.; Cheng, Y.C.; Kang, W.; Hu, S.H.; Liu, D.; Xiao, C.; Wang, H.D. A Novel Oral Astaxanthin Nanoemulsion from Haematococcus pluvialis Induces Apoptosis in Lung Metastatic Melanoma. *Oxid. Med. Cell Longev.* **2020**, *2020*, 2647670. [CrossRef]
8. Chen, Y.C.; Chen, Y.H.; Pan, B.S.; Chang, M.M.; Huang, B.M. Functional study of Cordyceps sinensis and cordycepin in male reproduction: A review. *J. Food Drug Anal.* **2017**, *25*, 197–205. [CrossRef]
9. Jędrejko, K.J.; Lazur, J.; Muszyńska, B. Cordyceps militaris: An Overview of Its Chemical Constituents in Relation to Biological Activity. *Foods* **2021**, *10*, 2634. [CrossRef]
10. Ashraf, S.A.; Elkhalifa, A.E.O.; Siddiqui, A.J.; Patel, M.; Awadelkareem, A.M.; Snoussi, M.; Ashraf, M.S.; Adnan, M.; Hadi, S. Cordycepin for Health and Wellbeing: A Potent Bioactive Metabolite of an Entomopathogenic Cordyceps Medicinal Fungus and Its Nutraceutical and Therapeutic Potential. *Molecules* **2020**, *25*, 2735. [CrossRef]
11. Rosenberg, S.A.; Restifo, N.P.; Yang, J.C.; Morgan, R.A.; Dudley, M.E. Adoptive cell transfer: A clinical path to effective cancer immunotherapy. *Nat. Rev. Cancer* **2008**, *8*, 299–308. [CrossRef] [PubMed]
12. Zheng, R.; Zhu, R.; Li, X.; Li, X.; Shen, L.; Chen, Y.; Zhong, Y.; Deng, Y. N6-(2-Hydroxyethyl) Adenosine From Cordyceps cicadae Ameliorates Renal Interstitial Fibrosis and Prevents Inflammation via TGF-β1/Smad and NF-κB Signaling Pathway. *Front. Physiol.* **2018**, *9*, 1229. [CrossRef]
13. Das, G.; Shin, H.S.; Leyva-Gómez, G.; Prado-Audelo, M.L.D.; Cortes, H.; Singh, Y.D.; Panda, M.K.; Mishra, A.P.; Nigam, M.; Saklani, S.; et al. Cordyceps spp.: A Review on Its Immune-Stimulatory and Other Biological Potentials. *Front. Pharmacol.* **2020**, *11*, 602364. [CrossRef] [PubMed]
14. Rupa, E.J.; Li, J.F.; Arif, M.H.; Yaxi, H.; Puja, A.M.; Chan, A.J.; Hoang, V.A.; Kaliraj, L.; Yang, D.C.; Kang, S.C. Cordyceps militaris Fungus Extracts-Mediated Nanoemulsion for Improvement Antioxidant, Antimicrobial, and Anti-Inflammatory Activities. *Molecules* **2020**, *25*, 5733. [CrossRef] [PubMed]
15. Lin, Y.Y.; Chen, C.Y.; Ma, D.L.; Leung, C.H.; Chang, C.Y.; Wang, H.D. Cell-derived artificial nanovesicle as a drug delivery system for malignant melanoma treatment. *Biomed. Pharmacother.* **2022**, *147*, 112586. [CrossRef]
16. Voss, K.; Larsen, S.E.; Snow, A.L. Metabolic reprogramming and apoptosis sensitivity: Defining the contours of a T cell response. *Cancer Lett.* **2017**, *408*, 190–196. [CrossRef]
17. Tseng, C.C.; Lin, Y.J.; Liu, W.; Lin, H.Y.; Chou, H.Y.; Thia, C.; Wu, J.H.; Chang, J.S.; Wen, Z.H.; Chang, J.J.; et al. Metabolic engineering probiotic yeast produces 3S, 3′S-astaxanthin to inhibit B16F10 metastasis. *Food Chem. Toxicol.* **2020**, *135*, 110993. [CrossRef]
18. Li, J.; Chen, C.Y.; Huang, J.Y.; Wang, L.; Xu, Z.; Kang, W.; Lin, M.H.; Wang, H.D. Isokotomolide A from Cinnamomum kotoense Induce Melanoma Autophagy and Apoptosis In Vivo and In Vitro. *Oxid. Med. Cell Longev.* **2020**, *2020*, 3425147. [CrossRef]
19. Kao, C.J.; Chou, H.Y.; Lin, Y.C.; Liu, Q.; David Wang, H.M. Functional Analysis of Macromolecular Polysaccharides: Whitening, Moisturizing, Anti-Oxidant, and Cell Proliferation. *Antioxidants* **2019**, *8*, 533. [CrossRef]
20. Wang, H.M.; Fu, L.; Cheng, C.C.; Gao, R.; Lin, M.Y.; Su, H.L.; Belinda, N.E.; Nguyen, T.H.; Lin, W.H.; Lee, P.C.; et al. Inhibition of LPS-Induced Oxidative Damages and Potential Anti-Inflammatory Effects of Phyllanthus emblica Extract via Down-Regulating NF-κB, COX-2, and iNOS in RAW 264.7 Cells. *Antioxidants* **2019**, *8*, 270. [CrossRef]
21. Li, P.H.; Chiu, Y.P.; Shih, C.C.; Wen, Z.H.; Ibeto, L.K.; Huang, S.H.; Chiu, C.C.; Ma, D.L.; Leung, C.H.; Chang, Y.N.; et al. Biofunctional Activities of Equisetum ramosissimum Extract: Protective Effects against Oxidation, Melanoma, and Melanogenesis. *Oxid. Med. Cell Longev.* **2016**, *2016*, 2853543. [CrossRef] [PubMed]
22. Berthenet, K.; Castillo Ferrer, C.; Fanfone, D.; Popgeorgiev, N.; Neves, D.; Bertolino, P.; Gibert, B.; Hernandez-Vargas, H.; Ichim, G. Failed Apoptosis Enhances Melanoma Cancer Cell Aggressiveness. *Cell Rep.* **2020**, *31*, 107731. [CrossRef] [PubMed]
23. Costantini, F.; Di Sano, C.; Barbieri, G. The Hydroxytyrosol Induces the Death for Apoptosis of Human Melanoma Cells. *Int. J. Mol. Sci.* **2020**, *21*, 8074. [CrossRef] [PubMed]

24. Karpathiou, G.; Mihailidis, V.; Nakou, E.; Anevlavis, S.; Tzouvelekis, A.; Kouliatsis, G.; Ntolios, P.; Bouros, D.; Kotsianidis, I.; Froudarakis, M.E. Chemotherapy-induced changes in bronchoalveolar lavage fluid CD4 + and CD8 + cells of the opposite lung to the cancer. *Sci. Rep.* **2020**, *10*, 19927. [CrossRef]
25. Khan, M.; Arooj, S.; Wang, H. NK Cell-Based Immune Checkpoint Inhibition. *Front. Immunol.* **2020**, *11*, 167. [CrossRef]

MDPI
St. Alban-Anlage 66
4052 Basel
Switzerland
www.mdpi.com

Antioxidants Editorial Office
E-mail: antioxidants@mdpi.com
www.mdpi.com/journal/antioxidants

www.ingramcontent.com/pod-product-compliance
Lightning Source LLC
Chambersburg PA
CBHW061835210926
51850CB00006B/898
9783036599656